Functional Integration

Basics and Applications

NATO ASI Series

Advanced Science Institutes Series

A series presenting the results of activities sponsored by the NATO Science Committee, which aims at the dissemination of advanced scientific and technological knowledge, with a view to strengthening links between scientific communities.

The series is published by an international board of publishers in conjunction with the NATO Scientific Affairs Division

A	**Life Sciences**	Plenum Publishing Corporation
B	**Physics**	New York and London
C	**Mathematical and Physical Sciences**	Kluwer Academic Publishers Dordrecht, Boston, and London
D	**Behavioral and Social Sciences**	
E	**Applied Sciences**	
F	**Computer and Systems Sciences**	Springer-Verlag
G	**Ecological Sciences**	Berlin, Heidelberg, New York, London,
H	**Cell Biology**	Paris, Tokyo, Hong Kong, and Barcelona
I	**Global Environmental Change**	

PARTNERSHIP SUB-SERIES

1. Disarmament Technologies	Kluwer Academic Publishers
2. Environment	Springer-Verlag
3. High Technology	Kluwer Academic Publishers
4. Science and Technology Policy	Kluwer Academic Publishers
5. Computer Networking	Kluwer Academic Publishers

The Partnership Sub-Series incorporates activities undertaken in collaboration with NATO's Cooperation Partners, the countries of the CIS and Central and Eastern Europe, in Priority Areas of concern to those countries.

Recent Volumes in this Series:

Volume 359 — Correlations and Clustering Phenomena in Subatomic Physics
edited by M. N. Harakeh, J. H. Koch, and O. Scholten

Volume 360 — Surface Diffusion: Atomistic and Collective Processes
edited by M. C. Tringides

Volume 361 — Functional Integration: Basics and Applications
edited by Cecile DeWitt-Morette, Pierre Cartier, and Antoine Folacci

Volume 362 — Low-Dimensional Applications of Quantum Field Theory
edited by Laurent Baulieu, Vladimir Kazakov, Marco Picco, and Paul Windey

Series B: Physics

Functional Integration

Basics and Applications

Edited by

Cecile DeWitt-Morette

The University of Texas at Austin
Austin, Texas

Pierre Cartier

Ecole Normale Supérieure
Paris, France

and

Antoine Folacci

Université de Corse
Corte, France

Springer Science+Business Media, LLC

Proceedings of a NATO Advanced Study Institute on
Functional Integration: Basics and Applications,
held September 1 – 14, 1996,
in Cargèse, France

NATO-PCO-DATA BASE

The electronic index to the NATO ASI Series provides full bibliographical references (with keywords and/or abstracts) to about 50,000 contributions from international scientists published in all sections of the NATO ASI Series. Access to the NATO-PCO-DATA BASE is possible in two ways:

—via online FILE 128 (NATO-PCO-DATA BASE) hosted by ESRIN, Via Galileo Galilei, I-00044 Frascati, Italy

—via CD-ROM "NATO Science and Technology Disk" with user-friendly retrieval software in English, French, and German (©WTV GmbH and DATAWARE Technologies, Inc. 1989). The CD-ROM contains the AGARD Aerospace Database.

The CD-ROM can be ordered through any member of the Board of Publishers or through NATO-PCO, Overijse, Belgium.

Library of Congress Cataloging-in-Publication Data

Functional integration : basics and applications / edited by Cecile DeWitt-Morette, Pierre Cartier and Antoine Folacci.
p. cm. -- (NATO ASI series. B, Physics ; v. 361)
"Published in cooperation with NATO Scientific Affairs Division."
"Proceedings of a NATO Advanced Study Institute on Functional Integration: Basics and Applications, held September 1-14, 1996, Cargèse, France"--T.p. verso.
Includes bibliographical references and index.

1. Integration, Functional--Congresses. I. DeWitt-Morette, Cécile. II. Cartier, P. (Pierre) III. Folacci, Antoine. IV. North Atlantic Treaty Organization. Scientific Affairs Division. V. NATO Advanced Study Institute on Functional Integration: Basics and Applications (1996 : Cargèse, France) VI. Series.
QC20.7.F85F86 1997
530.15'574--dc21 97-22007
CIP

DOI 10.1007/978-1-4899-0319-8

Originally published by Plenum Press, New York in 1997
MyCopy version of the original edition 1997

http://www.plenum.com

10 9 8 7 6 5 4 3 2 1

PREFACE

The program of the Institute covered several aspects of functional integration - from a robust mathematical foundation to many applications, heuristic and rigorous, in mathematics, physics, and chemistry. It included analytic and numerical computational techniques. One of the goals was to encourage cross-fertilization between these various aspects and disciplines.

The first week was focused on quantum and classical systems with a finite number of degrees of freedom; the second week on field theories.

During the first week the basic course, given by P. Cartier, was a presentation of a recent rigorous approach to functional integration which does not resort to discretization, nor to analytic continuation. It provides a definition of functional integrals simpler and more powerful than the original ones. Could this approach accommodate the works presented by the other lecturers? Although much remains to be done before answering "Yes," there seems to be no major obstacle along the road.

The other courses taught during the first week presented:

a) a solid introduction to functional numerical techniques (A. Sokal) and their applications to functional integrals encountered in chemistry (N. Makri).

b) integrals based on Poisson processes and their applications to wave propagation (S. K. Foong), in particular a wave-restorer or wave-designer algorithm yielding the initial wave profile when one can only observe its distortion through a dissipative medium.

c) the formulation of a quantum equivalence principle (H. Kleinert) which, given the flat space theory, yields a well-defined quantum theory in spaces with curvature and torsion.

d) a variational perturbation expansion for functional integrals which converts divergent perturbation series into exponentially fast-converging ones (H. Kleinert).

e) the use of functional integrals in studying physics on and near caustics (C. DeWitt-Morette).

During the second week the basic course was given by L. Kauffman on the theory of knots and its interfaces with polymers, quantum groups, and functional integration, thus increasing the scope of functional integration in many directions and opening up promising areas for future investigations.

Functional integrals in polymer and membrane theory were presented heuristically (B. Duplantier) and rigorously (J. Magnen). Highlights included:

a) renormalization theory, in the form of the ε expansion, of the Edwards model for self-avoiding polymers.

b) rigorous results on nongaussian path integrals with a singular (quartic) weight (Edwards model).

c) extension of perturbative renormalization to interacting and self-avoiding membranes, via the introduction of generalized Feynman diagrams for extended objects.

d) The use of polymers to model $\lambda\phi^4$ theory.

New results in basic quantum field theory (B. DeWitt) and in topological quantum field theory (M. Blau) using functional integration included:

a) a formulation of gauge theory without ghosts.

b) a discussion of a variety of formal functional integral techniques for low-dimensional gauge theories, including, in particular, a Weyl integral formula for functional integrals, i.e., a formula reexpressing an integral over group-valued maps as an integral over maps taking their values in the maximal torus of the group.

Needless to say, many aspects of functional integration were not covered in the formal lectures; some of them were discussed in informal presentations and discussions by a number of participants. All of them have been invited to contribute a short paragraph to the Proceedings stating either an open problem or a recently solved one in their fields of interest.

ACKNOWLEDGMENTS

This volume is an outcome of the NATO Advanced Study Institute held at the Institut d'Etudes Scientifiques de Cargèse in Corsica from September 1 to 14, 1997. It is not simply a set of lecture notes but a combination of lecture notes and articles organized in a book for a larger public than the Institute participants. It includes texts which can serve as an introduction to more specialized topics, namely:

- A. Sokal, *Monte Carlo Methods in Statistical Mechanics* (updated version of his Lausanne course) by permission of the Cours de Troisième Cycle de la Physique en Suisse Romande,
- P. Cartier, *Développements récents sur les groupes de tresses* by permission of l'Association des Collaborateurs de Nicolas Bourbaki,
- M. Blau, *Localization and Diagonalization* (updated version of his J. Math. Phys. article) by permission of the American Institute of Physics.

We are very grateful for the permissions to use these texts.

An Advanced Study Institute is more than a series of lectures, it is a community of participants who live together for a couple of weeks. The quality of life during this period contributes immeasurably to the success of the Institute. The financial support of the Collectivité Territoriale de Corse and the Staff of the Institut d'Etudes Scientifiques de Cargèse were instrumental in making the session fruitful. We wish to thank the administrator Annie Touchant for her impeccable and friendly management and the support staff, Joseph-Antoine Ariano, Chantal Ariano, Brigitte Cassegrain, Claudine Conforto and Vittoria Conforto for their excellent services, and personal care.

NSF-NATO travel grants made possible the participations of Mark Byrd, Mark Mims, Xiaorong Morrow and Alexander Wurm. A Tubitak-NATO travel grant made possible the participation of Nuri Unal.

The organizing committee
P. Cartier
C. DeWitt-Morette
A. Folacci
C.J. Isham

CONTENTS

A RIGOROUS MATHEMATICAL FOUNDATION OF FUNCTIONAL INTEGRATION

P. Cartier[1]* and C. DeWitt-Morette[2]

[1]Ecole Normale Supérieure
45 rue d'Ulm
F-75005 Paris, France
[2]Department of Physics and Center for Relativity
University of Texas at Austin
Austin, TX

Based on Lecture Notes by
Alexander Wurm[3,4]

with an appendix on Functional Integration over Complex paths prepared by
David Collins[3]

ABSTRACT

Due to the growing interest in path integrals as a valuable tool in theoretical physics, a rigorous mathematical foundation of functional integration is needed more than ever. In these lectures we will present a new approach that is in part a synthesis of what has been accomplished over the past decades and in part an extension of functional integration to a larger class of functionals. After a discussion of integration theory and its shortcomings in infinite-dimensional, non-compact spaces, we will present the new approach in detail and give examples of its application.

CONTENTS

*Lectures given by P. Cartier
[3]University of Texas at Austin, Texas, USA
[4]and University of Wurzburg, Germany

Functional Integration: Basics and Applications
Edited by Cécile DeWitt-Morette, Plenum Press, New York, 1997

1. INTRODUCTION

What have we learned since Feynman introduced functional integrals into quantum physics? First of all, the domain of integration is an infinite-dimensional, and not locally compact, function space. To establish the notion of integration in such a space we need a volume element. But this is where the problems begin. As we will see, Lebesgue's conventional integration theory does not work as expected on these spaces, so what can we do?

The starting point for our approach is actually a complete change in perspective. Functional integrals in quantum physics have usually been *defined* by expressions of the following type

$$\langle \ldots | e^{-iHt/\hbar} | \ldots \rangle =: \int \mathcal{D}x \ldots, \tag{1.1}$$

where the $\mathcal{D}x$ was an ill-defined, mysterious object.

A functional integral should be defined as a mathematical object in its own right, independently of the physical system which it is meant to solve and *independently of ambiguous discretization procedures*. This is not to say that discretization is not important! Most functional integrals cannot be evaluated in closed form, so one has to rely on approximation methods. But these should only be calculational tools and not an essential part of the definition.

Our aim is to discover a mathematically coherent definition of integrals over function spaces and robust methods for computing them. Various approaches have been followed. We make connections between some of them and recover well-established results within our framework. The new approach is essentially axiomatic. It can be seen as a synthesis of ideas that have taken shape over the years.

To motivate the new approach, we will first study volume elements and integration theory on well-understood spaces before we give a detailed description of the new formalism. There are several reasons for that: we want to understand exactly why conventional integration theory fails in the cases we want to consider, and why it is necessary to give up the notion of a measure and replace it by the more general notion of a volume element (or, as we will call it later, an "integrator"). The study of well-understood cases will suggest general properties of integrals and volume elements necessary for practical calculations.

In the next section we will discuss various volume elements on spaces that arise in physical problems. But first, we want to give a very simple mathematical example which already illustrates some of the key issues of defining a volume element on an infinite-dimensional space. It will also indicate to the reader the type of information we will extract from the next sections on the way to functional integration.

Consider the integral

$$\int_{\mathbb{R}^D} d^D x\, e^{-\frac{\pi}{a}|x|^2} = a^{D/2} \quad , \text{ where } \quad |x|^2 := \sum_{\alpha=1}^{D} (x^\alpha)^2 . \tag{1.2}$$

Now we would like to take the limit $D = \infty$, but the result fails to be continuous in the parameter a, as should be reasonably desired:

$$a^\infty = \begin{cases} 0 & : \quad 0 < a < 1 \\ 1 & : \quad a = 1 \\ \infty & : \quad 1 < a \end{cases} . \tag{1.3}$$

A way out of this scaling problem is to introduce an integrator $\mathcal{D}_a x$

$$\mathcal{D}_a x = a^{-D/2} dx^1 \ldots dx^D, \text{ for } x \in \mathbb{R}^D . \tag{1.4}$$

Eq. (1.2) can be rewritten in a form which is independent of the dimension:

$$\int_{\mathbb{R}^D} \mathcal{D}_a x\, e^{-\frac{\pi}{a}|x|^2} = 1. \tag{1.5}$$

This tells us, that we will need a *dimensionless* (in the sense of dimensional analysis) volume element when we deal with infinite-dimensional spaces.

2. REVIEW: VOLUME ELEMENTS IN FINITE-DIMENSIONAL SPACES

2.1. Volume Elements in Probability Space

Let Ω be a probability space of dimension δ, with elements given in coordinate form as $\omega = (\omega^1, \ldots, \omega^\delta)$. The natural volume element in Ω has the form $d^\delta \omega = d\omega^1 \ldots d\omega^\delta$. The normalization property

$$\int_\Omega dP(\omega) = 1 \tag{2.1}$$

of the probability measure $dP(\omega)$ can be achieved by introducing a weight factor $W(\omega) \geq 0$ and by defining

$$dP(\omega) = W(\omega) d^\delta \omega \tag{2.2}$$

provided that

$$\int_\Omega d^\delta \omega \, W(\omega) = 1. \tag{2.3}$$

But such a normalized $W(\omega)$ is not always provided offhand in a physical model. So one introduces a $W_0(\omega) \geq 0$ without insisting on normalization and defines

$$d\mu_0(\omega) = W_0(\omega) d^\delta \omega. \tag{2.4}$$

Nevertheless, normalization can still be achieved by introducing the notion of the partition function

$$\mathcal{Z} := \int_\Omega d\mu_0(\omega) \tag{2.5}$$

and defining

$$W(\omega) = \mathcal{Z}^{-1} W_0(\omega). \tag{2.6}$$

The mean value (expectation value) of a physical quantity $F(\omega)$ with respect to the probability measure is then given by

$$\langle F \rangle = \int_\Omega F(\omega) dP(\omega) = \frac{\int_\Omega d^\delta \omega \, F(\omega) W_0(\omega)}{\int_\Omega d^\delta \omega \, W_0(\omega)}. \tag{2.7}$$

One application in physics of this procedure is Statistical Mechanics (in the sense of Boltzmann and Gibbs):

Given a configuration space $\mathcal{M}^D$ with coordinates

$$(q^1, \ldots, q^D) \in \mathcal{M}^D, \tag{2.8}$$

its phase space is the cotangent bundle $\Phi = T^* \mathcal{M}^D$ with canonical coordinates

$$(q^1, \ldots, q^D, p_1, \ldots, p_D) \in \Phi. \tag{2.9}$$

The Liouville volume element dV is defined on Φ as

$$dV = d^D q \, d^D p \tag{2.10}$$

where $d^D q = dq^1 \ldots dq^D$ and $d^D p = dp_1 \ldots dp_D$. This dV is invariant under a canonical change of coordinates.

So in this case $\Omega = \Phi$ and $d^\delta \omega = dV$. Moreover $W_0(\omega)$ is the Boltzmann weight:

$$W_0(q,p) = e^{-\beta H(q,p)} \tag{2.11}$$

where $H(q,p)$ is the Hamiltonian, k_B the Boltzmann constant and $\beta = 1/k_B T$ the inverse temperature. Thus the partition function is given by

$$\mathcal{Z}(\beta) = \int_\Phi d^D q \, d^D p \, e^{-\beta H(q,p)} \tag{2.12}$$

and the normalized weight is

$$W(q,p) = \mathcal{Z}^{-1}(\beta) e^{-\beta H(q,p)}. \tag{2.13}$$

2.2. Infinite Volume

There are cases where the natural volume is infinite. One such example is what could be called *Geometrical Optics in the Plane*.

Let $\mathbf{L}$ be the space of all oriented straight lines in the plane (viewed as light rays). To describe a line $L \in \mathbf{L}$ we

1. fix a base point O in the plane
2. draw a unit vector $\vec{v}$ parallel to L through O as origin.

For coordinates of L in $\mathbf{L}$ we can choose

i) the angle θ from a chosen base line to $\vec{v}$ (taken modulo 2π)

ii) the (oriented) distance p from the base point to L (a real number).

One can see that $\mathbf{L}$ is a two-dimensional manifold with coordinates (p,θ). The space $\mathbf{L}$ is invariant under the group G of rigid motions of the plane. We call *natural* the volume element on $\mathbf{L}$ invariant under the same group G as $\mathbf{L}$. Modulo a multiplicative constant, $dL = dp d\theta$ is the only natural volume element on $\mathbf{L}$. The volume of $\mathbf{L}$, namely $\int_{\mathbf{L}} dL = \int_{-\infty}^{+\infty} dp \int_0^{2\pi} d\theta$, is infinite.

We cannot define a notion of probability on $\mathbf{L}$ *which would be invariant under the group* G. Indeed, the weight $W(L)$ should be constant, but we cannot achieve normalization since $\int_L dL\, W(L)$ is infinite. We can still define the *conditional probability* of A relative to B as

$$P[A|B] = \frac{\int_A dL}{\int_B dL} \tag{2.14}$$

for $A \subset B \subset \mathbf{L}$ whenever $\int_B dL$ is finite. *This conditional probability is invariant under our group* G *of motions.*

Returning to the general situation described in 2.1, the conditional probability of A relative to B is given by

$$P[A|B] = \frac{\int_A d^\delta\omega\, W_0(\omega)}{\int_B d^\delta\omega\, W_0(\omega)} \tag{2.15}$$

for $A \subset B \subset \Omega$ provided the denominator be finite. If $\int_\Omega d^\delta\omega\, W_0(\omega)$ is finite, the (absolute) probability is equal to $P[A|\Omega]$. The conditional mean value is given by

$$\langle F\rangle_B = \frac{\int_B d^\delta\omega\, F(\omega) W_0(\omega)}{\int_B d^\delta\omega\, W_0(\omega)}. \tag{2.16}$$

Introducing the *characteristic function*

$$\chi_A(\omega) = \begin{cases} 1 & : \ \omega \in A \\ 0 & : \ \omega \notin A \end{cases}, \tag{2.17}$$

we can rewrite the previous definitions using integrals over the whole space Ω

$$P[A|B] = \frac{\int_\Omega d^\delta\omega\, \chi_A(\omega) W_0(\omega)}{\int_\Omega d^\delta\omega\, \chi_B(\omega) W_0(\omega)}, \tag{2.18}$$

$$\langle F\rangle_B = \frac{\int_\Omega d^\delta\omega\, \chi_B(\omega) W_0(\omega) F(\omega)}{\int_\Omega d^\delta\omega\, \chi_B(\omega) W_0(\omega)}. \tag{2.19}$$

2.3. Quadratic Forms

Quadratic forms enter, via Pythagoras' theorem, in the measure of lengths, *hence of volumes*. For example, in a finite-dimensional vector space $\mathbf{X}$ with a positive definite quadratic form Q, the natural volume element is given by

$$d_Q x = d\bar{x}^1 \ldots d\bar{x}^D, \tag{2.20}$$

for every cartesian coordinate system $(\bar{x}^1, \ldots, \bar{x}^D)$ orthonormal with respect to Q, namely $Q = \sum_{\alpha=1}^{D} (\bar{x}^\alpha)^2$. We recall that in a general linear coordinate system $(x^\alpha)_{1 \leq \alpha \leq D}$ where

$$Q = \sum_{\alpha,\beta} g_{\alpha\beta} x^\alpha x^\beta \tag{2.21}$$

then

$$d_Q x = \sqrt{\det g}\, dx^1 \ldots dx^D. \tag{2.22}$$

The formula remains valid for the natural volume element in a Riemannian manifold $(\mathcal{M}, g)$ with line element ds given by

$$ds^2 = \sum_{\alpha,\beta} g_{\alpha\beta} dx^\alpha dx^\beta. \tag{2.23}$$

3. REVIEW: BOREL MEASURES AND INTEGRATION

3.1. A Brief Tutorial on Lebesgue's Integration

For completeness, we must recall some important definitions of Lebesgue integration theory.

1. A **σ-field** $\mathcal{A}$ in a set $\mathbf{X}$ is a class of subsets of $\mathbf{X}$ such that
   (a) $A, B \in \mathcal{A} \Rightarrow A \setminus B \in \mathcal{A}$ and $A \cup B \in \mathcal{A}$;
   (b) $\mathbf{X} \in \mathcal{A}$, $\emptyset \in \mathcal{A}$;
   (c) $\mathcal{A}$ is closed under countable unions of subsets, $A_i \in \mathcal{A}, i \in \mathbb{N} \Rightarrow \bigcup_{i=1}^{\infty} A_i \in \mathcal{A}$.
2. A **positive measure** μ on the set $\mathbf{X}$ is a countably additive positive set function from the σ-field $\mathcal{A}$ in $\mathbf{X}$ into $\mathbb{R}_+ \cup \{+\infty\}$.

In most cases, the set $\mathbf{X}$ to be endowed with a measure is already a topological space. It is then desirable to have the measure related, in a certain sense, to the topology.

3. The **Borel σ-field** of a topological space $\mathbf{X}$ is the smallest σ-field containing the open sets of $\mathbf{X}$ (or, equivalently, the closed sets of $\mathbf{X}$). An element of the Borel σ-field is called a **Borel set**. All countable unions and countable intersections of open and closed sets are Borel sets.
4. Let $\mathbf{X}$ be a locally compact Hausdorff space. Assume $\mathbf{X}$ to be separable: there exists a countable basis for the open sets. A **Radon measure** on $\mathbf{X}$ is a positive measure μ on the Borel σ-field such that $\mu(K)$ be finite whenever $K \subset \mathbf{X}$ is compact.

Let be given a set $\mathbf{X}$, a σ-field $\mathcal{A}$ of subsets of $\mathbf{X}$ and a positive measure μ. One defines the **measurable functions** $f : \mathbf{X} \longrightarrow \mathbb{R}$ by the property that the set

$$\{f \geq a\} := \{x \in \mathbf{X} | f(x) \geq a\} \tag{3.1}$$

should belong to $\mathcal{A}$ for every real number a. Denoting by $\mathcal{M}_+$ the class of non-negative measurable functions, there is a unique way to define an integral $\int_\mathbf{X} f d\mu = \int_\mathbf{X} f(x) d\mu(x)$ for the functions f in $\mathcal{M}_+$ satisfying the following properties:

a) $0 \leq \int_\mathbf{X} f d\mu \leq +\infty$.

b) *If* $f(x) = \sum_{n=1}^{\infty} f_n(x)$ *for all* x *in* $\mathbf{X}$, *then* $\int_\mathbf{X} f d\mu = \sum_{n=1}^{\infty} \int_\mathbf{X} f_n d\mu$ ("term-by-term integration of a series").

c) *For every set* B *in* $\mathcal{A}$, *the characteristic function* χ_B *is measurable, and* $\int_\mathbf{X} \chi_B d\mu = \mu(B)$.

To complete the definition of the integral, one introduces the space L^1 of measurable functions f on $\mathbf{X}$ such that $\int_\mathbf{X} |f| d\mu$ be finite. This is a vector space, and the integral is extended by linearity from $L^1 \cap \mathcal{M}_+$ to L^1. More generally, we define the spaces L^p by

$$f \in L^p \quad \text{iff } f \text{ is measurable and } \int_\mathbf{X} |f|^p d\mu < +\infty. \tag{3.2}$$

3.2. Riesz–Markov Theorem

In this section, we state the main result and refer the reader for more details to any standard textbook on functional analysis (e.g., [Reed and Simon, 1980]).

We consider a separable compact Hausdorff space Ω. We denote by $\mathcal{C}^0(\Omega)$ the set of all continuous real-valued functions $f : \Omega \to \mathbb{R}$. We provide $\mathcal{C}^0(\Omega)$ with a Banach space structure by using the pointwise addition

$$(f_1 + f_2)(\omega) = f_1(\omega) + f_2(\omega), \tag{3.3}$$

and the sup norm

$$\| f \| := \sup_{\omega \in \Omega} |f(\omega)|. \tag{3.4}$$

Let P be a Radon measure on Ω. Hence P is a countably additive set function defined on the Borel subsets of Ω with finite values in $\mathbb{R}_+$: since Ω is compact, $P(\Omega)$ is finite, and for any Borel subset B of Ω, one has

$$0 \leq P(B) \leq P(\Omega), \tag{3.5}$$

hence $P(B)$ is finite. It follows that every measurable *bounded* function is integrable, hence $\mathcal{C}^0(\Omega)$ is contained in L^1. To the measure P we can therefore associate a linear form

$$\gamma : \mathcal{C}^0(\Omega) \longrightarrow \mathbb{R} \tag{3.6}$$

by

$$\gamma(f) = \int_\Omega f(\omega) dP(\omega). \tag{3.7}$$

We have the following properties:

a) $\gamma(f) \geq 0$ for $f \geq 0$ in $C^0(\Omega)$,

b) $|\gamma(f)| \leq P(\Omega) \parallel f \parallel$ for any f in $C^0(\Omega)$.

Conversely:

Riesz–Markov theorem: *To every linear form γ on the Banach space $C^0(\Omega)$ such that $\gamma(f) \geq 0$ for $f \geq 0$, there exists a unique Radon measure P on Ω such that*

$$\gamma(f) = \int_\Omega f(\omega) dP(\omega) \tag{3.8}$$

for all f in $C^0(\Omega)$. Moreover, P is a probability law, that is $P(\Omega) = 1$, iff γ is normalized, that is $\gamma(1) = 1$.

We describe the main steps of the construction of P:

a) Let K be a compact subset of Ω. We can construct a sequence of functions f_n in $C^0(\Omega)$ such that

$$f_0(\omega) \geq f_1(\omega) \geq \ldots \geq f_n(\omega) \geq f_{n+1}(\omega) \geq \ldots \tag{3.9}$$

for every ω in Ω and furthermore

$$\lim_{n \longrightarrow \infty} f_n(\omega) = \begin{cases} 1 & : \quad \text{if } \omega \in K \\ 0 & : \quad \text{if } \omega \in \Omega \backslash K \end{cases} . \tag{3.10}$$

Then the measure of K is given by

$$P(K) = \lim_{n \longrightarrow \infty} \gamma(f_n) \tag{3.11}$$

independently of the chosen sequence (f_n).

b) If $B \subset \Omega$ is a Borel subset, set

$$P(B) = \sup_K P(K) \tag{3.12}$$

where K runs over the compact subsets of B. This relation (3.12) expresses the so-called *inner regularity* of the measure P.

c) It remains to check the relation $\gamma(f) = \int_\Omega f dP$ for any function f in $C^0(\Omega)$.

3.3. Spin System on a Lattice

Consider a d-dimensional lattice L, isomorphic to $\mathbb{Z}^d$. At each lattice point $x = (x^1, \ldots, x^d) \in L$, with $x^i \in \mathbb{Z}$ there is associated a spin which can take the values +1 or -1. Define Ω, the space of all possible spin configurations with elements

$$\omega : L \longrightarrow \{-1, 1\}. \tag{3.13}$$

The task will be to integrate over Ω, i.e., to choose a measure on Ω. In the same spirit as in section 2.2 we hope to choose a measure which is invariant under a certain group of operations on Ω.

A natural group Γ acting on Ω is the group of *spin flips*. Let

$$s : \Omega \times L \longrightarrow \{-1, 1\} \tag{3.14}$$

$$(\omega, x) \mapsto \omega(x). \tag{3.15}$$

So we can also define the spin variables s_x (for x in L) as the functions

$$\begin{aligned} s_x : \Omega &\longrightarrow \{-1,1\} \\ \omega &\mapsto s(\omega,x) = \omega(x). \end{aligned} \tag{3.16}$$

Now define the transformation σ_x which flips the spin s_x at x and keeps the other spins s_y (for $y \neq x$) fixed. More precisely:

$$\sigma_x : \Omega \longrightarrow \Omega \tag{3.17}$$

is given by

$$(\sigma_x \omega)(y) = \begin{cases} s_y(\omega) & : \ y \neq x \\ -s_x(\omega) & : \ y = x \end{cases}. \tag{3.18}$$

It is obvious that $\sigma_x^2 = 1$ and $\sigma_x \sigma_y = \sigma_y \sigma_x$. We denote by Γ the group of transformations of Ω generated by the σ_x's. For each finite subset Λ of L, with $|\Lambda|$ elements, the spin flips σ_x for x in Λ generate a finite subgroup Γ_Λ of Γ, of order $2^{|\Lambda|}$, and isomorphic to $\mathbb{Z}_2^\Lambda$. The group Γ is the union of these subgroups Γ_Λ.

Given the group Γ, one tries to find a probability measure μ_0 on Ω, invariant under the operations of the group. A general theorem states that this measure exists and is unique (Haar measure). One then fixes μ_0 which corresponds to using a certain *a priori* distribution.

In statistical mechanics, it means fixing the high-temperature regime. Indeed, by

$$\beta := \frac{1}{k_B T} \quad , \quad d\mu_\beta := e^{-\beta H} d\mu_0 \quad , \quad Z(\beta) := \int d\mu_\beta \tag{3.19}$$

we define the Gibbs probability measure $Z(\beta)^{-1} d\mu_\beta$ at temperature T, and this has a limit proportional to $d\mu_0$ when T is large.

We begin by constructing integrals of the form

$$\int_\Omega f d\mu_0 \quad \text{where} \quad f \in C^0(\Omega). \tag{3.20}$$

The idea is to define a measure on finite approximations of Ω, then to define integration over functions on them and to take the limit as the subspaces approach Ω.

3.3.1. Finite subspaces. Let $\Lambda \subset L$ be a finite set of lattice sites of L. Define

$$\omega_\Lambda : \Lambda \longrightarrow \{-1,1\} \tag{3.21}$$

which is the restriction of $\omega \in \Omega$ to Λ, i.e., ω_Λ takes into account the values of the spins in Λ only. Introduce now the finite set

$$\Omega_\Lambda := \{1,-1\}^\Lambda \tag{3.22}$$

of "configurations in Λ." The number of elements in each Ω_Λ is finite, namely $2^{|\Lambda|}$. The measure we use on Ω_Λ will assign an equal weight $2^{-|\Lambda|}$ to each element of Ω_Λ.

3.3.2. Local functionals. Given a finite subset Λ of L and the space of local configurations Ω_Λ on Λ one can consider so-called *local* functionals on Ω_Λ. They are simply functions

$$f_\Lambda : \Omega_\Lambda \longrightarrow \mathbb{R}. \tag{3.23}$$

They form a finite-dimensional vector space $\mathbb{R}^{\Omega_\Lambda}$, that we denote by $C^0(\Omega_\Lambda)$ for consistency. Combining the local functionals with the projection

$$\begin{aligned} \Pi_\Lambda : \Omega &\longrightarrow \Omega_\Lambda \\ \omega &\mapsto \omega_\Lambda \end{aligned} \tag{3.24}$$

results in the map

$$\begin{aligned} f_\Lambda \circ \Pi_\Lambda : \Omega &\longrightarrow \mathbb{R} \\ \omega &\mapsto f_\Lambda(\omega_\Lambda). \end{aligned} \tag{3.25}$$

This map is continuous on Ω by definition of the (Tychonov) topology on the space $\Omega = \{1,-1\}^L$. By Tychonov's theorem, this space Ω *is a separable compact Hausdorff space.*

By identifying f_Λ with $f_\Lambda \circ \Pi_\Lambda$, we can consider that $C^0(\Omega_\Lambda)$ is a finite-dimensional subspace of $C^0(\Omega)$. The local functionals in $C^0(\Omega_\Lambda)$ are exactly the linear combinations of the form:

$$\sum_{\textit{all comb.}} c(x_1,\ldots,x_p)\, s_{x_1}\ldots s_{x_p} \tag{3.26}$$

with summation over all subsets $\{x_1,\ldots,x_p\}$ of Λ. The products $s_{x_1}\ldots s_{x_p}$ are functionals taking values ± 1, defined by

$$s_{x_1}\ldots s_{x_p}(\omega_\Lambda) = \omega(x_1)\ldots\omega(x_p). \tag{3.27}$$

The space of local functionals on Ω is obtained by varying Λ

$$C^0(\Omega) \supset C^0_{loc} := \bigcup_\Lambda C^0(\Omega_\Lambda). \tag{3.28}$$

It is dense in $C^0(\Omega)$.

3.3.3. Approximation of a general function. Any function f in $C^0(\Omega)$ can therefore be described as the uniform limit of a sequence of local functionals $f_{\Lambda_n} : \Omega_{\Lambda_n} \to \mathbb{R}$:

$$\lim_{n \longrightarrow \infty} f_{\Lambda_n} = f \tag{3.29}$$

where $\Lambda_1 \subset \Lambda_2 \subset \Lambda_3 \subset \ldots$, $\bigcup_n \Lambda_n = L$. Our aim is to integrate as follows:

$$\int_\Omega f\, d\mu_0 = \lim_{n \longrightarrow \infty} \int_\Omega f_{\Lambda_n} d\mu_0 \tag{3.30}$$

3.3.4. Definition of μ_0. On each Λ there are $2^{|\Lambda|}$ configurations. We assign to each element of Ω_Λ the same weight $2^{-|\Lambda|}$. This amounts to a definition of μ_0 on Ω_Λ. So if $g \in \mathcal{C}^0(\Omega_\Lambda)$ we define

$$\int_{\Omega_\Lambda} g\, d\mu_0 := \frac{1}{2^{|\Lambda|}} \sum_{\omega_\Lambda \in \Omega_\Lambda} g(\omega_\Lambda). \tag{3.31}$$

Now we can extend this to integrals of certain functions on Ω. So if: $f_\Lambda = g \circ \Pi_\Lambda$, where $g \in \mathcal{C}^0(\Omega_\Lambda)$ and $\Pi_\Lambda : \Omega \longrightarrow \Omega_\Lambda$ we define

$$\int_\Omega f_\Lambda\, d\mu_0 = \frac{1}{2^{|\Lambda|}} \sum_{\omega_\Lambda} g(\omega_\Lambda). \tag{3.32}$$

Given a sequence of functions f_{Λ_n} as in Eq. (3.29), each of them integrable in the above way, we can define $\int f d\mu_0$ as the limit of the sequence of the integrals. It is easy to show that this integral is independent of the choice of the limiting sequence (f_n).

To conclude, we now apply the Riesz–Markov theorem. From the linear functional defined above on $\mathcal{C}^0(\Omega)$ we can construct a Radon measure on Ω; it is obviously normalized, hence μ_0 is a probability measure. Furthermore, since the elements in Ω_Λ carry the same weight, our probability law μ_0 is obviously invariant under spin flips.

3.4. Integration on $\mathbb{R}^\infty$

Next we want to consider integration on $\mathbb{R}^\infty$ as an example of an infinite-dimensional space where the construction of a Borel measure is still possible. Let $\Omega = \mathbb{R}^\infty$ whose elements are sequences $x = (x_1, x_2, \ldots)$ with $x_n \in \mathbb{R}$. Define a metric

$$d(x,y) = \sum_{n=1}^{\infty} \inf(|x_n - y_n|, 2^{-n}). \tag{3.33}$$

(2^{-n} could be replaced by any constant ε_n converging to 0 for $n \longrightarrow \infty$.) The result is a non-compact Polish space Ω.

By definition, a space Ω is said to be a *Polish* space, if

1. Ω is a metric space with topology inherited from the metric;
2. Ω is complete;
3. there exists a countable dense subset in Ω.

We will first consider *compact subspaces* of Ω. For instance, let I be a finite closed interval in $\mathbb{R}$, for instance $I = [0,1]$. The subspace I^∞ of $\mathbb{R}^\infty$ consists of the sequences $x = (x_1, x_2, \ldots)$ with $x_n \in I$ for every index n. It is a compact metric space. To construct a Radon measure on I^∞, we shall use Riesz–Markov theorem; we have therefore to construct the integral of the continuous functions $F : I^\infty \to \mathbb{R}$. We use the same strategy as in the previous section. For a given integer $n \geq 1$, define the space I^n as the unit cube in $\mathbb{R}^n$ consisting of vectors $x = (x_1, \ldots, x_n)$ with

$$0 \leq x_1 \leq 1\,,\; 0 \leq x_2 \leq 1\,,\; \ldots,\; 0 \leq x_n \leq 1. \tag{3.34}$$

We use the projection map $\Pi_n : I^\infty \longrightarrow I^n$ which associates to an infinite sequence $x = (x_1, x_2, \ldots)$ its truncation

$$\Pi_n(x) = (x_1, \ldots, x_n). \tag{3.35}$$

If $F_n : I^n \to \mathbb{R}$ is a continuous function, the composition $F_n \circ \Pi_n = f_n$ is a continuous function on I^∞. In this way, we identify $C^0(I^n)$ to a subspace of $C^0(I^\infty)$ and define *local functionals* on I^∞ by

$$C^0_{loc} := \bigcup_{n\geq 1} C^0(I^n). \tag{3.36}$$

This is a dense subspace of $C^0(I^\infty)$. That is, any continuous functional F on I^∞ is the uniform limit of local functionals; the approximation can be done explicitly as follows:

$$F(x) = \lim_{n\longrightarrow\infty} F(x_1,\ldots,x_n,0,0,\ldots). \tag{3.37}$$

We then define explicitly the integral of F by

$$\int_{I^\infty} d^\infty x\, F(x) := \lim_{n\longrightarrow\infty} \int_{I^n} dx_1 \ldots dx_n\, F(x_1,\ldots,x_n,0,0,\ldots). \tag{3.38}$$

Using the Riesz–Markov theorem again, we construct a probability measure μ on I^∞, to be denoted as a volume element $d^\infty x$. This is essentially the way Daniell used around 1920 to define an integral of functions of infinitely many variables.

For this construction the normalization of the measure $\int_0^1 dx = 1$ was crucial. But finite intervals $J = [a,b]$ can also be dealt with by re-scaling of the measure: $dx \longrightarrow \frac{dx}{b-a}$ since $\int_a^b \frac{dx}{b-a} = 1$.

But this is still not satisfactory for our purposes because the restriction to a finite interval I remains. One way to get integration on $\mathbb{R}$ is to introduce a *probability density* $p(x)$ with the properties:

1. $p(x) \geq 0$,
2. $\int_{-\infty}^{+\infty} dx\, p(x) = 1$.

From this one defines a probability measure on $\mathbb{R}^n$

$$d\gamma_n(\vec{x}) := p(x_1)\ldots p(x_n) dx_1 \ldots dx_n \tag{3.39}$$

(for $\vec{x} = (x_1,\ldots,x_n)$ in $\mathbb{R}^n$). If there exists a finite interval I in $\mathbb{R}$ such that $p(x) = 0$ for $x \notin I$, we can imitate Daniell's procedure and define a probability measure γ_∞ on $\mathbb{R}^\infty$ such that the integral of a continuous functional F on $\mathbb{R}^\infty$ can be defined explicitly by

$$\int_{\mathbb{R}^\infty} d\gamma_\infty\, F = \lim_{n\longrightarrow\infty} \int_{\mathbb{R}^n} dx_1 \ldots dx_n\, p(x_1)\ldots p(x_n)\, F(x_1,\ldots,x_n,0,0,\ldots). \tag{3.40}$$

Unfortunately, our condition $p(x) = 0$ for $x \notin I$ is not satisfied in the basic example of a Gaussian distribution

$$p(x) = \frac{1}{\sigma\sqrt{2\pi}} e^{-\frac{x^2}{2\sigma^2}} \tag{3.41}$$

with standard deviation σ. Since we have to consider the whole space $\mathbb{R}^\infty$, and not its compact subspace I^∞, we cannot apply Daniell's method directly. Here is a familiar trick: we compactify $\mathbb{R}$ to the extended real line

$$\overline{\mathbb{R}} = \mathbb{R} \cup \{-\infty, +\infty\} = [-\infty, +\infty]. \tag{3.42}$$

Assume that $p(x) > 0$ for all x in $\mathbb{R}$, as in the case of Eq. (3.41) for instance. Using the probability density $p(x)$, define a homeomorphism Φ of $\overline{\mathbb{R}}$ onto $I = [0,1]$ by

$$\Phi(x) = \int_{-\infty}^{x} dt\, p(t). \tag{3.43}$$

We extend it to a homeomorphism Φ^∞ of $\overline{\mathbb{R}}^\infty$ onto I^∞ component-wise. The Daniell measure $d^\infty x$ on I^∞ can be transported back to a measure $\overline{\gamma}_\infty$ on $\overline{\mathbb{R}}^\infty$. Now it can be shown that the spurious sequences in $\overline{\mathbb{R}}^\infty \setminus \mathbb{R}^\infty$ (with at least one infinite coordinate) form a null set N, that is $\overline{\gamma}_\infty(N) = 0$. It is then possible to restrict $\overline{\gamma}_\infty$ to a probability measure γ_∞ on $\mathbb{R}^\infty$. It is characterized by the following property

$$\int_{\mathbb{R}^\infty} F(x) d\gamma_\infty(x) = \int_{\mathbb{R}^n} dx_1 \ldots dx_n\, p(x_1) \ldots p(x_n)\, F_n(x_1, \ldots, x_n) \tag{3.44}$$

if F is a local functional, namely

$$F(x) = F_n(x_1, \ldots, x_n) \quad \text{for } x = (x_1, x_2, \ldots). \tag{3.45}$$

In more abstract terms, any measure μ defined on the Borel σ-algebra of a Polish space Ω such that $\mu(\Omega)$ be finite has the *tightness* property. Namely, we can decompose Ω as the union of a null set N (with $\mu(N) = 0$) and disjoint *compact* subsets $K_1, K_2, \ldots$. For any function F on Ω, we have

$$\int_\Omega F d\mu = \sum_n \int_{K_n} f d\mu_n \tag{3.46}$$

where μ_n is a Radon measure on K_n. From this property, one can deduce a generalization of the Riesz–Markov theorem valid for an arbitrary Polish space; the measures on Ω can be identified with the so-called *tight linear forms* on the Banach space $C_b^0(\Omega)$ of bounded continuous functions on Ω (see [Bourbaki, 1982], p.58 for instance).

3.5. Problems with Infinite Dimensions

Our final aim is to define a volume element in the infinite-dimensional space $\mathbb{R}^\infty$. In the D-dimensional space $\mathbb{R}^D$ consisting of vectors $\vec{x} = (x_1, \ldots, x_D)$, the volume element is $d^D\vec{x} = \prod_{n=1}^D dx_n$ and an integral like $\int_A d^D x$ extended over the domain A defined by

$$a_1 \leq x_1 \leq b_1, \ldots, a_D \leq x_D \leq b_D \tag{3.47}$$

is equal to $\prod_{n=1}^D (b_n - a_n)$. By analogy, the volume element $d^\infty x = \prod_{n=1}^\infty dx_n$ in $\mathbb{R}^\infty$ should satisfy

$$\int_A d^\infty x = \prod_{n=1}^\infty (b_n - a_n) \tag{3.48}$$

when A is the domain in $\mathbb{R}^\infty$ defined by the inequalities $a_n \leq x_n \leq b_n$ for all $n \geq 1$. By Daniell's method described in section 3.4, we can make sense of this when $A = I^\infty$, or more generally when $b_n - a_n = 1$ for all $n \geq 1$ (or slightly more generally, when $\prod_{n=1}^\infty (b_n - a_n)$ is a convergent infinite product). We cannot go beyond this because of the scaling problem mentioned in the introduction: replacing x_n by Cx_n for all n transforms $d^D x$ into $C^D d^D x$ for finite D, but the limit $D = \infty$ is untractable (for $C \neq 1$) since by any means we ought to have

$$C^\infty = \begin{cases} 0 & : \quad 0 < C < 1 \\ 1 & : \quad C = 1 \\ \infty & : \quad C > 1 \end{cases} . \tag{3.49}$$

A way out of this difficulty is to introduce a damping factor as we have seen in the previous section. If we replace dx_n by $p(x_n)\,dx_n$ where $p(x)$ is a probability density, the probability distribution $d\gamma_D(\vec{x}) = \prod_{n=1}^{D} p(x_n)\,dx_n$ on $\mathbb{R}^D$ has an analogue on $\mathbb{R}^\infty$: it is a measure γ_∞ on the Borel σ-algebra of the Polish space $\mathbb{R}^\infty$ such that

$$\int_{\mathbb{R}^\infty} F(x)\,d\gamma_\infty(x) = \lim_{n\to\infty}\int_{\mathbb{R}^n} dx_1\dots dx_n\,p(x_1)\dots p(x_n)\,F(x_1,\dots,x_n,0,0,\dots) \tag{3.50}$$

for an arbitrary bounded continuous functional F on $\mathbb{R}^\infty$. Because of this relation, we can write this measure in the form

$$d\gamma_\infty(x) = \prod_{n=1}^{\infty} p(x_n)\,dx_n. \tag{3.51}$$

To be specific, consider the Gaussian case where

$$p_\sigma(x) = \frac{1}{\sigma\sqrt{2\pi}}e^{-\frac{1}{2}\left(\frac{x}{\sigma}\right)^2}. \tag{3.52}$$

To bypass the scaling problem of manipulating expressions like $\left(\frac{1}{\sigma\sqrt{2\pi}}\right)^\infty$, we limit ourselves to the case $\sigma := \frac{1}{\sqrt{2\pi}}$, hence

$$p(x) = e^{-\pi x^2}. \tag{3.53}$$

Notice the normalization property:

$$\int_{-\infty}^{\infty} dx\,e^{-\pi x^2} = 1. \tag{3.54}$$

On the space $\mathbb{R}^\infty$, there exists a *bona fide* probability measure

$$d\Gamma_\infty(x) = \prod_{n=1}^{\infty}\left(e^{-\pi x_n^2}dx_n\right). \tag{3.55}$$

On the finite-dimensional space $\mathbb{R}^D$, we can split the corresponding measure

$$d\Gamma_D(\vec{x}) := \prod_{n=1}^{D}\left(e^{-\pi x_n^2}dx_n\right) \tag{3.56}$$

as a product

$$d\Gamma_D(\vec{x}) = \prod_{n=1}^{D}\left(e^{-\pi x_n^2}\right)\cdot d^D\vec{x}. \tag{3.57}$$

We cannot split $d\Gamma_\infty(x)$ *as the product of* $\prod_{n=1}^{\infty} e^{-\pi x_n^2}$ *and some volume element* $d^\infty x$: the reason is that, except for a null set N of sequences x with $\Gamma(N) = 0$, one has $\sum_{n=1}^{\infty}(x_n)^2 = +\infty$, hence

$$\prod_{n=1}^{\infty} e^{-\pi x_n^2} = e^{-\pi\sum_{n=1}^{\infty}(x_n)^2} = 0. \tag{3.58}$$

3.6. Invariance Properties of Gaussian Measures

We consider $\mathbb{R}^\infty$ as a vector space with the vector addition

$$(x_1,x_2,\ldots)+(y_1,y_2,\ldots)=(x_1+y_1,x_2+y_2,\ldots). \tag{3.59}$$

There is a vector subspace ℓ^2 of $\mathbb{R}^\infty$ consisting of the sequences x such that $\sum_{n=1}^{\infty}(x_n)^2<+\infty$, with the standard dot product

$$x\cdot y=\sum_{n=1}^{\infty}x_ny_n \tag{3.60}$$

for $x=(x_1,x_2,\ldots)$ and $y=(y_1,y_2,\ldots)$ in ℓ^2.

A fundamental invariance property of the Gaussian measure $d\Gamma_\infty(x)$ defined by Eq. (3.55) is supplied by the so-called *Cameron–Martin formula*:

$$\int_{\mathbb{R}^\infty} d\Gamma_\infty(x)\,F(x)=e^{-\pi c\cdot c}\int_{\mathbb{R}^\infty} d\Gamma_\infty(x)\,e^{-2\pi c\cdot x}F(x+c) \tag{3.61}$$

for any sequence c in ℓ^2. The subtle point in it is the definition of $c\cdot x$ for c in ℓ^2 and x in $\mathbb{R}^\infty$. Indeed, the dot product $c\cdot x$ is *a priori* defined only for x in ℓ^2, but ℓ^2 is a null set in $\mathbb{R}^\infty$, namely $\Gamma_\infty(\ell^2)=0$. But it can be proven that, for any c in ℓ^2, the series $\sum_{n=1}^{\infty}c_nx_n$ converges (not absolutely!) for all x in $\mathbb{R}^\infty\setminus N_c$ where N_c is a null set, namely $\Gamma_\infty(N_c)=0$.

Life would be easier if we could split $d\Gamma_\infty(x)$ in the form

$$d\Gamma_\infty(x)=e^{-\pi x\cdot x}\mathcal{D}x, \tag{3.62}$$

where $\mathcal{D}x$ corresponds to our previously denoted $d^\infty x$. In a symbolic way, the Cameron–Martin formula can be written as

$$d\Gamma_\infty(x+c)=e^{-\pi(c\cdot c+2c\cdot x)}d\Gamma_\infty(x). \tag{3.63}$$

Using Eq. (3.62) in a formal way, we rewrite this as follows:

$$e^{-\pi(x+c)\cdot(x+c)}\mathcal{D}(x+c)=e^{-\pi(c\cdot c+2c\cdot x)}e^{-\pi x\cdot x}\mathcal{D}x. \tag{3.64}$$

From the obvious relation:

$$(x+c)\cdot(x+c)=c\cdot c+2c\cdot x+x\cdot x, \tag{3.65}$$

one concludes that Eq. (3.64) can be rewritten as

$$\mathcal{D}(x+c)=\mathcal{D}x. \tag{3.66}$$

Hence, at a formal level, *Cameron–Martin's formula expresses the translational invariance of the mythical integrator* $\mathcal{D}x=\prod_{n=1}^{\infty}dx_n$ *under any element* c *of* ℓ^2.

Rotational invariance can be treated in a similar way. Let U be an orthogonal operator in the (real) Hilbert space ℓ^2: there exists an infinite matrix $(u_{mn})_{m,n\geq 1}$ with real entries such that

$$\sum_{n=1}^{\infty}u_{mn}^2=1 \tag{3.67}$$

$$(Uc)_m=\sum_{n=1}^{\infty}u_{mn}c_n \tag{3.68}$$

for all integers $m \geq 1$ and all c in ℓ^2. By the convergence result stated above, there exists a null set N in $\mathbb{R}^\infty$ such that, for any x in $\mathbb{R}^\infty \setminus N$, the transform $\hat{U}x$ is defined as an element of $\mathbb{R}^\infty$, in such a way that

$$(\hat{U}x)_m = \sum_{n=1}^{\infty} u_{mn}x_n \tag{3.69}$$

for all integers $m \geq 1$. Defining arbitrarily $\hat{U}x$ for the exceptional x's in N, we have constructed a measurable map $\hat{U}$ from $\mathbb{R}^\infty$ to $\mathbb{R}^\infty$, extending U. *This transformation $\hat{U}$ leaves the measure Γ_∞ invariant.*

Since the orthogonality of U is expressed by

$$Ux \cdot Ux = x \cdot x \tag{3.70}$$

it follows from Eq. (3.62) that the *invariance of Γ_∞ under $\hat{U}$ can be restated as an invariance of $\mathcal{D}x$ under the rotation U of ℓ^2.*

4. REVIEW: INTEGRATION ON SPACES OF PATHS

A space of paths has more structure than $\mathbb{R}^\infty$.

4.1. Theory of Oscillatory Integrals

4.1.1. Motivation. We will be concerned with integrands of the form $e^{i\pi Q(x)}$ where Q is a (real) quadratic form. Why is there a problem to integrate these?

The main problem is that the function to be integrated is of modulus one, hence the integral is not absolutely convergent over an infinite domain. In one dimension, we define the *improper integral*

$$\gamma := \lim_{A \to \infty} \int_{-A}^{A} dx\, e^{i\pi x^2} \tag{4.1}$$

as a limit. That the limit exists can be proved by an integration by parts:

$$\int_1^A dx\, e^{i\pi x^2} = \frac{1}{2\pi i} \int_1^A dx\, e^{i\pi x^2}/x^2 + \left.\frac{e^{i\pi x^2}}{2\pi i x}\right|_{x=1}^{x=A}; \tag{4.2}$$

this trick reduces the improper integral $\int_1^{+\infty} dx\, e^{i\pi x^2}$ to the *absolutely convergent* integral $\int_1^{+\infty} dx\, e^{i\pi x^2}/x^2$.

In the many-dimensional case, there is no convenient method to define an improper integral in a way that is independent of the choice of a coordinate system. For instance, from Eq. (4.1), one gets

$$\gamma^2 = \lim_{R \to \infty} \int_{-R}^{R} dx \int_{-R}^{R} dy\, e^{i\pi(x^2+y^2)}, \tag{4.3}$$

providing a definition of the improper integral over the plane $\mathbb{R}^2$ of the function $e^{i\pi\left(x^2+y^2\right)}$. But if we change to polar coordinates, we get the integral

$$\int_0^\infty dr \int_0^{2\pi} d\theta\, r e^{i\pi r^2} = 2\pi \int_0^\infty dr\, r e^{i\pi r^2}$$

$$= \lim_{R \to \infty} \int_0^R dr\, 2\pi r e^{i\pi r^2} \tag{4.4}$$
$$= i \lim_{R \to \infty} \left(1 - e^{i\pi R^2}\right)$$

and the limit does not exist.

Let us remark that both integrals in Eq. (4.3) and Eq. (4.4) are obtained via a *sharp cut-off* of the form

$$\lim_{R \to \infty} \int\int_{\mathbb{R}^2} dx\, dy\, \phi(x/R, y/R)\, e^{i\pi(x^2+y^2)} \tag{4.5}$$

with the following choices of ϕ

$$\phi(x,y) = \begin{cases} 1 & : \text{ if } |x| \leq 1, |y| \leq 1 \\ 0 & : \text{ otherwise} \end{cases}, \tag{4.6}$$

in the first case, and

$$\phi(x,y) = \begin{cases} 1 & : \text{ if } x^2 + y^2 \leq 1 \\ 0 & : \text{ otherwise} \end{cases} \tag{4.7}$$

in the second case. Notice that $e^{i\pi R^2}$ oscillates and does not have a limit for $R \to \infty$. But in a heuristic sense, $e^{i\pi R^2}$ oscillates so rapidly for large R that it vanishes at a macroscopic level. On a more mathematical side, the statement is substantiated by using the well-known

Riemann–Lebesgue lemma: *For any function f in $L^1(\mathbb{R})$, one has*

$$\lim_{\omega \to \infty} \int_{\mathbb{R}} dx\, f(x) e^{i\omega x} = 0. \tag{4.8}$$

Assuming this, the limit in Eq. (4.4) is equal to i; hence $\gamma^2 = i$.

To ensure a limit in the case of polar coordinates, we have therefore to smear over the radius R, and this means to replace the sharp plateau function ϕ in Eq. (4.7) by a smooth function. Let us make the following assumptions:

a) *the function ϕ is continuous with continuous first derivatives $\partial_x \phi$ and $\partial_y \phi$;*

b) *the integral*

$$\int\int dxdy \left(|\partial_x \phi| + |\partial_y \phi| + |\phi|\right) \tag{4.9}$$

is finite;

c) *ϕ is normalized by $\phi(0,0) = 1$.*

For instance, we can take $\phi(x,y) = e^{-(x^2+y^2)}$.

As cut-off function, we use the scaled function

$$\phi_R(x,y) = \phi(x/R, y/R) \tag{4.10}$$

for $R > 0$. Hence, $\phi_R(x,y)$ tends to 1 as $R \to \infty$ (by assumption c)). Then, *for any $R > 0$, the function $e^{i\pi(x^2+y^2)} \phi_R(x,y)$ is absolutely integrable, and the limit*

$$\lim_{R \to \infty} \int_{\mathbb{R}^2} dxdy\, e^{i\pi(x^2+y^2)} \phi_R(x,y) \tag{4.11}$$

exists and is independent of the cut-off function ϕ. The limit will be used as the *regularized integral* of $e^{i\pi(x^2+y^2)}$ over the plane $\mathbb{R}^2$.

This example provides us with an idea of how to define the integral of non-absolutely integrable functions. The way we will introduce smooth cut-offs will be via Fourier transforms and the Parseval formula. This method will later be generalized to the infinite-dimensional case and will enable us to define so-called *oscillatory integrals* that play an important part in our new approach. These *oscillatory integrals* are essentially, apart from notational changes, the same as introduced by Albeverio and Høegh-Krohn (see [Albeverio and Høegh-Krohn, 1976]).

4.1.2. A brief reminder on Fourier transforms. Let V be a finite-dimensional real vector space of dimension d. Using a linear system of coordinates, we identify V to the space $\mathbb{R}^d$ of column vectors $x = (x^1, \dots, x^d)^T$. The elements of the dual space V' are the linear forms on V, and are represented by row vectors $\xi = (\xi_1, \dots, \xi_d)$. The natural duality between V and V' is defined by

$$\langle \xi, x \rangle_V = \sum_{\alpha=1}^{d} \xi_\alpha x^\alpha. \tag{4.12}$$

The *Schwartz space* $S(V)$ consists of the C^∞ functions $\phi(x)$ on V such that $p(x, \partial_x)\phi(x)$ be bounded over V for all differential operators with polynomial coefficients, of the form

$$p(x, \partial_x) = \sum_{\alpha_1=1}^{N} \cdots \sum_{\alpha_d=1}^{N} \sum_{\beta_1=1}^{N} \cdots \sum_{\beta_d=1}^{N} c_{\alpha\beta} (x^1)^{\alpha_1} \dots (x^d)^{\alpha_d} \left(\frac{\partial}{\partial x^1}\right)^{\beta_1} \cdots \left(\frac{\partial}{\partial x^d}\right)^{\beta_d}. \tag{4.13}$$

We define similarly the Schwartz space $S(V')$.

For any function $\phi(x)$ in $S(V)$, its *Fourier transform* is defined by an absolutely convergent integral

$$\hat{\phi}(\xi) = \int_V d_V x \, \phi(x) e^{2\pi i \langle \xi, x \rangle}, \tag{4.14}$$

with the volume element $d_V x = dx^1 \dots dx^d$. With this definition, the function $\hat{\phi}(\xi)$ belongs to $S(V')$ and the *Fourier inversion formula* holds, namely

$$\phi(x) = \int_{V'} d_{V'} \xi \, \hat{\phi}(\xi) e^{-2\pi i \langle \xi, x \rangle} \tag{4.15}$$

with the volume element $d_{V'} \xi = d\xi_1 \dots d\xi_d$ (with no constants!).

Considering the functions in $S(V)$ as test functions, one introduces a dual space $S'(V)$ of generalized functions ("tempered distributions") consisting of the linear forms T on $S(V)$ satisfying an estimate of the form

$$|T(\phi)| \le \sup_{x \in V} |p(x, \partial_x)\phi(x)|, \tag{4.16}$$

for a suitable differential operator p. The tempered distributions on V' are defined similarly. We abuse notation by writing $T(\phi)$ in the form

$$T(\phi) = \int_V d_V x \, T(x)\phi(x), \tag{4.17}$$

despite the fact that $T(x)$ does not make sense for an individual point x. The so-called Dirac function $\delta(x)$ is a particular case, with the standard definition

$$\int_V d_V x\, \phi(x)\delta(x) = \phi(0). \tag{4.18}$$

For a tempered distribution T in $S'(V)$, its Fourier transform is a tempered distribution $\hat{T}$ in $S'(V')$ characterized by the *Parseval formula*

$$T(\bar{\phi}) = \hat{T}(\bar{\hat{\phi}}), \tag{4.19}$$

where $\bar{\phi}$ denotes the complex-conjugate of ϕ. For instance, if $T(x)$ is a function such that $\int_V d_V x\, |T(x)|$ is finite, it can be considered as a tempered distribution by using Eq. (4.17) literally, and its Fourier transform is the bounded continuous function $\hat{T}(\xi)$ defined by

$$\hat{T}(\xi) = \int_V d_V x\, e^{2\pi i\langle\xi,x\rangle} T(x). \tag{4.20}$$

4.1.3. Oscillatory distributions. Definition: *An element T in $S'(V)$ is said to be an oscillatory distribution if its Fourier transform $\hat{T}$ in $S'(V')$ is a smooth function in $C^\infty(V')$.*

More explicitly, $\hat{T}(\xi)$ is a smooth function on V, necessarily of *polynomial growth*, namely

$$|\hat{T}(\xi)| \leq |P(\xi)| \tag{4.21}$$

holds for a suitable polynomial P and every ξ in V'. Furthermore, for any test function $\phi(x)$ in $S(V)$, one has

$$T(\phi) = \int_{V'} d_{V'}\xi\, \hat{T}(\xi) \int_V d_V x\, e^{-2\pi i\langle\xi,x\rangle} \phi(x). \tag{4.22}$$

Here is an important remark: If $\hat{T}$ is integrable in Lebesgue's sense, that is if $\int_{V'} d_{V'}\xi\, |\hat{T}(\xi)|$ is finite, we can interchange the order of integrations in Eq. (4.22) and we get

$$T(\phi) = \int_V d_V x\, T(x)\phi(x) \tag{4.23}$$

with

$$T(x) = \int_{V'} d_{V'}\xi\, e^{-2\pi i\langle\xi,x\rangle} \hat{T}(\xi). \tag{4.24}$$

It means that $T(x)$ is a function, the inverse Fourier transform of the function $\hat{T}(\xi)$.

We give a specific example. Consider the function

$$T(x) = e^{i\pi\left((x^1)^2+\ldots+(x^d)^2\right)}. \tag{4.25}$$

It can be shown that its Fourier transform is similar in form

$$\hat{T}(\xi) = \gamma^d e^{-i\pi\left((\xi_1)^2+\ldots+(\xi_d)^2\right)} \tag{4.26}$$

with the constant γ defined in Eq. (4.1). It is well-known that γ is equal to $e^{i\pi/4}$. More generally, let $Q(x)$ be a nondegenerate quadratic form on V, of the form

$$Q(x) = \sum_{\alpha=1}^{d} \sum_{\beta=1}^{d} q_{\alpha\beta}\, x^\alpha x^\beta. \tag{4.27}$$

Introduce the inverse quadratic form W on V' by

$$W(\xi) = \sum_{\alpha=1}^{d} \sum_{\beta=1}^{d} q^{\alpha\beta} \xi_\alpha \xi_\beta \tag{4.28}$$

where the matrices $(q_{\alpha\beta})$ and $(q^{\alpha\beta})$ are inverse to each other. Then the Fourier transform of $T(x) = e^{i\pi Q(x)}$ is $\hat{T}(\xi) = \gamma^I |\Delta|^{-1/2} e^{-i\pi W(\xi)}$, where $\Delta = \det(q_{\alpha\beta})$ is the discriminant and I is the index of the quadratic form Q: that is $I = I_+ - I_-$ where I_+ (I_-) is the number of positive (negative) eigenvalues of the real symmetric matrix $(q_{\alpha\beta})$. To conclude, *$e^{i\pi Q}$ is an oscillatory distribution.*

4.1.4. Oscillatory integrals. Definition: *If T is an oscillatory distribution on V, with Fourier transform $\hat{T}$ in $C^\infty(V')$, we define*

$$\tilde{\int_V} d_V x \, T(x) := \hat{T}(0). \tag{4.29}$$

Notice that if $T(x)$ is an integrable function, its Fourier transform is given by an absolutely convergent integral

$$\hat{T}(\xi) = \int_V d_V x \, e^{2\pi i \langle \xi, x \rangle} T(x). \tag{4.30}$$

Letting $\xi = 0$, we see that $\hat{T}(0)$ is the integral of $T(x)$ over V, hence *our definition generalizes the standard definition of an integral.*

In our basic example $T(x) = e^{i\pi Q(x)}$, we get $\hat{T}(\xi) = \gamma^I |\Delta|^{-1/2} e^{-i\pi W(\xi)}$ and letting $\xi = 0$, we calculate the oscillatory integral

$$\tilde{\int_V} d_V x \, e^{i\pi Q(x)} = \gamma^I |\det(q_{\alpha\beta})|^{-1/2} \tag{4.31}$$

and in particular

$$\tilde{\int_{\mathbb{R}^d}} dx^1 \dots dx^d \, e^{i\pi((x^1)^2 + \dots + (x^d)^2)} = e^{i\pi d/4}. \tag{4.32}$$

We add a few remarks:

a) Let $T(x)$ be an oscillatory function (or distribution). Then
   - $U(x) = T(x) e^{2\pi i \langle \xi_0, x \rangle}$ is again oscillatory. (If we perform the Fourier transformation on both sides, we get $\hat{U}(\xi) = \hat{T}(\xi + \xi_0)$).
   - $\tilde{\int_V} d_V x \, T(x) e^{2\pi i \langle \xi_0, x \rangle} = \hat{T}(\xi_0)$ since $\tilde{\int_V} d_V x \, U(x) = \hat{U}(0)$.

   So the formula

$$\hat{T}(\xi) = \tilde{\int_V} d_V x \, T(x) e^{2\pi i \langle \xi, x \rangle} \tag{4.33}$$

   holds.

b) Introduce a scaling parameter $R > 0$ and the cut-off function $\phi_R(x) = \phi(x/R)$ for a given function ϕ in $S(V)$ with $\phi(0) = 1$. Then

$$\tilde{\int_V} d_V x \, T(x) = \lim_{R \to \infty} T(\phi_R) \tag{4.34}$$

holds. If T is an oscillatory *function*, we get in particular

$$\tilde{\int}_V d_V x\, T(x) = \lim_{R\longrightarrow\infty} \int_V d_V x\, \phi(x/R)T(x), \tag{4.35}$$

where the right-hand side is a limit of absolutely convergent integrals. To prove Eq. (4.34), use first Parseval's formula, hence

$$\begin{aligned} T(\phi_R) &= \int_{V'} d_{V'}\xi\, R^d \hat{\phi}(R\xi)\hat{T}(-\xi) \\ &= \int_{V'} d_{V'}\eta\, \hat{\phi}(\eta)\hat{T}(-\eta/R), \end{aligned} \tag{4.36}$$

and then the Lebesgue's dominated convergence theorem

$$\begin{aligned} \lim_{R\longrightarrow\infty} \int_{V'} d_{V'}\eta\, \hat{\phi}(\eta)\hat{T}(-\eta/R) &= \hat{T}(0) \int_{V'} d_{V'}\eta\, \hat{\phi}(\eta) \\ &= \hat{T}(0)\phi(0) \\ &= \hat{T}(0) \\ &= \tilde{\int}_V d_V x\, T(x). \end{aligned} \tag{4.37}$$

From now on, we shall abuse notation by writing an oscillatory integral $\tilde{\int}$ as an ordinary integral $\int$.

To summarize what has been achieved: Given a finite-dimensional vector space V, with dual space V', we defined a class of objects, the "oscillatory distributions" on V. Such a distribution T is introduced via the Fourier transform $\hat{T}$, a smooth (C^∞) function on V', with polynomial growth (see Eq. (4.21)). By convention, we define the oscillatory integral $\int_V d_V x\, T(x)$ (or more pedantically $\tilde{\int}_V d_V x\, T(x)$) as $\hat{T}(0)$. We want also to consider T as a linear functional on a space of "test functions" on V. The standard way is to consider the Schwartz space $S(V)$ consisting of the functions of the form

$$\phi(x) = \int_{V'} d_{V'}\xi\, \psi(\xi) e^{2\pi i\langle\xi,x\rangle} \tag{4.38}$$

and to state a definition

$$\langle T, \phi\rangle := \int_{V'} d_{V'}\xi\, \psi(\xi)\hat{T}(\xi). \tag{4.39}$$

In Eq. (4.38), the function ψ belongs to the Schwartz space $S(V')$, but the crucial property is that the product $P(\xi)\psi(\xi)$ should be (absolutely) integrable on V' for every polynomial $P(\xi)$ on V'. Because of the polynomial growth of $\hat{T}(\xi)$, the integral in Eq. (4.39) is then absolutely convergent for such ψ. This suggests to introduce a larger space of test functions, to be denoted $\Sigma(V)$, consisting of the Fourier–Stieltjes transforms

$$\phi(x) = \int_{V'} d\mu(\xi)\, e^{2\pi i\langle\xi,x\rangle}. \tag{4.40}$$

Here μ is a (complex) bounded Radon measure on the locally compact space V' such that the integral $\int_{V'} |d\mu(\xi)||P(\xi)|$ should be finite for every polynomial function $P(\xi)$ on V'. Equivalently, we assume that

$$\int_{V'} |d\mu(\xi)|\, \| \xi \|^N < +\infty \tag{4.41}$$

holds for every integer $N \geq 0$, where $\| \xi \|$ is a norm on V' (arbitrarily chosen). Under these conditions, we can state the following definition

$$\langle T, \phi \rangle := \int_{V'} d\mu(\xi)\, \hat{T}(\xi), \tag{4.42}$$

which generalizes Eq. (4.39) since $S(V)$ is contained in $\Sigma(V)$.

There are two main advantages in this new definition:

a) Taking μ to be a point mass at 0, so that

$$\int_{V'} d\mu(\xi)\, F(\xi) = F(0) \tag{4.43}$$

for every measurable function $F(\xi)$, we see that the constant function 1 on V is a test function in $\Sigma(V)$, and that $\langle T, 1 \rangle = \hat{T}(0)$, hence

$$\langle T, 1 \rangle = \int_V d_V x\, T(x). \tag{4.44}$$

b) Given an oscillatory distribution T, defined via its Fourier transform $\hat{T}(\xi)$, and a test function $\phi(x)$ in $\Sigma(V)$ as in Eq. (4.40), we can define a new oscillatory distribution $\phi \cdot T$ via its Fourier transform

$$\widehat{\phi \cdot T}(\xi) = \int_{V'} d\mu(\eta)\, \hat{T}(\xi + \eta). \tag{4.45}$$

We have an associativity rule, namely

$$\langle T, \phi_1 \phi_2 \rangle = \langle \phi_1 \cdot T, \phi_2 \rangle \tag{4.46}$$

for ϕ_1 and ϕ_2 in $\Sigma(V)$, hence

$$\langle T, \phi \rangle = \langle \phi \cdot T, 1 \rangle. \tag{4.47}$$

That is $\langle T, \phi \rangle$ is the oscillatory integral of $\phi \cdot T$ over V. Things are consistent.

4.1.5. Infinite-dimensional case. We are ready to consider the infinite-dimensional case. Let $\mathbf{X}$ be any separable Banach space, with norm $\| x \|$. We consider the dual Banach space $\mathbf{X}'$ consisting of the continuous linear forms ξ on $\mathbf{X}$, and denote by $\langle \xi, x \rangle$ the value taken by ξ at x. The norm on $\mathbf{X}$ induces a norm on $\mathbf{X}'$ by

$$\| \xi \| = \sup_{x \neq 0} |\langle \xi, x \rangle| / \| x \| . \tag{4.48}$$

Hence $\mathbf{X}'$ is a Banach space; *we assume* $\mathbf{X}'$ *to be separable.* Then both $\mathbf{X}$ and $\mathbf{X}'$ are Polish spaces, and integration theory works on these spaces as explained in section 3. A **complex bounded measure** on $\mathbf{X}'$ is a set function $\mu(A)$ assigning complex values to the Borel subsets A in $\mathbf{X}'$ such that

$$\mu(A) = \sum_{n=1}^{\infty} \mu(A_n) \tag{4.49}$$

whenever A is a disjoint union of the Borel subsets A_n. It is a linear combination with complex coefficients c_i of the form $\mu(A) = \sum_i c_i \mu_i(A)$ where each μ_i is a positive measure in the sense of section 3.1, with $\mu_i(\mathbf{X}')$ finite.

We now repeat the definition of the space of test functions: $\Sigma(\mathbf{X})$ consists of the functions $\phi(x)$ given by Eq. (4.40) where the complex bounded measure μ on $\mathbf{X}'$ satisfies the restriction (4.41) for every integer $N \geq 0$. The oscillatory distribution T is introduced as the linear functional on the space $\Sigma(\mathbf{X})$ of the form

$$\langle T, \phi \rangle = \int_{\mathbf{X}'} d\mu(\xi)\, \hat{T}(\xi) \tag{4.50}$$

for ϕ given by Eq. (4.40). Here $\hat{T}$ is a continuous function on $\mathbf{X}'$ of polynomial growth, namely the inequality

$$|\hat{T}(\xi)| \leq C\,(1+\parallel \xi \parallel)^M \tag{4.51}$$

(for all ξ in $\mathbf{X}'$) holds for a suitable integer $M \geq 0$ and a suitable constant $C > 0$. This assumption guarantees that $\hat{T}$ is integrable for the measure μ. As before, the constant function $\phi \equiv 1$ on $\mathbf{X}$ is a test function in $\Sigma(\mathbf{X})$ and we define the integral $\int_{\mathbf{X}} T$ as $\langle T, 1 \rangle$.

We will not repeat the remarks made at the end of subsection 4.1.4, but we conclude by two more remarks.

a) For every ξ in $\mathbf{X}'$, the function $e^{2\pi i \xi}$ on $\mathbf{X}$, taking the value $e^{2\pi i \langle \xi, x \rangle}$ at the point x, is a test function in $\Sigma(\mathbf{X})$. Moreover $\hat{T}(\xi)$ is equal to $\left\langle T, e^{2\pi i \xi} \right\rangle$. Using an integral notation, we get the equation

$$\hat{T}(\xi) = \int_{\mathbf{X}} e^{2\pi i \xi} \cdot T \tag{4.52}$$

which substantiates the claim that $\hat{T}(\xi)$ is the Fourier transform of the distribution T.

b) As explained in section 3.5, we cannot define in a naive way the volume element $d_{\mathbf{X}}x$ in the infinite-dimensional space $\mathbf{X}$. That is why we write the integral of T in the form $\int_{\mathbf{X}} T$, and not in the form $\int_{\mathbf{X}} d_{\mathbf{X}}x\, T(x)$. This difficulty will be addressed again in section 5.

4.2. Promeasures and Prodistributions

Promeasures (so named after [Bourbaki, 1982]) have been introduced under various names, as cylindrical measures or canonical systems. They provide a — somewhat limited — method of integration in certain probabilistic questions. From the beginning, the importance of Fourier transformation was clearly recognized. DeWitt-Morette introduced in [DeWitt-Morette, 1972, 1974] a more general notion of prodistributions, aimed at solving explicit physical problems. Prodistributions were then called "pseudomeasures"; they were "baptized" prodistributions by J. Dieudonné because they are families of *distributions* defined on *pro*jective systems of quotient spaces of infinite-dimensional spaces. They precede the more mathematically oriented work of Albeverio and Høegh-Krohn on oscillatory integrals. A skeleton of the theories of promeasures and prodistributions, and their relationship to oscillatory distributions (Eq. (4.60)) is presented in this section. It sheds light on our general theory in section 5.

4.2.1. General setup. Feynman's heuristic procedure consists in approximating the trajectory $x(t)$ of a particle by giving the position at a finite number of times. This discretization procedure is explained in subsection 4.2.4. More generally, the dynamical states of a system being represented by points in an infinite-dimensional space $\mathbf{X}$, we can approximate it by a system $\{\mathbf{X}_\alpha\}$ of finite-dimensional spaces.

For simplicity, we consider a separable Banach space $\mathbf{X}$. We denote by $\mathcal{F}(\mathbf{X})$ the collection of closed subspaces V of $\mathbf{X}$ for which the quotient space $\mathbf{X}/V$ is finite-dimensional. Such a subspace V can be described by a set of equations of the form

$$\langle \xi_1, x\rangle = \ldots = \langle \xi_n, x\rangle = 0 \tag{4.53}$$

where $\xi_1, \ldots \xi_n$ belong to the dual $\mathbf{X}'$ of $\mathbf{X}$. For such a V denote by P_V the canonical projection of $\mathbf{X}$ onto $\mathbf{X}/V$. The set $\mathcal{F}(\mathbf{X})$ is partially ordered by inclusion, $V \subset W$. There exists then a unique linear map

$$P_{WV} : \mathbf{X}/V \to \mathbf{X}/W \tag{4.54}$$

such that

$$P_W = P_{WV} \circ P_V. \tag{4.55}$$

The quotient spaces $\mathbf{X}/V, \mathbf{X}/W, \ldots$ together with the maps P_{WV} form the projective system of finite-dimensional quotient spaces of $\mathbf{X}$ indexed by $\mathcal{F}(\mathbf{X})$.

A *promeasure* μ_Q is a family of (complex) bounded measures $\{\mu_V\}$ on $\{\mathbf{X}/V\}$ such that:

a) the number $\mu_V(\mathbf{X}/V)$ is independent of V;

b) whenever $V \subset W$, the projection of μ_V under P_{WV} is equal to μ_W.

We can develop an integration theory based on μ_Q as follows. A function $F : \mathbf{X} \to \mathbb{C}$ is called *tame* (or sometimes *cylindrical*) if there exists a subspace V in $\mathcal{F}(\mathbf{X})$ and a function f on $\mathbf{X}/V$ such that $F = f \circ P_V$. Assuming that F is bounded and measurable (that is, f is bounded and measurable), it can be shown that the number $\int_{\mathbf{X}/V} d\mu_V\, f$ is independent of V. We then define

$$\int_{\mathbf{X}} d\mu_Q\, F := \int_{\mathbf{X}/V} d\mu_V\, f. \tag{4.56}$$

We are now in a position to define the Fourier transform $\hat{\mu}_Q$ of a promeasure μ_Q by

$$\hat{\mu}_Q(\xi) = \int_{\mathbf{X}} d\mu_Q\, e^{2\pi i \xi} \tag{4.57}$$

for ξ in $\mathbf{X}'$. Indeed, the function ξ is tame, hence $e^{2\pi i \xi}$ is tame. According to standard properties of the Fourier transformation, the promeasure μ_Q is *uniquely characterized by its Fourier transform* $\hat{\mu}_Q$. Furthermore, the function $\hat{\mu}_Q$ has a continuous restriction to each finite-dimensional subspace of $\mathbf{X}'$; in standard situations, the function $\hat{\mu}_Q$ itself is continuous on the whole space $\mathbf{X}'$.

Let us consider a promeasure μ_Q for which the Fourier transform $\hat{\mu}_Q$ is continuous. The function $\hat{\mu}_Q$ is bounded that is, it satisfies the estimate

$$|\hat{\mu}_Q(\xi)| \leq C \cdot \| \xi \|^0 . \tag{4.58}$$

Comparing this with Eq. (4.51), we conclude that there exists an oscillatory distribution T on $\mathbf{X}$ with the same Fourier transform as μ_Q, namely $\hat{T} = \hat{\mu}_Q$. Furthermore, it can be shown that if a function ϕ is at the same time a test function in the space $\Sigma(\mathbf{X})$ *and* a tame function, then

$\int_{\mathbf{X}} d\mu_Q\, \phi = \langle T, \phi \rangle$. Henceforth, we shall not distinguish between μ_Q and T: *promeasures are therefore oscillatory distributions of order* 0 (because of Eq. (4.58)).

We compare now the oscillatory distributions to the prodistributions of DeWitt-Morette. Let T be an oscillatory distribution on $\mathbf{X}$, with Fourier transform $\hat{T}(\xi)$. For V in $\mathcal{F}(\mathbf{X})$, the linear map $P_V : \mathbf{X} \to \mathbf{X}/V$ has a transposed map $\tilde{P}_V : (\mathbf{X}/V)' \to \mathbf{X}'$ which takes a linear form λ on $\mathbf{X}/V$ into the linear form $\lambda \circ P_V$ on $\mathbf{X}$. To the distribution T on $\mathbf{X}$, there corresponds a oscillatory distribution T_V on $\mathbf{X}/V$ whose Fourier transform is given by

$$\hat{T}_V(\lambda) = \hat{T}(\lambda \circ P_V), \tag{4.59}$$

that is $\hat{T}_V = \hat{T} \circ \tilde{P}_V$. Generalizing from the case of bounded measures, we say that T_V is the image of T under the linear map $P_V : \mathbf{X} \to \mathbf{X}/V$. It is then tautological that, for $V \subset W$, the image of T_V (on $\mathbf{X}/V$) by the linear map $P_{WV} : \mathbf{X}/V \to \mathbf{X}/W$ is the oscillatory distribution T_W. The collection $\{T_V\}$ indexed by $\mathcal{F}(\mathbf{X})$ is therefore a prodistribution. We conclude: *any oscillatory distribution on* $\mathbf{X}$ *is a prodistribution.*

$$\{\text{promeasures}\} \subset \{\text{oscillatory distributions}\} \subset \{\text{prodistributions}\} \tag{4.60}$$

4.2.2. Gaussian integrators. In the prodistribution scheme, a pervasive example is furnished by the so-called Gaussian prodistributions. These are distributions w on $\mathbf{X}$ with Fourier transform of the form

$$\hat{w}(\xi) = w(\mathbf{X}) e^{-\pi s W(\xi,\xi)}. \tag{4.61}$$

Here s is equal to 1 or i, $w(\mathbf{X})$ is a constant, equal to $\int_{\mathbf{X}} w$, and W is a continuous symmetric bilinear form on $\mathbf{X}'$ (positive definite if $s = 1$). The letter w reminds us of the famous prototype, the Wiener measure. We shall see in section 5 how to give a meaning to the equality

$$dw(x) = Ce^{-\pi Q(x)/s} \mathcal{D}x \tag{4.62}$$

where $Q(x)$ is a quadratic form on $\mathbf{X}$, inverse in a suitable sense to the quadratic form $W(\xi, \xi)$ on $\mathbf{X}'$ (the so-called variance).

4.2.3. Linear methods. Consider two separable Banach spaces $\mathbf{X}$ and $\mathbf{Y}$ and a continuous linear map $P : \mathbf{X} \to \mathbf{Y}$. The transpose $\tilde{P}$ is the linear map from $\mathbf{Y}'$ to $\mathbf{X}'$ such that

$$\langle \eta, Px \rangle_{\mathbf{Y}} = \langle \tilde{P}\eta, x \rangle_{\mathbf{X}} \tag{4.63}$$

for x in $\mathbf{X}$ and η in $\mathbf{Y}'$, where $\langle\ ,\ \rangle_{\mathbf{X}}$ is the duality between $\mathbf{X}'$ and $\mathbf{X}$, and similarly for $\langle\ ,\ \rangle_{\mathbf{Y}}$.

Let T be an oscillatory distribution on $\mathbf{X}$, with Fourier transform $\hat{T}(\xi)$ (for ξ in $\mathbf{X}'$). The function $\hat{T} \circ \tilde{P}$, taking η into $\hat{T}(\tilde{P}\eta)$ is continuous on $\mathbf{Y}'$, and satisfies an estimate like in Eq. (4.51). Hence there exists an oscillatory distribution S on $\mathbf{Y}$ such that $\hat{S}(\eta) = \hat{T}(\tilde{P}\eta)$ for η in $\mathbf{Y}'$. Generalizing a definition given at the end of subsection 4.2.1, we call *S the image of T under the map $P : \mathbf{X} \to \mathbf{Y}$.*

We can now state a very general principle of change of variable in our oscillatory integrals. Let ϕ be a test function in $\Sigma(\mathbf{Y})$ of the form

$$\phi(y) = \int_{\mathbf{Y}'} d\nu(\eta)\, e^{2\pi i \langle \eta, y \rangle}, \tag{4.64}$$

where ν is a complex bounded measure on $\mathbf{Y}'$. Under the map $\tilde{P}: \mathbf{Y}' \to \mathbf{X}'$, the measure ν on $\mathbf{Y}'$ transforms into a measure $\mu = \tilde{P}\nu$ on $\mathbf{X}'$, and we get

$$\begin{aligned} \phi(Px) &= \int_{\mathbf{Y}'} d\nu(\eta)\, e^{2\pi i \langle \eta, Px \rangle} \\ &= \int_{\mathbf{Y}'} d\nu(\eta)\, e^{2\pi i \langle \tilde{P}\eta, x \rangle} \\ &= \int_{\mathbf{X}'} d\mu(\xi)\, e^{2\pi i \langle \xi, x \rangle}. \end{aligned} \tag{4.65}$$

Hence, the composite function $\phi \circ P$ is a test function in $\Sigma(\mathbf{X})$. Our *change of variable formula* is therefore

$$\int_{\mathbf{X}} \phi(Px)\, T(x) = \int_{\mathbf{Y}} \phi(y)\, S(y). \tag{4.66}$$

This is especially useful in the case of Gaussian integrals.

4.2.4. Example: Discretization. Here is a specific application of the previous result. We suppose that $\mathbf{X}$ is a space of paths $x = \{x(t)\}$, where t runs over a time interval $[t_a, t_b]$. Select a sequence of intermediate times t_i (for $1 \leq i \leq n$), namely

$$t_a = t_1 < t_2 < \ldots < t_{n-1} < t_n = t_b. \tag{4.67}$$

We consider the linear map $P_n : \mathbf{X} \to \mathbb{R}^n$ mapping the path x into the vector $u = \left(u^1, \ldots, u^n\right)$ with

$$u^i = x(t_i) = \langle \delta_{t_i}, x \rangle \tag{4.68}$$

(where δ_t is the Dirac measure at t). In many cases a functional $F[x]$ of the path x can be approximated by functionals of the form

$$F_n[x] = f_n\left(x(t_1), \ldots, x(t_n)\right) \tag{4.69}$$

for a suitable function f_n of n real variables. Then a path integral of the form $\int_{\mathbf{X}} F_n[x] T(x)$ can be reduced to the finite-dimensional integral

$$\int \ldots \int du^1 \ldots du^n\, f_n\left(u^1, \ldots, u^n\right) T_n\left(u^1, \ldots, u^n\right) \tag{4.70}$$

where the oscillatory distribution T_n on $\mathbb{R}^n$ is the image of T under the linear map $P_n : \mathbf{X} \to \mathbb{R}^n$.

5. A NEW APPROACH

5.1. The General Setup

5.1.1. Domain of integration. The primary component of the proposed axiomatic for functional integration is the domain of integration, an infinite-dimensional space $\mathbf{X}$ of functions or maps. A point x in $\mathbf{X}$ is a map specified by its domain, its range, its behavior on the boundary of its domain and its analytic properties, i.e., a continuous map, an element of a specified Sobolev space, a Grassmann variable etc.

5.1.2. Gaussian volume elements. There is no universal expression for a volume element on an infinite-dimensional space, but given $\mathbf{X}$ there are choices of volume elements on which one can build a theory of integration.

We denote by $\mathbf{X}$ a real, separable, Banach space, by $\mathbf{X}'$ its dual and by $\langle \xi, x\rangle$ the duality between $\mathbf{X}$ and $\mathbf{X}'$. We assume the existence of a continuous linear map $D : \mathbf{X} \longrightarrow \mathbf{X}'$ with the following properties:

- *symmetry*: $\langle Dx, y\rangle = \langle Dy, x\rangle$ for x, y in $\mathbf{X}$;
- *invertibility*: there exists a continuous linear map $G : \mathbf{X}' \to \mathbf{X}$ inverse of D, i.e.,

$$DG = 1 \quad \text{and} \quad GD = 1. \tag{5.1}$$

Out of these data one constructs two quadratic forms, Q on $\mathbf{X}$ and W on $\mathbf{X}'$, by the rules

$$\boxed{Q(x) = \langle Dx, x\rangle \, , \; W(\xi) = \langle \xi, G\xi\rangle .} \tag{5.2}$$

They are related to each other as follows:

$$Q(x) = W(Dx) \, , \; W(\xi) = Q(G\xi) \tag{5.3}$$

for x in $\mathbf{X}$ and ξ in $\mathbf{X}'$.

We denote by s a parameter equal to 1 or i, to treat in a unified way the oscillatory integrators $e^{i\pi Q(x)}\mathcal{D}x$ and the integrators of type $e^{-\pi Q(x)}\mathcal{D}x$. As in subsection 4.2.2, both will be called *Gaussian integrators*. The function $e^{-\pi s W}$ on $\mathbf{X}'$ is continuous. It is bounded in the following cases:

- $s = i$;
- $s = 1$ and $W(\xi) > 0$ for $\xi \neq 0$ in $\mathbf{X}'$. Equivalently, $s = 1$ and $Q(x) > 0$ for $x \neq 0$ in $\mathbf{X}$.

The oscillatory case ($s = i$) has been treated extensively in the last section. The justification for the following equations in the case $s = 1$ will be sketched in 5.1.5.

We denote by $\mathcal{D}_{s,Q}x$ the integrator characterized by

$$\boxed{\int_{\mathbf{X}} \mathcal{D}_{s,Q}x \cdot e^{-\frac{\pi}{s}Q(x)} e^{+2\pi i\langle \xi, x\rangle} = e^{-\pi s W(\xi)} .} \tag{5.4}$$

Formally, this means that the integrator $\mathcal{D}w(x)$ defined by

$$\mathcal{D}w(x) = e^{-\frac{\pi}{s}Q(x)} \mathcal{D}_{s,Q}x \tag{5.5}$$

has a Fourier transform equal to $e^{-\pi s W(\xi)}$.

5.1.3. General case. Gaussian integrators are not the only possible integrators. In [Cartier and DeWitt-Morette, 1993] we have developed an axiomatic for functional integrals on a Banach space $\mathbf{X}$ expressed in terms of integrators $\mathcal{D}_{\Theta,Z}$ defined by

$$\boxed{\int_{\mathbf{X}} \mathcal{D}_{\Theta,Z}x \, \Theta(x, \xi) = Z(\xi)} \tag{5.6}$$

for $x \in \mathbf{X}$ and $\xi \in \mathbf{X}'$, where Θ and Z are two given continuous bounded functionals

$$\Theta : \mathbf{X} \times \mathbf{X}' \longrightarrow \mathbb{C} \, , \; Z : \mathbf{X}' \longrightarrow \mathbb{C}. \tag{5.7}$$

Gaussian integrators are a particular example of this general formalism.

5.1.4. Integrands (oscillatory case). We want to introduce a suitable space $\mathcal{F}_Q(\mathbf{X})$ of functionals on $\mathbf{X}$ integrable by $\mathcal{D}_{i,Q}x$ and a norm on $\mathcal{F}_Q(\mathbf{X})$, and then compute integrals of the type

$$I = \int_{\mathbf{X}} \mathcal{D}_{i,Q}x\, F(x) \tag{5.8}$$

for $F \in \mathcal{F}_Q(\mathbf{X})$. By the results of 4.1.5, we have:

$$\hat{T}(\xi) = e^{-i\pi W(\xi)} \tag{5.9}$$

$$\phi(x) = \int_{\mathbf{X}'} d\mu(\xi)\, e^{2\pi i\langle \xi, x\rangle}. \tag{5.10}$$

So we know how to calculate

$$\langle T, \phi\rangle = \int_{\mathbf{X}'} d\mu(\xi)\, e^{-i\pi W(\xi)}. \tag{5.11}$$

We take for $\mathcal{F}_Q(\mathbf{X})$ the space of all functions of the form

$$F(x) := e^{i\pi Q(x)}\phi(x) = e^{i\pi Q(x)} \int_{\mathbf{X}'} d\mu(\xi)\, e^{2\pi i\langle \xi, x\rangle} \tag{5.12}$$

where μ is a complex bounded measure on $\mathbf{X}'$. Let us denote by $Var(\mu)$ the total variation of μ, that is the l.u.b. of numbers of the form $\sum_{i=1}^{n} |\mu(A_i)|$ where $(A_1, \ldots, A_n)$ is a partition of $\mathbf{X}'$ into Borel subsets. For a given function F in $\mathcal{F}_Q(\mathbf{X})$, the measure μ in Eq. (5.12) is uniquely determined. Hence we are justified to define the *norm* of F and its *integral*

$$\| F \|_{\mathcal{F}} := Var(\mu) \tag{5.13}$$

$$\int_{\mathbf{X}} \mathcal{D}x\, F(x) := \int_{\mathbf{X}'} d\mu(\xi) e^{-i\pi W(\xi)}. \tag{5.14}$$

We write also $\mathcal{D}(F)$ for the integral of F.

We can show that:

- $\mathcal{F}_Q(\mathbf{X})$ is a Banach space
- $\mathcal{D}$ is a linear form on $\mathcal{F}_Q(\mathbf{X})$ bounded by $|\mathcal{D}(F)| \leq \| F \|_{\mathcal{F}}$
- $\mathcal{F}_Q(\mathbf{X})$ is invariant under translations
- $\mathcal{D}$ is invariant under translations.

As a consequence we obtain:

a) The function $\Phi_0(x) := e^{i\pi Q(x)}$ belongs to $\mathcal{F}_Q(\mathbf{X})$.

b) $\int \mathcal{D}x\, \Phi_0(x) = 1$ by definition.

c) A general function F in $\mathcal{F}_Q(\mathbf{X})$ can be written as $\int_{\mathbf{X}} d\nu(y)\, \Phi_0(x+y)$ with a suitable complex bounded measure ν on $\mathbf{X}$, that is as an average of translates of Φ_0.

5.1.5. The case $s = 1$. For the case $s = 1$ and Q positive definite, Eq. (5.4) becomes

$$\int_{\mathbf{X}} \mathcal{D}x\, e^{-\pi Q(x)+2\pi i\langle \xi, x\rangle} = e^{-\pi W(\xi)}. \tag{5.15}$$

One can interpret this relation as in the oscillatory case. The conclusion is the integration formula for

$$F(x) := \int_{\mathbf{X}'} d\mu(\xi)\, e^{-\pi Q(x)+2\pi i\langle \xi, x\rangle}, \tag{5.16}$$

namely

$$\int_{\mathbf{X}} \mathcal{D}x\, F(x) := \int_{\mathbf{X}'} d\mu(\xi)\, e^{-\pi W(\xi)}. \tag{5.17}$$

But the space $\mathcal{F}_Q(\mathbf{X})$ is no longer invariant under translations. To restore this invariance, one has to replace Fourier–Stieltjes transforms by Laplace–Stieltjes transforms. For that purpose, one has to consider the complex dual space $\mathbf{X}'_{\mathbb{C}}$ consisting of the continuous real-linear maps $\xi : \mathbf{X} \longrightarrow \mathbb{C}$ and measures μ on $\mathbf{X}'_{\mathbb{C}}$.

5.2. What is New?

5.2.1. Minor changes in bookkeeping open up new perspectives.

a) Formulae in $\mathbb{R}^d$ which do not depend on d suggest formulae valid in infinite-dimensional spaces. For instance

$$\int_{\mathbb{R}^d} dv\, e^{-\frac{\pi}{a}|x|^2+2\pi i\langle \xi, x\rangle} = e^{-\pi a|\xi|^2} \tag{5.18}$$

with dv the dimensionless volume element given by Eq. (1.4). Note the position of π in Eq. (5.18); the definition of Fourier transforms which leads to equations independent of d has long been advocated by L. Schwartz.

Eq. (5.18) suggests Eq. (5.4), i.e., the definition of a volume element on an infinite-dimensional space $\mathbf{X}$.

b) The integrator $\mathcal{D}w(x)$ is broken up into two components $e^{-\frac{\pi}{s}Q(x)}$ and $\mathcal{D}x$ (Eq. (5.5)). Bringing out the quadratic form Q on $\mathbf{X}$ brings back the action (Eq. (6.1)) which does not show explicitly in the prodistribution scheme. For a Banach space $\mathbf{X}$, one can choose translation invariant $\mathcal{D}x$, and see at a glance the change of $Q(x)$ under translation. Transformations of $\mathcal{D}w(x)$ under linear and affine maps on $\mathbf{X}$ have given, via Fourier transform, a wealth of simple and compact results; but physics problems are often easier to analyze in terms of functionals of $x \in \mathbf{X}$ than in terms of functionals of $\xi \in \mathbf{X}'$, such as (functional) Fourier transforms.

c) We give equal status to the two quadratic forms Q on $\mathbf{X}$ and W on $\mathbf{X}'$, inverse of each other in the sense of Eq. (5.1) and Eq. (5.2). In works done with prodistributions, Q, whenever it is needed, is called W^{-1}. The new notation emphasizes the interplay of $\mathbf{X}$ and $\mathbf{X}'$.

5.2.2. A greater emphasis on domains of integration. The original use of $\mathbb{E}$, or any other average symbol, which meant "integral over Brownian paths" was not conducive to a full theory of integration; it has lead to some cumbersome results, discouraging non-specialists.

A major progress was made by considering rigged Hilbert spaces $\mathcal{H}$ (Gelfand triples) as suitable domains of integration

$$S \subset \mathcal{H} \subset S' \tag{5.19}$$

where S is the Schwartz space of test functions decreasing rapidly at infinity (see subsection 4.1.2), and its topological dual S' is the space of tempered distributions. The work of Albeverio and Høegh-Krohn is done on Hilbert spaces of functions; the white noise calculus developed by de Faria, Hida, Kuo, Potthoff, Streit [Hida et al., 1993] is done on the space S' of distributions.

But $\mathcal{H}$ is not the only Banach space which can be identified in a space of distributions. Sobolev has defined a collection of them, their norms and their duals which are appropriate to physical systems (see section 6). This collection is very useful in locating solutions of partial differential equations. It is also very useful in "visualizing" the domain of integration $\mathbf{X}$ appropriate for a given physical system. Therefore we have generalized the triple (Eq. (5.19)) to a quadruple

$$\boxed{S \subset \mathbf{X} \rightleftarrows \mathbf{X}' \subset S'}, \tag{5.20}$$

where the arrows represent the maps $D : \mathbf{X} \to \mathbf{X}'$ and $G : \mathbf{X}' \to \mathbf{X}$. In the following section $\mathbf{X}$ is an $L^{2,1}$ Sobolev space, and $\mathbf{X}'$ its dual $L^{2,-1}$. Continuity arguments can be used to express an integral on $\mathbf{X}$ as an integral on S'.

5.2.3. A unified scheme for oscillatory (Fresnel) and Gaussian volume elements. The same defining equation (Eq. (5.4)) can be used for both cases with an extra condition imposed on Q when $s = 1$ (see 5.1.2). Introducing $s \in \{1, i\}$ and claiming results to be true for both values of s had been justified in the prodistribution scheme. Nevertheless, it was still often considered as a heuristic statement. The new formalism confirms the prodistribution proof and extends its domain of application. Therefore in following sections we call *Gaussian* what has previously been distinguished as *oscillatory* and *Gaussian*.

5.2.4. Spaces of pointed paths, (dynamical) vector fields, and Lie derivatives w.r.t. such vector fields in Schrödinger equations. An arbitrary manifold $\mathcal{M}^d$ cannot be mapped globally into $\mathbb{R}^d$, but a space $\mathcal{P}_a\mathcal{M}^d$ of pointed paths on $\mathcal{M}^d$ is contractible and can be mapped globally into a space $\mathcal{P}_0\mathbb{R}^d$ of pointed paths on $\mathbb{R}^d$ (see section 7.1).

A map $\mathcal{P}_a\mathcal{M}^d \longrightarrow \mathcal{P}_0\mathbb{R}^d$ was used first by Eells and Elworthy (see [Eells and Elworthy, 1971]), and more extensively by Elworthy (see [Elworthy, 1982]) in constructing Wiener-type functional integrals over $\mathcal{P}_a\mathcal{M}^d$ for $\mathcal{M}^d$ a Riemannian space.

By generalizing this idea we have solved a number of problems (see section 7 and [Cartier and DeWitt-Morette, 1995]) which have, over the years, been sources of many controversies. The map $\mathcal{P}_a\mathcal{M}^d \longrightarrow \mathcal{P}_0\mathbb{R}^d$ by $x \mapsto z$ is often expressed in terms of one or several ordinary differential equations. Vector fields enter these ODEs (see section 7.1.1), and a general theorem relates a functional integral on $\mathcal{P}_a\mathcal{M}^d$ to a generalized Schrödinger equation stated in terms of Lie derivatives w.r.t such vector fields.

5.2.5. A greater role for the action functional. As we shall see in the following section, the action functional dictates the analytical properties of the paths (i.e., properties of the domain of integration) and defines a quadratic form appropriate to the system (i.e., an appropriate volume element).

6. FUNCTIONAL INTEGRALS IN PHYSICS

6.1. The Action

Functional integrals in physics are not limited to systems with classical action functionals. And even if a system has a classical action functional, it is not necessary to consider an integral, schematically written

$$\boxed{\int_{\mathbf{X}} \mathcal{D}x\, e^{2\pi i \frac{S(x)}{h}} F(x)} \tag{6.1}$$

as the primary functional integral yielding the quantum properties of the system. Indeed, there is a scheme (see e.g., [DeWitt-Morette, 1983]) in which the primary object is a stochastic process on a fibre bundle; the classical action function is obtained as the phase of the WKB approximation of the functional integral defined by the chosen stochastic process; the classical action functional is obtained from the classical action function (see [DeWitt-Morette, 1990] for explicit calculations).

This being said, functional integrals of type of Eq. (6.1) play a dominant role in quantum physics and these are the ones we consider in this section.

6.2. The Action Defines the Analytical Properties of the Paths

The key requirement for the existence of Eq. (6.1) is the finiteness of the action

$$|\mathcal{S}(x)| < \infty. \tag{6.2}$$

For systems with a finite number of degrees of freedom, one considers the standard action

$$\mathcal{S}(x) = \int_{t_a}^{t_b} dt\, L(x(t),\dot{x}(t),t) \quad , x(t) \in \mathcal{M}^d. \tag{6.3}$$

We assume that the Lagrangian L includes a kinetic energy term

$$K(x) = \frac{1}{2}\int_{t_a}^{t_b} dt\, g_{jk}(x(t))\dot{x}^j(t)\dot{x}^k(t)\ , x \in \mathbf{X}; \tag{6.4}$$

or, after a parametrization $x \mapsto z$ defined in 7.1,

$$K_E(z) = \frac{1}{2}\int_{t_a}^{t_b} dt\, h_{jk}\dot{z}^j(t)\dot{z}^k(t)\ , z \in \mathbf{Z}. \tag{6.5}$$

K, or K_E, are finite if x, or z, belongs to a Sobolev space $L^{2,1}$, i.e., to a space of paths whose first derivative, in the sense of distributions, is square integrable.

$$\boxed{\mathbf{X} \text{ and } \mathbf{Z} \text{ are Sobolev spaces } L^{2,1}.} \tag{6.6}$$

6.3. Choosing the Boundary Values of the Paths

We consider paths

$$x : \mathbf{T} = [t_a, t_b] \longrightarrow \mathcal{M}^d, \tag{6.7}$$

where $\mathcal{M}^d$ is a d-dimensional manifold. We keep the time interval $\mathbf{T}$ finite, not because it is necessary but because it simplifies the presentation considerably.

Since integrals over spaces of pointed paths (paths with one fixed point) are easy to compute, and since the Schrödinger equation is a partial differential equation satisfied by a function of (t_b, x_b), we shall consider here domains of integration $\mathbf{X}_\beta$ of pointed paths x such that

$$x(t_b) = x_b \in \mathcal{M}^d, \text{ a fixed point of } \mathcal{M}^d; \tag{6.8}$$

or domains of integration $\mathbf{Z}_\beta$ of pointed paths z such that

$$z(t_b) = 0 \in \mathbb{R}^d, \text{ the origin of } \mathbb{R}^d. \tag{6.9}$$

If Eq. (6.1) is to represent the probability amplitude for a transition due to a Hamiltonian $\mathbf{H}$ from an initial state $|\alpha\rangle$ to a final state $|\beta\rangle$, then a correspondence must be established between the boundary values of the paths and the states, for instance, in a position-to-position transition

$$x(t_a) \text{ corresponds to } |\alpha\rangle \tag{6.10}$$

$$x(t_b) \text{ corresponds to } |\beta\rangle, \tag{6.11}$$

but the initial and final states are not necessarily position states. In Eq. (6.8), we have chosen $|\beta\rangle$ to be the state of the system which is at the point x_b of $\mathcal{M}^d$ at time t_b. The space $\mathcal{M}^d$ is not necessarily the configuration space, or a polarization of phase space. In general it is a fibre bundle describing the system; for instance the frame bundle over a configuration space, as in 7.3. The information contained in the state $|\alpha\rangle$ will then be encoded in an initial wave function: for instance, a Dirac δ-function $\delta(x(t_a) - x_a)$ for a position-to-position transition.

6.4. The Action Defines Gaussian Volume Elements

6.4.1. Choosing a quadratic form $Q_\mathbf{X}$. A Gaussian volume element on $\mathbf{X}_\beta$, or $\mathbf{Z}_\beta$, is defined by a quadratic form $Q_\mathbf{X}(x)$ or $Q_\mathbf{Z}(z)$. The best quadratic form defined by the action S is the Hessian at a critical point of the action (see for instance in this volume "Physics on and near Caustics"). In some cases it may even be the only meaningful one. For instance for an anharmonic oscillator with action

$$S(x) = \int_{t_a}^{t_b} dt \left(\frac{1}{2}\dot{x}^2(t) + \frac{1}{2}x^2(t) + \frac{\lambda}{4}x^4(t) \right), \tag{6.12}$$

there are three obvious quadratic forms: the kinetic energy, the action of a harmonic oscillator, the Hessian of the action computed at a solution of the Euler–Lagrange equation of the system. The last choice gives a functional integral which approaches smoothly the functional integral of a harmonic oscillator in the limit $\lambda \longrightarrow 0$; the others do not. Explicit calculations for the quantum anharmonic oscillator can be found in the PhD thesis of M. M Mizrahi ([Mizrahi, 1975]) and are summarized in ([Mizrahi, 1979]).

Mizrahi's example: A simple example due to Mizrahi shows that the space of solutions of the anharmonic oscillator tends to the space of solutions of the harmonic oscillator, but an arbitrarily chosen solution in one space does not tend to an arbitrarily chosen solution in the other space.

The solution of $\ddot{x}(t) + k\dot{x}(t) + g = 0$ with arbitrary constants of integration, namely $x(t) = -gtk^{-1} + Ae^{-kt} + B$, where A and B are constants of integration, blows up when k tends to zero. But the solution of $\ddot{x}(t) + k\dot{x}(t) + g = 0$ tends to the solution of $\ddot{y}(t) + g = 0$ if both systems have the same initial conditions.

6.4.2. Invertibility of the quadratic form $Q_{\mathbf{X}}$. To illustrate the main features, we take the elementary example of a free particle moving on a line, with position $x(t)$ at time t. The quadratic form is defined as

$$Q(x) = \int_{\mathbf{T}} dt\, \dot{x}(t)^2 \tag{6.13}$$

if we normalize the mass by $m = 2$. the choice of the space $\mathbf{X}$ of paths is dictated by the following considerations. First, the integral $Q(x)$ should be finite, that is, the derivative $\dot{x}(t)$ should be a square-integrable function of t in $\mathbf{T}$. Notice that $Q(x) \geq 0$ and that according to our assumptions. we should have $Q(x) = 0$ for $x = 0$ only. To preclude the constant maps to belong to $\mathbf{X}$, we should impose a boundary condition, for instance $x(t_a) = 0$. The natural choice is therefore the space $L^{2,1}(\mathbf{T})$ *of functions* $x : \mathbf{T} \to \mathbb{R}$ *with square-integrable derivative* $\dot{x}$, *and boundary condition* $x(t_a) = 0$. This space shall also be denoted as $\mathbf{X}_a$ to refer to the boundary condition at t_a; the space $\mathbf{X}_b$ is defined similarly by $x(t_b) = 0$, and the intersection $\mathbf{X}_{a,b} := \mathbf{X}_a \cap \mathbf{X}_b$ is characterized by $x(t_a) = x(t_b) = 0$.

The space $\mathbf{X}_{a,b}$ is used in the calculation of the position-to-position amplitudes by

$$\langle x_b t_b | x_a t_a \rangle = \int \mathcal{D}x\, e^{iS(x)/\hbar}. \tag{6.14}$$

Here, the action $S(x)$ is equal to $Q(x)$ and the paths in the domain of integration are pinned at both sides $x(t_a) = x_a, x(t_b) = x_b$. The *classical path* is defined by

$$x_{cl}(t) = (x_a(t_b - t) + x_b(t - t_a)) / (t_b - t_a) \tag{6.15}$$

and we make the change of variable

$$x(t) = x_{cl}(t) + z(t) \tag{6.16}$$

in Eq. (6.14), where the *quantum fluctuation* $z(t)$ satisfies $z(t_a) = z(t_b) = 0$, hence belongs to $\mathbf{X}_{a,b}$. Therefore, in Eq. (6.14), the integral reduces to an integral over $\mathbf{X}_{a,b}$.

On the analytical side, the space $\mathbf{X}_{a,b}$ is the Sobolev space[*,†] $L_0^{2,1}(\mathbf{T})$ (denoted $H_0^1(\mathbf{T})$). We denote by $L^{2,0}(\mathbf{T})$ the Hilbert space of (real-valued) square-integrable functions on $\mathbf{T}$, by $L_0^{2,0}(\mathbf{T})$ the subspace of $L^{2,0}(\mathbf{T})$ consisting of the functions $u(t)$ with $\int_{\mathbf{T}} dt\, u(t) = 0$. Finally, we denote by $L^{2,-1}(\mathbf{T})$ the space of distributions on the open interval $]t_a, t_b[$ which are derivatives of square-integrable functions. One has inclusions of spaces

$$\begin{array}{ccccc} L^{2,1}(\mathbf{T}) & \subset & L^{2,0}(\mathbf{T}) & \subset & L^{2,-1}(\mathbf{T}) \\ \cup & & \cup & & \\ L_0^{2,1}(\mathbf{T}) & & L_0^{2,0}(\mathbf{T}) & & \end{array} . \tag{6.17}$$

Moreover, the derivation d/dt defines isomorphisms of $L_0^{2,1}$ with $L_0^{2,0}$, of $L^{2,1}$ with $L^{2,0}$, and of $L_0^{2,0}$ with $L^{2,-1}$.

It follows that the operator $D = -d^2/dt^2$ *defines an isomorphism of* $L_0^{2,1}(\mathbf{T})$ with $L^{2,-1}(\mathbf{T})$. Furthermore, these two spaces are in natural duality by

$$\langle \xi, x \rangle = \int_{\mathbf{T}} dt\, \xi(t) x(t), \tag{6.18}$$

[*] For the basic properties of Sobolev spaces, we refer the reader to chapter IV in [Dautray-Lions, 1988].

[†] The exponent "2" refers to the integrability of the *square*, and the exponent "1" to the *first* derivative. The subscript "0" is here to remind us that the functions *vanish* at the boundary of the time interval, consisting of t_a and t_b.

and the relation

$$\langle Dx,y\rangle = \int_{\mathbf{T}} dt\, \dot{x}(t)\dot{y}(t) \tag{6.19}$$

follows by an integration by parts.

To summarize: the dual $\mathbf{X}'_{a,b}$ of the space $\mathbf{X}_{a,b} = L^{2,1}_0(\mathbf{T})$ *is the space* $L^{2,-1}(\mathbf{T})$, *and* $D = -d^2/dt^2$ *is an isomorphism of* $\mathbf{X}_{a,b}$ *with* $\mathbf{X}'_{a,b}$, *satisfying the relations*:

$$\langle Dx,y\rangle = \langle Dy,x\rangle \quad , \quad Q(x) = \langle Dx,x\rangle \tag{6.20}$$

We add two remarks:

a) The space $\mathbf{X}_a = L^{2,1}(\mathbf{T})$ is the direct sum of the subspace $\mathbf{X}_{a,b} = L^{2,1}_0(\mathbf{T})$ and the one-dimensional space generated by the linear function $L(t) = t - t_a$. As dual $\mathbf{X}'_a$ we can take the space of distributions of the form

$$T(t) = \frac{d}{dt}u(t) + \delta(t - t_b) \tag{6.21}$$

with $\int_{\mathbf{T}} dt\, u(t)^2 < +\infty$. The operator $D_a : \mathbf{X}_a \to \mathbf{X}'_a$ coincides with $D = -d^2/dt^2$ on $L^{2,1}_0(\mathbf{T})$ and takes $t - t_a$ into $\delta(t - t_b)$. Notice that the C^1 functions are dense in $L^{2,1}(\mathbf{T})$ and that D_a is given by

$$(D_a x)(t) = -\frac{d^2}{dt^2}x(t) + \dot{x}(t_b)\delta(t - t_b) \tag{6.22}$$

for any function in $C^2(\mathbf{T})$. The formula

$$\langle D_a x,y\rangle = \int_{\mathbf{T}} dt\, \dot{x}(t)\dot{y}(t) \tag{6.23}$$

follows by an integration by parts.

b) In the basic integration formula

$$\int_{\mathbf{X}} \mathcal{D}x\, e^{-\frac{\pi}{s}Q(x) + 2\pi i\langle\xi,x\rangle} = e^{-\pi s W(\xi)}, \tag{6.24}$$

we have to define the two Banach spaces $\mathbf{X}$ and $\mathbf{X}'$ in duality such that $Q(x)$ makes sense for x in $\mathbf{X}$ and $W(\xi)$ for ξ in $\mathbf{X}'$. this means that we have limitations on the sizes of both $\mathbf{X}$ and $\mathbf{X}'$. But narrowing $\mathbf{X}$, for instance, to a dense subspace $\mathbf{Y}$ enlarges the dual: $\mathbf{X}'$ is contained as a subspace in $\mathbf{Y}'$. We have therefore no freedom, and once the quadratic form $Q(x)$ (in physical terms, the action) and $W(\xi)$ (in physical terms, the Green's function) are given, there is a unique choice of $\mathbf{X}$ and $\mathbf{X}'$.

6.4.3. The quadratic form $W_{\mathbf{X}'}$ and Green's function. We discuss now the meaning of the operator $G : \mathbf{X}' \to \mathbf{X}$ inverse to $D : \mathbf{X} \to \mathbf{X}'$ and the *covariance* $\langle\xi, G\eta\rangle$. The basic relation is the following

$$\int_{\mathbf{X}} \mathcal{D}x\, e^{-\frac{\pi}{s}Q(x)} \langle\xi,x\rangle\langle\eta,x\rangle = \frac{s}{2\pi}\langle\xi, G\eta\rangle. \tag{6.25}$$

Let x be any function in $L^{2,1}(\mathbf{T})$. From the relation

$$x(t) = \int_{t_a}^{t} ds\, \dot{x}(s) \tag{6.26}$$

and Cauchy–Schwarz formula, it follows that x is continuous and obeys the estimate:

$$\begin{aligned}|x(t_0)|^2 &\leq \int_{t_a}^{t_0} ds\, |\dot{x}(s)|^2 \cdot (t_0 - t_a) \\ &\leq Q(x) \cdot (t_0 - t_a)\end{aligned} \tag{6.27}$$

for any point t_0 in $\mathbf{T}$. Hence the Dirac's "function" δ_{t_0} defined by

$$\langle \delta_{t_0}, x\rangle = x(t_0) \tag{6.28}$$

belongs to the dual of $L^{2,1}(\mathbf{T}) = \mathbf{X}_a$ for $t_a < t_0 \leq t_b$ while δ_{t_a} degenerates to the 0 element in $\mathbf{X}'_a$ since $x(t_a) = 0$ by definition. Similarly, δ_{t_0} belongs to the dual $\mathbf{X}'_{a,b} = L^{2,-1}(\mathbf{T})$ of $\mathbf{X}_{a,b} = L_0^{2,1}(\mathbf{T})$ and degenerates to 0 for $t_0 = t_a$ or $t_0 = t_b$.

Since the appropriate Dirac's functions belong to $\mathbf{X}'$, we can define the kernel of the operator G inverse of D, by

$$G_a(t,s) = \langle \delta_t, G_a\delta_s\rangle \tag{6.29}$$

$$G_{a,b}(t,s) = \langle \delta_t, G_{a,b}\delta_s\rangle. \tag{6.30}$$

Here $G_{a,b} : L^{2,1}(\mathbf{T}) \to L_0^{2,1}(\mathbf{T})$ is the inverse of the operator $D_{a,b} = -d^2/dt^2$ which takes isomorphically $L_0^{2,1}(\mathbf{T})$ into $L^{2,-1}(\mathbf{T})$. Similarly, $G_a : \mathbf{X}'_a \to \mathbf{X}_a$ is inverse to the isomorphism $D_a : \mathbf{X}_a \to \mathbf{X}'_a$. By specializing Eq. (6.25) we get the familiar equations

$$\frac{s}{2\pi} G_a(t,s) = \int_{\mathbf{X}_a} \mathcal{D}x\, e^{-\frac{\pi}{s}Q(x)} x(t)x(s) \tag{6.31}$$

$$\frac{s}{2\pi} G_{a,b}(t,s) = \int_{\mathbf{X}_{a,b}} \mathcal{D}x\, e^{-\frac{\pi}{s}Q(x)} x(t)x(s) \tag{6.32}$$

We shall not repeat the standard calculations and content ourselves with stating the final results:

$$G_a(t,s) = \inf(t - t_a, s - t_a) \tag{6.33}$$

$$G_{a,b}(t,s) = \inf(t - t_a, s - t_a) - (t - t_a)(s - t_a)/(t_b - t_a). \tag{6.34}$$

These are symmetric functions as expected:

$$G_a(t,s) = G_a(s,t) \quad , \quad G_{a,b}(t,s) = G_{a,b}(s,t). \tag{6.35}$$

Furthermore, the boundary conditions

$$G_a(t_a, s) = 0, \tag{6.36}$$

$$G_{a,b}(t_a, s) = G_{a,b}(t_b, s) = 0, \tag{6.37}$$

follow from the integral representations given in Eq. (6.31) and Eq. (6.32). Indeed the boundary condition $x(t_a) = 0$ and also $x(t_b) = 0$ in the second case) is encoded in the definition of the integration domain $\mathbf{X}_a$ (or $\mathbf{X}_{a,b}$). That $\partial G_a(t,s)/\partial t$ should vanish for $t = t_b$ is less obvious since the functions x in $\mathbf{X}_a$ do not in general satisfy the relation $\dot{x}(t_b) = 0$. They do it in *a statistical sense* only. But notice that $G_a(t,s)$ is the Green kernel for the operator D_a given by Eq. (6.22), which includes a δ term.

7. APPLICATIONS

The proofs and the details of the material presented in this section can be found in [Cartier and DeWitt-Morette, 1995], henceforth referred to as [4].

7.1. The Functional Integral Solution of a Schrödinger Equation on Tensors

Let ϕ be a given tensor field (possibly an initial wave function) on a manifold $\mathcal{N}$; let $\mathcal{P}_{x_0}\mathcal{N}$ be the space of pointed $L^{2,1}$ paths on $\mathcal{N}$

$$x : \mathbf{T} \to \mathcal{N} \quad \text{s.t.} \quad x(t_0) = x_0\ ,\ t_0 \in \mathbf{T}; \tag{7.1}$$

let $\mathcal{P}_0\mathbb{R}^d \equiv \mathbf{Z}_0$ be the space of pointed $L^{2,1}$ paths on $\mathbb{R}^d$

$$z : \mathbf{T} \to \mathbb{R}^d \quad \text{s.t.} \quad z(t_0) = 0; \tag{7.2}$$

let the transformation of paths

$$P : \mathbf{Z}_0 \to \mathcal{P}_{x_0}\mathcal{N} \quad \text{by } z \mapsto x \tag{7.3}$$

be such that

$$\begin{cases} dx^i(t,z) = X^i_{(\alpha)}(x(t,z))dz^\alpha + Y^i(x(t,z))dt \\ x(t_0,z) = x_0 \end{cases} . \tag{7.4}$$

This is an equation between $L^{2,1}$ paths, *not a stochastic equation on continuous paths*. It depends on the choice of the vector fields $X_{(1)},\ldots,X_{(d)},Y$ on $\mathcal{N}$, and it is but an example of a *nonholonomic transformation of paths*. Let the solution of Eq. (7.4) be written

$$x(t,z) = x_0 \cdot \Sigma(t,z), \tag{7.5}$$

x is a function of t and a functional of z; not a function of t and $z(t)$. Let

$$Q_0(z) = \int_{\mathbf{T}} dt\ h_{\alpha\beta}\dot{z}^\alpha(t)\dot{z}^\beta(t) \tag{7.6}$$

be the quadratic form defining the volume element $\mathcal{D}_{s,Q_0}z$ by Eq. (5.4). Then the integral

$$\int_{\mathbf{Z}_b} \mathcal{D}_{s,Q_0}z \cdot e^{-\frac{\pi}{s}Q_0(z)}\phi(x_0 \cdot \Sigma(t,z)) =: \Psi(t,x_0) \tag{7.7}$$

is the solution of the generalized Schrödinger equation on tensors:

$$\begin{cases} \frac{\partial\Psi}{\partial t} = \frac{s}{4\pi}h^{\alpha\beta}\mathcal{L}_{X_{(\alpha)}}\mathcal{L}_{X_{(\beta)}}\Psi + \mathcal{L}_Y\Psi \\ \Psi(t_0,\cdot) = \phi(\cdot) \end{cases} , \tag{7.8}$$

where $\mathcal{L}_X$ denotes the Lie derivative at x_0 w.r.t $X(x_0)$, and $h^{\alpha\beta}h_{\beta\gamma} = \delta^\alpha_\gamma$.

A word of caution is in order. When we write $\phi(x_0 \cdot \Sigma(t,z))$, we mean the value at x_0 of the tensor field ϕ transported by the diffeomorphism $\Sigma(t,z)$ of $\mathcal{N}$ with itself. The infinitesimal form of such a diffeomorphism is a Lie derivative acting on tensor fields. There is no parallel transport here, no connection. But of course, a variant exists of this result, which refers to parallel transport of ϕ, with $\mathcal{L}_X$ replaced by a covariant derivative, *once a connection has been chosen*.

7.2. The Feynman–Kac Formula

If, in the previous section, we choose ϕ to be a scalar function (i.e., $\mathcal{L}_X = X$), $X_{(\alpha)}$ to be constant, $t_0 = t_b$ (the endpoint of $\mathbf{T} = [t_a, t_b]$), and $s = i$, then we recover the Feynman formula. If, with the same conditions for ϕ and $X_{(\alpha)}$ we choose $t_0 = t_a$ and $s = 1$, we recover the Kac formula.

Feynman formula: The pointed paths are pinned down at t_b, $x(t_b) = x_b$, hence, $x_b \cdot \Sigma(t_a, z) = x(t_a)$. Set $X_{(\alpha)} = \gamma \frac{\partial}{\partial x^\alpha}$ for $1 \leq \alpha \leq d$, where (x^α) is a global coordinate system on $\mathcal{N} = \mathbb{R}^d$. Then

$$\Psi(t_b, x_b) = \int_{\mathbf{Z}_b} \mathcal{D}_{i,Q_0} z \cdot \exp\left(i\pi \int_{\mathbf{T}} dt \, |\dot{z}(t)|^2 \right) \phi(x(t_a)) \tag{7.9}$$

is the solution of

$$\frac{\partial \Psi}{\partial t} = \frac{i}{4\pi} \gamma^2 \Delta \Psi \quad , \quad \Psi(t_a, \cdot) = \phi(\cdot). \tag{7.10}$$

With $\gamma^2 = h/m$ we recover the Schrödinger equation for a free particle. If a potential is present we use Eq. (7.8) with $Y \neq 0$ (for details see [4]).

Kac formula: The pointed paths are pinned down at t_a, $x(t_a) = x_a$, hence $x_a \cdot \Sigma(t_b, z) = x(t_b)$. Then

$$\Psi(t_b, x_a) = \int_{\mathbf{Z}_a} \mathcal{D}_{1,Q_0} z \cdot \exp\left(-\pi \int_{\mathbf{T}} dt \, |\dot{z}(t)|^2 \right) \phi(x(t_b)) \tag{7.11}$$

is the solution of

$$\frac{\partial \Psi}{\partial t} = \frac{1}{4\pi} \gamma^2 \Delta \Psi \quad , \quad \Psi(t_a, \cdot) = \phi(\cdot). \tag{7.12}$$

With $\gamma^2/4\pi = D$ we recover the diffusion equation with diffusion coefficient D..

So far we have only recovered old results, however the blueprint of section 7.1 has many more applications (in addition to be defined on non-scalar functions). A simple one is the solution of parabolic (Schrödinger or diffusion equations) in non-cartesian coordinates. For example in polar coordinates (r, θ) where

$$\Delta = \frac{\partial^2}{\partial r^2} + \frac{1}{r^2} \frac{\partial^2}{\partial \theta^2} + \frac{1}{r} \frac{\partial}{\partial r}, \tag{7.13}$$

we use the mapping (7.3), $P : z \mapsto x$, to express $x(t)$ in polar coordinates and $z(t)$ in cartesian coordinates:

$$z^1 = r \cdot \cos\theta \quad , \quad z^2 = r \cdot \sin\theta. \tag{7.14}$$

Eq. (7.4) is obtained by inverting the matrix mapping $(dr, d\theta)$ into (dz^1, dz^2), namely

$$\begin{aligned} dr(t,z) &= \cos\theta(t,z) \cdot dz^1(t) + \sin\theta(t,z) \cdot dz^2(t) \\ d\theta(t,z) &= -\frac{\sin\theta(t,z)}{r} \cdot dz^1(t) + \frac{\cos\theta(t,z)}{r} \cdot dz^2(t). \end{aligned} \tag{7.15}$$

Therefore the vector fields $\{X_{(\alpha)}\}$ can be read off this equation:

$$\begin{aligned} X_{(1)} &= \cos\theta \cdot \frac{\partial}{\partial r} - \frac{\sin\theta}{r} \cdot \frac{\partial}{\partial \theta}, \\ X_{(2)} &= \sin\theta \cdot \frac{\partial}{\partial r} + \frac{\cos\theta}{r} \cdot \frac{\partial}{\partial \theta}. \end{aligned} \tag{7.16}$$

One deduces

$$X_{(1)}^2 + X_{(2)}^2 = \frac{\partial^2}{\partial r^2} + \frac{1}{r^2}\frac{\partial^2}{\partial \theta^2} + \frac{1}{r}\frac{\partial}{\partial r} = \Delta, \tag{7.17}$$

as expected.

Generalizations to the case of n-dimensional spherical coordinates can be found in [Wurm, 1995].

7.3. Spaces of Paths Taking their Values in a Riemannian Manifold

Notation:

- $\mathcal{M}$ a Riemannian manifold with metric g, $x_b \in \mathcal{M}$
- $\mathcal{N}$ the orthonormal frame bundle over $\mathcal{M}$, $\rho_b \in \mathcal{N}$, $\rho_b = (x_b, u_b)$, u_b is a frame
- Connection on $\mathcal{N}$: the Riemannian connection defined by g. $\mathcal{M}$ is d-dimensional, $\mathcal{N}$ is D-dimensional ($D = \frac{1}{2}(d^2 + d)$)
- $T\mathcal{N}$ tangent bundle to $\mathcal{N}$; H_{ρ_b} horizontal subspace of $T_{\rho_b}\mathcal{N}$, therefore H_{ρ_b} is d-dimensional and $T_{\rho_b}\mathcal{N}$ is D-dimensional
- $\mathcal{P}_{x_b}\mathcal{M}$ space of pointed paths on $\mathcal{M}$, $x(t_b) = x_b$
- $\mathcal{P}_{\rho_b}\mathcal{N}$ space of pointed paths on $\mathcal{N}$, $\rho(t_b) = \rho_b$
- $\mathcal{P}^H_{\rho_b}\mathcal{N}$ space of pointed paths on $\mathcal{N}$ which are the horizontal lifts of paths in $\mathcal{P}_{x_b}\mathcal{M}$
- $\mathbf{Z}_b$ a vector space of pointed paths in $\mathbb{R}^d$, $z(t_b) = 0$.

One goal is to find the vector fields $\{X_{(\alpha)}\}$ such that we can parametrize $\mathcal{P}_{x_b}\mathcal{M}$ by $\mathbf{Z}_b$. We need a bijection between $\mathbf{Z}_b$ and $\mathcal{P}_{x_b}\mathcal{M}$, and, on the basis of previous works (see references in [4]), we expect it to be the composition

$$\mathbf{Z}_b \longrightarrow \mathcal{P}^H_{\rho_b}\mathcal{N} \longrightarrow \mathcal{P}_{x_b}\mathcal{M} \tag{7.18}$$

by $z \mapsto \rho \mapsto x$.

The ultimate goal is to find the solution of a parabolic equation (with given initial function) on $\mathcal{M}$,

$$\begin{aligned} \frac{\partial}{\partial t_b}\psi(t_b, x_b) &= \frac{s}{4\pi} g^{\lambda\mu} D_\lambda D_\mu \psi(t_b, x_b) \\ &= \frac{s}{4\pi}\Delta\psi(t_b, x_b), \\ \psi(t_a, \cdot) &= \phi(\cdot), \end{aligned} \tag{7.19}$$

where D_λ is the covariant derivative defined by the Riemannian connection, therefore Δ is the *Laplace–Beltrami* operator on $\mathcal{M}$.

It has been shown [see for instance Choquet-Bruhat and DeWitt-Morette, 1994] that the covariant Laplacian Δ at a point x_b of $\mathcal{M}$ can be lifted to a sum of products of (Lie) derivatives $h^{\alpha\beta} X_{(\alpha)}(\rho_b) X_{(\beta)}(\rho_b)$; the vectors $X_{(\alpha)}(\rho_b) \in H_{\rho_b}$ are the horizontal lifts of the orthonormal basis $\{e_\alpha\}$ of $T_{x_b}\mathcal{M}$ corresponding to the frame ρ_b.

Therefore Eq. (7.19) can be lifted to a parabolic equation on $\mathcal{N}$,

$$\begin{cases} \frac{\partial}{\partial t_b}\Psi(t_b,\rho_b) = \frac{s}{4\pi}h^{\alpha\beta}X_{(\alpha)}X_{(\beta)}\Psi(t_b,\rho_b) \\ \Psi(t_a,\cdot) = \Phi(\cdot) \end{cases} . \tag{7.20}$$

with

$$h_{\alpha\beta} = g_{x_b}(e_\alpha, e_\beta). \tag{7.21}$$

Given a trivialization $\rho(t) = (x(t), u(t))$ at each point $\rho(t) \in \mathcal{N}$ we can construct a basis $\{X_{(\alpha)}(\rho(t)\}$ in $H_{\rho(t)} \subset T_{\rho(t)}\mathcal{N}$; this basis is the horizontal lift of the basis defined by $u(t)$.

In order to construct the functional integral solution of Eq. (7.20) we define a field of quadratic forms $h(\rho(t)) \equiv h_{\rho(t)}$ by the equation

$$h_{\rho(t)}\left(X_{(\alpha)}(\rho(t)), X_{(\beta)}(\rho(t))\right) = h_{\alpha\beta}. \tag{7.22}$$

Let A be an action on the space of horizontal pointed paths $\mathcal{P}^H_{\rho_b}\mathcal{N}$ on $\mathcal{N}$

$$A(\rho) = \int_{\mathbf{T}} dt\, h_{\rho(t)}\left(\dot\rho(t), \dot\rho(t)\right). \tag{7.23}$$

Since the vectors $X_{(\alpha)}(\rho(t))$ (for $1 \le \alpha \le d$) form a basis of $H_{\rho(t)}$, and since the path x is of class $L^{2,1}$, we infer that there exist functions $\dot z_\alpha$ in $L^2(\mathbf{T})$ such that

$$\begin{cases} \dot\rho(t) = X_{(\alpha)}(\rho(t))\dot z^\alpha(t) \\ \rho(t_b) = \rho_b \end{cases} . \tag{7.24}$$

The function $\dot z^\alpha$ is the derivative of a function z^α in $L^{2,1}(\mathbf{T})$ normalized by $z^\alpha(t_b) = 0$. The vector function $z = \left(z^1, \ldots, z^d\right)$ is an element of the space denoted by $\mathbf{Z}_b$.

This construction associates to a path ρ in $\mathcal{P}^H_{x_b}\mathcal{N}$ a path z in $\mathbf{Z}_b$ with conservation of the action:

$$A(\rho) = Q_0(z), \tag{7.25}$$

where

$$Q_0(z) = \int_{\mathbf{T}} dt\, h_{\alpha\beta}\dot z^\alpha(t)\dot z^\beta(t). \tag{7.26}$$

Assume now that $\mathcal{M}$ is compact. By using the theory of $L^{2,1}$ differential equations (see [4]), it can be shown that we can invert the transformation $\rho \mapsto z$. Given any z in $\mathbf{Z}_b$, the differential equation Eq. (7.4) has a unique solution ρ in $\mathcal{P}^H_{x_b}\mathcal{N}$, and we obtain a *parametrization*

$$P : \mathbf{Z}_b \longrightarrow \mathcal{P}^H_{x_b}\mathcal{N}. \tag{7.27}$$

The desired bijection $\mathbf{Z}_b \longrightarrow \mathcal{P}^H_{x_b}\mathcal{M}$ is now complete and the solution of Eq. (7.20) on $\mathcal{P}^H_{\rho_b}\mathcal{N}$ is

$$\Psi(t_b,\rho_b) = \int_{\mathbf{Z}_b} \mathcal{D}z \cdot e^{-\frac{\pi}{s}Q_0(z)}\Phi\left(\rho(t_a,z)\right) \tag{7.28}$$

with Q_0 given by Eq. (7.26) and $\rho(t_a,z)$ solution of Eq. (7.24).

The projection of Eq. (7.28) on $\mathcal{P}_{x_b}\mathcal{M}$ is most simply stated by giving the geometrical interpretation of Eq. (7.24). It can be shown that (see [4], p.2266)

$$\Pi \circ \rho(t,z) = (\mathrm{Dev}\, z)(t) \tag{7.29}$$

where Dev z is the Cartan development map for $L^{2,1}$ paths; in other words, the bijection $\mathbf{Z}_b \to \mathcal{P}_{x_b}\mathcal{M}$ associates a path z to a path x which is the Cartan development of z. Therefore the projection of Eq. (7.28) is

$$\psi(t_b,x_b) = \int_{\mathbf{Z}_b} \mathcal{D}z \cdot e^{-\frac{\pi}{s}Q_0(z)} \phi((\mathrm{Dev}\, z)(t_a)). \tag{7.30}$$

Remark: The parametrization of $\mathcal{P}^H_{p_b}\mathcal{N}$ by $\mathbf{Z}_b$ is valid for spaces of paths on any subbundle of $T\mathcal{N}$.

ACKNOWLEDGMENTS

Many thanks to Christiane Koch, Xiaorong Wu-Morrow, Mark Byrd, Mark Mims, Jon Urrestilla and Bruce Jensen for discussions and sharing of lecture notes. Also, the preliminary lecture notes written by Jens Lang during a two-week lecture series given by P. Cartier in spring 1996 at the University of Munich [Cartier, 1996] have been helpful.

REFERENCES

[1] Albeverio, S. A. and Høegh-Krohn, R. J., *Mathematical Theory of Feynman Path Integrals*, Springer Verlag Lecture Notes in Mathematics **523** (1976).

[2] Bourbaki, N., *"Intégration, chapitre 9"*, Masson, Paris (1982).

[3] Cartier,P. and DeWitt-Morette, C., "Intégration fonctionnelle; éléments d'axiomatique," C. R. Acad. Sci. Paris 316, 733–738 (1993).

[4] Cartier, P. and DeWitt-Morette, C., "A new perspective on functional integration," J. Math. Phys. **36**, 2237–2312 (1995).

[5] Cartier, P., "A New Perspective on Path Integrals: Rigorous Mathematical Foundations," Lectures at the University of Munich Spring 1996, Notes taken by Lang, J., preprint (1996).

[6] Choquet-Bruhat, Y. and DeWitt-Morette, C., "Supplement to *Analysis, Manifolds and Physics*," Armadillo preprint, Center for Relativity, University of Texas, Austin, TX 78712 (1994).

[7] Dautray, R. and Lions, J.-L., *Mathematical Analysis and Numerical Methods for Science and Technology*, Springer Verlag (1988).

[8] DeWitt-Morette, C., "Feynman's Path Integral: Definition without Limiting Procedure," Commun. Math. Phys. **28**, 47–67 (1972).

[9] DeWitt-Morette, C., "Feynman Path Integrals: I. Linear and Affine Techniques, II. The Feynman–Green Function," Commun. Math. Phys. **37**, 63–81 (1974).

[10] DeWitt-Morette, C., Maheshwari, A. and Nelson, B., "Path Integration in Non-Relativistic Quantum Mechanics," Phys. Rep **50**, 266–372 (1979).

[11] DeWitt-Morette, C., "Path integration at the crossroad of stochastic and differential calculus," pp.166–70, in *Gauge Theory and Gravitation* Eds. K. Kikkawa, N. Nakanishi, and H. Nariai, Springer Verlag Lecture Notes in Physics **176** (1983).

[12] DeWitt-Morette, C., "Quantum Mechanics in curved spacetimes; stochastic processes on frame bundles," pp.49–87 in *Quantum Mechanics in Curved Space–Time* Eds. J. Audretsch and V. de Sabbata, Plenum Press, New York (1990).

[13] Eells, J. and Elworthy, K. D., "Wiener integration on certain manifolds," pp. 67–94, in *Problems in non-linear analysis*, C. I. M. E IV (1971).

[14] Elworthy, K. D., *Stochastic Differential Equations on Manifolds*, Cambridge University Press, Cambridge U. K. (1982).

[15] Hida, T., Kuo, H.-H., Potthoff, J. and Streit, L., *White Noise — an Infinite Dimensional Calculus*, Kluwer-Academic, New York (1993).

[16] Mizrahi, M. M., Ph. D. Thesis, University of Texas at Austin (1975).

[17] Mizrahi, M. M., "The semiclassical expansion of the anharmonic-oscillator propagator," J. Math. Phys. **20**, 844–855 (1979).

[18] Reed, M. and Simon, B., *Functional Analysis* (Methods of Modern Mathematical Physics I), Academic Press, New York (1980).
[19] Young, A. and DeWitt-Morette, C., "Time substitutions in stochastic Processes as a Tool in Path Integration," Ann. of Phys. **69**, 140–166 (1986).
[20] Wurm, A., Master Thesis, University of Texas at Austin (1995).

8. APPENDIX: FUNCTIONAL INTEGRATION OVER COMPLEX POISSON PROCESSES

A.1. Motivation for Considering Poisson Processes

One of the shortcomings of the various approaches to functional integration including that proposed in section 5 has been that they do not offer methods for the solution of quantum systems for which the state space is discrete (for instance the dimension of the Hilbert space is finite). An example is a two-state system, such as a nucleus of spin $\frac{1}{2}$ in a magnetic field. The Schrödinger equation for this system has the form:

$$i\hbar\frac{\partial}{\partial t}\begin{pmatrix}\psi_1(t)\\ \psi_2(t)\end{pmatrix}=\begin{pmatrix}-\varepsilon & \Delta\\ \Delta & \varepsilon\end{pmatrix}\begin{pmatrix}\psi_1(t)\\ \psi_2(t)\end{pmatrix}. \tag{A.1}$$

where $\varepsilon, \Delta \in \mathbb{R}$. Usually a functional integral approach is not necessary as matrix methods may be used to solve this equation. However, in the case where a two-state system is coupled to an environment one of the methods for extracting information about the influence of the environment is the influence functional method proposed by Feynman and Vernon [Feynman and Vernon, 1963]. This method makes use of the Feynman path integral formalism and has been implemented by several authors ([Niu, 1991] and [Leggett et al, 1987]).

The method offered in section 5 breaks down in this case as several requirements are not satisfied:

(i) There is no classical action from which to construct a quadratic form of the sort frequently used in defining the integrator.

(ii) The space of paths which most naturally correspond to possible histories of the system, those which flip from one state to the other, do not form a vector space.

This forces a search for alternative methods of functional integration.

The Schrödinger equation for the two-state system is closely related to the Weyl representation of the Dirac equation in one space and one time dimension (henceforth called the $1+1$ Dirac equation). With $\hbar = c = 1$ this has the form:

$$\frac{\partial}{\partial t}\begin{pmatrix}\psi_+(x,t)\\ \psi_-(x,t)\end{pmatrix}=\begin{pmatrix}-\partial_x & -im\\ -im & \partial_x\end{pmatrix}\begin{pmatrix}\psi_+(x,t)\\ \psi_-(x,t)\end{pmatrix} \tag{A.2}$$

and this is related to the equation for the two-state system by means of a Fourier transform. The beginnings of a functional integral solution to the $1+1$ Dirac equation have been provided [Gaveau et al, 1984] by noting its similarity to a master equation for a certain random walk process considered by Kac [Kac, 1956]. In this random walk process an object moves in one space dimension at a constant speed v. At time intervals Δt it experiences an interaction which does one of the following:

(i) it reverses direction with probability $b\Delta t$ or

(ii) it maintains its direction with probability $1 - b\Delta t$

where $b \in \mathbb{R}_+$. Let

$$P_+(x, n\Delta t) := \begin{cases} \text{probability that the object is located at } x \text{ at time } t = n\Delta t \\ \text{and is moving right immediately after the interaction} \end{cases} \tag{A.3}$$

$$P_-(x, n\Delta t) := \begin{cases} \text{probability that the object is located at } x \text{ at time } t = n\Delta t \\ \text{and is moving left immediately after the interaction.} \end{cases} \tag{A.4}$$

These satisfy the following master equations:

$$\begin{aligned} P_+(x,(n+1)\Delta t) = (1 - b\Delta t)P_+(x - v\Delta t, n\Delta t) + \\ b\Delta t P_-(x + v\Delta t, n\Delta t) \end{aligned} \tag{A.5}$$

$$\begin{aligned} P_-(x,(n+1)\Delta t) = (1 - b\Delta t)P_-(x + v\Delta t, n\Delta t) + \\ b\Delta t P_+(x - v\Delta t, n\Delta t). \end{aligned} \tag{A.6}$$

In the limit as $\Delta t \to 0$ these difference equations supply the following differential equations:

$$\frac{\partial}{\partial t}\begin{pmatrix} P_+(x,t) \\ P_-(x,t) \end{pmatrix} = \begin{pmatrix} -v\partial_x - b & b \\ b & v\partial_x - b \end{pmatrix}\begin{pmatrix} P_+(x,t) \\ P_-(x,t) \end{pmatrix}. \tag{A.7}$$

These master equations have similar forms to the $1+1$ Dirac equation. The principal difference, the appearance of b in the diagonals of the matrix, can be removed by considering $e^{bt}P_\pm(x,t)$. Kac [Kac, 1956] used the fact that the limit of this random walk (for $\Delta t \to 0$) is a Poisson process to supply solutions to these equations in the form of averages over Poisson processes. This was used [Gaveau et al., 1984] to provide solutions to the $1+1$ Dirac equation although these required that b be purely imaginary. One of the objectives of this Appendix is to obtain solutions to such equations in this case without resorting to analytic continuation.

With this as motivation, we shall present a rigorous method for integrating over Poisson paths. The method will differ from those presented in section 5, principally in that Poisson processes rather than Gaussian processes provide its foundation.

A.2. Measures on the Set of Poisson Paths

A Poisson process is a stationary random process consisting of a sequence of events, called reversals, occurring at certain times and which has the Markov property that the occurrence of an event is independent of what happened in the past. In what follows observation of the process will begin at an initial time, t_0.

A.2.1. Description in terms of waiting times. Let

$$t_n := \text{time of } n^{th} \text{ direction reversal.} \tag{A.8}$$

One method of parametrizing a Poisson path is to specify the set of all times $\{t_n\}$ at which reversals occur. However, for the purposes of stating the probability rule it is more convenient to parametrize it in terms of the waiting times between successive reversals, defined as follows:

$$u_n := t_n - t_{n-1}. \tag{A.9}$$

Clearly $u_n > 0$ and thus any particular Poisson path, denoted ω, may be represented as follows:

$$\omega = (u_1, u_2, u_3, \ldots)$$

where $u_n \in \mathbb{R}_+$. The set of all Poisson paths is

$$\Omega = \{(u_1, u_2, u_3, \ldots) \mid u_n \in \mathbb{R}_+\}.$$

A probability measure $d\Gamma(\omega)$ on Ω may be constructed by imposing the following requirements:

(i) the waiting times are stochastically independent random variables,

(ii) for all u_n the probability distributions are identical and

(iii) for the u_1 the probability distribution is independent of the time at which observation begun.

The first requirement implies that

$$d\Gamma(\omega) = \prod_{n=1}^{\infty} d\gamma_n(u_n) \tag{A.10}$$

where $d\gamma_n(u_n)$ is a probability measure on $\mathbb{R}_+$.

The second requirement implies that

$$d\gamma_n(u_n) = d\gamma(u_n) \tag{A.11}$$

where $d\gamma(u_n)$ is a probability measure on $\mathbb{R}_+$.

The third requirement implies that

$$d\gamma_a(u_n) = ae^{-au_n} du_n \tag{A.12}$$

where $a \in \mathbb{R}_+$ and the subscript a indicates the dependence of the measure on this choice. This is called the decay rate and a simple calculation shows that a^{-1} is the average lifespan between two reversals.

Consequently a probability measure on Ω which satisfies the abovementioned requirements has the form

$$d\Gamma_a(\omega) = \prod_{n=1}^{\infty} ae^{-au_n} du_n. \tag{A.13}$$

Some of the features of this measure are:

(i) It depends on the choice of a and this is indicated via the subscript a.

(ii) It is bounded provided that $a \in \mathbb{R}_+$. If we attempt to extend this method to a case where a is purely imaginary $d\Gamma_a(\omega)$ will no longer be bounded.

(iii) A separation of the form

$$d\Gamma_a(\omega) = (\lim_{n\to\infty} a^n) e^{-a(u_1+u_2+\ldots)} \prod_{n=1}^{\infty} du_n \tag{A.14}$$

is meaningless since on average $e^{-a(u_1+u_2+\ldots)} = 0$ and $\lim_{n\to\infty} a^n$ behaves wildly with respect to a.

Thus this method of parametrization and the resulting construction of the probability measure are ill suited for situations in which the decay rate will have to be imaginary such as appears to be the case for the $1+1$ Dirac equation.

A.2.2. Poisson processes over a finite interval. An alternative method of description of Poisson processes may be provided by considering Poisson processes over a finite time interval $[t_0, T]$. In this case two sets of quantities are needed to specify a sample path:

(i) the number of reversals in $[t_0, T]$, say n, and

(ii) the times at which reversals occur: $t_1, t_2, \ldots, t_n \in [t_0, T]$.

Let

$$P_n := \{\text{paths with } n \text{ reversals in } [t_0, T]\}. \tag{A.15}$$

The set of all paths is

$$P := \bigcup_{n=0}^{\infty} P_n. \tag{A.16}$$

It follows that P_n may be represented as the following subset of $[t_0, T]^n$:

$$P_n = \{(t_1, \ldots, t_n) \mid t_0 < t_1 < t_2 < \ldots < t_n < T\}. \tag{A.17}$$

A path in P_n may be represented graphically by using the reversal-counting function:

$$N(t) := \text{ number of reversals in } [t_0, t]\,. \tag{A.18}$$

An example for P_4 on $[0, T]$ is provided in Fig. 1.

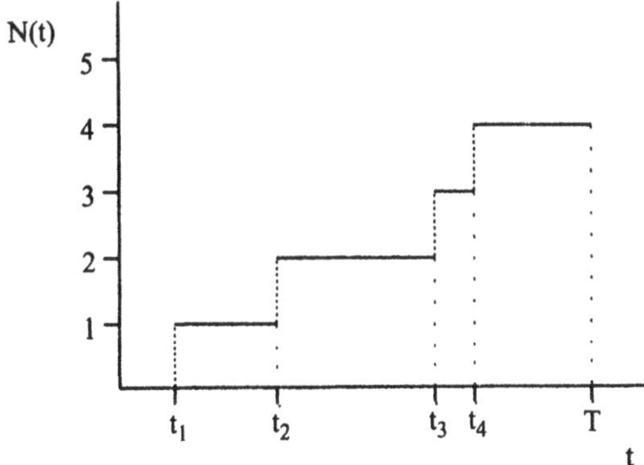

Figure 1. A path $N \in \mathcal{P}_4$.

The probability distribution on P_n, denoted $d\mu_a^n$, can be computed from that provided in Eq. (A.13). This gives:

$$d\mu_a^n(t_1, \ldots, t_n) = e^{-a(T-t_0)} dv_a(t_1) \ldots dv_a(t_n) \tag{A.19}$$

where

$$dv_a(t_j) := a dt_j \tag{A.20}$$

and for $n = 0$

$$d\mu_a^0 = e^{-a(T-t_0)}. \tag{A.21}$$

By construction, P_0 consists of a single path ("no reversal in the time interval $[t_0, T]$").

The probability that a process has n reversals is the volume, $\mu_a^n(P_n)$ of P_n and a straightforward calculation shows that to be:

$$\mu_a^n(P_n) = \frac{a^n(T-t_0)^n}{n!} e^{-a(T-t_0)}. \tag{A.22}$$

Hence the random variable $N(T)$ measuring the number of reversals in the total time interval $[t_0, T]$ follows a Poisson law with parameter $c = a(T-t_0)$. That is, the probability that $N(T) = n$ is $e^{-c}\frac{c^n}{n!}$. More generally, if we subdivide the time interval into subintervals $\Delta_1, \ldots, \Delta_k$, then $N(T)$ is the sum of the random variables $N(\Delta_1), \ldots, N(\Delta_k)$ measuring the number of reversals as occurring during the various time intervals. These variables are stochastically independent, and each one follows a Poisson law: the parameter c for a variable $N(\Delta)$ corresponding to an interval Δ is

$$c = \int_\Delta d\nu_a(t) = a \times \text{length}(\Delta). \tag{A.23}$$

The probability measure on P is the collection:

$$\{d\mu_a^0, d\mu_a^1(t_1), d\mu_a^2(t_1,t_2), \ldots\}.$$

A.3. Integration over Poisson Paths

A.3.1. Functions of Poisson paths. Any function F on P may be represented in terms of a collection of functions, one on each component of P i.e.,

$$F = \{F_0, F_1(t_1), F_2(t_1,t_2), \ldots\} \tag{A.24}$$

where

$$F_n := F\,|_{P_n}\,. \tag{A.25}$$

A.3.2. Integral of a function on P. The average of $F = \{F_0, F_1(t_1), F_2(t_1,t_2), \ldots\}$ is

$$\langle F \rangle_a := \sum_{n=0}^{\infty} \int_{P_n} F_n(t_1, \ldots, t_n) d\mu_a^n(t_1, \ldots, t_n) \tag{A.26}$$

where in the $n = 0$ term integration is replaced by multiplication. Explicitly this gives:

$$\langle F \rangle_a = e^{-a(T-t_0)} F_0 + \sum_{n=1}^{\infty} e^{-a(T-t_0)} a^n \int_{t_0}^{T} dt_1 \int_{t_1}^{T} dt_2 \ldots \int_{t_{n-1}}^{T} dt_n F_n(t_1, \ldots, t_n). \tag{A.27}$$

This allows for the following definition of an integrator on P:

$$\int_P \mathcal{D}_a N\, F(N) := \langle F \rangle_a \tag{A.28}$$

where we now represent a point in P by the corresponding counting function $N(t)$ as defined in Eq. (A.18). Some of the features of this integrator are:

(i) *The integrator is defined for any complex value of a, not only for a in $\mathbb{R}_+$.*

(ii) *$\mathcal{D}_a N$ is a complex bounded measure on the locally compact space P.* This is a consequence of the fact that $d\mu_a^n(t_1, \ldots, t_n)$ is a measure on P_n and of the absolute convergence of the series $\sum_{n=0}^{\infty} e^{-a(T-t_0)} \frac{a^n(T-t_0)^n}{n!}$.

(iii) *The integrator depends on the choice of* a and this is indicated by the subscript a.

A characterization of $\mathcal{D}_a N$ in terms of its *Fourier transform* may be provided by setting:

$$\begin{cases} F_n(t_1,\ldots,t_n) & := & \exp\left(\sum_{j=1}^{n} u(t_j)\right) \\ F_0 & := & 1 \end{cases} \tag{A.29}$$

where u is any (measurable) bounded complex-valued function on $[t_0,T]$. Hence $F(N)$ may be reexpressed as:

$$F(N) = \exp\left(\int_{t_0}^{T} u(t)dN(t)\right) \tag{A.30}$$

where $dN(t)$ is the Stieltjes integral. A straightforward calculation shows that

$$\int_P \mathcal{D}_a N \exp\left(\int_{t_0}^{T} udN\right) = \exp\left(\int_{t_0}^{T} (e^{u(t)}-1)dv_a(t)\right). \tag{A.31}$$

A.4. Solution of Differential Equations

Suppose that V is a finite-dimensional vector space, and that $Q(t)$ is a linear operator on V for $t \in [t_0,T]$. Then consider the operator-valued function F on P for which

$$\begin{cases} F_n(t_1,\ldots,t_n) & = & Q(t_n)\ldots Q(t_1) \\ F_0 & = & 1 \end{cases} \tag{A.32}$$

In this case we shall denote $F(N)$ by $Q(N)$. Let

$$U_a(T,t_0) := \langle Q \rangle_a := \int_P \mathcal{D}_a N\, Q(N). \tag{A.33}$$

Clearly $U_a(T,t_0)$ is a linear map on V. Differentiating with respect to T and making use of the definition of the integral over P yields:

$$\begin{cases} \frac{\partial U_a}{\partial T}(T,t_0) & = & aQ(T)U_a(T,t_0) - aU_a(T,t_0) \\ U_a(t_0,t_0) & = & 1 \end{cases}. \tag{A.34}$$

For any $v_0 \in V$ let

$$v(T) := U_a(T,t_0)v_0 \tag{A.35}$$

which can now be shown to satisfy

$$\begin{cases} \frac{\partial v}{\partial T} & = & aQ(T)v(T) - av(T) \\ v(t_0) & = & v_0 \end{cases}. \tag{A.36}$$

The operator $U_a(T,t_0)$, attained directly by an integration over P, is simply the evolution operator for the differential equation (A.36). A related and useful equation can be solved in a similar fashion by defining:

$$w(T) := e^{a(T-t_0)}v(T), \tag{A.37}$$

which satisfies

$$\frac{\partial w}{\partial T} = aQ(T)w(T) \tag{A.38}$$

$$w(t_0) = v_0 \tag{A.39}$$

and whose solution is given by:

$$w(T) = e^{a(T-t_0)} \int_P \mathcal{D}_a N\, Q(N)\, w(t_0). \tag{A.40}$$

A.5. Applications to Physical Problems

The general methods for solving certain types of differential equations in terms of integrals over Poisson processes may now be applied to particular physical problems.

A.5.1. Random walk (Kac process). Equation (A.7) shows that the distribution functions satisfy:

$$\frac{\partial}{\partial t}\begin{pmatrix} P_+(x,t) \\ P_-(x,t) \end{pmatrix} = (Q_0 - bI + Q_1)\begin{pmatrix} P_+(x,t) \\ P_-(x,t) \end{pmatrix}, \tag{A.41}$$

where

$$Q_0 = \begin{pmatrix} -v\partial_x & 0 \\ 0 & v\partial_x \end{pmatrix} \tag{A.42}$$

$$Q_1 = \begin{pmatrix} 0 & b \\ b & 0 \end{pmatrix} \tag{A.43}$$

and $b \in \mathbb{R}_+$. There are a variety of approaches for providing a functional integral solution of the form given in Eq. (A.40). One such method, similar to that used in the interaction picture of quantum mechanics, is to consider

$$\begin{pmatrix} R_+(x,t) \\ R_-(x,t) \end{pmatrix} := e^{-Q_0(t-t_0)+b(t-t_0)}\begin{pmatrix} P_+(x,t) \\ P_-(x,t) \end{pmatrix} \tag{A.44}$$

which can easily be shown to satisfy

$$\frac{\partial}{\partial t}\begin{pmatrix} R_+(x,t) \\ R_-(x,t) \end{pmatrix} = bS(t)\begin{pmatrix} R_+(x,t) \\ R_-(x,t) \end{pmatrix} \tag{A.45}$$

where

$$S(t) = \begin{pmatrix} 0 & \exp(2v(t-t_0)\partial_x) \\ \exp(-2v(t-t_0)\partial_x) & 0 \end{pmatrix}. \tag{A.46}$$

This clearly has the same form as equation (A.38) and hence

$$\begin{pmatrix} R_+(x,t) \\ R_-(x,t) \end{pmatrix} = e^{b(t-t_0)}\int_P \mathcal{D}_b N\, S(N)\begin{pmatrix} R_+(x,t_0) \\ R_-(x,t_0) \end{pmatrix}. \tag{A.47}$$

Finally this yields the following solution for the distribution functions:

$$\begin{pmatrix} P_+(x,t) \\ P_-(x,t) \end{pmatrix} = e^{Q_0(t-t_0)}\int_P \mathcal{D}_b N\, S(N)\begin{pmatrix} P_+(x,t_0) \\ P_-(x,t_0) \end{pmatrix}. \tag{A.48}$$

In this case P is the set of all Poisson paths on $[t_0,t]$ with decay rate b. The evolution operator for this equation

$$U(t,t_0) := e^{Q_0(t-t_0)}\int_P \mathcal{D}_b N\, S(N) \tag{A.49}$$

is a 2×2 matrix. Using the definition of the functional integral over P, equation (A.27), its matrix elements may be expressed as:

$$U^i_j = \left\langle \exp\left((-1)^j v\partial_x \int_{t_0}^t (-1)^{N(\tau)} d\tau\right) \mid (-1)^{N(t)} = (-1)^{i+1}\right\rangle, \tag{A.50}$$

where $N(\tau)$ is the reversal counting function and $\langle A|B\rangle$ indicates the average of A over all Poisson paths which satisfy condition B.

A.5.2. 1 + 1 Dirac equation. The Dirac equation for $1+1$ spacetime takes on the following form in the Weyl representation (with $\hbar = c = 1$):

$$\frac{\partial}{\partial t}\begin{pmatrix} \psi_+(x,t) \\ \psi_-(x,t) \end{pmatrix} = \begin{pmatrix} -\partial_x & -im \\ -im & \partial_x \end{pmatrix}\begin{pmatrix} \psi_+(x,t) \\ \psi_-(x,t) \end{pmatrix}. \tag{A.51}$$

A similar approach to that used to solve the random walk equation yields:

$$\begin{pmatrix} \psi_+(x,t) \\ \psi_-(x,t) \end{pmatrix} = e^{Q_0(t-t_0)-im(t-t_0)} \int_P \mathcal{D}_{-im}N\, S(N) \begin{pmatrix} \psi_+(x,t_0) \\ \psi_-(x,t_0) \end{pmatrix} \tag{A.52}$$

where

$$S(t) = \begin{pmatrix} 0 & \exp(2(t-t_0)\partial_x) \\ \exp(-2(t-t_0)\partial_x) & 0 \end{pmatrix} \tag{A.53}$$

and

$$Q_0 = \begin{pmatrix} -\partial_x & 0 \\ 0 & \partial_x \end{pmatrix}. \tag{A.54}$$

Notice that in Eq. (A.52), the complex integrator $\mathcal{D}_{-im}N$ could be rewritten as $(-i)^N \mathcal{D}_m N$, using the integrator $\mathcal{D}_m N$ with real parameter m.

A.5.3. Two state system. The two state system satisfies:

$$\frac{\partial}{\partial t}\begin{pmatrix} \psi_1(t) \\ \psi_2(t) \end{pmatrix} = -\frac{i}{\hbar}\begin{pmatrix} -\varepsilon & \Delta \\ \Delta & \varepsilon \end{pmatrix}\begin{pmatrix} \psi_1(t) \\ \psi_2(t) \end{pmatrix}. \tag{A.55}$$

A variety of solutions in terms of functional integrals over Poisson paths to this equation may be provided. For any $\alpha \neq 0$ let

$$Q(t) := \frac{1}{\alpha}\begin{pmatrix} \varepsilon & \Delta \\ \Delta & -\varepsilon \end{pmatrix}. \tag{A.56}$$

Then it can be shown that

$$\begin{pmatrix} \psi_1(t) \\ \psi_2(t) \end{pmatrix} = e^{\frac{-i\alpha(t-t_0)}{\hbar}} \int_P \mathcal{D}_{\frac{-i\alpha}{\hbar}}N\, Q(N) \begin{pmatrix} \psi_1(t_0) \\ \psi_2(t_0) \end{pmatrix}. \tag{A.57}$$

Again, $\mathcal{D}_{-\frac{i\alpha}{\hbar}}N$ can be replaced by $(-i)^N \mathcal{D}_{\frac{\alpha}{\hbar}}N$, where the parameter $\frac{\alpha}{\hbar}$ is real.

Alternatively let

$$Q(t) := \begin{pmatrix} 0 & \exp\left(-\frac{2i\varepsilon(t-t_0)}{\hbar}\right) \\ \exp\left(-\frac{2i\varepsilon(t-t_0)}{\hbar}\right) & 0 \end{pmatrix}. \tag{A.58}$$

Then it can be shown that

$$\begin{pmatrix} \psi_1(t) \\ \psi_2(t) \end{pmatrix} = \exp\left[\frac{-i\Delta(t-t_0)}{\hbar}\begin{pmatrix} \varepsilon-\Delta & 0 \\ 0 & -\varepsilon-\Delta \end{pmatrix}\right] \int_P \mathcal{D}_{\frac{-i\Delta}{\hbar}}N\, Q(N) \begin{pmatrix} \psi_1(t_0) \\ \psi_2(t_0) \end{pmatrix}. \tag{A.59}$$

In both cases the Poisson paths are over $[t_0, t]$ but the decay constants differ as do the integrands.

A.6. Conclusion

A.6.1. Coupled parabolic equations and wave equations. We have studied certain systems of coupled parabolic equations, each of which is equivalent to a wave equation. One standard technique for solving the wave equation makes use of "light cone" coordinates:

$$x_{\pm} = x \pm vt. \tag{A.60}$$

These also simplify both the $1+1$ Dirac equation as well as those governing the Kac process. For instance the Kac process equations become:

$$2v\frac{\partial P_+}{\partial x_+} = b(P_- - P_+) \tag{A.61}$$

$$2v\frac{\partial P_-}{\partial x_-} = b(P_- - P_+). \tag{A.62}$$

But the same coordinates are naturally suitable for the description of any sample path in P; this may be specified by alternating constant values for x_+ and x_-.

A.6.2. Three blueprints for parabolic equations. Parabolic equations may be solved using three different blueprints:

$\frac{\partial P}{\partial t} = QP$	$\Leftrightarrow$	$P(t) = e^{Q(t-t_0)}P(t_0)$	$\Leftrightarrow$	difference equation (A.6)
$\downarrow$		$\downarrow$		$\downarrow$
Dyson series (time ordered exponential)		Trotter formula (product integral)		Kac Solution (average over Poisson processes)
$\searrow$		$\downarrow$		$\swarrow$
		$\int \mathcal{D}_a N\, Q(N)$		

In this Appendix we have generalized the Kac route to **complex** Poisson processes. Furthermore we have worked out the Dyson route in terms of **Poisson processes**. The links to the method involving the Trotter [Nelson,1964] formula are provided by the diagram above.

REFERENCES FOR APPENDIX

[1] Feynman, R. P. and Vernon, F. L., "The Theory of a General Quantum System Interacting with a Linear Dissipative System," Ann. Phys. **24**, 118–173 (1963).

[2] Gaveau, B., Jacobson, T., Kac, M. and Schulman, L. S., "Relativistic Extension of the Analogy between Quantum Mechanics and Brownian Motion," Phys. Rev. Lett. **53**, 419–422 (1984).

[3] Kac, M., "A Stochastic Model Related to the Telegrapher's Equation," Rocky Mountain J. of Math. **4**, 497–509 (1974). Reprinted from the Magnolia Petroleum Company Colloquium Lectures in the Pure and Applied Sciences, No 2. October 1956, *"Some Stochastic Problems in Physics and Mathematics"*.

[4] Leggett, A., Chakravarty, S., Dorsey, A. T., Fisher, M. P. A., Garg, A. and Zwerger, W., "Dynamics of the Dissipative Two-state System," Rev. Mod. Phys. **59**, 1–85 (1987).

[5] Nelson, E., "Feynman Integrals and the Schrödinger Equation," J. Math. Phys. **5**, 332–43 (1964).

[6] Niu, Q., "Quantum Coherence of a Narrow-Band Particle Interacting with Phonons and Static Disorder," J. Stat. Phys. **65**, 317–361 (1991).

2

PHYSICS ON AND NEAR CAUSTICS

C. DeWitt-Morette[1] (with P. Cartier[2])

[1]Department of Physics and Center for Relativity
University of Texas
Austin, TX
[2]Ecole Normale Supérieure
F-75005 Paris

ABSTRACT

Physics *on* caustics is obtained by studying physics *near* caustics. Physics on caustics is at the cross-roads of calculus of variation and functional integration. More precisely consider a space $\mathcal{P}\mathbf{M}^d$ of paths $x : [t_a, t_b] \to \mathbf{M}^d$, on a d-dimensional manifold $\mathbf{M}^d$. Consider two of its subspaces:

i) The $2d$-dimensional space $\mathcal{U}$ of critical points of an action functional $S : \mathcal{P}\mathbf{M}^d \to \mathbb{R}$; $q \in \mathcal{U} \Leftrightarrow \frac{\delta S}{\delta q} q = 0$.

ii) The space $\mathcal{P}_{\mu,\nu}\mathbf{M}^d$ of paths satisfying d initial conditions (μ) and d final conditions (ν).

The nature of the intersection $\mathcal{P}_{\mu,\nu}\mathbf{M}^d \cap \mathcal{U}$ is the key to identifying and analyzing caustics. Caustics occur when the roots of $S'(q) = 0$, for S restricted to $\mathcal{P}_{\mu,\nu}\mathbf{M}^d$, are not isolated points in $\mathcal{U}$. We consider two cases:

i) The roots define a subspace of non zero dimension ℓ of $\mathcal{U}$. For example, the classical system with action S is constrained by conservation laws.

ii) There is a multiple root of $S'(q) = 0$. For example the classical flow has an envelope.

In both situations there is, at least, one nonzero Jacobi field along q with d initial vanishing boundary conditions (μ) and d final vanishing boundary conditions (ν), i.e., $q(t_a)$ and $q(t_b)$ are conjugate points along q. In both cases we say "$q(t_b)$ is on the caustics."

The strict WKB approximations of functional integrals for the system S "break down" on caustics, but the full semiclassical expansion, including contributions from $S''(q)$ and $S'''(q)$ (and possibly higher derivatives) yield the physical properties of the system S on and near the caustics. Caustics display classical physics as a limit of quantum mechanics.

We work out glory scattering cross-sections because, there, both situations occur simultaneously: the system is constrained by conservation laws, and the classical flow is caustic forming. The cross-section is given in closed form in section 5.

Functional Integration: Basics and Applications
Edited by Cécile DeWitt-Morette, Plenum Press, New York, 1997

1. INTRODUCTION

Interesting phenomena occur when a flow of classical paths is caustic forming. Glory scattering, rainbows, orbiting, etc... are only a few examples of physics on and near caustics. Conservation laws are, as we shall see, another example.

Physics on caustics is at the cross-roads of calculus of variation and functional integration. More precisely consider a space $\mathcal{P}\mathbf{M}^d$ of $(L^{2,1})$ paths

$$x : \mathbf{T} \to \mathbf{M}^d \quad , \quad \mathbf{T} = [t_a, t_b]$$

on a d-dimensional manifold $\mathbf{M}^d$. Consider two of its subspaces:

i) the $2d$-dimensional subspace $\mathcal{U}$ of critical points of an action functional

$$S : \mathcal{P}\mathbf{M}^d \to \mathbb{R} \tag{1.1}$$

$$q \in \mathcal{U} \Leftrightarrow \frac{\delta S}{\delta q} = 0; \tag{1.2}$$

ii) the space $\mathcal{P}_{\mu,\nu}\mathbf{M}^d$ of paths satisfying d boundary conditions (μ) at $t = t_a$, and d boundary conditions (ν) at $t = t_b$.

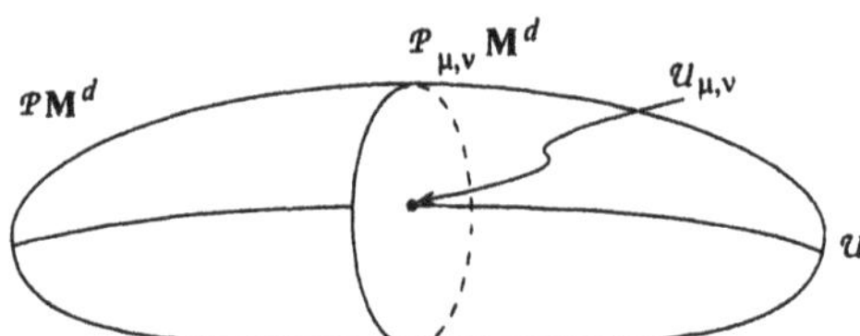

Figure 1. $\mathcal{U}$ is the space of classical motions (critical points of the action functional S); $\mathcal{P}_{\mu,\nu}\mathbf{M}^d$ is the space of paths satisfying d initial conditions and d final conditions; $\mathcal{U}_{\mu,\nu}$ is their intersection.

The nature of the intersection

$$\mathcal{U}_{\mu,\nu} := \mathcal{P}_{\mu,\nu}\mathbf{M}^d \cap \mathcal{U} \tag{1.3}$$

is the key to identifying and analyzing caustics.

In the context of functional integration, the action functional S is restricted to $\mathcal{P}_{\mu,\nu}\mathbf{M}^d$; unless otherwise specified *we assume here S to be restricted to $\mathcal{P}_{\mu,\nu}\mathbf{M}^d$.* Caustics can be analyzed by studying the expansion of S around $S(q)$ with $q \in \mathcal{U}_{\mu,\nu}$, namely

$$S(x) = S(q) + S'(q) \cdot \xi + \frac{1}{2!} S''(q) \cdot \xi\xi + \frac{1}{3!} S'''(q) \cdot \xi\xi\xi + \cdots \tag{1.4}$$

$$x = q + \xi \tag{1.5}$$

where ξ has d vanishing boundary conditions at t_a, and d vanishing boundary conditions at t_b. For q in $\mathcal{P}_{\mu,\nu}\mathbf{M}^d$ we have

$$q \in \mathcal{U}_{\mu,\nu} \Leftrightarrow S'(q) \cdot \xi = 0 \text{ for all } \xi \text{ in } T_q \mathcal{P}_{\mu,\nu}\mathbf{M}^d . \tag{1.6}$$

We can consider 4 cases:

i) $\mathcal{U}_{\mu,\nu}$ consists of isolated points: no caustics, e.g., the anharmonic oscillator (provided (μ,ν) does not define a conjugate point).

ii) $\mathcal{U}_{\mu,\nu}$ is a space of dimension $\ell > 0$: conservation laws.

iii) $q \in \mathcal{U}_{\mu,\nu}$ is a multiple root: the classical flow has an envelope.

iv) $\mathcal{U}_{\mu,\nu}$ is an empty set.

In case i) the critical points are not degenerate; in cases ii) and iii) the critical points are degenerate.

In case i)] we need to keep only the second variation.

In case ii) we need to keep also the first variation.

In case iii) we need to keep (at least) the first three variations.

In this paper, we shall not consider the case in which $\mathcal{U}_{\mu,\nu}$ is an empty set; this case is interesting but does not belong to the study of caustics. A typical example is the knife edge problem solved by L. S. Schulman [1 and references therein].

There is another wording to distinguish cases ii) and iii) from case i): "The WKB approximation of the functional integral for the transition $K(\nu,t_b;\mu,t_a)$ of the system governed by S breaks down in cases ii) and iii)."

These different wordings reflect the different ways one approaches the intersection $\mathcal{U}_{\mu,\nu}$, namely from $\mathcal{U}$ (calculus of variation) or from $\mathcal{P}_{\mu,\nu}\mathbf{M}^d$ (functional integration). They are equivalent because an eigenvector of the Jacobi operator with zero eigenvalue is also a nonzero Jacobi field with vanishing boundary conditions at t_a and t_b.

2. DEGENERATE CRITICAL POINTS

Fixing the boundary conditions (μ) and (ν), the point q in $\mathcal{P}_{\mu,\nu}\mathbf{M}^d$ is said to be a *critical point of S* if

$$S'(q)\cdot\xi = 0 \text{ for all } \xi \text{ in } T_q\,\mathcal{P}_{\mu,\nu}\mathbf{M}^d\,. \tag{2.1}$$

It is said to be degenerate if the hessian form of S is degenerate; this condition means that, for some $\eta \neq 0$ in $T_q\,\mathcal{P}_{\mu,\nu}\mathbf{M}^d$, one has

$$S''(q)\cdot\xi\eta = 0 \text{ for all } \xi \text{ in } T_q\,\mathcal{P}_{\mu,\nu}\mathbf{M}^d\,. \tag{2.2}$$

The integral kernel of the hessian is the (functional) Jacobi operator

$$\mathcal{J}_{\alpha\beta}(q,s,t) = \frac{1}{2}\,\frac{\delta^2}{\delta\xi^\alpha(s)\,\delta\xi^\beta(t)}\,S''(q)\cdot\xi\xi\,. \tag{2.3}$$

The hessian defines the (differential) Jacobi operator on $T_q\,\mathcal{P}_{\mu,\nu}\mathbf{M}^d$ by

$$S''(q)\cdot\xi\xi = \langle\xi,\mathcal{J}(q)\,\xi\rangle = \int_{\mathbf{T}} dt\,\xi^\alpha(t)\,\mathcal{J}_{\alpha\beta}(q(t))\,\xi^\beta(t)\,. \tag{2.4}$$

Here ξ is a $L^{2,1}$ vector field along q and $\langle\ ,\ \rangle$ is the $L^{2,1}$ duality. See in the Appendix the explicit expression for $S''(q)\cdot\xi\xi$ when S is not restricted to $\mathcal{P}_{\mu,\nu}\mathbf{M}^d$. It is useful when one explores $\mathcal{U}_{\mu,\nu}$ by varying the (μ) and/or the (ν) boundary conditions.

A *Jacobi field* $h(q)$ is a vector field along q in the space $T_q\,\mathcal{U}$; it is obtained by a variation through classical paths around q. It is therefore a solution of the Jacobi equation

$$\mathcal{J}_{\alpha\beta}(q(t))\,(h^{\beta}(q))(t) = 0 \qquad h \in T_q\,\mathcal{U} \tag{2.5}$$

often abbreviated to

$$\mathcal{J}_{\alpha\beta}(q)\,h^{\beta}(t) = 0\,.$$

To say that the hessian is degenerate is to say that there is at least one nonzero Jacobi field $h \in T_q\,\mathcal{P}_{\mu,\nu}\mathbf{M}^d$, that is at least one nonzero Jacobi field with d vanishing boundary conditions at t_a and d at t_b. Equivalently, the hessian is degenerate if the Jacobi operator has one or more eigenvectors in $T_q\,\mathcal{P}_{\mu,\nu}\mathbf{M}^d$ with zero eigenvalue.

The eigenvectors of the Jacobi operator form a convenient basis of $T_q\,\mathcal{P}_{\mu,\nu}\mathbf{M}^d$ because they diagonalize the hessian, and the hessian is the best quadratic form for defining a volume element on $T_q\,\mathcal{P}_{\mu,\nu}\mathbf{M}^d$. Explicitly let $\{\psi_k\}$ be a complete set of orthonormal eigenvectors of the Jacobi operator in the space $T_q\,\mathcal{P}_{\mu,\nu}\mathbf{M}^d$:

$$\mathcal{J}(q)\,\psi_k(t) = \alpha_k\,\psi_k(t) \qquad k \in \{0,1,\ldots\}\,. \tag{2.6}$$

There may be ℓ zero eigenvalues,

$$\alpha_k = 0 \qquad k \in \{0,\ldots,\ell-1\}\,;$$

$$\int_{\mathbf{T}} \mathrm{d}t\,(\psi_k(t)\mid\psi_j(t)) = \delta_{kj}\,. \tag{2.7}$$

We expand ξ in this basis

$$\xi^{\alpha}(t) = \sum_{k=0}^{\infty} u^k\,\psi_k^{\alpha}(t) \tag{2.8}$$

$$S''(q)\cdot\xi\xi = \langle\xi,\mathcal{J}(q)\,\xi\rangle = \sum_{k=0}^{\infty}\alpha_k(u^k)^2 = \sum_{k=\ell}^{\infty}\alpha_k(u^k)^2\,. \tag{2.9}$$

We call ℓ^2(hessian) the space of points u such that $\Sigma\,\alpha_k(u^k)^2 < \infty$. On the subspace $\mathbb{X} \subset T_q\,\mathcal{P}_{\mu,\nu}\mathbf{M}^d$ spanned by the eigenvectors $\{\psi_k\}$ with nonzero eigenvalues, $S''(q)\cdot\xi\xi$ defines a nondegenerate quadratic form $\mathcal{Q}(u)$

$$\mathcal{Q}(u) := S''(q)\cdot\xi\xi\,|_{\mathbb{X}} = \sum_{k=\ell}^{\infty}\alpha_k(u^k)^2 \tag{2.10}$$

$\mathbb{X}$ is parametrized by $\{u^k\}$ for $k \in \{\ell,\ldots\}$.

With ξ expanded in the eigenvector basis $\{\psi_k\}$, the first, second, and third variations of $S(x)$ around $S(q)$ read

$$\begin{aligned} S'(q)\cdot\xi &= \int_{\mathbf{T}}\mathrm{d}t\,\frac{\delta S}{\delta q^{\alpha}(t)}\sum_{k=0}^{\infty}u^k\,\psi_k^{\alpha}(t) \\ &= \sum_{k=0}^{\infty}\left(\int_{\mathbf{T}}\mathrm{d}t\,\frac{\delta S}{\delta q^{\alpha}(t)}\,\psi_k^{\alpha}(t)\right)u^k =: \sum_{k=0}^{\infty}c_k\,u^k \end{aligned} \tag{2.11}$$

$$S''(q)\cdot\xi\xi = \sum_{k=0}^{\infty}\alpha_k(u^k)^2,\ \text{with } \alpha_k \text{ the eigenvalues of (2.6)} \tag{2.12}$$

$$S'''(q)\cdot\xi\xi\xi = \sum_{k,\ell,m}\left(\int_{\mathrm{T}}\mathrm{d}r\int_{\mathrm{T}}\mathrm{d}s\int_{\mathrm{T}}\mathrm{d}t\,\frac{\delta^3 S}{\delta q^\alpha(r)\,\delta q^\beta(s)\,\delta q^\gamma(t)}\,\psi_k^\alpha(r)\,\psi_\ell^\beta(s)\,\psi_m^\gamma(t)\right)\cdot u^k u^\ell u^m. \tag{2.13}$$

For k between 0 and $\ell-1$, ψ_k is a Jacobi field with d vanishing boundary conditions at t_a and d vanishing boundary conditions at t_b.

2.1. Explicit Calculations of an Eigenvector of the Jacobi Operator with Zero Eigenvalue

Jacobi fields, with or without vanishing boundary conditions, are derivatives of one parameter families of classical solutions. For instance let $q \in \mathcal{U}$ be specified by its *initial* position and *initial* momentum

$$q(t_a,a,p_a) = a,\ p(t_a,a,p_a) = p_a \tag{2.14}$$

then the set of $2d$ Jacobi fields

$$\begin{cases} \partial q^\alpha(t,a,p_a)/\partial p_{a\beta} =: j^{\alpha(\beta)}(t) \\ \partial q^\alpha(t,a,p_a)/\partial a^\beta =: k^\alpha_{(\beta)}(t) \end{cases} \tag{2.15}$$

is a particularly useful basis for $T_q\,\mathcal{U}$. Also useful in phase space, and in the study of caustics is the set

$$\begin{cases} \partial p_\alpha(t,a,p_a)/\partial p_{a\beta} =: \tilde{k}_\alpha^{(\beta)}(t) \\ \partial p_\alpha(t,a,p_a)/\partial a^\beta =: \ell_{\alpha(\beta)}(t) \end{cases} \tag{2.16}$$

$\tilde{k}^{(\beta)}$ and $\ell_{(\beta)}$ are not Jacobi fields in configuration space but the matrices $J,K,\tilde{K},L$, whose columns consist of the components of $j^{(\beta)}$, $k_{(\beta)}$, $\tilde{k}^{(\beta)}$, $\ell_{(\beta)}$ respectively,

$$J^{\alpha\beta}(t,t_a) = j^{\alpha(\beta)}(t) \tag{2.17}$$

$$K^\alpha_\beta(t,t_a) = k^\alpha_{(\beta)}(t) \tag{2.18}$$

$$\tilde{K}^\beta_\alpha(t,t_a) = \tilde{k}^{(\beta)}_\alpha(t) \tag{2.19}$$

$$L_{\alpha\beta}(t,t_a) = \ell_{\alpha(\beta)}(t), \tag{2.20}$$

together make a $2d\times 2d$ matrix $\mathbf{J}$ of Jacobi fields in phase space, namely;

$$\mathbf{J}(t,t_a) := \begin{pmatrix} J^{\alpha\beta}(t,t_a) & K^\alpha_\beta(t,t_a) \\ \tilde{K}^\beta_\alpha(t,t_a) & L_{\alpha\beta}(t,t_a) \end{pmatrix}. \tag{2.21}$$

Therefore we call $J,K,\tilde{K},L$ "Jacobi matrices." The properties of the Jacobi matrices are simple and powerful tools for solving problems [2–5]. Here we shall use them to express a Jacobi field ψ_0 with vanishing boundary conditions:

$$\text{If } \psi_0(t_a) = 0, \text{ then } \psi_0(t) = J(t,t_a)\,\dot{\psi}_0(t_a). \tag{2.22}$$

$$\text{If } \dot{\psi}_0(t_a) = 0, \text{ then } \psi_0(t) = K(t,t_a)\,\psi_0(t_a). \tag{2.23}$$

$$\text{If } \dot{\psi}_0(t_b) = 0, \text{ then } \psi_0(t) = \psi_0(t_b)\,\tilde{K}(t_b,t). \tag{2.24}$$

The case of derivatives vanishing both at t_a and t_b is best discussed in phase space path integrals (see reference [6] and [4]). To prove (2.22)–(2.24), note that both sides of the equations satisfy the same differential equation and the same boundary values at t_a (for eq. II.22, 23) or at t_b (for eq. II.24).

To construct a Jacobi field ψ_0 such that $\psi_0(t_a) = 0$ *and* $\psi_0(t_b) = 0$, it suffices therefore by (2.22) to choose $\dot{\psi}_0(t_a)$ in the kernel of the linear map $J(t_b, t_a)$. A similar procedure applies for the two other cases $\dot{\psi}_0(t_a) = 0$, $\psi_0(t_b) = 0$ in (2.23) and $\psi_0(t_a) = 0$, $\dot{\psi}_0(t_b) = 0$ in (2.24).

3. THE INTERSECTION $\mathcal{U}_{\mu,\nu}$ IS OF DIMENSION $\ell > 0$; CONSTANTS OF THE MOTION

This situation has a finite-dimensional analog when S is a function, rather than a functional, namely the critical points of S are not isolated and Morse's lemma needs to be generalized.

3.1. Generalized Morse Lemma; Finite Dimensional Case

Let $S : \mathbf{M}^d \to \mathbb{R}$ be a smooth function on a d-dimensional manifold $\mathbf{M}^d$. Its Taylor expansion around a point x_0 reads

$$S(x) = S(x_0) + S'_\alpha(x_0) y^\alpha + \frac{1}{2} S''_{\alpha\beta}(x_0) y^\alpha y^\beta + O(|y|^3) \tag{3.1}$$

with coordinates y^α vanishing at x_0. We consider the case in which there exists a submanifold $\mathbf{V}^\ell \subset \mathbf{M}^d$ of dimension ℓ, such that

$$T_{x_0}\mathbf{M}^d = T_{x_0}\mathbf{V}^\ell \oplus T_{x_0}\mathbf{F}^{d-\ell} \tag{3.2}$$

for some supplementary manifold $\mathbf{F}^{d-\ell}$ of dimension $d - \ell$ through x_0. We assume that $S'(x_0)$ induces 0 on $T_{x_0}\mathbf{F}^{d-\ell}$ but that the hessian

$$S''(x_0) \text{ is not degenerate on } T_{x_0}\mathbf{F}^{d-\ell}, \tag{3.3}$$

and induces 0 on $T_{x_0}\mathbf{V}^\ell$.

Possibly after a linear change of coordinates, we shall have two sets of equations

$$\begin{cases} S'_a = g_a & \text{constant on } \mathbf{V}^\ell; \\ S'_A(x_0) = 0 & \text{determine } d-\ell \text{ coordinates of } x_0. \end{cases} \tag{3.4}$$

In physical terms, the first set of equations corresponds to a constrained system, or to conservation laws. The second set of equations defines a *relative critical point*, under the constraints $S'_a(x_0) = g_a$. In the new coordinates, labelled y^a and y^A, with $1 \le a \le \ell$, $\ell + 1 \le A \le d$, one gets

$$S(x) = S(x_0) + g_a y^a + \frac{1}{2} S''_{AB}(x_0) y^A y^B + O(|y|^3). \tag{3.5}$$

A non linear change of the variables $\{y^A\} \mapsto \{\bar{y}^A\}$ *can now be used to remove the terms of order greater than* 2, according to the Morse lemma.

We express the situation in more geometrical terms. The space $\mathbf{M}^d$ is fibered by $(d-\ell)$-dimensional fibers $\mathbf{F}^{d-\ell}(g_a)$; each fiber is defined by the values of $\{g_a\}$. On each fiber there is a non degenerate critical point. Such critical points will be said to be *relative to a given fiber* because we take the variations of S along the fibers. The set of relative critical points define a (in general local) section of the fibered space $\mathbf{M}^d$.

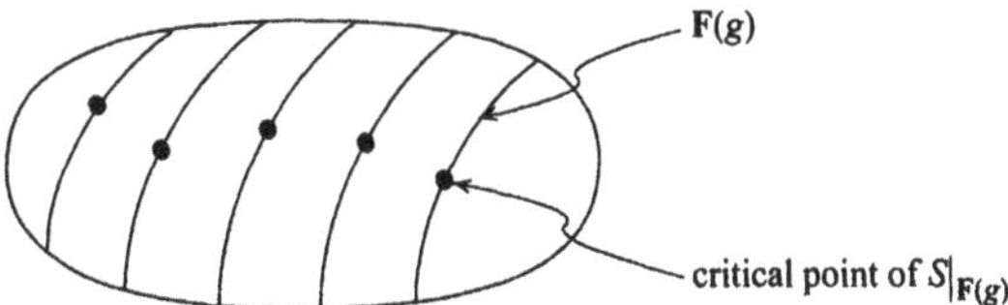

Figure 2. The space $\mathbf{M}^2$ is fibered by the fibers $\mathbf{F}(g)$. The function S restricted to $\mathbf{F}(g)$ has a nondegenerate relative critical point.

In conclusion, if the (relative) first variation $S'(x_0) = 0$ defines a submanifold $\mathbf{V}^\ell$, and if the second variation S'' restricted to the complement of $\mathbf{V}^\ell$ in $\mathbf{M}^d$ is not degenerate, the generalized Morse lemma says that there exists a system of coordinates $(\bar{y}^a, \bar{y}^A)$ in the neighborhood of a relative critical point x_0 such that

$$S(x) = S(x_0) + \sum_{a=1}^{\ell} g_a \bar{y}^a + \sum_{A,B=\ell+1}^{d} g_{AB}(x_0)\bar{y}^A \bar{y}^B . \tag{3.6}$$

Remark. The extreme cases $\ell = 0$ (isolated critical points) and $\ell = d$ (no critical point) are well known.

3.2. Generalized Morse Lemma; Infinite Dimensional Case; Lagrange Multipliers

Equation (1.4) is the infinite dimensional version of (3.1). Formally the argument leading from (3.1) to the generalized Morse lemma (3.6) applies to (1.4). But the contents is much richer in the infinite dimensional case:

- The algebraic equations for the critical points $x_0 \in \mathbf{M}^d$ become equations for q in $\mathcal{P}_{\mu,\nu}\mathbf{M}^d$, i.e., Euler–Lagrange differential equations.
- The set of Euler–Lagrange equations splits into 2 sets:

$$\begin{cases} S'_a = g_a & \text{is a set of } \ell \text{ constraints on the system} \\ S'_A(q) = 0 & \text{determine the components } q^A \text{ of the path } q: T \to \mathbf{M}^d . \end{cases} \tag{3.7}$$

Equation (3.5) now reads, with the notation of section 1

$$S(x) = S(q) + g_a \cdot \xi^a + \frac{1}{2} S''_{AB}(q) \cdot \xi^A \xi^B + O(|\xi|^3) . \tag{3.8}$$

The variables $\{\xi^a\}$ play the role of Lagrange multipliers of the system.

In a functional integral symbolically written

$$\int_\Xi \exp(2\pi i S(x)/h)\, \mathcal{D}_\Xi \xi \tag{3.9}$$

with

$$\mathcal{D}_{\Xi}\,\xi = \mathcal{D}\xi^{a}\,\mathcal{D}\xi^{A} \tag{3.10}$$

the integration with respect to $\mathcal{D}\xi^a$ contributes δ-functionals in g_a. In equation (3.23) we shall give explicitly the linear change of variables, $\xi \mapsto u$, which brings the action in the form (3.5). We do not have a general prescription for the non linear change of variable which would remove $O(|\xi|^3)$.

Remark. The nonlinear change of variable $x \mapsto z$, exploited in [8], which brings the second variation into a quadratic form on $\mathcal{P}\mathbb{R}^d$ is valid only on spaces of pointed paths. It can be used here if we replace $\mathcal{P}_{\mu,\nu}\mathbf{M}^d$ by a space of pointed paths, say $\mathcal{P}_{\mu}\mathbf{M}^d$, and insert in the integrand an index function which kills the paths which are not in $\mathcal{P}_{\mu,\nu}\mathbf{M}^d$.

Remark. The generalized Morse lemma in finite dimensions was derived in [3]; it was used for the infinite dimensional case by constructing the WKB approximation of a propagator when the final state is restricted by a conservation law. For example, with obvious notations, one writes, K being the WKB approximation of the propagator,

$$K(p_b,t_b;p_a,t_a) = \lim_{h=0} \int_{\mathbf{M}^d} dx\, K(p_b,t_b;x,t)\, K(x,t;p_a,t_a)$$

and looks for the critical points of the sum $F(x)$ of the action functions $\mathcal{S}$ (not the action functionals S)

$$F(x) = \mathcal{S}(p_b,t_b;x,t) + \mathcal{S}(x,t;p_a,t_a).$$

The analysis of the degenerate critical points of the function $F(x)$ reproduces the results obtained in Section 3.1, and Section 3.2. If $F(x)$ has no critical point, i.e., if there is no classical path between (p_a,t_a) and (p_b,t_b), then $K(p_b,t_b;p_a,t_a) = O(h^n)$ for n an arbitrary integer. It follows that *conservation laws* (here conservation of momentum) *appear only in the classical limit of quantum physics.*

3.3. Functional Integral; Semiclassical Approximation

The functional integral representation of the probability amplitude $K(\nu,t_b;\mu,t_a)$ for the transition of the system S from the state μ at t_a to the state ν at t_b can be written schematically

$$K(\nu,t_b;\mu,t_a) = \int_{\mathcal{P}_{\mu,\nu}\mathbf{M}^d} \mathcal{D}x \exp\left(\frac{2\pi i}{h} S(x)\right) \Phi(x(t_a)). \tag{3.11}$$

The domain of integration $\mathbb{X}$ is the space of $L^{2,1}$ paths

$$x : \mathbf{T} \to \mathbf{M}^d$$

satisfying boundary conditions (μ) at t_a, and (ν) at t_b.

The semiclassical approximation is a path integral over the tangent space $T_q\,\mathcal{P}_{\mu,\nu}\mathbf{M}^d$. To fit within the general framework of [8], we recall the expressions as written there for semiclassical approximations when the second variation is not degenerate. We recall a specific case: "momentum-to-position" transition [8, Sect. III.2], because it is more illuminating than a generic case. One of the important points of [8] is to integrate over a space of pointed paths (paths with one fixed point) because a space of pointed paths on any manifold $\mathbf{M}^d$ is contractible and can be parametrized by a space of pointed paths on $\mathbb{R}^d$, vanishing at the origin of $\mathbb{R}^d$. In a momentum-to-position transition, the obvious parametrizing space is the space $\mathcal{P}_0 T_b \mathbf{M}^d$ of paths on $T_b \mathbf{M}^d$ where $x(t_b) = b$.

The (v) boundary conditions are $x(t_b) = b$.

The (μ) boundary conditions at t_a are encoded in the functional integral by the choice of initial "wave function" Φ in (3.11) — a plane wave of momentum $p(t_a) = p_a$ if $\mathbf{M}^d$ is flat, or its "generalization" [8, eq. (3.2)] if $\mathbf{M}^d$ is riemannian.

If the hessian of S is not degenerate, the strict WKB approximation [8, eq. III.17] of $K(b,t_b;p_a,t_a)$ is [9]

$$K_{\text{WKB}}(b,t_b;p_a,t_a) = \exp\left(\frac{2\pi i}{h}\mathcal{S}(t_b,x_b)\right)\left(\text{Det}\,\frac{Q_0}{Q+Q_0}\right)^{1/2} \tag{3.12}$$

where $\mathcal{S}$ is the action function, solution of the Hamilton–Jacobi equation of the system and where by [8, eq. (B.30)]

$$Q_0(\xi) + Q(\xi) = \frac{2}{h}\mathcal{S}''(q)\cdot\xi\xi \tag{3.13}$$

with

$$Q_0(\xi) = \int_{\mathbf{T}} dt\, \frac{\partial^2 L}{\partial\dot{q}^\alpha(t)\,\partial\dot{q}^\beta(t)}\,\dot{\xi}^\alpha(t)\,\dot{\xi}^\beta(t)\,.$$

The infinite determinant in (3.12) is the ratio of finite determinants of Jacobi matrices (see [8, section 3.2] [10 and references therein]. The determinant in (3.12) together with (3.13) comes from [8, eq. (B.34)]

$$\int \mathcal{D}_{s,Q_0}\xi \exp\left(-\frac{2\pi}{sh}\mathcal{S}''(q)\cdot\xi\xi\right) = \left(\text{Det}\,\frac{Q_0}{Q_0+Q}\right)^{1/2},\ s\in\{1,i\}. \tag{3.14}$$

This equation can be thought of as defining $\mathcal{D}_{s,Q_0}$ on $T_q\mathcal{P}_{\mu,v}\mathbf{M}^d$. Anticipating the case in which the hessian of S is degenerate, we make the linear change of variable defined by (2.8)

$$L : T_q\mathcal{P}_{\mu,v}\mathbf{M}^d \to \ell^2(\text{hessian}) \text{ by } \xi \mapsto u \tag{3.15}$$

with ℓ^2(hessian) the space of points u such that $\Sigma\,\alpha_k(u^k)^2$ is finite (see II.9). Moreover, assume for simplicity that there is only one eigenvector ψ_0 of the Jacobi operator which has a zero eigenvalue.

We recall briefly the transformation of a functional integral under a linear change of variable of integration [see 8, Sect. A2.2]. Let

$$L : \mathbb{X} \to \mathbb{Y},$$

let $\mathcal{D}x$ on $\mathbb{X}$ be defined by

$$\int_{\mathbb{X}} \mathcal{D}x \exp\left(-\frac{\pi}{s}Q_{\mathbb{X}}(x) - 2\pi i\langle x',x\rangle\right) = \exp(-\pi s W_{\mathbb{X}'}(x')) \tag{3.16}$$

$Q_{\mathbb{X}}$ a quadratic form on $\mathbb{X}$, $W_{\mathbb{X}'}$ its "inverse" in the dual $\mathbb{X}'$ of $\mathbb{X}$, whose precise definition [8, (1.6)] is not needed here. Let $\mathcal{D}y$ on $\mathbb{Y}$ be defined similarly by

$$\int_{\mathbb{Y}} \mathcal{D}y \exp\left(-\frac{\pi}{s}Q_{\mathbb{Y}}(y) - 2\pi i\langle y',y\rangle\right) = \exp(-\pi s W_{\mathbb{Y}'}(y')) \tag{3.17}$$

where, assuming L invertible

$$Q_{\mathbb{Y}} = Q_{\mathbb{X}} \circ L^{-1}, \text{ i.e. } Q_{\mathbb{Y}}(y) = Q_{\mathbb{X}}(x(y)). \tag{3.18}$$

Then,

$$\int_{\mathbb{X}} \mathcal{D}x \exp\left(-\frac{\pi}{s} Q_{\mathbb{X}}(x)\right) \cdot g(Lx) = \int_{\mathbb{Y}} \mathcal{D}y \exp\left(-\frac{\pi}{s} Q_{\mathbb{Y}}(y)\right) \cdot g(y). \tag{3.19}$$

Therefore, (3.14) in the u variable reads

$$\begin{aligned}
&\int_{T_q \mathcal{P}_{\mu,\nu}\mathbf{M}^d} \mathcal{D}_{s,Q_0}\xi \exp\left(-\frac{\pi}{s} Q_0(\xi)\right) \exp\left(-\frac{\pi}{s} Q(\xi)\right) \\
&= \int \mathcal{D}_{s,Q_0 \circ L^{-1}} u \exp\left(-\frac{\pi}{s} Q_0(\xi(u))\right) \exp\left(-\frac{\pi}{s} Q(\xi(u))\right) \\
&= \int \mathcal{D}_{s,Q_0 \circ L^{-1}} u \exp\left(-\frac{2\pi}{sh} S''(q) \cdot \xi(u)\,\xi(u)\right).
\end{aligned} \tag{3.20}$$

The domain of integration ℓ^2(hessian) is spanned by the complete set of eigenvectors $\{\psi_k\}$, $k \in \{0, \ldots\}$, and can be decomposed into a one-dimensional space $\mathbb{X}^1$ spanned by ψ_0 and an infinite dimensional space $\mathbb{X}^\infty$ of codimension 1, spanned by $\{\psi_k\}$, $k \in \{1, \ldots\}$

$$\ell^2(\text{hessian}) = \mathbb{X}^1 \times \mathbb{X}^\infty. \tag{3.21}$$

The volume element on ℓ^2(hessian) is the product of the volume elements on $\mathbb{X}^1$ and on $\mathbb{X}^\infty$ respectively. We write

$$\mathcal{D}_{s,Q_0 \circ L^{-1}} u = \mathcal{D}_{\mathbb{X}^1} u^0 \, \mathcal{D}_{\mathbb{X}^\infty} u. \tag{3.22}$$

Under the change (3.15) defined by (2.8), the dominating terms in the expansion of $S(x)$ around $S(q)$ read (3.8)

$$S(x) \simeq S(q) + c_0 u^0 + \frac{1}{2} \sum_{k=1}^{\infty} \alpha_k (u^k)^2 \tag{3.23}$$

with

$$c_0 = \int_{\mathbb{T}} dt \frac{\delta S}{\delta q^\alpha(t)} \psi_0^\alpha(t) \tag{3.24}$$

and the functional integral, with $s = i$,

$$\begin{aligned}
&\int \mathcal{D}_{i,Q_0 \circ L^{-1}} u \exp\left(\frac{2\pi i}{h} c_0 u^0 + \frac{\pi i}{h} \sum_{k=1}^{\infty} \alpha_k (u^k)^2\right) \cdot \Phi(x(t_a)) \\
&= \int_{\mathbb{X}^1} \mathcal{D}_{\mathbb{X}^1} u^0 \exp\left(\frac{2\pi i}{h} c_0 u^0\right) \int_{\mathbb{X}^\infty} \mathcal{D}_{\mathbb{X}^\infty} u \exp\left(\frac{\pi i}{h} \sum_{k=1}^{\infty} \alpha_k (u^k)^2\right) \cdot \Phi(x(t_a)).
\end{aligned} \tag{3.25}$$

In the space $\mathbb{X}^\infty$, the quadratic form

$$S''(q) \cdot \xi\xi \,|_{\mathbb{X}^\infty} = \sum_{k=1}^{\infty} \alpha_k (u^k)^2$$

is invertible and the integral over $\mathbb{X}^\infty$ proceeds as before. The integral over $\mathbb{X}^1$ contributes a δ-function to the propagator

$$\delta\left(\frac{1}{h} \int_{\mathbb{T}} dt \frac{\delta S}{\delta q^\alpha(t)} \psi_0^\alpha(t)\right)$$

which says that the propagator vanishes, unless the conservation law

$$\frac{1}{h}\int_{\mathrm{T}} \mathrm{d}t \,\frac{\delta S}{\delta q^{\alpha}(t)}\,\psi_0^{\alpha}(t) = 0$$

is satisfied.

Explicit expressions for a variety of cases can be found in [3]. In conclusion, if the action is invariant under automorphisms of $\mathcal{U}_{\mu,\nu}$, then the dominating terms of the semiclassical expansion of $S(x)$ around $S(q)$ imply conservation laws. Conservation laws appear in the classical limit of quantum physics.

4. THE INTERSECTION $\mathcal{U}_{\mu,\nu}$ IS A MULTIPLE ROOT OF $S'(Q)\cdot\xi = 0$

In this situation the classical flows are caustic forming. Four examples are treated in references [11] and [4]. Two of them enter into well-known problems.

i) The soap bubble problem [12]. The "paths" are the curves defining (by rotation around an axis) the surface of a soap bubble held by two rings. The "classical flow" is a family of catenaries with one fixed point. The caustic is the envelope of the catenaries.

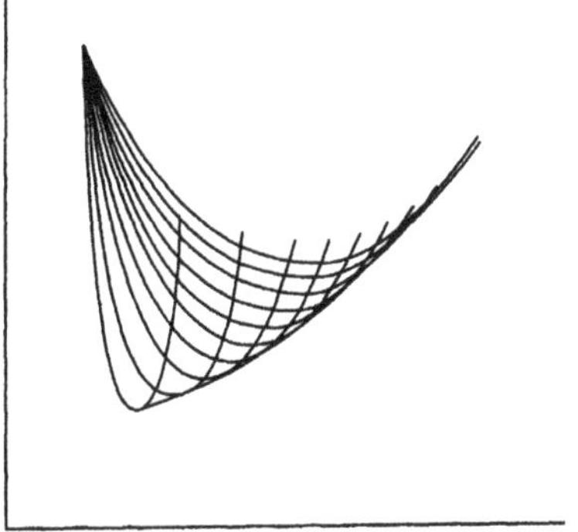

Figure 3. For a point in the "dark" side of the caustic there is no classical path; for a point on the "bright" side there are two classical paths which coalesce into a single one as the intersection of the two paths approaches the caustic. Note that the paths do not arrive at an intersection at the same time, the paths do not intersect in a space time diagram.

ii) The scattering of particles by a repulsive Coulomb potential. The flow is a family of Coulomb paths with fixed initial momentum. Its envelope is a parabola.

The two other examples are not readily identified as caustic problems because the flows do not have an envelope in the physical space. The vanishing boundary conditions of the Jacobi field at the caustic is the vanishing of its first derivative. In phase space the projection of the flow on the momentum space has an envelope.

iii) Rainbow scattering from a point source.

iv) Rainbow scattering from a source at infinity.

The relevant features can be analyzed on a specific example. We choose the scattering of particles by a repulsive Coulomb potential since we have already discussed momentum-to-position transitions in Section 3. For other examples see [7].

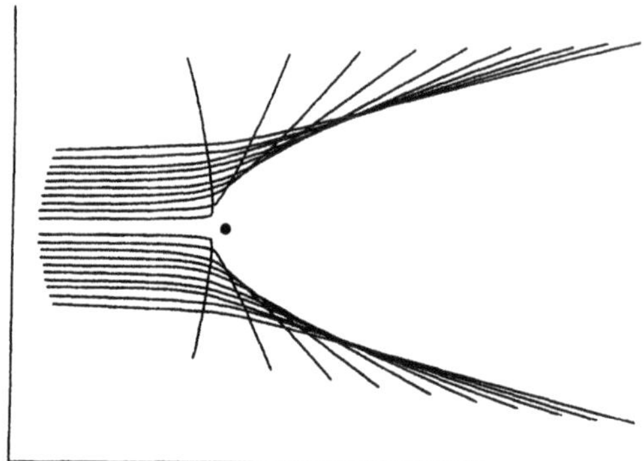

Figure 4. A flow on configuration space of charged particles in a repulsive Coulomb potential.

Let q and q^Δ be two solutions of the same Euler–Lagrange equation with slightly different boundary conditions at t_b

$$p(t_a) = p_a \qquad q(t_b) = b$$

$$p^\Delta(t_a) = p_a \qquad q^\Delta(t_b) = b^\Delta .$$

Assume p_a and b to be conjugate along q with multiplicity 1; i.e., the Jacobi fields h along q such that

$$\dot{h}(t_a) = 0, \quad h(t_b) = 0$$

form a one-dimensional space. Assume (p_a, b^Δ) not conjugate along q^Δ.

We shall compute the probability amplitude $K(b^\Delta, t_b; p_a, t_a)$ when b^Δ is close to the caustic on the "bright" side or on the "dark" side. We shall *not* compute K by expanding S around q^Δ for the following reasons:

- If b^Δ is on the dark side, q^Δ does not exist.
- If b^Δ is on the bright side, one could consider K to be the limit of the sum of two contributions corresponding to the two paths q and q^Δ intersecting at b^Δ

$$K(b, t_b; p_a, t_a) = \lim_{\Delta=0} K_q(b^\Delta, t_b + \Delta t; p_a, t_a) + K_{q^\Delta}(b^\Delta, t_b; p_a, t_a)$$

 but, at b^Δ, q has touched the caustic and "picked up" an additional phase equal to $-\pi/2$; both limits are infinite and their sum is not defined.

We compute $K(b^\Delta, t_b; p_a, t_a)$ by expanding S around q, using (1.4)—and possibly higher derivatives if the third variation is singular. The calculation requires some care [see reference [7] for details] because $q(t_b) \neq b^\Delta$; in other words q is not a critical point of the action restricted to the space of paths such that $x(t_b) = b^\Delta$.

As before we make the change (2.8) of variable $\xi \mapsto u$ which diagonalizes $S''(q) \cdot \xi\xi$. In the u variable, the expansion of $S(x)$ is given by (2.11)–(2.13). Again we decompose the domain of integration in the u-variable

$$\ell^2(\text{hessian}) = \mathbb{X}^1 \times \mathbb{X}^\infty .$$

Again the second variation restricted to $\mathbb{X}^\infty$ is non singular, and calculating the integral over $\mathbb{X}^\infty$ proceeds as usual for the strict WKB approximation. The integral over $\mathbb{X}^1$ is

$$I(\nu,c) = \int_{\mathbb{R}} du^0 \exp\left(i\left(cu^0 - \frac{\nu}{3}(u^0)^3\right)\right) = \nu^{-1/3}\,\mathrm{Ai}(\nu^{-1/3}c) \tag{4.1}$$

where

$$\nu = \frac{\pi}{h}\int_{\mathrm{T}} \mathrm{d}r \int_{\mathrm{T}} \mathrm{d}s \int_{\mathrm{T}} \mathrm{d}t \,\frac{\delta^3 S}{\delta q^\alpha(r)\,\delta q^\beta(s)\,\delta q^\gamma(t)}\,\psi_0^\alpha(r)\,\psi_0^\beta(s)\,\psi_0^\gamma(t) \tag{4.2}$$

$$c = -\frac{2\pi}{h}\int_{\mathrm{T}} \mathrm{d}t\,\frac{\delta S}{\delta q(t)}\cdot\psi_0(t)\,(b^\Delta - b) \tag{4.3}$$

Ai is the Airy function. The leading contribution of the Airy function when h tends to zero can be computed by the stationary phase method. At $v^2 = \nu^{-1/3}c$

$$\mathrm{Ai}(\nu^{-1/3}c) \simeq \begin{cases} 2\sqrt{\pi}v^{-1/4}\cos\left(\frac{2}{3}v^2 - \frac{\pi}{4}\right) & \text{for } v > 0 \\ \sqrt{\pi}(-v)^{-1/4}\exp\left(-\frac{2}{3}v^3\right) & \text{for } v < 0 \end{cases} \tag{4.4}$$

v is the critical point of the phase in the integrand of the Airy function; it is of order $h^{-1/3}$. For $v > 0$, b^Δ is in the illuminated region and the probability amplitude oscillates rapidly as h tends to zero. For $v < 0$, b^Δ is in the shadow region and the probability amplitude decays exponentially.

The probability amplitude $K(b^\Delta, t_b; a, t_a)$ does not blow up when b^Δ tends to b. Quantum mechanics softens up the caustics.

Remark. The normalization and the argument of the Airy function can be expressed solely in terms of the Jacobi fields.

Remark. Other cases, such as position-to-momentum, position-to-position, momentum-to-momentum, angular momentum transitions have been treated explicitly in references [3] and [7].

5. GLORY SCATTERING

Backward scattering of light, very close to the direction of the incoming rays has a long and interesting history (see for instance [13] and references therein). It creates a bright halo around one's shadow, and is usually called glory scattering. Early derivations of glory scattering were cumbersome, and used several approximations. It has been computed from first principles by functional integration using only the expansion in powers of the square root of Planck's constant [11,4].

The classical cross-section for the scattering of a beam of particles in a solid angle $d\Omega = 2\pi\sin\theta\,d\theta$ by an axisymmetric potential is

$$d\sigma_{c\ell}(\Omega) = 2\pi B(\theta)\,dB(\theta) \tag{5.1}$$

where the deflection function $\Theta(B)$ giving the scattering angle θ as a function of the impact parameter B is assumed to have a unique inverse $B(\Theta)$. We can write

$$d\sigma_{c\ell}(\Omega) = B(\Theta)\,\frac{dB(\Theta)}{d\Theta}\,|_{\Theta=\theta}\,\frac{d\Omega}{\sin\theta} \tag{5.2}$$

abbreviated henceforth

$$d\sigma_{c\ell}(\Omega) = B(\theta)\frac{dB(\theta)}{d\theta}\frac{d\Omega}{\sin\theta}. \tag{5.3}$$

It can happen that for a certain value of B, say B_g (g for glory), the deflection function vanishes,

$$\theta = \Theta(B_g) \text{ is } 0 \text{ or } \pi, \tag{5.4}$$

implying $\sin\theta = 0$, and making (5.3) useless.

The classical glory scattering cross-section is infinite because glory scattering is a caustic problem on two accounts.

i) There is a conservation law: the final momentum $p_b = -p_a$ the initial momentum.

ii) Near glory, particles with impact parameter $B_g + \delta B$ and $-B_g + \delta B$ exist with approximately the same angles, namely $\pi +$ terms of order $(\delta B)^3$.

The glory cross-section can be computed [11,4] using the methods presented in sections 3 and 4. The result is

$$d\sigma(\Omega) = 4\pi^2 h^{-1}|p_a| B^2(\theta)\frac{dB(\theta)}{d\theta} J_0(2\pi h^{-1}|p_a| B(\theta)\sin\theta)^2 d\Omega \tag{5.5}$$

where J_0 is the Bessel function of order 0.

A similar calculation [14,4,13] gives the WKB cross-section for polarized glories of massless waves in curved space–times

$$d\sigma(\Omega) = 4\pi^2\lambda^{-1} B_g^2\frac{dB}{d\theta} J_{2s}(2\pi\lambda^{-1} B_g\sin\theta)^2 d\Omega \tag{5.6}$$

$s = 0$ for scalar waves; at glory $J_0(0)^2 \neq 0$

$s = 1$ for electromagnetic waves; at glory $J_2(0)^2 \neq 0$

$s = 2$ for gravitational waves; at glory $J_4(0)^2 \neq 0$.

λ is the wave length of the incoming wave.

Equation (5.6) matches perfectly with the numerical calculations [13] of R. Matzner based on the partial wave decomposition method.

6. CONCLUSION

Caustics are classical limits of quantum physics. They are best characterized by the existence of nonvanishing Jacobi fields with vanishing boundary conditions. They include conservation laws as well as caustics in the strict sense of the term.

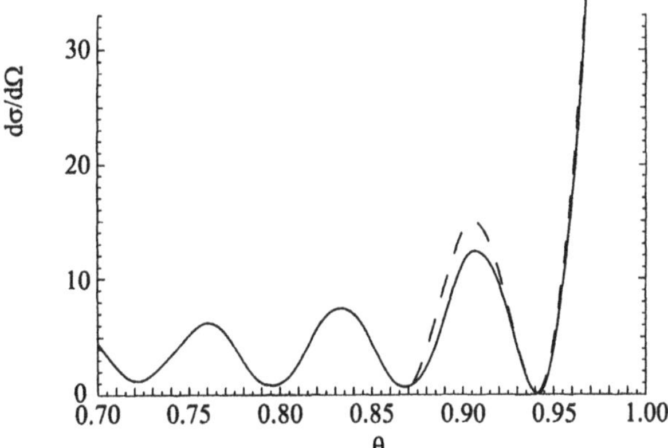

Figure 5. Glory cross-section:
$\frac{d\sigma}{d\Omega} = 2\pi\omega B_g^2 \left|\frac{dB}{d\theta}\right|_{\theta=\pi} J_{2s}(\omega B_g \sin\theta)^2$
$B_g = B(\pi)$ is the glory impact parameter.
$\omega = 2\pi\lambda^{-1}$; $s = 2$ for gravitational wave.
Analytic cross section: dashed line.
Numerical cross section: solid line.

APPENDIX

Although it is often easier to work with eq. II.4 rather than with the differential form of the Jacobi operator obtained via integration by parts, we record the differential form below, not assuming vanishing boundary conditions for ξ at t_a nor at t_b. This equation can be useful in particular when varying the boundary conditions.

Let $S(q) = \int_{\mathbf{T}} dt\, L(q(t), \dot{q}(t), t)$; then, the Jacobi operator defined by

$$S''(q)\cdot\xi\xi = \int_{\mathbf{T}} dt\, \xi^\alpha(t)\, \mathcal{J}_{\alpha\beta}(q(t))\, \xi^\beta(t) + \Sigma(t_b) - \Sigma(t_a)$$

can be written (abbreviating $q(t)$ by q and $\xi(t)$ by ξ)

$$\mathcal{J}_{\alpha\beta}(q) = -\frac{\partial^2 L}{\partial\dot{q}^\alpha \partial\dot{q}^\beta}\frac{d^2}{dt^2} + \left(\frac{\partial^2 L}{\partial q^\alpha \partial\dot{q}^\beta} - \frac{\partial^2 L}{\partial\dot{q}^\alpha \partial q^\beta} - \frac{d}{dt}\frac{\partial^2 L}{\partial\dot{q}^\alpha \partial\dot{q}^\beta}\right)\frac{d}{dt} - \frac{d}{dt}\frac{\partial^2 L}{\partial\dot{q}^\alpha \partial q^\beta} + \frac{\partial^2 L}{\partial q^\alpha \partial q^\beta}$$

and the boundary terms

$$\Sigma(\cdot) = \left(\frac{\partial^2 L}{\partial\dot{q}^\alpha \partial q^\beta} - \frac{1}{2}\frac{d}{dt}\frac{\partial^2 L}{\partial\dot{q}^\alpha \partial\dot{q}^\beta}\right)\xi^\alpha\xi^\beta + \frac{1}{2}\frac{d}{dt}\left(\frac{\partial^2 L}{\partial\dot{q}^\alpha \partial\dot{q}^\beta}\xi^\alpha\xi^\beta\right),$$

equivalently

$$\Sigma(t) = \frac{\partial^2 L}{\partial\dot{q}^\alpha \partial q^\beta}\xi^\alpha\xi^\beta + \frac{\partial^2 L}{\partial\dot{q}^\alpha \partial\dot{q}^\beta}\dot{\xi}^\alpha\xi^\beta.$$

REFERENCES

[1] C. DeWitt-Morette, G. Low, L. S. Schulman, A. Y. Shiekh: "Wedges I," *Foundations of Physics* **16**, 311–349 (1986).

[2] The properties of Jacobi fields in configuration space and phase space, and the corresponding WKB approximations are scattered in several papers of Cécile DeWitt-Morette, John La Chapelle, Maurice Mizrahi, Bruce Nelson, Benny Sheeks, Alice Young, and Tian-Rong Zhang. For a summary of results obtained prior to 1984, see references [3,4]. For more recent results, see reference [5].

[3] Cécile DeWitt-Morette and Tian-Rong Zhang: "Path integrals and conservation laws," *Phys. Rev. D* **28**, 2503–2516 (1983).

[4] Cécile DeWitt-Morette: "Feynman path integrals, from the prodistribution definition to the calculation of glory scattering," *Acta Physica Austriaca Suppl.* **XXVI**, 101–170 (1984).

[5] John La Chapelle: "Functional Integration on Symplectic Manifolds," (Ph. D. Dissertation, The University of Texas at Austin, May 1995).

[6] Cécile DeWitt-Morette and Tian-Rong Zhang: "Feynman–Kac formula in phase space with applications to coherent state transitions," *Phys. Rev. D* **28**, 2517–2525 (1983).

[7] Cécile DeWitt-Morette, Bruce Nelson, and Tian-Rong Zhang: "Caustic problems in quantum mechanics with applications to scattering theory," *Phys. Rev. D* **28**, 2526–2546 (1983).

[8] Pierre Cartier and Cécile DeWitt-Morette: "A new perspective on functional integration," *J. Math. Phys.* **36**, 2237–2312 (1995). For strict WKB approximations, see the calculation leading to (III.55).

[9] We use h rather than $\hbar = h/2\pi$, in order that expressions such as Gaussian, Fourier transforms, etc... on $\mathbb{R}^d$ do not depend on d.

[10] C. DeWitt-Morette, A. Maheshwari, and B. Nelson: "Path integration in non-relativistic quantum mechanics," *Phys. Rep.* **50**, 266–372 (1979). This article includes a summary of earlier articles.

[11] Cécile DeWitt-Morette and Bruce L. Nelson: "Glories — and other degenerate points of the action," *Phys. Rev. D* **29**, 1663–1668 (1984).

[12] Cécile DeWitt-Morette: "Catastrophes in Lagrangian systems," and C. DeWitt-Morette and P. Tshumi: "Catastrophes in Lagrangian systems. An example," in *Long Time Prediction in Dynamics* Eds. V. Szebehely and B. D. Tapley (D. Reidel Pub. Co. 1976) pp. 57–69. Also Y. Choquet-Bruhat and C. DeWitt-Morette *Analysis, Manifolds and Physics* Vol. I pp. 105–109, and pp. 277–281 (North Holland 1982–1996).

[13] R. A. Matzner, C. DeWitt-Morette, B. Nelson, and T.-R. Zhang: "Glory scattering by black holes," *Phys. Rev. D* **31**, 1869–1878 (1985).

[14] Tian-Rong Zhang and Cécile DeWitt-Morette: "WKB cross-section for polarized glories of massless waves in curved space–times," *Phys. Rev. Letters* **52**, 2313–2316 (1984).

References Added in Proof

Ph. Choquard and F. Steiner: "The story of Van Vleck's and Morette–Van Hove's determinants" *Helv. Phys. Acta* **69**, 636–654 (1969) analyzes carefully the contributions of Cécile Morette ("On the definition and approximation of Feynman's path integral," *Phys. Rev.* **81**, 848–852 (1951)), Leon Van Hove ("annexe" to his doctorate thesis) and Wolfang Pauli (unpublished notes) to the strict WKB approximation of Feynman's path integral. The appearance of singular determinants ("WKB breaks down") was first noted by Ph. Choquard in *Helv. Phys. Acta* **28**, 89–157 (1955) on pages 114–119.

3

QUANTUM EQUIVALENCE PRINCIPLE

H. Kleinert*

Institut für Theoretische Physik
Freie Universität Berlin
Arnimallee 14
D-14195 Berlin, Germany

ABSTRACT

A simple mapping procedure is presented by which classical orbits and path integrals for the motion of a point particle in flat space can be transformed directly into those in curved space with torsion. Our procedure evolved from well-established methods in the theory of plastic deformations, where crystals with defects are described mathematically as images of ideal crystals under *active nonholonomic* coordinate transformations.

Our mapping procedure may be viewed as a natural extension of Einstein's famous *equivalence principle.* When applied to time-sliced path integrals, it gives rise to a new *quantum equivalence principle* which determines short-time action and measure of fluctuating orbits in spaces with curvature and torsion. The nonholonomic transformations possess a nontrivial Jacobian in the path integral measure which produces in a curved space an additional term proportional to the curvature scalar R, thus canceling a similar term found earlier by DeWitt. This cancellation is important for correctly describing semiclassically and quantum mechanically various systems such as the hydrogen atom, a particle on the surface of a sphere, and a spinning top. It is also indispensable for the process of bosonization, by which Fermi particles are redescribed by those fields.

1. INTRODUCTION

In 1957, Bryce DeWitt [1] proposed a path integral formula for a point particle in a curved space using a specific generalization of Feynman's time-sliced formula in Cartesian coordinates. Surprisingly, his amplitude turned out to satisfy a Schrödinger equation different from what had previously been considered as correct [2]. In addition to the Laplace–Beltrami

*E-mail: kleinert@physik.fu-berlin.de;
URL: http://www.physik.fu-berlin.de/~kleinert;
Phone/Fax: 0049/30/8383034
Source and postscript available from eprint archive (quant-ph/9612040)

operator for the kinetic term, his Hamilton operator contained an extra effective potential proportional to the curvature scalar R. At the time of his writing, DeWitt could not think of any argument to rule out the presence of such an extra term.

Since DeWitt's pioneering work, the time-sliced path integral in curved spaces has been reformulated by many people in a variety of ways [3]. The basic problem is the freedom in time slicing the functional integral. Literature offers prepoint, midpoint, and postpoint prescriptions which in the Schrödinger equation correspond to different orderings of the momentum operators $\hat{p}_\mu$ with respect to the position variables q^λ in the Hamiltonian operator $\hat{H} = g^{\mu\nu}(q)\hat{p}_\mu\hat{p}_\nu/2$. Similar ambiguities are well known in the theory of stochastic differential equations where different algorithms have been developed by Itô and Stratonovich based on different time discretization procedures [4]. In the stochastic context, covariant versions of the Fokker–Planck equation in curved spaces have been derived by Graham [5]. The mathematical approach to path integrals uses techniques [6] similar to the stochastic one. The inherent ambiguities can be removed by demanding a certain form for the Schrödinger equation of the system, which in curved space has the Laplace–Beltrami operator as an operator for the kinetic energy [2], without an additional curvature scalar.

It has often been repeated that a Hamiltonian whose kinetic term depends on the position variable has in principle many different operator versions. For an arbitrary model Hamiltonian, this is of course, true. A specific physical system, however, must have a unique Hamilton operator. If a system has a high symmetry, it is often possible to find its correct form on the basis of group theory. Recall that in standard textbooks on quantum mechanics [7], a spinning top is quantized by expressing its Hamiltonian in terms of the generators of the rotation group, and quantizing these via the well-known commutation rules, rather than canonical variables. This procedure avoids the ordering problem by avoiding canonical variables. The resulting Hamilton operator contains only a Laplace–Beltrami operator and *no extra term* proportional to R. A particle on the surface of a sphere is quantized similarly. This procedure forms the basis of the so-called *group quantization* or *geometric quantization* [8] which corresponds to Schrödinger equations containing only the Laplace–Beltrami operator and no extra curvature terms.

Until recently, geometric quantization was the only procedure which *predicted* the form of the Schrödinger equation uniquely on the basis of symmetry, with generally accepted results. Moreover, canonical quantization in flat space is a special case since it corresponds to a geometric quantization of the generators of the euclidean group. Unfortunately, it is quite difficult to generalize this procedure to systems in more general geometries without symmetry. In particular, it makes no prediction as to the form of the Schrödinger equation in spaces with curvature and torsion. In DeWitt's time-sliced approach and its various successors, Schrödinger equations are found which contain many possible selections of scalar combinations of curvature and torsion tensor. As a consequence, there is a definite need for a principle capable of predicting the correct Schrödinger equation in such spaces.

In the context of gravity, this may seem a somewhat academic question since nobody has ever experimentally observed a scalar term in the Schrödinger equation of gravitating matter for a point particle in a curved space, even if torsion is neglected, and it is not even clear, whether gravity will generate torsion outside spinning matter. The simplest generalization of Einstein's theory to the *Einstein–Cartan theory* [9] does not permit propagating torsion.

Fortunately, there exist fields other than gravitational physics and accessible to experiment, where torsion enters geometry. Most notable is the field of defect physics, where geometric methods have been used successfully for a long time to describe the plastic properties of materials [9]. Defects are described mathematically by means of *active nonholonomic*

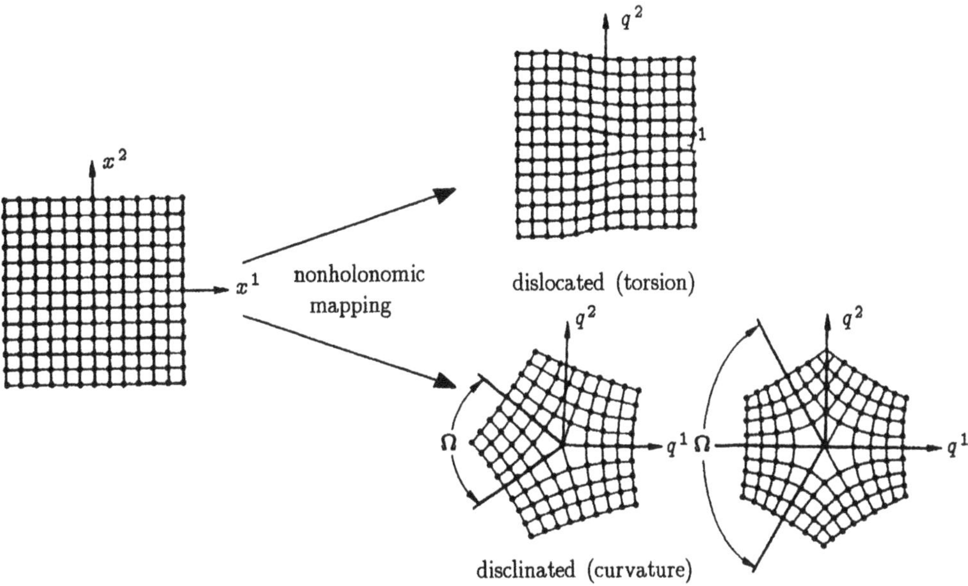

Figure 1. Crystal with dislocation and disclination generated by nonholonomic coordinate transformations from an ideal crystal. Geometrically, the former transformation introduces torsion and no curvature, the latter curvature and no torsion.

coordinate transformations.* They will be described in detail in Section 2. In Fig. 1 we show two typical elementary defects in two dimensions which can be generated by such transformations. It has been understood a long time ago that, geometrically, crystals with defects correspond to spaces with curvature and torsion [10].

In the context of path integrals, such transformations are of crucial importance. They provided us with a key to finding the resolvent of the most elementary atomic system, the hydrogen atom [11]. Two such transformations brought it to a harmonic form. Only recently was it recognized that one of these transformations may be interpreted as leading to a space with torsion [12]. If DeWitt's construction rules for a path integral in curved space would be adapted to this case, the resulting path integral, would produce the wrong atomic spectrum.

The resolution of this puzzle has led to the discovery of a simple rule for correctly transforming Feynman's time-sliced path integral formula from its well-known Cartesian form to spaces with curvature and torsion [12, 13, 14]. The rule plays the same fundamental role in quantum physics as Einstein's *equivalence principle* does within classical physics, where it governs the form of the equations of motion in curved spaces. It is therefore called *quantum equivalence principle* (QEP) [12].

The crucial place where this principle makes a nontrivial statement is in the measure of the path integral. The nonholonomic nature of the differential coordinate transformation gives rise to an additional term with respect to the naive DeWitt measure, and this cancels precisely the bothersome term proportional to R found by DeWitt.

It is the purpose of these lectures to demonstrate the power of the new quantum equivalence principle and to discuss its consequences also at the classical level, where the familiar

*Note that passive nonholonomic coordinate transformations lead to an alternative, usually inconvenient description of ideal crystals. The role of torsion is then played by so-called *objects of anholonomity*.

action principle breaks down and requires an important modification [12, 15, 16]. The geometric reason for this lies in the fact that infinitesimal variations can no longer be taken as closed curves; they possess a defect analogous to the Burgers vector in crystal physics. This surprising result has been verified by deriving the Euler equations for the motion of a spinning top from an action principle formulated *within* the body-fixed reference frame, where the geometry of the nonholonomic coordinates possesses torsion [18].

2. CLASSICAL MOTION OF A MASS POINT IN A SPACE WITH TORSION

We begin by recalling that Einstein formulated the rules for finding the classical laws of motion in a gravitational field on the basis of his famous equivalence principle. He assumed the space to be free of torsion since otherwise his geometric principle was not able to determine the classical equations of motion uniquely. Since our nonholonomic mapping principle is free of this problem, we do not need to restrict the geometry in this way. The correctness of the resulting laws of motion is exemplified by several physical systems with well-known experimental properties. Basis for these "experimental verifications" will be the fact that classical equations of motion are invariant under nonholonomic coordinate transformations. Since it is well known that active versions of such transformations introduce curvature and torsion into a parameter space, such redescriptions of standard mechanical systems provide us with sample systems in general metric-affine spaces.

To be as specific and as simple as possible, we restrict ourselves to the theory for a nonrelativistic massive point particle in a general metric-affine space. The entire discussion may easily be extended to relativistic particles in space–time.

2.1. Equations of Motion

Consider the action of the particle along the orbit $\mathbf{x}(t)$ in a flat space parametrized with D rectilinear, Cartesian coordinates:

$$\mathcal{A} = \int_{t_a}^{t_b} dt \frac{M}{2} (\dot{x}^i)^2. \tag{1}$$

It may be transformed to curvilinear coordinates q^μ, $\mu = 1,2,3$, via some functions

$$x^i = x^i(q), \tag{2}$$

leading to

$$\mathcal{A} = \int_{t_a}^{t_b} dt \frac{M}{2} g_{\mu\nu}(q) \dot{q}^\mu \dot{q}^\nu, \tag{3}$$

where

$$g_{\mu\nu}(q) = \partial_\mu x^i(q) \partial_\nu x^i(q) \tag{4}$$

is the *induced metric* for the curvilinear coordinates. Repeated indices are understood to be summed over, as usual. For Cartesian coordinates, upper and lower indices i are the same. The indices μ, ν of the curvilinear coordinates, on the other hand, are lowered by contraction with the metric $g_{\mu\nu}$ or raised with the inverse metric $g^{\mu\nu} \equiv (g_{\mu\nu})^{-1}$.

The length of the orbit in the flat space is given by

$$l = \int_{t_a}^{t_b} dt \sqrt{g_{\mu\nu}(q)\dot{q}^\mu \dot{q}^\nu}. \tag{5}$$

Both the action (3) and the length (5) are invariant under arbitrary *reparametrizations of space* $q^\mu \to q'^\mu$.

Einstein's equivalence principle amounts to the postulate that the transformed action (3) describes directly the motion of the particle in the presence of a gravitational field caused by other masses. The forces caused by the field are all a result of the geometric properties of the metric tensor.

The equations of motion are obtained by extremizing the action in Eq. (3) with the result

$$\partial_t(g_{\mu\nu}\dot{q}^\nu) - \frac{1}{2}\partial_\mu g_{\lambda\nu}\dot{q}^\lambda \dot{q}^\nu = g_{\mu\nu}\ddot{q}^\nu + \bar{\Gamma}_{\lambda\nu\mu}\dot{q}^\lambda \dot{q}^\nu = 0. \tag{6}$$

Here

$$\bar{\Gamma}_{\lambda\nu\mu} \equiv \frac{1}{2}(\partial_\lambda g_{\nu\mu} + \partial_\nu g_{\lambda\mu} - \partial_\mu g_{\lambda\nu}) \tag{7}$$

is the *Riemann connection* or *Christoffel symbol* of the *first kind*. Defining also the Christoffel symbol of the *second kind*

$$\bar{\Gamma}_{\lambda\nu}{}^\mu \equiv g^{\mu\sigma}\bar{\Gamma}_{\lambda\nu\sigma}, \tag{8}$$

we can write

$$\ddot{q}^\mu + \bar{\Gamma}_{\lambda\nu}{}^\mu \dot{q}^\lambda \dot{q}^\nu = 0. \tag{9}$$

The solutions of these equations are the classical orbits. They coincide with the extrema of the length l of a curve in (5). Thus, in a curved space, classical orbits are the *shortest curves*, called *geodesics*. The reason for the name *shortest lines* is that they minimize the invariant length (5) of all lines connecting two given points $q_a^\mu = q^\mu(t_a)$ and $q_b^\mu = q^\mu(t_b)$.

The same equations can also be obtained directly by transforming the equation of motion from

$$\ddot{x}^i = 0 \tag{10}$$

to curvilinear coordinates q^μ, which gives

$$\ddot{x}^i = \frac{\partial x^i}{\partial q^\mu}\ddot{q}^\mu + \frac{\partial^2 x^i}{\partial q^\lambda \partial q^\nu}\dot{q}^\lambda \dot{q}^\nu = 0. \tag{11}$$

At this place it is useful to employ the so-called *basis triads*

$$e^i{}_\mu(q) \equiv \frac{\partial x^i}{\partial q^\mu} \tag{12}$$

and the *reciprocal basis triads*

$$e_i{}^\mu(q) \equiv \frac{\partial q^\mu}{\partial x^i}, \tag{13}$$

which satisfy the orthogonality and completeness relations

$$e_i{}^\mu e^i{}_\nu = \delta^\mu{}_\nu, \tag{14}$$

$$e_i{}^\mu e^j{}_\mu = \delta_i{}^j. \tag{15}$$

The induced metric can then be written as

$$g_{\mu\nu}(q) = e^i{}_\mu(q) e^i{}_\nu(q). \tag{16}$$

Using the basis triads, Eq. (11) can be rewritten as

$$\frac{d}{dt}(e^i{}_\mu \dot{q}^\mu) = e^i{}_\mu \ddot{q}^\mu + e^i{}_{\mu,\nu}\dot{q}^\mu \dot{q}^\nu = 0, \tag{17}$$

or as

$$\ddot{q}^\mu + e_i{}^\mu e^i{}_{\kappa,\lambda}\dot{q}^\kappa \dot{q}^\lambda = 0. \tag{18}$$

The subscript λ separated by a comma denotes the partial derivative $\partial_\lambda = \partial/\partial q^\lambda$, i.e., $f_{,\lambda} \equiv \partial_\lambda f$. The quantity in front of $\dot{q}^\kappa \dot{q}^\lambda$ is called the *affine connection*:

$$\Gamma_{\lambda\kappa}{}^\mu = e_i{}^\mu e^i{}_{\kappa,\lambda}. \tag{19}$$

Due to (14), it can also be written as

$$\Gamma_{\lambda\kappa}{}^\mu = -e^i{}_\kappa e_i{}^\mu{}_{,\lambda}. \tag{20}$$

Thus we arrive at the transformed flat-space equation of motion

$$\ddot{q}^\mu + \Gamma_{\kappa\lambda}{}^\mu \dot{q}^\kappa \dot{q}^\lambda = 0. \tag{21}$$

The solutions of this equation are called the *straightest lines* or *autoparallels*.

If the coordinate transformation $x^i(q)$ is smooth and single-valued, it is integrable. This property is expressed by Schwarz's integrability condition, according to which derivatives in front of such a function $x^i(q)$ commute:

$$(\partial_\lambda \partial_\kappa - \partial_\kappa \partial_\lambda) x^i(q) = 0. \tag{22}$$

Then the triads satisfy the identity

$$e^i_{\kappa,\lambda} = e^i_{\lambda,\kappa}, \tag{23}$$

implying that the connection $\Gamma_{\mu\nu}{}^\lambda$ is symmetric in the lower indices. In this case it coincides with the Riemann connection, the Christoffel symbol $\bar{\Gamma}_{\mu\nu}{}^\lambda$. This follows immediately after inserting $g_{\mu\nu}(q) = e^i{}_\mu(q) e^i{}_\nu(q)$ into (7) and working out all derivatives using (23). Thus, for a space with curvilinear coordinates q^μ which can be reached by an integrable coordinate transformation from a flat space, the autoparallels coincide with the geodesics.

2.2. Nonholonomic Mapping to Spaces with Torsion

It is possible to map the x^i-space locally into a q-space via an infinitesimal transformation

$$dx^i = e^i{}_\mu(q) dq^\mu, \tag{24}$$

with coefficient functions $e^i{}_\mu(q)$ which are not integrable in the sense of Eq. (22), i.e.,

$$\partial_\mu e^i{}_\nu(q) - \partial_\nu e^i{}_\mu(q) \neq 0. \tag{25}$$

Such a mapping will be called *nonholonomic*. There exists no single-valued function $x^i(q)$ for which $e^i{}_\mu(q) = \partial x^i(q)/\partial q^\mu$. Nevertheless, we shall write (25) in analogy to (22) as

$$(\partial_\lambda \partial_\kappa - \partial_\kappa \partial_\lambda) x^i(q) \neq 0, \tag{26}$$

since this equation involves only the differential dx^i. This violation of mathematical conventions will not cause any problems.

From Eq. (25) we see that the image space of a nonholonomic mapping carries torsion. The connection $\Gamma_{\lambda\kappa}{}^\mu = e_i{}^\mu e^i{}_{\kappa,\lambda}$ has a nonzero antisymmetric part, called the *torsion tensor* [17]:

$$S_{\lambda\kappa}{}^\mu = \frac{1}{2}(\Gamma_{\lambda\kappa}{}^\mu - \Gamma_{\kappa\lambda}{}^\mu). \tag{27}$$

In contrast to $\Gamma_{\lambda\kappa}{}^\mu$, the antisymmetric part $S_{\lambda\kappa}{}^\mu$ is a proper tensor under holonomic coordinate transformations. The contracted tensor

$$S_\mu \equiv S_{\mu\lambda}{}^\lambda \tag{28}$$

transforms like a vector, whereas the contracted connection $\Gamma_\mu \equiv \Gamma_{\mu\nu}{}^\nu$ does not. Even though $\Gamma_{\mu\nu}{}^\lambda$ is not a tensor, we shall freely lower and raise its indices using contractions with the metric or the inverse metric, respectively: $\Gamma^\mu{}_\nu{}^\lambda \equiv g^{\mu\kappa}\Gamma_{\kappa\nu}{}^\lambda$, $\Gamma_\mu{}^{\nu\lambda} \equiv g^{\nu\kappa}\Gamma_{\mu\kappa}{}^\lambda$, $\Gamma_{\mu\nu\lambda} \equiv g_{\lambda\kappa}\Gamma_{\mu\nu}{}^\kappa$. The same thing will be done with $\bar{\Gamma}_{\mu\nu}{}^\lambda$.

In the presence of torsion, the connection is no longer equal to the Christoffel symbol. In fact, by rewriting $\Gamma_{\mu\nu\lambda} = e_{i\lambda}\partial_\mu e^i{}_\nu$ trivially as

$$\begin{aligned}\Gamma_{\mu\nu\lambda} &= \frac{1}{2}\left\{e_{i\lambda}\partial_\mu e^i{}_\nu + \partial_\mu e_{i\lambda} e^i{}_\nu + e_{i\mu}\partial_\nu e^i{}_\lambda + \partial_\nu e_{i\mu} e^i{}_\lambda - e_{i\mu}\partial_\lambda e^i{}_\nu - \partial_\lambda e_{i\mu} e^i{}_\nu\right\} \\ &+ \frac{1}{2}\left\{\left[e_{i\lambda}\partial_\mu e^i{}_\nu - e_{i\lambda}\partial_\nu e^i{}_\mu\right] - \left[e_{i\mu}\partial_\nu e^i{}_\lambda - e_{i\mu}\partial_\lambda e^i{}_\nu\right] + \left[e_{i\nu}\partial_\lambda e^i{}_\mu - e_{i\nu}\partial_\mu e^i{}_\lambda\right]\right\}\end{aligned}$$

and using $e^i{}_\mu(q) e^i{}_\nu(q) = g_{\mu\nu}(q)$, we find the decomposition

$$\Gamma_{\mu\nu}{}^\lambda = \bar{\Gamma}_{\mu\nu}{}^\lambda + K_{\mu\nu}{}^\lambda, \tag{29}$$

where the combination of torsion tensors

$$K_{\mu\nu\lambda} \equiv S_{\mu\nu\lambda} - S_{\nu\lambda\mu} + S_{\lambda\mu\nu} \tag{30}$$

is called the *contortion tensor*. It is antisymmetric in the last two indices so that

$$\Gamma_{\mu\nu}{}^\nu = \bar{\Gamma}_{\mu\nu}{}^\nu. \tag{31}$$

In Einstein's theory of gravitation, torsion is assumed to be absent, i.e., the integrability condition for $x^i(q)$ is not violated as in (26). The main effect of matter in Einstein's theory of gravitation manifests itself in the violation of the integrability condition for the *derivative* of the coordinate transformation $x^i(q)$, namely,

$$(\partial_\mu \partial_\nu - \partial_\nu \partial_\mu)\partial_\lambda x^i(q) \neq 0. \tag{32}$$

A transformation for which $x^i(q)$ itself is integrable, while the first derivatives $\partial_\mu x^i(q) = e^i{}_\mu(q)$ are not, carries a flat-space region into a purely curved one. The quantity which records the nonintegrability is the *Cartan curvature tensor*

$$R_{\mu\nu\lambda}{}^\kappa = e_i{}^\kappa(\partial_\mu \partial_\nu - \partial_\nu \partial_\mu) e^i{}_\lambda. \tag{33}$$

Working out the derivatives using (19) we see that $R_{\mu\nu\lambda}{}^{\kappa}$ can be written as a covariant curl of the connection,

$$R_{\mu\nu\lambda}{}^{\kappa} = \partial_\mu \Gamma_{\nu\lambda}{}^{\kappa} - \partial_\nu \Gamma_{\mu\lambda}{}^{\kappa} - [\Gamma_\mu, \Gamma_\nu]_\lambda{}^{\kappa}. \tag{34}$$

In the last term we have used a matrix notation for the connection. The tensor components $\Gamma_{\mu\lambda}{}^{\kappa}$ are viewed as matrix elements $(\Gamma_\mu)_\lambda{}^{\kappa}$, so that we can use the matrix commutator

$$[\Gamma_\mu, \Gamma_\nu]_\lambda{}^{\kappa} \equiv (\Gamma_\mu \Gamma_\nu - \Gamma_\nu \Gamma_\mu)_\lambda{}^{\kappa} = \Gamma_{\mu\lambda}{}^{\sigma} \Gamma_{\nu\sigma}{}^{\kappa} - \Gamma_{\nu\lambda}{}^{\sigma} \Gamma_{\mu\sigma}{}^{\kappa}. \tag{35}$$

Einstein's original theory of gravity assumes the absence of torsion. The space properties are completely specified by the *Riemann curvature tensor* formed from the Riemann connection (the Christoffel symbol)

$$\bar{R}_{\mu\nu\lambda}{}^{\kappa} = \partial_\mu \bar{\Gamma}_{\nu\lambda}{}^{\kappa} - \partial_\nu \bar{\Gamma}_{\mu\lambda}{}^{\kappa} - [\bar{\Gamma}_\mu, \bar{\Gamma}_\nu]_\lambda{}^{\kappa}. \tag{36}$$

The relation between the two curvature tensors is

$$R_{\mu\nu\lambda}{}^{\kappa} = \bar{R}_{\mu\nu\lambda}{}^{\kappa} + \bar{D}_\mu K_{\nu\lambda}{}^{\kappa} - \bar{D}_\nu K_{\mu\lambda}{}^{\kappa} - [K_\mu, K_\nu]_\lambda{}^{\kappa}. \tag{37}$$

In the last term, the $K_{\mu\lambda}{}^{\kappa}$'s are viewed as matrices $(K_\mu)_\lambda{}^{\kappa}$. The symbols $\bar{D}_\mu$ denote the *covariant derivatives* formed with the Christoffel symbol. Covariant derivatives act like ordinary derivatives if they are applied to a scalar field. When applied to a vector field, they act as follows:

$$\begin{aligned} \bar{D}_\mu v_\nu &\equiv \partial_\mu v_\nu - \bar{\Gamma}_{\mu\nu}{}^{\lambda} v_\lambda, \\ \bar{D}_\mu v^\nu &\equiv \partial_\mu v^\nu + \bar{\Gamma}_{\mu\lambda}{}^{\nu} v^\lambda. \end{aligned} \tag{38}$$

The effect upon a tensor field is the generalization of this; every index receives a corresponding additive $\bar{\Gamma}$ contribution.

In the presence of torsion, there exists another covariant derivative formed with the affine connection $\Gamma_{\mu\nu}{}^{\lambda}$ rather than the Christoffel symbol which acts upon a vector field as

$$\begin{aligned} D_\mu v_\nu &\equiv \partial_\mu v_\nu - \Gamma_{\mu\nu}{}^{\lambda} v_\lambda, \\ D_\mu v^\nu &\equiv \partial_\mu v^\nu + \Gamma_{\mu\lambda}{}^{\nu} v^\lambda. \end{aligned} \tag{39}$$

The two derivatives (38) and (39) are equally covariant under holonomic coordinate transformations. Thus, in conventional differential geometry it is not clear which of them should play a more fundamental role in physics. They do differ, however, in their transformation behavior under nonholonomic transformations, and there (39) is definitely the preferred object for reasons of simplicity.

From either of the two curvature tensors, $R_{\mu\nu\lambda}{}^{\kappa}$ and $\bar{R}_{\mu\nu\lambda}{}^{\kappa}$, one can form the once-contracted tensors of rank 2, the *Ricci tensor*

$$R_{\nu\lambda} = R_{\mu\nu\lambda}{}^{\mu}, \tag{40}$$

and the *curvature scalar*

$$R = g^{\nu\lambda} R_{\nu\lambda}. \tag{41}$$

The celebrated Einstein equation for the gravitational field postulates that the tensor

$$G_{\mu\nu} \equiv R_{\mu\nu} - \frac{1}{2} g_{\mu\nu} R, \tag{42}$$

the so-called *Einstein tensor*, is proportional to symmetric energy-mo-mentum tensor of all matter fields. This postulate was made only for spaces with no torsion, in which case $R_{\mu\nu} = \bar{R}_{\mu\nu}$ and $R_{\mu\nu}, G_{\mu\nu}$ are both symmetric. As mentioned in the Introduction, it is not yet clear how Einstein's field equations should be generalized in the presence of torsion since the experimental consequences are as yet too small to be observed. In this paper, we are not concerned with the generation of curvature and torsion but only with their consequences upon the motion of point particles.

2.3. Simple Nonholonomic Sample Mappings

The generation of defects illustrated in Fig. 1 provides us with two simple examples for nonholonomic mappings which show us in which way these mappings are capable of generating a space with curvature and torsion from a euclidean space. The reader not familiar with this subject is advised to consult the standard literature on this subject quoted in Ref. [9, 10].

Consider first the upper example in Fig. 1, in which a dislocation is generated, characterized by a missing or an additional layer of atoms. In two dimensions, it may be described differentially by the transformation

$$dx^i = \begin{cases} dq^1 & \text{for } i=1, \\ dq^2 + \varepsilon\partial_\mu\phi(q)dq^\mu & \text{for } i=2, \end{cases} \tag{43}$$

with infinitesimal ε and the multi-valued function

$$\phi(q) \equiv \arctan(q^2/q^1). \tag{44}$$

The triads reduce to dyads, with the components

$$\begin{aligned} e^1{}_\mu &= \delta^1{}_\mu, \\ e^2{}_\mu &= \delta^2{}_\mu + \varepsilon\partial_\mu\phi(q), \end{aligned} \tag{45}$$

and the torsion tensor has the components

$$e^1{}_\lambda S_{\mu\nu}{}^\lambda = 0, \qquad e^2{}_\lambda S_{\mu\nu}{}^\lambda = \frac{\varepsilon}{2}(\partial_\mu\partial_\nu - \partial_\nu\partial_\mu)\phi. \tag{46}$$

If we differentiate (44) formally, we find $(\partial_\mu\partial_\nu - \partial_\nu\partial_\mu)\phi \equiv 0$. This, however, is incorrect at the origin. Using Stokes' theorem we see that

$$\int d^2q(\partial_1\partial_2 - \partial_2\partial_1)\phi = \oint dq^\mu\partial_\mu\phi = \oint d\phi = 2\pi \tag{47}$$

for any closed circuit around the origin, implying that there is a δ-function singularity at the origin with

$$e^2{}_\lambda S_{12}{}^\lambda = \frac{\varepsilon}{2}2\pi\delta^{(2)}(q). \tag{48}$$

By a linear superposition of such mappings we can generate an arbitrary torsion in the q-space. The mapping introduces no curvature. When encircling a dislocation along a closed path C, its counter image C' in the ideal crystal does not form a closed path. The closure failure is called the *Burgers vector*

$$b^i \equiv \oint_{C'} dx^i = \oint_C dq^\mu e^i{}_\mu. \tag{49}$$

It specifies the direction and thickness of the layer of additional atoms. With the help of Stokes' theorem, it is seen to measure the torsion contained in any surface S spanned by C:

$$b^i = \oint_S d^2 s^{\mu\nu} \partial_\mu e^i{}_\nu = \oint_S d^2 s^{\mu\nu} e^i{}_\lambda S_{\mu\nu}{}^\lambda, \tag{50}$$

where $d^2 s^{\mu\nu} = -d^2 s^{\nu\mu}$ is the projection of an oriented infinitesimal area element onto the plane $\mu\nu$. The above example has the Burgers vector

$$b^i = (0, \varepsilon). \tag{51}$$

A corresponding closure failure appears when mapping a closed contour C in the ideal crystal into a crystal containing a dislocation. This defines a Burgers vector:

$$b^\mu \equiv \oint_{C'} dq^\mu = \oint_C dx^i e_i{}^\mu. \tag{52}$$

By Stokes' theorem, this becomes a surface integral

$$\begin{aligned} b^\mu &= \oint_S d^2 s^{ij} \partial_i e_j{}^\mu = \oint_S d^2 s^{ij} e_i{}^\nu \partial_\nu e_j{}^\mu \\ &= - \oint_S d^2 s^{ij} e_i{}^\nu e_j{}^\lambda S_{\nu\lambda}{}^\mu, \end{aligned} \tag{53}$$

the last step following from (20).

The second example is the nonholonomic mapping in the lower part of Fig. 1 generating a disclination which corresponds to an entire section of angle Ω missing in an ideal atomic array. For an infinitesimal angel Ω, this may be described, in two dimensions, by the differential mapping

$$x^i = \delta^i{}_\mu [q^\mu + \Omega \varepsilon^\mu{}_\nu q^\nu \phi(q)], \tag{54}$$

with the multi-valued function (44). The symbol $\varepsilon_{\mu\nu}$ denotes the antisymmetric Levi-Civita tensor. The transformed metric

$$g_{\mu\nu} = \delta_{\mu\nu} - \frac{2\Omega}{q^\sigma q_\sigma} \varepsilon_{\mu\lambda} \varepsilon_{\nu\kappa} q^\lambda q^\kappa. \tag{55}$$

is single-valued and has commuting derivatives. The torsion tensor vanishes since $(\partial_1 \partial_2 - \partial_2 \partial_1) x^{1,2}$ is proportional to $q^{2,1} \delta^{(2)}(q) = 0$. The local rotation field $\omega(q) \equiv \frac{1}{2}[\partial_1 x^2(q) - \partial_2 x^1(q)]$, on the other hand, is equal to the multi-valued function $-\Omega\phi(q)$, thus having the noncommuting derivatives:

$$(\partial_1 \partial_2 - \partial_2 \partial_1) \omega(q) = -2\pi\Omega\delta^{(2)}(q). \tag{56}$$

To lowest order in Ω, this determines the curvature tensor, which in two dimensions posses only one independent component, for instance R_{1212}. Using the fact that $g_{\mu\nu}$ has commuting derivatives, R_{1212} can be written as

$$R_{1212} = (\partial_1 \partial_2 - \partial_2 \partial_1) \omega(q). \tag{57}$$

2.4. Straightest versus Shortest Particle Trajectories

We have seen in Eqs. (38) and (39) that there exist two different types of covariant derivatives. Thus there are two types of parallel vector fields, $v_{\rm a}^{\mu}$ and $v_{\rm g}^{\mu}$, defined by

$$D_{\nu}v_{\rm a}^{\mu}(q)=0, \qquad \bar{D}_{\nu}v_{\rm g}^{\mu}(q)=0. \tag{58}$$

The stream lines of these vector fields are found by introducing an arbitrary parameter s and searching for a function $q^{\mu}(s)$ whose tangent is given by these vector fields:

$$\frac{dq_{\rm a}^{\mu}(q)}{ds}=v_{\rm a}^{\mu}(q), \qquad \frac{dq_{\rm g}^{\mu}(q)}{ds}=v_{\rm g}^{\mu}(q). \tag{59}$$

By forming one more derivative with respect to s, we find the differential equations for these stream lines

$$\ddot{q}_{\rm a}^{\mu}+\Gamma_{\kappa\lambda}{}^{\mu}\dot{q}_{\rm a}^{\kappa}\dot{q}_{\rm a}^{\lambda}=0, \tag{60}$$

$$\ddot{q}_{\rm g}^{\mu}+\bar{\Gamma}_{\kappa\lambda}{}^{\mu}\dot{q}_{\rm g}^{\kappa}\dot{q}_{\rm g}^{\lambda}=0. \tag{61}$$

The first are the autoparallels (21), the second are the shortest lines or geodesics. In the presence of torsion, the shortest and straightest lines are no longer equal. This keeps surprising people, since by (27), torsion is the asymmetric part of the connection and the asymmetric part of $\Gamma_{\kappa\lambda}{}^{\mu}$ certainly drops out of the equation of motion (60). However, the decomposition (29) with the contortion tensor (30) shows, that $\Gamma_{\kappa\lambda}{}^{\mu}$ contains a contribution from torsion also in its symmetric part:

$$\Gamma_{\{\kappa\lambda\}}{}^{\mu}=\bar{\Gamma}_{\kappa\lambda}{}^{\mu}+2S^{\mu}{}_{\kappa\lambda}. \tag{62}$$

Since the two types of lines play geometrically an equally favored role, the question arises as to which of them describes the correct classical particle orbits. The answer will be given in the rest of these lectures. Both types of curves are a priori equally good candidates for particle trajectories in a theory of gravitation in which all particles move along geometrically determined paths.

From our nonholonomic mapping principle, a free-particle trajectory in Euclidean space is mapped into the autoparallel. Since we know that, in classical mechanics, equations of motion remain correct under nonholonomic coordinate transformations, we conclude that nature must have chosen the autoparallels as the geometrically distinguished curves along which particles move.

However, this conclusion might be too hasty. The fundamental *Hamilton principle* of classical mechanics states that particle trajectories should emerge from a variational approach, in which an action $\mathcal{A}[q]$ which is a functional of arbitrary possible paths $q^{\mu}(t)$ is minimized with respect to small changes $\delta q^{\mu}(t)$. If we take as an action the nonholonomic image (3) of the flat-space action (1), and minimize this without varying the endpoints, i.e. with the boundary conditions

$$\delta q^{\mu}(t_a)=\delta q^{\mu}(t_b)=0, \tag{63}$$

we find for the particle trajectories the geodesic differential equations (61), rather than the autoparallel ones (60).

Which conclusion is physically correct? At first sight, the nonholonomic mapping principle seems to be inconsistent. In Section 2.5 we shall see that consistency can be ensured by a proper extension of Hamilton's principle to particles in spaces with torsion.

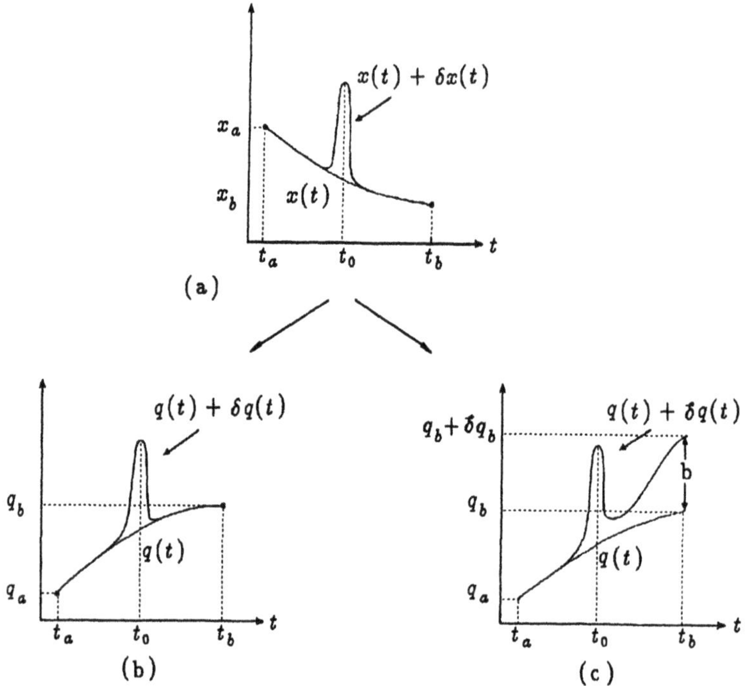

Figure 2. Images under a holonomic and a nonholonomic mapping of a fundamental path variation. In the holonomic case, the paths $x(t)$ and $x(t)+\delta x(t)$ in (a) turn into the paths $q(t)$ and $q(t)+\delta q(t)$ in (b). In the nonholonomic case with $S_{\mu\nu}{}^{\lambda} \neq 0$, they go over into $q(t)$ and $q(t)+\delta q(t)$ shown in (c) with a closure failure b^{μ} at t_b analogous to the Burgers vector b^{μ} in a solid with dislocations.

2.5. Classical Action Principle for Spaces with Curvature and Torsion

We have seen in the last section that for a unique consistent theory in spaces with torsion we must reexamine the Hamilton action principle for the classical motion of a spinless point particle. We must make sure that autoparallels emerge as the extremals of an action (3) that involves only the metric tensor $g_{\mu\nu}$. The action is independent of the torsion and carries only information on the Riemann part of the space geometry. Torsion can therefore enter the equations of motion only via some novel feature of the variation procedure. Since we know how to perform variations of an action in the euclidean x^i-space, we deduce the correct procedure in the general metric-affine space by transferring the variations $\delta x^i(t)$ under the nonholonomic mapping

$$\dot{q}^{\mu} = e_i{}^{\mu}(q)\dot{x}^i \tag{64}$$

into the q^{μ}-space. Their images are quite different from ordinary variations as illustrated in Fig. 2(a). The variations of the Cartesian coordinates $\delta x^i(t)$ are done at fixed end points of the paths. Thus they form *closed paths* in the x-space. Their images, however, lie in a space with defects and thus possess a closure failure indicating the amount of torsion introduced by the mapping. This property will be emphasized by writing the images $\delta q^{\mu}(t)$ and calling them *nonholonomic variations.*

Let us calculate them explicitly. The paths in the two spaces are related by the integral equation

$$q^\mu(t) = q^\mu(t_a) + \int_{t_a}^{t} dt' e_i{}^\mu(q(t'))\dot{x}^i(t'). \tag{65}$$

For two neighboring paths in x-space differing from each other by a variation $\delta x^i(t)$, Eq. (65) determines the nonholonomic variation $\boldsymbol{\delta} q^\mu(t)$:

$$\boldsymbol{\delta} q^\mu(t) = \int_{t_a}^{t} dt' \delta[e_i{}^\mu(q(t'))\dot{x}^i(t')]. \tag{66}$$

A comparison with (64) shows that the variations $\boldsymbol{\delta} q^\mu$ and the time derivative of q^μ are independent of each other

$$\boldsymbol{\delta}\dot{q}^\mu(t) = \frac{d}{dt}\boldsymbol{\delta} q^\mu(t), \tag{67}$$

just as for ordinary variations δx^i.

Let us introduce *auxiliary holonomic variations* in q-space:

$$\delta q^\mu \equiv e_i{}^\mu(q)\delta x^i. \tag{68}$$

In contrast to $\boldsymbol{\delta} q^\mu(t)$, these vanish at the endpoints,

$$\delta q(t_a) = \delta q(t_b) = 0, \tag{69}$$

i.e., they form closed paths with the unvaried orbits.

Using (68) we derive from (66) the relation

$$\begin{aligned} \frac{d}{dt}\boldsymbol{\delta} q^\mu(t) &= \delta e_i{}^\mu(q(t))\dot{x}^i(t) + e_i{}^\mu(q(t))\delta\dot{x}^i(t) \\ &= \delta e_i{}^\mu(q(t))\dot{x}^i(t) + e_i{}^\mu(q(t))\frac{d}{dt}[e^i{}_\nu(t)\delta q^\nu(t)]. \end{aligned} \tag{70}$$

After inserting

$$\delta e_i{}^\mu(q) = -\Gamma_{\lambda\nu}{}^\mu \delta q^\lambda e_i{}^\nu, \qquad \frac{d}{dt}e^i{}_\nu(q) = \Gamma_{\lambda\nu}{}^\mu \dot{q}^\lambda e^i{}_\mu, \tag{71}$$

this becomes

$$\frac{d}{dt}\boldsymbol{\delta} q^\mu(t) = -\Gamma_{\lambda\nu}{}^\mu \delta q^\lambda \dot{q}^\nu + \Gamma_{\lambda\nu}{}^\mu \dot{q}^\lambda \delta q^\nu + \frac{d}{dt}\delta q^\mu. \tag{72}$$

It is useful to introduce the difference between the nonholonomic variation $\boldsymbol{\delta} q^\mu$ and the auxiliary holonomic variation δq^μ:

$$\delta b^\mu \equiv \boldsymbol{\delta} q^\mu - \delta q^\mu. \tag{73}$$

Then we can rewrite (72) as a first-order differential equation for δb^μ:

$$\frac{d}{dt}\delta b^\mu = -\Gamma_{\lambda\nu}{}^\mu \delta b^\lambda \dot{q}^\nu + 2S_{\lambda\nu}{}^\mu \dot{q}^\lambda \delta q^\nu. \tag{74}$$

Under an arbitrary nonholonomic variation $\boldsymbol{\delta} q^\mu = \delta q^\mu + \delta b^\mu$, the action (3) changes by

$$\boldsymbol{\delta}\mathcal{A} = M\int_{t_a}^{t_b} dt\left(g_{\mu\nu}\dot{q}^\nu \boldsymbol{\delta}\dot{q}^\mu + \frac{1}{2}\partial_\mu g_{\lambda\kappa}\boldsymbol{\delta} q^\mu \dot{q}^\lambda \dot{q}^\kappa\right). \tag{75}$$

We use (67), (69) for a partial integration of the $\delta\dot{q}$-term, and apply the identity $\partial_\mu g_{\nu\lambda} \equiv \Gamma_{\mu\nu\lambda} + \Gamma_{\mu\lambda\nu}$, which follows from the definitions $g_{\mu\nu} \equiv e^i{}_\mu e^i{}_\nu$ and $\Gamma_{\mu\nu}{}^\lambda \equiv e_i{}^\lambda \partial_\mu e^i{}_\nu$, to obtain

$$\delta\mathcal{A} = M\int_{t_a}^{t_b} dt\Big[-g_{\mu\nu}\left(\ddot{q}^\nu + \bar{\Gamma}_{\lambda\kappa}{}^\nu\dot{q}^\lambda\dot{q}^\kappa\right)\delta q^\mu + \left(g_{\mu\nu}\dot{q}^\nu\frac{d}{dt}\delta b^\mu + \Gamma_{\mu\lambda\kappa}\delta b^\mu\dot{q}^\lambda\dot{q}^\kappa\right)\Big]. \tag{76}$$

To derive the equation of motion we first vary the action in a space without torsion. Then $\delta b^\mu(t) \equiv 0$, and we obtain

$$\delta\mathcal{A} = \delta\mathcal{A} = -M\int_{t_a}^{t_b} dt g_{\mu\nu}(\ddot{q}^\nu + \bar{\Gamma}_{\lambda\kappa}{}^\nu\dot{q}^\lambda\dot{q}^\kappa)q^\nu. \tag{77}$$

Thus, the action principle $\delta\mathcal{A} = 0$ produces the equation for the geodesics (9), which are the correct particle trajectories in the absence of torsion.

In the presence of torsion where $\delta b^\mu \neq 0$, the equation of motion receives a contribution from the second parentheses in (76). After inserting (74), the nonlocal terms proportional to δb^μ cancel and the total nonholonomic variation of the action becomes

$$\begin{aligned}\delta\mathcal{A} &= -M\int_{t_a}^{t_b} dt g_{\mu\nu}\left[\ddot{q}^\nu + \left(\bar{\Gamma}_{\lambda\kappa}{}^\nu + 2S^\nu{}_{\lambda\kappa}\right)\dot{q}^\lambda\dot{q}^\kappa\right]\delta q^\mu \\ &= -M\int_{t_a}^{t_b} dt g_{\mu\nu}\left(\ddot{q}^\nu + \Gamma_{\lambda\kappa}{}^\nu\dot{q}^\lambda\dot{q}^\kappa\right)\delta q^\mu.\end{aligned} \tag{78}$$

The second line follows from the first after using the identity $\Gamma_{\lambda\kappa}{}^\nu = \bar{\Gamma}_{\{\lambda\kappa\}}{}^\nu + 2S^\nu{}_{\{\lambda\kappa\}}$. The curly brackets indicate the symmetrization of the enclosed indices. Setting $\delta\mathcal{A} = 0$ gives the autoparallels (21) as the equations of motions, which is what we wanted to show.

In order appreciate the geometric significance of the differential equation (74), we introduce the matrices

$$G^\mu(t)_\lambda \equiv \Gamma_{\lambda\nu}{}^\mu(q(t))\dot{q}^\nu(t) \tag{79}$$

and

$$\Sigma^\mu{}_\nu(t) \equiv 2S_{\lambda\nu}{}^\mu(q(t))\dot{q}^\lambda(t), \tag{80}$$

and rewrite Eq. (74) as a differential equation for a vector

$$\frac{d}{dt}\delta b = -G\delta b + \Sigma(t)\,\delta q^\nu(t). \tag{81}$$

The solution is

$$\delta b(t) = \int_{t_a}^{t} dt' U(t,t')\,\Sigma(t')\,\delta q(t'), \tag{82}$$

with the matrix

$$U(t,t') = T\exp\left[-\int_{t'}^{t} dt'' G(t'')\right]. \tag{83}$$

In the absence of torsion, $\Sigma(t)$ vanishes identically and $\delta b(t) \equiv 0$, and the variations $\delta q^\mu(t)$ coincide with the holonomic $\delta q^\mu(t)$ [see Fig. 2(b)]. In a space with torsion, the variations $\delta q^\mu(t)$ and $\delta q^\mu(t)$ are different from each other [see Fig. 2(c)].

2.6. Alternative Formulation of Action Principle with Torsion

The above variational treatment of the action is still somewhat complicated and calls for a simpler procedure which was found recently [16].

Let us vary the paths $q^\mu(t)$ in the usual holonomic way, i.e., with fixed endpoints, and consider the associated variations $\delta x^i = e^i{}_\mu(q)\delta q^\mu$ of the Cartesian coordinates. Taking their time derivative $d_t \equiv d/dt$ we find

$$d_t\,\delta x^i = e^i{}_\lambda(q)d_t\delta q^\lambda + \partial_\mu e^i{}_\lambda(q)\dot{q}^\mu\delta q^\lambda. \tag{84}$$

On the other hand, we may write the relation (24) in the form $d_t x^i = e^i{}_\mu(q)d_t q^\mu$ and vary this to yield

$$\delta d_t x^i = e^i{}_\lambda(q)\delta\dot{q}^\lambda + \partial_\mu e^i{}_\lambda(q)\dot{q}^\lambda\delta q^\mu. \tag{85}$$

Using now the fact that time derivatives d_t and variations δ commute for Cartesian paths,

$$\delta d_t x^i - d_t\delta x^i = 0, \tag{86}$$

we deduce from (84) and (85) that this is no longer true in the presence of torsion, where

$$\delta d_t q^\lambda - d_t\delta q^\lambda = 2S_{\mu\nu}{}^\lambda(q)\,\dot{q}^\mu\,\delta q^\nu. \tag{87}$$

In other words, the variations of the velocities $\dot{q}^\mu(t)$ no longer coincide with the time derivatives of the variations of $q^\mu(t)$.

This failure to commute is responsible for shifting the trajectory from geodesics to autoparallels. Indeed, let us vary an action

$$\mathcal{A} = \int_{t_a}^{t_b} dt L\left(q^\lambda(t), \dot{q}^\lambda(t)\right) \tag{88}$$

by $\delta q^\lambda(t)$ and impose (87), we find

$$\delta\mathcal{A} = \int_{t_a}^{t_b} dt\left\{\frac{\partial L}{\partial q^\lambda}\delta q^\lambda + \frac{\partial L}{\partial\dot{q}^\lambda}\frac{d}{dt}\delta q^\lambda + 2S_{\mu\nu}{}^\lambda\frac{\partial L}{\partial\dot{q}^\lambda}\dot{q}^\mu\delta q^\nu\right\}. \tag{89}$$

After a partial integration of the second term using the vanishing $\delta q^\lambda(t)$ at the endpoints, we obtain the Euler–Lagrange equation

$$\frac{\partial L}{\partial q^\lambda} - \frac{d}{dt}\frac{\partial L}{\partial\dot{q}^\lambda} = 2S_{\lambda\mu}{}^\nu\dot{q}^\mu\frac{\partial L}{\partial\dot{q}^\nu}. \tag{90}$$

This differs from the standard Euler–Lagrange equation by an additional contribution due to the torsion tensor. For the action (3) we thus obtain the equation of motion

$$M\left[\ddot{q}^\lambda + g^{\lambda\kappa}\left(\partial_\mu g_{\nu\kappa} - \frac{1}{2}\partial_\kappa g_{\mu\nu}\right) + 2S^\lambda{}_{\mu\nu}\right]\dot{q}^\mu\dot{q}^\nu = 0, \tag{91}$$

which is once more Eq. (21) for autoparallels.

3. PATH INTEGRAL IN SPACES WITH CURVATURE AND TORSION

We now turn to the quantum mechanics of a point particle in a general metric-affine space. We first consider the path integral in a flat space with Cartesian coordinates

$$(\mathbf{x}t|\mathbf{x}'t') = \frac{1}{\sqrt{2\pi i\varepsilon\hbar/M}^D} \prod_{n=1}^{N} \left[\int_{-\infty}^{\infty} dx_n\right] \prod_{n=1}^{N+1} K_0^{\varepsilon}(\Delta\mathbf{x}_n), \tag{92}$$

where $K_0^{\varepsilon}(\Delta\mathbf{x}_n)$ is an abbreviation for the short-time amplitude

$$K_0^{\varepsilon}(\Delta\mathbf{x}_n) \equiv \langle\mathbf{x}_n|\exp\left(-\frac{i}{\hbar}\varepsilon\hat{H}\right)|\mathbf{x}_{n-1}\rangle = \frac{1}{\sqrt{2\pi i\varepsilon\hbar/M}^D} \exp\left[\frac{i}{\hbar}\frac{M}{2}\frac{(\Delta\mathbf{x}_n)^2}{\varepsilon}\right] \tag{93}$$

with $\Delta\mathbf{x}_n \equiv \mathbf{x}_n - \mathbf{x}_{n-1}, \mathbf{x} \equiv \mathbf{x}_{N+1}, \mathbf{x}' \equiv \mathbf{x}_0$. A possible external potential has been omitted since this would contribute in an additive way, uninfluenced by the space geometry.

Our basic postulate is that the path integral in a general metric-affine space should be obtained by an appropriate nonholonomic transformation of the amplitude (92) to a space with curvature and torsion.

3.1. Nonholonomic Transformation of the Action

The short-time action contains the square distance $(\Delta\mathbf{x}_n)^2$ which we have to transform to q-space. For an infinitesimal coordinate difference $\Delta\mathbf{x}_n \approx d\mathbf{x}_n$, the square distance is obviously given by $(d\mathbf{x})^2 = g_{\mu\nu}dq^\mu dq^\nu$. For a finite $\Delta\mathbf{x}_n$, however, it is well known that we must expand $(\Delta\mathbf{x}_n)^2$ up to the fourth order in $\Delta q_n{}^\mu = q_n{}^\mu - q_{n-1}{}^\mu$ to find all terms contributing to the relevant order ε.

It is important to realize that with the mapping from dx^i to dq^μ not being holonomic, the finite quantity Δq^μ is not uniquely determined by Δx^i. A unique relation can only be obtained by integrating the functional relation (65) along a specific path. The preferred path is the classical orbit, i.e., the autoparallel in the q-space. It is characterized by being the image of a straight line in the x-space. There $\dot{x}^i(t)$ =const and the orbit has the linear time dependence

$$\Delta x^i(t) = \dot{x}^i(t_0)\Delta t, \tag{94}$$

where the time t_0 can lie anywhere on the t-axis. Let us choose for t_0 the final time in each interval (t_n, t_{n-1}). At that time, $\dot{x}^i_n \equiv \dot{x}^i(t_n)$ is related to $\dot{q}^\mu_n \equiv \dot{q}^\mu(t_n)$ by

$$\dot{x}^i_n = e^i{}_\mu(q_n)\dot{q}^\mu_n. \tag{95}$$

It is easy to express $\dot{q}^\mu_n$ in terms of $\Delta q^\mu_n = q^\mu_n - q^\mu_{n-1}$ along the classical orbit. First we expand $q^\mu(t_{n-1})$ into a Taylor series around t_n. Dropping the time arguments, for brevity, we have

$$\Delta q \equiv q^\lambda - q'^\lambda = \varepsilon\dot{q}^\lambda - \frac{\varepsilon^2}{2!}\ddot{q}^\lambda + \frac{\varepsilon^3}{3!}\dddot{q}^\lambda + \ldots, \tag{96}$$

where $\varepsilon = t_n - t_{n-1}$ and $\dot{q}^\lambda, \ddot{q}^\lambda, \ldots$ are the time derivatives at the final time t_n. An expansion of this type is referred to as a *postpoint expansion*. Due to the arbitrariness of the choice of the time t_0 in Eq. (95), the expansion can be performed around any other point just as well, such as t_{n-1} and $\bar{t}_n = (t_n + t_{n-1})/2$, giving rise to the so-called *prepoint* or *midpoint* expansions of Δq.

Now, the term $\ddot{q}^\lambda$ in (96) is given by the equation of motion (21) for the autoparallel

$$\ddot{q}^\lambda = -\Gamma_{\mu\nu}{}^\lambda \dot{q}^\mu \dot{q}^\nu. \tag{97}$$

A further time derivative determines

$$\dddot{q}^\lambda = -(\partial_\sigma \Gamma_{\mu\nu}{}^\lambda - 2\Gamma_{\mu\nu}{}^\tau \Gamma_{\{\sigma\tau\}}{}^\lambda)\dot{q}^\mu \dot{q}^\nu \dot{q}^\sigma. \tag{98}$$

Inserting these expressions into (96) and inverting the expansion, we obtain $\dot{q}^\lambda$ at the final time t_n expanded in powers of Δq. Using (94) and (95) we arrive at the mapping of the finite coordinate differences:

$$\begin{aligned}
\Delta x^i &= e^i{}_\lambda \dot{q}^\lambda \Delta t \\
&= e^i{}_\lambda \left[\Delta q^\lambda - \frac{1}{2!}\Gamma_{\mu\nu}{}^\lambda \Delta q^\mu \Delta q^\nu \right. \\
&\quad \left. + \frac{1}{3!}(\partial_\sigma \Gamma_{\mu\nu}{}^\lambda + \Gamma_{\mu\nu}{}^\tau \Gamma_{\{\sigma\tau\}}{}^\lambda)\Delta q^\mu \Delta q^\nu \Delta q^\sigma + \ldots\right],
\end{aligned} \tag{99}$$

where $e^i{}_\lambda$ and $\Gamma_{\mu\nu}{}^\lambda$ are evaluated at the postpoint. Inserting this into the short-time amplitude (93), we obtain

$$K_0^\varepsilon(\Delta \mathbf{x}) = \langle \mathbf{x}| \exp\left(-\frac{i}{\hbar}\varepsilon \hat{H}\right)|\mathbf{x} - \Delta \mathbf{x}\rangle = \frac{1}{\sqrt{2\pi i \varepsilon \hbar / M}^D} \exp\left[\frac{i}{\hbar}\mathcal{A}_>^\varepsilon(q, q - \Delta q)\right] \tag{100}$$

with the short-time postpoint action

$$\begin{aligned}
\mathcal{A}_>^\varepsilon(q, q-\Delta q) &= \frac{M}{2\varepsilon}(\Delta x^i)^2 = \varepsilon \frac{M}{2} g_{\mu\nu}\dot{q}^\mu \dot{q}^\nu \\
&= \frac{M}{2\varepsilon}\left\{ g_{\mu\nu}\Delta q^\mu \Delta q^\nu - \Gamma_{\mu\nu\lambda}\Delta q^\mu \Delta q^\nu \Delta q^\lambda \right. \\
&\quad + \left[\frac{1}{3} g_{\mu\tau}(\partial_\kappa \Gamma_{\lambda\nu}{}^\tau + \Gamma_{\lambda\nu}{}^\delta \Gamma_{\{\kappa\delta\}}{}^\tau) \right. \\
&\quad \left.\left. + \frac{1}{4}\Gamma_{\lambda\kappa}{}^\sigma \Gamma_{\mu\nu\sigma}\right] \Delta q^\mu \Delta q^\nu \Delta q^\lambda \Delta q^\kappa + \ldots \right\}.
\end{aligned} \tag{101}$$

Separating the affine connection into Christoffel symbol and torsion, this can also be written as

$$\begin{aligned}
\mathcal{A}_>^\varepsilon(q, q-\Delta q) &= \frac{M}{2\varepsilon}\left\{ g_{\mu\nu}\Delta q^\mu \Delta q^\nu - \bar{\Gamma}_{\mu\nu\lambda}\Delta q^\mu \Delta q^\nu \Delta q^\lambda \right. \\
&\quad + \left[\frac{1}{3} g_{\mu\tau}(\partial_\kappa \bar{\Gamma}_{\lambda\nu}{}^\tau + \bar{\Gamma}_{\lambda\nu}{}^\delta \bar{\Gamma}_{\delta\kappa}{}^\tau) + \frac{1}{4}\bar{\Gamma}_{\lambda\kappa}{}^\sigma \bar{\Gamma}_{\mu\nu\sigma} \right. \\
&\quad \left.\left. + \frac{1}{3} S^\sigma{}_{\lambda\kappa} S_{\sigma\mu\nu} + \ldots \right]\right\} \Delta q^\mu \Delta q^\nu \Delta q^\lambda \Delta q^\kappa.
\end{aligned} \tag{102}$$

Note that the right-hand side contains only quantities *intrinsic* to the q-space. For the systems treated there (which all live in a euclidean space parametrized with curvilinear coordinates), the present intrinsic result reduces to the previous one.

At this point we observe that the final short-time action (101) could also have been introduced without any reference to the flat coordinates x^i. Indeed, the same action is obtained by evaluating the continuous action (3) for the small time interval $\Delta t = \varepsilon$ along the classical orbit between the points q_{n-1} and q_n. Due to the equations of motion (21), the Lagrangian

$$L(q,\dot{q}) = \frac{M}{2} g_{\mu\nu}(q(t))\, \dot{q}^\mu(t)\dot{q}^\nu(t) \tag{103}$$

is independent of time (this is true for autoparallels as well as geodesics). The short-time action

$$\mathcal{A}^\varepsilon(q,q') = \frac{M}{2}\int_{t-\varepsilon}^{t} dt'\, g_{\mu\nu}(q(t'))\dot{q}^\mu(t')\dot{q}^\nu(t') \tag{104}$$

can therefore be written in either of the three forms

$$\mathcal{A}^\varepsilon = \frac{M}{2}\varepsilon g_{\mu\nu}(q)\dot{q}^\mu\dot{q}^\nu = \frac{M}{2}\varepsilon g_{\mu\nu}(q')\dot{q}'^\mu\dot{q}'^\nu = \frac{M}{2}\varepsilon g_{\mu\nu}(\bar{q})\dot{\bar{q}}^\mu\dot{\bar{q}}^\nu, \tag{105}$$

where $q^\mu, q'^\mu, \bar{q}^\mu$ are the coordinates at the final time t_n, the initial time t_{n-1}, and the average time $(t_n + t_{n-1})/2$, respectively. The first expression obviously coincides with (101). The others can be used as a starting point for deriving equivalent prepoint or midpoint actions. The prepoint action $\mathcal{A}^\varepsilon_<$ arises from the postpoint one $\mathcal{A}^\varepsilon_>$ by exchanging Δq by $-\Delta q$ and the postpoint coefficients by the prepoint ones. The midpoint action has the most simple-looking appearance:

$$\bar{\mathcal{A}}^\varepsilon(\bar{q}+\frac{\Delta q}{2}, \bar{q}-\frac{\Delta q}{2}) = \frac{M}{2\varepsilon}\Big[g_{\mu\nu}(\bar{q})\Delta q^\mu \Delta q^\nu + \frac{1}{12} g_{\kappa\tau}(\partial_\lambda \Gamma_{\mu\nu}{}^\tau \tag{106}$$

$$+ \Gamma_{\mu\nu}{}^\delta \Gamma_{\{\lambda\delta\}}{}^\tau)\Delta q^\mu \Delta q^\nu \Delta q^\lambda \Delta q^\kappa + \dots\Big],$$

where the affine connection can be evaluated at any point in the interval (t_{n-1}, t_n). The precise position is irrelevant to the amplitude producing only changes beyond the relevant order epsilon.

In the textbook [12], the postpoint action turned out to be the most useful one since it gives ready access to the time evolution of amplitudes. The prepoint action is completely equivalent to it and useful if one wants to describe the time evolution backwards. Some authors favor the midpoint action because of its symmetry and intimate relation to an ordering prescription in operator quantum mechanics which was advocated by H. Weyl. This prescription is, however, only of historic interest since it does not lead to the correct physics. In the following, the action $\mathcal{A}^\varepsilon$ without subscript will always denote the preferred postpoint expression (101):

$$\mathcal{A}^\varepsilon \equiv \mathcal{A}^\varepsilon_>(q, q - \Delta q). \tag{107}$$

3.2. The Measure of Path Integration

We now turn to the integration measure in the Cartesian path integral (92)

$$\frac{1}{\sqrt{2\pi i\varepsilon\hbar/M}^D}\prod_{n=1}^{N} d^D x_n.$$

This has to be transformed to the general metric-affine space. We imagine evaluating the path integral starting out from the latest time and performing successively the integrations over

$x_N, x_{N-1}, \ldots$, i.e., in each short-time amplitude we integrate over the earlier position coordinate, the prepoint coordinate. For the purpose of this discussion, we relabel the product $\prod_{n=1}^{N} d^D x_n^i$ by $\prod_{n=2}^{N+1} dx_{n-1}^i$, so that the integration in each time slice (t_n, t_{n-1}) with $n = N+1, N, \ldots$ runs over dx_{n-1}^i.

In a flat space parametrized with curvilinear coordinates, the transformation of the integrals over $d^D x_{n-1}^i$ into those over $d^D q_{n-1}^\mu$ is obvious:

$$\prod_{n=2}^{N+1} \int d^D x_{n-1}^i = \prod_{n=2}^{N+1} \left\{ \int d^D q_{n-1}^\mu \det\left[e_\mu^i(q_{n-1})\right] \right\}. \tag{108}$$

The determinant of $e^i{}_\mu$ is the square root of the determinant of the metric $g_{\mu\nu}$:

$$\det(e^i{}_\mu) = \sqrt{\det g_{\mu\nu}(q)} \equiv \sqrt{g(q)}, \tag{109}$$

and the measure may be rewritten as

$$\prod_{n=2}^{N+1} \int d^D x_{n-1}^i = \prod_{n=2}^{N+1} \left[\int d^D q_{n-1}^\mu \sqrt{g(q_{n-1})} \right]. \tag{110}$$

This expression is not directly applicable. When trying to do the $d^D q_{n-1}^\mu$-integrations successively, starting from the final integration over dq_N^μ, the integration variable q_{n-1} appears for each n in the argument of $\det\left[e_\mu^i(q_{n-1})\right]$ or $g_{\mu\nu}(q_{n-1})$. To make this q_{n-1}-dependence explicit, we expand in the measure (108) $e_\mu^i(q_{n-1}) = e^i{}_\mu(q_n - \Delta q_n)$ around the postpoint q_n into powers of Δq_n. This gives

$$dx^i = e_\mu^i(q - \Delta q) dq^\mu = e_\mu^i dq^\mu - e^i{}_{\mu,\nu} dq^\mu \Delta q^\nu + \frac{1}{2} e^i{}_{\mu,\nu\lambda} dq^\mu \Delta q^\nu \Delta q^\lambda + \ldots, \tag{111}$$

omitting, as before, the subscripts of q_n and Δq_n. Thus the Jacobian of the coordinate transformation from dx^i to dq^μ is

$$J_0 = \det(e^i{}_\kappa) \det\left[\delta^\kappa{}_\mu - e_i{}^\kappa e^i{}_{\mu,\nu} \Delta q^\nu + \frac{1}{2} e_i{}^\kappa e^i{}_{\mu,\nu\lambda} \Delta q^\nu \Delta q^\lambda\right], \tag{112}$$

giving the relation between the infinitesimal integration volumes $d^D x^i$ and $d^D q^\mu$:

$$\prod_{n=2}^{N+1} \int d^D x_{n-1}^i = \prod_{n=2}^{N+1} \left\{ \int d^D q_{n-1}^\mu J_{0n} \right\}. \tag{113}$$

The well-known expansion formula

$$\det(1+B) = \exp \operatorname{tr} \log(1+B) = \exp \operatorname{tr}(B - B^2/2 + B^3/3 - \ldots) \tag{114}$$

allows us now to rewrite J_0 as

$$J_0 = \det(e^i{}_\kappa) \exp\left(\frac{i}{\hbar} \mathcal{A}_{J_0}^\varepsilon\right), \tag{115}$$

with the determinant $\det(e_\mu^i) = \sqrt{g(q)}$ evaluated at the postpoint. This equation defines an effective action associated with the Jacobian, for which we obtain the expansion

$$\frac{i}{\hbar} \mathcal{A}_{J_0}^\varepsilon = -e_i{}^\kappa e^i{}_{\kappa,\mu} \Delta q^\mu + \frac{1}{2} \left[e_i{}^\mu e^i_{\mu,\nu\lambda} - e_i{}^\mu e^i{}_{\kappa,\nu} e_j{}^\kappa e^j{}_{\mu,\lambda}\right] \Delta q^\nu \Delta q^\lambda + \ldots. \tag{116}$$

To express this in terms of the affine connection, we use (19) and derive the relations

$$\frac{1}{4}e_{i\nu,\mu}e^i{}_{\kappa,\lambda} = \frac{1}{4}e_i{}^\sigma e^i{}_{\nu,\mu}e_{j\sigma}e^j{}_{\kappa,\lambda} = \frac{1}{4}\Gamma_{\mu\nu}{}^\sigma,\Gamma_{\lambda\kappa\sigma} \tag{117}$$

$$\begin{aligned}\frac{1}{3}e_{i\mu}e^i{}_{\nu,\lambda\kappa} &= \frac{1}{3}g_{\mu\tau}[\partial_\kappa(e_i{}^\tau e^i{}_{\nu,\lambda}) - e^{i\sigma}e^i{}_{\nu,\lambda}e^j{}_\sigma e^{j\tau}{}_{,\kappa}] \\ &= \frac{1}{3}g_{\mu\tau}(\partial_\kappa\Gamma_{\lambda\nu}{}^\tau + \Gamma_{\lambda\nu}{}^\sigma\Gamma_{\kappa\sigma}{}^\tau).\end{aligned} \tag{118}$$

With these, the Jacobian action becomes

$$\frac{i}{\hbar}\mathcal{A}^\varepsilon_{J_0} = -\Gamma_{\mu\nu}{}^\nu\Delta q^\mu + \frac{1}{2}\partial_\mu\Gamma_{\nu\kappa}{}^\kappa\Delta q^\nu\Delta q^\mu + \dots. \tag{119}$$

The same result would, of course, be obtained by writing the Jacobian in accordance with (110) as

$$J_0 = \sqrt{g(q-\Delta q)}, \tag{120}$$

which leads to the alternative formula for the Jacobian action

$$\exp\left(\frac{i}{\hbar}\mathcal{A}^\varepsilon_{J_0}\right) = \frac{\sqrt{g(q-\Delta q)}}{\sqrt{g(q)}}. \tag{121}$$

An expansion in powers of Δq gives

$$\exp\left(\frac{i}{\hbar}\mathcal{A}^\varepsilon_{J_0}\right) = -\frac{1}{\sqrt{g(q)}}\sqrt{g(q)}_{,\mu}\Delta q^\mu + \frac{1}{2\sqrt{g(q)}}\sqrt{g(q)}_{,\mu\nu}\Delta q^\mu\Delta q^\nu + \dots. \tag{122}$$

Using the formula

$$\frac{1}{\sqrt{g}}\partial_\mu\sqrt{g} = \frac{1}{2}g^{\sigma\tau}\partial_\mu g_{\sigma\tau} = \bar{\Gamma}_{\mu\nu}{}^\nu, \tag{123}$$

this becomes

$$\exp\left(\frac{i}{\hbar}\mathcal{A}^\varepsilon_{J_0}\right) = 1 - \bar{\Gamma}_{\mu\nu}{}^\nu\Delta q^\mu + \frac{1}{2}(\partial_\mu\bar{\Gamma}_{\nu\lambda}{}^\lambda + \bar{\Gamma}^\sigma_{\mu\sigma}\bar{\Gamma}_{\nu\lambda}{}^\lambda)\Delta q^\mu\Delta q^\nu + \dots, \tag{124}$$

so that

$$\frac{i}{\hbar}\mathcal{A}^\varepsilon_{J_0} = -\bar{\Gamma}_{\mu\nu}{}^\nu\Delta q^\mu + \frac{1}{2}\partial_\mu\bar{\Gamma}_{\nu\lambda}{}^\lambda\Delta q^\mu\Delta q^\nu + \dots. \tag{125}$$

In a space without torsion where $\bar{\Gamma}^\lambda_{\mu\nu} \equiv \Gamma_{\mu\nu}{}^\lambda$, the Jacobian actions (119) and (125) are trivially equal to each other. But the equality holds also in the presence of torsion. Indeed, when inserting the decomposition (29), $\Gamma_{\mu\nu}{}^\lambda = \bar{\Gamma}_{\mu\nu}{}^\lambda + K_{\mu\nu}{}^\lambda$, into (119), the contortion tensor drops out since it is antisymmetric in the last two indices and these are contracted in both expressions.

In terms of $\mathcal{A}^\varepsilon_{J_{0n}}$, we can rewrite the transformed measure (108) in the more useful form

$$\prod_{n=2}^{N+1}\int d^Dx^i_{n-1} = \prod_{n=2}^{N+1}\left\{\int d^Dq^\mu_{n-1}\det\left[e^i_\mu(q_n)\right]\exp\left(\frac{i}{\hbar}\mathcal{A}^\varepsilon_{J_{0n}}\right)\right\}. \tag{126}$$

In a flat space parametrized in terms of curvilinear coordinates, the right-hand sides of (108) and (126) are related by an ordinary coordinate transformation, and both give the correct measure for a time-sliced path integral. In a general metric-affine space, however, this is no longer true. Since the mapping $dx^i \to dq^\mu$ is nonholonomic, there are in principle infinitely many ways of transforming the path integral measure from Cartesian coordinates to a noneuclidean space. Among these, there exists a preferred mapping which leads to the correct quantum-mechanical amplitude in all known physical systems. It is this mapping which led to the correct solution of the path integral of the hydrogen atom [11].

The clue for finding the correct mapping is offered by an unesthetic feature of Eq. (111): The expansion contains both differentials dq^μ and differences Δq^μ. This is somehow inconsistent. When time-slicing the path integral, the differentials dq^μ in the action are increased to finite differences Δq^μ. Consequently, the differentials in the measure should also become differences. A relation such as (111) containing simultaneously differences and differentials should not occur.

It is easy to achieve this goal by changing the starting point of the nonholonomic mapping and rewriting the initial flat space path integral (92) as

$$(\mathbf{x}t|\mathbf{x}'t') = \frac{1}{\sqrt{2\pi i\varepsilon\hbar/M}^D}\prod_{n=1}^{N}\left[\int_{-\infty}^{\infty} d\Delta x_n\right]\prod_{n=1}^{N+1} K_0^\varepsilon(\Delta\mathbf{x}_n). \tag{127}$$

Since x_n are Cartesian coordinates, the measures of integration in the time-sliced expressions (92) and (127) are certainly identical:

$$\prod_{n=1}^{N} d^D x_n \equiv \prod_{n=2}^{N+1} d^D \Delta x_n. \tag{128}$$

Their images under a nonholonomic mapping, however, are different so that the initial form of the time-sliced path integral is a matter of choice. The initial form (127) has the obvious advantage that the integration variables are precisely the quantities Δx_n^i which occur in the short-time amplitude $K_0^\varepsilon(\Delta x_n)$.

Under a nonholonomic transformation, the right-hand side of Eq. (128) leads to the integral measure in a general metric-affine space

$$\prod_{n=2}^{N+1}\int d^D\Delta x_n \to \prod_{n=2}^{N+1}\left[\int d^D\Delta q_n J_n\right], \tag{129}$$

with the Jacobian following from (99) (omitting n)

$$J = \frac{\partial(\Delta x)}{\partial(\Delta q)} = \det(e^i{}_\kappa)\,\det\Big[\delta_\mu{}^\lambda - \Gamma_{\{\mu\nu\}}{}^\lambda\Delta q^\nu \tag{130}$$
$$+\frac{1}{2}(\partial_\sigma\Gamma_{\mu\nu}{}^\lambda + \Gamma_{\{\mu\nu}{}^\tau\Gamma_{\{\tau|\sigma\}\}}{}^\lambda)\Delta q^\nu\Delta q^\sigma + \ldots\Big].$$

In a space with curvature and torsion, the measure on the right-hand side of (129) replaces the flat-space measure on the right-hand side of (110). The curly double brackets around the indices ν,κ,σ,μ indicate a symmetrization in τ and σ followed by a symmetrization in μ,ν, and σ. With the help of formula (114) we now calculate the Jacobian action

$$\frac{i}{\hbar}\mathcal{A}_J^\varepsilon = -\Gamma_{\{\mu\nu\}}{}^\mu\Delta q^\nu \tag{131}$$
$$+\frac{1}{2}\left[\partial_{\{\mu}\Gamma_{\nu\kappa\}}{}^\kappa + \Gamma_{\{\nu\kappa}{}^\sigma\Gamma_{\{\sigma|\mu\}\}}{}^\kappa - \Gamma_{\{\nu\kappa\}}{}^\sigma\Gamma_{\{\sigma\mu\}}{}^\kappa\right]\Delta q^\nu\Delta q^\mu + \ldots.$$

The curly double brackets around the indices ν, κ, σ, μ indicate a symmetrization in τ and σ followed by a symmetrization in μ, ν, and σ (here the index μ is excluded as indicated by the bar). This expression differs from the earlier Jacobian action (119) by the symmetrization symbols. Dropping them, the two expressions coincide. This is allowed if q^μ are curvilinear coordinates in a flat space. Since then the transformation functions $x^i(q)$ and their first derivatives $\partial_\mu x^i(q)$ are integrable and possess commuting derivatives, the two Jacobian actions (119) and (131) are identical.

There is a further good reason for choosing (128) as a starting point for the nonholonomic transformation of the measure. According to Huygens' principle of wave optics, each point of a wave front is a center of a new spherical wave propagating from that point. Therefore, in a time-sliced path integral, the differences Δx_n^i play a more fundamental role than the coordinates themselves. Intimately related to this is the observation that in the canonical form, a short-time piece of the action reads

$$\int \frac{dp_n}{2\pi\hbar} \exp\left[\frac{i}{\hbar} p_n (x_n - x_{n-1}) - \frac{ip_n^2}{2M\hbar} t\right]. \tag{132}$$

Each momentum is associated with a coordinate difference $\Delta x_n \equiv x_n - x_{n-1}$. Thus, we should expect the spatial integrations conjugate to p_n to run over the coordinate differences $\Delta x_n = x_n - x_{n-1}$ rather than the coordinates x_n themselves, which makes the important difference in the subsequent nonholonomic coordinate transformation.

We are thus led to postulate the following time-sliced path integral in q-space:

$$\langle q|\exp\left[-\frac{i}{\hbar}(t-t')\hat{H}\right]|q'\rangle = \frac{1}{\sqrt{2\pi i\hbar\varepsilon/M}^D} \prod_{n=2}^{N+1}\left[\int d^D\Delta q_n \frac{\sqrt{g(q_n)}}{\sqrt{2\pi i\varepsilon\hbar/M}^D}\right] \times \exp\left[\frac{i}{\hbar}\sum_{n=1}^{N+1}(\mathcal{A}^\varepsilon + \mathcal{A}_J^\varepsilon)\right], \tag{133}$$

where the integrals over Δq_n may be performed successively from $n = N$ down to $n = 1$.

Let us emphasize that this expression has not been *derived* from the flat space path integral. It is the result of a specific new *quantum equivalence principle* which rules how a flat space path integral behaves under nonholonomic coordinate transformations.

It is useful to reexpress our result in a different form which clarifies best the relation with the naively expected measure of path integration (110), the product of integrals

$$\prod_{n=1}^{N} \int d^D x_n = \prod_{n=1}^{N}\left[\int d^D q_n \sqrt{g(q_n)}\right]. \tag{134}$$

The measure in (133) can be expressed in terms of (134) as

$$\prod_{n=2}^{N+1}\left[\int d^D\Delta q_n \sqrt{g(q_n)}\right] = \prod_{n=1}^{N}\left[\int d^D q_n \sqrt{g(q_n)} e^{-i\mathcal{A}_{J_0}^\varepsilon/\hbar}\right].$$

The corresponding expression for the entire time-sliced path integral (133) in the metric-affine space reads

$$\langle q|\exp\left[-\frac{i}{\hbar}(t-t')\hat{H}\right]|q'\rangle = \frac{1}{\sqrt{2\pi i\hbar\varepsilon/M}^D} \prod_{n=1}^{N}\left[\int d^D q_n \frac{\sqrt{g(q_n)}}{\sqrt{2\pi i\hbar\varepsilon/M}^D}\right] \times \exp\left[\frac{i}{\hbar}\sum_{n=1}^{N+1}(\mathcal{A}^\varepsilon + \Delta\mathcal{A}_J^\varepsilon)\right], \tag{135}$$

where $\Delta\mathcal{A}_J^\varepsilon$ is the difference between the correct and the wrong Jacobian actions in Eqs. (119) and (131):

$$\Delta\mathcal{A}_J^\varepsilon \equiv \mathcal{A}_J^\varepsilon - \mathcal{A}_{J_0}^\varepsilon. \tag{136}$$

In the absence of torsion where $\Gamma_{\{\mu\nu\}}{}^\lambda = \bar{\Gamma}_{\mu\nu}{}^\lambda$, this simplifies to

$$\frac{i}{\hbar}\Delta\mathcal{A}_J^\varepsilon = \frac{1}{6}\bar{R}_{\mu\nu}\Delta q^\mu \Delta q^\nu, \tag{137}$$

where $\bar{R}_{\mu\nu}$ is the Ricci tensor associated with the Riemann curvature tensor, i.e., the contraction (40) of the Riemann curvature tensor associated with the Christoffel symbol $\bar{\Gamma}_{\mu\nu}{}^\lambda$.

Being quadratic in Δq, the effect of the additional action can easily be evaluated perturbatively using the methods explained in Chapter 8 of the textbook [12], according to which $\Delta q^\mu \Delta q^\nu$ may be replaced by its lowest order expectation

$$\langle \Delta q^\mu \Delta q^\nu \rangle_0 = i\varepsilon\hbar g^{\mu\nu}(q)/M.$$

Then $\Delta\mathcal{A}_J^\varepsilon$ yields the additional effective potential

$$V_{\text{eff}} = -\frac{\hbar^2}{6M}\bar{R}, \tag{138}$$

where $\bar{R}$ is the Riemann curvature scalar.* By including this potential in the action, the path integral in a curved space can be written down in the naive form (134) as follows:

$$\begin{aligned}\langle q|\exp\left[-\frac{i}{\hbar}(t-t')\hat{H}\right]|q'\rangle = \frac{1}{\sqrt{2\pi i\hbar\varepsilon/M}^D}\prod_{n=1}^{N}\left[\int d^D q_n \frac{\sqrt{g(q_n)}}{\sqrt{2\pi i\varepsilon\hbar/M}^D}\right] \\ \times\exp\left[\frac{i}{\hbar}\sum_{n=1}^{N+1}(\mathcal{A}^\varepsilon + \varepsilon V_{\text{eff}})\right].\end{aligned} \tag{139}$$

The integrals over q_n are conveniently performed successively downwards over $\Delta q_{n+1} = q_{n+1} - q_n$ at fixed q_{n+1}. The weights $\sqrt{g(q_n)} = \sqrt{g(q_{n+1} - \Delta q_{n+1})}$ require a postpoint expansion leading to the naive Jacobian J_0 of (112) and the Jacobian action $\mathcal{A}_{J_0}^\varepsilon$ of Eq. (119).

It goes without saying that the path integral (139) also has a phase space version. It is obtained by omitting all $(M/2\varepsilon)(\Delta q_n)^2$ terms in the short-time actions $\mathcal{A}^\varepsilon$ and extending the multiple integral by the product of momentum integrals

$$\prod_{n=1}^{N+1}\left[\frac{dp_n}{2\pi\hbar\sqrt{g(q_n)}}\right] e^{(i/\hbar)\sum_{n=1}^{N+1}\left[p_{n\mu}\Delta q^\mu - \varepsilon\frac{1}{2M}g^{\mu\nu}(q_n)p_{n\mu}p_{n\nu}\right]}. \tag{140}$$

When using this expression, all problems which were encountered in the literature with canonical transformations of path integrals disappear.

*This is one of the $\bar{R}$-terms of DeWitt. Another term with opposite sign and a factor -1/2 was found by him from the prefactor in the DeWitt-Morette semiclassical amplitude which he employed for the short-time propagator. See the discussion in Appendix 11B of [12].

4. CONCLUSION

It appears as though the new variational and quantum equivalence principles constitute the proper basis for a correct extension of our physical laws into geometries with torsion. In both principles, nonholonomic mappings play a fundamental role. When applied to classical paths, these mappings lead directly to the new variational principle and thus to the correct equations of motion. Their correctness is a consequence of the fact that classical equations of motion remain valid under nonholonomic coordinate transformations. An important application not discussed here is the derivation of the Euler–Lagrange equations of a spinning top within the rotating body-fixed frame of references from an extremum of the kinetic action [18].

The quantum equivalence principle adds to the nonholonomic mapping procedure the postulate that the measure of path integration which is to be mapped into a space with curvature and torsion contains the same time-sliced intervals Δx^i which appear in the short-time action [see Eq. (127)]. The most important theoretical evidence for the correctness of this principle comes from the solution of the path integral of the Coulomb problem. This was presented in the Lectures, but will not be repeated here, referring the reader to the textbook [12]. Only with this measure has it been possible to find the solution without undesirable time-slicing corrections.

Another theoretical evidence which was mentioned only briefly in the lectures comes from the bosonization of Fermi theories [19–28]. Only with the new measure is this bosonization possible [29] without errors in the energy spectrum.

REFERENCES

[1] B. S. DeWitt, Rev. Mod. Phys. **29**, 377 (1957).
[2] B. Podolsky, Phys. Rev. **32**, 812 (1928).
[3] K. S. Cheng, J. Math. Phys. **13**, 1723 (1972);
H. Kamo and T. Kawai, Prog. Theor. Phys. **50**, 680 (1973);
T. Kawai, Found. Phys. **5**, 143 (1975);
H. Dekker, Physica A **103**, 586 (1980);
G. M. Gavazzi, Nuovo Cimento A **101**, 241 (1981).
A good survey over similar attempts is given by
M. S. Marinov, Phys. Rep. **60**, 1 (1980).
[4] Among the most widely discussed procedures was a postpoint discretization due to Ito and a midpoint discretization due to Stratonovich, with different mathematical advantages. For a detailed discussion see the textbooks
H. Risken, *The Fokker–Planck Equation*, second edition, Springer, 1983, Vol. 18;
R. Kubo, M. Toda, and N. Hashitsume, *Statistical Physics II*, Springer, Berlin 1985.
A recent description of the relation between time slicing and Ito versus Stratonovich calculus can be found in
H. Nakazato, K. Okano, L. Schülke, and Y. Yamanaka, Nucl. Phys. B **346**, 611 (1990).
Stochastic differential equations in curved spaces are developed in
K. D. Elworthy, *Stochastic differential equations on manifolds*, Cambridge Univ. Press, 1982;
M. Emery, *Stochastic calculus in manifolds*, Springer, Berlin 1989.
[5] R. Graham, Z. Phys. B **26**, 397 (1977).
[6] K. D. Elworthy, *Path Integration on Manifolds*, in *Mathematical Aspects of Superspace*, eds. H.-J. Seifert, C. Clarke, and A. Rosenblum, Reidel, 1984.
[7] L. D. Landau and E. M. Lifshitz, *Quantum Mechanics*, Pergamon, New York, 1965.
[8] D. J. Simms and N. M. J. Woodhouse, *Lectures on geometric quantization*, Springer, Berlin 1976;
J. Sniatycki, *Geometric quantization and quantum mechanics*, Springer, Berlin 1980;
P. L. Robinson and J. H. Rawnsley, *The metaplectic representation, Mpc structures, and geometric quantization*, publ. by the American Mathematical Society in the series *Memoirs of the American Mathematical Society* no. 410 0065-9266, Providence, R. I., 1989.
[9] For details and more references see H. Kleinert, *Gauge Fields in Condensed Matter*, Vol. I *Superflow and Vortex Lines*, pp. 1–744, and Vol. II *Stresses and Defects*, World Scientific, Singapore 1989, pp. 744–1443.

[10] K. Kondo, in: Proc. 2nd Japan Nat. Congr. Applied Mechanics, Tokio, 1952
B. A. Bilby, R. Bullough and E. Smith, Proc. R. Soc. London A **231**, 263 (1955);
E. Kröner, in: *Physics of defects*, Les Houches summer school XXXV, North-Holland, Amsterdam 1981.

[11] H. Duru and H. Kleinert, Phys. Lett. B **84**, 185 (1979); Fortschr. d. Phys. **30**, 401 (1982).

[12] H. Kleinert, *Path Integrals in Quantum Mechanics, Statistics and Polymer Physics,*, second edition, World Scientific, Singapore 1995.

[13] H. Kleinert, Mod. Phys. Lett. A **4**, 2329 (1989).

[14] H. Kleinert, Phys. Lett. B **236**, 315 (1990).

[15] P. Fiziev and H. Kleinert, *New Action Principle for Classical Particle Trajectories In Spaces with Torsion*, Europh. Lett. **35**, 241 1996 (hep-th/9503074 and `http://www.physik.fu-berlin.de/~kleinert/kleiner_re219/newvar.html`).

[16] A. Pelster and H. Kleinert, FU-Berlin preprint, May 1996 (gr-qc/9605028 and `http://www.physik.fu-berlin.de/~kleinert/kleiner_re243/preprint.html`).

[17] Our notation for the geometric quantities in spaces with curvature and torsion is the same as in
J. A. Schouten, *Ricci Calculus*, Springer, Berlin 1954.

[18] P. Fiziev and H. Kleinert, *Euler Equations for Rigid-Body — A Case for Autoparallel Trajectories in Spaces with Torsion*, Berlin preprint 1995 (hep-th/9503075 and `http://www.physik.fu-berlin.de/~kleinert/kleiner_re224/euler.html`).

[19] H. Kleinert, *Collective Quantum Fields*,
Lectures presented at the First Erice Summer School on Low-Temperature Physics, 1977, Fortschr. Physik **26**, 565–671 (1978).
See also the predecessors:
H. Kleinert, *Field Theory of Collective Excitations — A Soluble Model*,
Phys. Lett. B **69**, 9 (1977),
as well as the derivation of an SU(3)×SU(3) chirally invariant field theory of mesons from a quark theory in
H. Kleinert, *Hadronization of Quark Theories and a Bilocal form of QED*, Phys. Lett. B **62**, 429 (1976);
H. Kleinert, *On the Hadronization of Quark Theories*, Lectures presented at the Erice Summer Institute 1976, in *Understanding the Fundamental Constituents of Matter*,
Plenum Press, New York, 1978, A. Zichichi ed., pp. 289–390.

[20] L. P. Gorkov, Sov. Phys. JETP **9**, 1364 (1959).

[21] V. L. Ginzburg and L. D. Landau, Eksp. Teor. Fiz. **20**, 1064 (1950).

[22] A. L. Leggett, Rev. Mod. Phys. **47**, 331 (1975).

[23] K. D. Schotte and U. Schotte, Phys. Rev. **182**, 479 (1969);
see also:
S. Tomonaga, Progr. Theor. Phys. **5**, 63 (1950).

[24] For a review see:
D. R. Bes, R. A. Broglia, Lectures delivered at "E. Fermi" Varenna Summer School, Varenna, Como Italy, 1976.
For recent studies: D. R. Bes, R. A. Broglia, R. Liotta, B. R. Mottelson, Phys. Letters B **52**, 253 (1974); B **56**, 109 (1975), Nuclear Phys. A **260**, 127 (1976).
See also:
R. W. Richardson, J. Math. Phys. **9**, 1329 (1968), Ann. Phys. (N. Y.) **65**, 249 (1971) and N. Y. U. Preprint 1977, as well as references therein.

[25] J. Hubbard, Phys. Rev. Letters **3**, 77 (1959); B. Mühlschlegel, J. Math. Phys. , **3**, 522 (1962); J. Langer, Phys. Rev. A **134**, 553 (1964); T. M. Rice, Phys. Rev. A **140** 1889 (1965); J. Math. Phys. **8**, 1581 (1967); A. V. Svidzinskij, Teor. Mat. Fiz. **9**, 273 (1971); D. Sherrington, J. Phys. C **4**, 401 (1971).

[26] E. Witten, Commun. Math. Phys. **92**, 455 (1984);
P. DiVecchia and P. Rossi, Phys. Lett. B **140**, 344 (1984);
P. DiVecchia, B. Durhuus and J. L. Petersen, Phys. Lett. B **144**, 245 (1984);
Y. Frishman, Phys. Lett. B **146**, 204 (1984);
E. Abdalla and M. C. B. Abdalla, Nucl. Phys. B **225**, 392 (1985);
D. Gonzales and A. N. Redlich, Phys. Lett. B **147**, 150 (1984);
C. M. Naón, Phys. Rev. D **31**, 2035 (1985);
See also the recent development by
P. H. Damgaard, H. B. Nielsen, and R. Sollacher, Nuclear Phys. B **385**, 227 (1992) (hep-th/9407022);
P. H. Damgaard and R. Sollacher, Cern preprint (hep-th/9407022);
A. N. Theron; F. A. Schaposnik, F. G. Scholtz and H. B. Geyer, Nucl. Phys. B **437**, 187 (1995) (hep-th/9410035);
C. P. Burgess and F. Quevedo, Phys. Lett. B **329** (1994) 457; Nucl. Phys. B **421**, 373 (1994);
C. P. Burgess, A. Lutkin, and F. Quevedo, Phys. Lett. B **336**, 18 (1994);
J. Fröhlich, R. Götschmann and P. A. Marchetti, preprint (hep-th/9406154).

[27] S. Coleman, Phys. Rev. D **11**, 2088 (1975);
S. Mandelstam, Phys. Rev. D **11**, 3026 (1975);
B. Schroer and T. T. Truong, Phys. Rev. D **15**, 1684 (1977).
[28] For a semiclassical study of the model at finite times see
H. Kleinert and H. Reinhardt, Nucl. Phys. A **332**, 33 (1979).
[29] H. Kleinert, *Nonabelian Bosonization as a Nonholonomic Transformations from Flat to Curved Field Space.* FU-Berlin preprint 1996
(`http://www.physik.fuberlin.de/~kleinert/kleiner_re239/preprint.html`).

4

VARIATIONAL PERTURBATION THEORY FOR PATH INTEGRALS

H. Kleinert*

Institut für Theoretische Physik
Freie Universität Berlin
Arnimallee 14
D-14195 Berlin, Germany

ABSTRACT

Variational Perturbation Theory (VPT) of path integrals arose from a 1986 collaboration with Feynman, combining ordinary divergent perturbation series with a variational procedure. It produces exponentially-fast convergent sequences of approximations, uniformly up to infinite coupling strengths. In this way, divergent *weak-coupling expansions* can be converted into convergent *strong-coupling expansions*, whose convergence radius can be deduced from details how the variational procedure converges. A method of interpolating between weak- and strong-coupling expansions follows naturally, with interesting consequences for polarons.

VPT can be continued to negative coupling constants to yield imaginary parts of amplitudes in the *sliding regime*. The precise imaginary parts obtained in this way can be inserted into a dispersion relation to find the large-order behavior of perturbation coefficients which turns out to be accurate down to the lowest order. A method of interpolating between weak-coupling and large-order information arises from this.

A combination of the variational approach with instanton calculus leads to a *Variational Tunneling Theory*. This can be applied to non-Borel-summable expansions making these convergent as well.

SUMMARY OF LECTURES

The content of these lectures is contained in the 1995 edition of my book on Path Integrals (PI) [1]. Chapter 5 of that book gives an introduction to the Feynman–Kleinert variational

*E-mail: kleinert@physik.fu-berlin.de;
URL: http://www.physik.fu-berlin.de/~kleinert;
Phone/Fax: 0049/30/8383034

Functional Integration: Basics and Applications
Edited by Cécile DeWitt-Morette, Plenum Press, New York, 1997

procedure for path integrals [2], from which the present approach arose by a systematic extension to arbitrary orders (Section 5.13 of PI), as proposed in [3] and continued in Ref. [4].

To lowest order, the method gives the same results as a Hartree–Fock–Bogoljubov approximation of many-body systems which is known to be applicable in the strong-coupling limit. This property survives to all orders with increasing accuracy, in contrast to previous variational approaches via Legendre transforms of generating functionals, where higher orders possess no strong-coupling limit [3].

The convergence behavior of the method is studied in Section 5.15 of PI for the energy eigenvalues of the quantum mechanical anharmonic oscillator, using the expansion coefficients obtained with the help of recursion relations of Bender and Wu [5]. It is found to be exponentially fast with superimposed oscillations. These can be related to the radius of convergence of the strong-coupling expansion [6].

By discovering a simple scaling law of the variational expression [7], the order was driven to 250, and a very accurate strong-coupling expansion was derived from the weak-coupling expansion [8].

The results can be continued analytically to give the correct imaginary parts of amplitudes in the sliding regime (see [9, 10] and Section 17.10.4 of PI). By a combination of the variational approach with instanton methods, also the tunneling regime can be reached with increasing accuracy. The imaginary parts of amplitudes obtained in this way can be inserted into a dispersion relation to find the large-order behavior of perturbation coefficients. Due to the information in the sliding regime the results are good even to lowest order. In the anharmonic oscillator, the first coefficient $3/4$ is reproduced to 0.1% (see Table 17.1 in PI).

An extension of this method leads to a resummation of non-Borel resummable series, such as those for the low-lying energy levels of a double-well potential (see [11] and Section 5.17 of PI).

The method can be applied to correlation functions and particle densities (see [12, 13] and Section 5.9 of PI).

Attempts to prove the convergence of this method were undertaken some time ago [14], and have recently been successful [6, 15] for the anharmonic oscillator.

The method can be improved further by introducing independent knowledge of strong-coupling [16] or large-order behavior [17]. It may also be used to calculate ordinary effective potentials, which always emerge as convex functions [18], in contrast to other methods.

REFERENCES

[1] H. Kleinert, *Path Integrals in Quantum Mechanics, Statistics and Polymer Physics*, 2nd edition (World Scientific, Singapore, 1995).

[2] R. P. Feynman and H. Kleinert, Phys. Rev. A **34**, 5080 (1986);
R. Giachetti and V. Tognetti, Phys. Rev. Lett. **55**, 912 (1985);
Phys. Rev. B **33**, 7647 (1986).

[3] H. Kleinert, Phys. Lett. A **174**, 332 (1993); A variational method based on higher Legendre transforms is investigated in
J. Jaenicke and H. Kleinert, Phys. Lett. A **176**, 409 (1993);

[4] H. Kleinert and H. Meyer, Phys. Lett. A **184**, 319 (1994).

[5] C. M. Bender and T. T. Wu, Phys. Rev. **184**, 1231 (1969);
Phys. Rev. D **7**, 1620 (1973).

[6] H. Kleinert and W. Janke, Phys. Lett. A **206**, 283 (1995).

[7] W. Janke and H. Kleinert, Phys. Lett. A **199**, 287 (1995).

[8] W. Janke and H. Kleinert, Phys. Rev. Lett. **75**, 2787 (1995).

[9] H. Kleinert, Phys. Lett. B **300**, 261 (1993).

[10] R. Karrlein and H. Kleinert, Phys. Lett. A **187**, 133 (1994).
[11] H. Kleinert, Phys. Lett. A **190**, 131 (1994).
[12] H. Kleinert, Phys. Lett. A **118**, 267 (1986).
[13] W. Janke and H. Kleinert, Phys. Lett. A **118**, 371 (1986).
[14] I. R. C. Buckley, A. Duncan, and H. F. Jones, Phys. Rev. D **47**, 2554 (1993); A. Duncan and H. F. Jones, Phys. Rev. D **47**, 2560 (1993); C. M. Bender, A. Duncan, and H. F. Jones, Phys. Rev. D **49**, 4219 (1994); C. Arvanitis, H. F. Jones, and C. S. Parker, Phys. Rev. D **52**, 3704 (1995); R. Guida, K. Konishi, and H. Suzuki, Ann. Phys. **241**, 152 (1995).
[15] R. Guida, K. Konishi, and H. Suzuki, Ann. Phys. **249**, 109 (1996).
[16] H. Kleinert, Phys. Lett. A **207**, 133 (1995).
[17] H. Kleinert, Phys. Lett. B **360**, 65 (1995).
[18] H. Kleinert, Phys. Lett. B **181**, 324 (1986); A **118**, 195 (1986).

5

FUNCTIONAL INTEGRATION AND WAVE PROPAGATION

S. K. Foong

Department of Natural Sciences
College of Education
University of the Ryukyus
Okinawa 903-01, Japan
e-mail address: `foong@edu.u-ryukyu.ac.jp`

ABSTRACT

We present a pedagogical introduction, including several new results, to Kac's path integral solution for the telegrapher equation, with emphases on: 1) wave propagation especially waveform (signal) restoration (reconstruction), designing and prediction; and 2) the underlying Poissonian random walk (essentially an asynchronous telegraph signal), especially the measure, and the connection to Brownian motion.

1. INTRODUCTION

The objects of study in these lectures are centered around wave propagations as described by the telegrapher equation, the simplest wave equation with dissipation. In particular, we are interested in waveform restoration, designing, and prediction. The treatment of the subject may be unconventional, and it is inspired by the path integral approach to the subject started by M. Kac [1], who was in turn inspired by S. Goldstein's [2] realization that a certain random walk, the Poissonian motion, is related to the telegrapher equation, just as the Brownian motion is related to the diffusion equation. We will also emphasize these random motions and their relation.

In 1956, M. Kac wrote down a path integral solution to the telegrapher equation in arbitrary space dimensions. After Kac's work, there have been quite a number of generalizations and applications (see for instance [3]–[28]). However, we will restrict our attention to wave propagation.

If one examines Kac's novel path integral solution (with subsequent refinements by others) of such wave equations, one is naturally led to ask the following fundamental questions in wave propagation: What would have been the waveform received if it had not suffered any

Functional Integration: Basics and Applications
Edited by Cécile DeWitt-Morette, Plenum Press, New York, 1997

distortion? And the closely related question: What should the to-be-transmitted waveform be such that the *distorted* distorted-waveform to be received becomes the intended one? In the third lecture, we will explain a solution to each of the problems inspired by Kac's approach, namely a waveform restorer and designer respectively for the case of distortions caused by the dissipation term in the telegrapher equation.

The organization of these lecture notes is as follows: In section 2, the telegrapher equation and several physical equations which have the same form are given. In section 3, the telegrapher equation is derived from the Poissonian motion, after reviewing how the diffusion equation is derived from the Brownian motion. In section 4, the Poissonian motion is examined further. In section 5, several examples are given for the application of the measure of the Poissonian motion. In section 6, the connection between the Poissonian motion and the Brownian motion is clarified. In section 7, Kac's solution is reviewed and proved for a special case. Section 8 is a brief experimental consideration for propagation in a co-axial cable. In section 9, a review and an extension of Kaplan's generalization of the Kac's solution are given. In section 10, an analytical scheme for calculating the density distribution of the displacement for the random walk is derived from a result of Kaplan. In section 11, two methods, Restorer No. 1 and No. 2 for restoration of waveform in arbitrary dimension are considered. In section 12, we show how the Restorer No. 1 can be used as a waveform designer in one dimension. Section 13 is on waveform prediction.

These written notes do not cover everything said in the lectures, especially on numerical simulation of the random walk, but references on published material are given. Several new results not covered in the lectures are included in this written version.

2. TELEGRAPHER EQUATION

The telegrapher equation we shall be concerned with is the following wave equation with dissipation with general initial conditions:

$$\frac{1}{v^2}\frac{\partial^2 F}{\partial t^2}+\frac{2a(t)}{v^2}\frac{\partial F}{\partial t}-O_{\mathbf{x}}F=0, \tag{1}$$

$$F(\mathbf{x},0)=\kappa(\mathbf{x}), \qquad \left.\frac{\partial F}{\partial t}\right|_{t=0}=\xi(\mathbf{x}),$$

where $a(t)$ is a non-negative and continuous function defined on $[0,\infty)$. $O_{\mathbf{x}}$ is any linear and spatial operator such as the Laplacian ∇^2. A somewhat generalized functional form of $a(t)$ will be stated in section 7.

Physical equations which have the form of the telegrapher equation are for example:

1. Telegrapher equation for a transmission line: The current and voltage both satisfy the telegrapher equation with $v=1/\sqrt{LC}$, $a=R/(2L)$, and $O_{\mathbf{x}}=\partial^2/\partial x^2$), where L is the inductance, R the resistance, and C the capacitance, all per unit length.

2. The Maxwell equations for the interior of a linear medium characterized by ε (permittivity), μ (permeability), σ (conductivity) reduce to

$$\nabla^2\mathbf{E}-\mu\varepsilon\frac{\partial^2\mathbf{E}}{\partial t^2}-\mu\sigma\frac{\partial\mathbf{E}}{\partial t}=0, \tag{2}$$

   and an identical equation for $\mathbf{B}$ [29]. Here, $v=1/\sqrt{\mu\varepsilon}$, and $a=\sigma/(2\varepsilon)$.

3. A flexible string under tension T vibrating in a dissipative medium such as water. Its amplitude y satisfies
$$\frac{\rho}{T}\frac{\partial^2 y}{\partial t^2}+\frac{R}{T}\frac{\partial y}{\partial t}-\frac{\partial^2 y}{\partial x^2}=0 \tag{3}$$
where ρ is the mass per unit length and R is the frictional force exerted by the medium per unit length per unit transverse velocity of the string.

4. Heat conduction equation according to Cattaneo's law [30] reads:
$$\frac{1}{v^2}\frac{\partial^2 T}{\partial t^2}+\frac{1}{v^2\tau}\frac{\partial T}{\partial t}-\nabla^2 T=0 \tag{4}$$
where T is the temperature, τ is a relaxation time associated with an average collision time between quanta (free electrons and phonons) responsible for heat conduction, and $v=(k/(\gamma\tau))^{1/2}$ where k is thermal conductivity and γ is heat capacity. This equation, in contrast to the usual heat diffusion equation, gives a *finite* propagation velocity for the temperature.

5. A more accurate description of the diffusion of chemical mixtures [31] requires
$$\nabla^2\rho=\frac{1}{v^2}\frac{\partial^2\rho}{\partial t^2}+\frac{1}{b^2}\frac{\partial\rho}{\partial t}, \tag{5}$$
where ρ is the concentration of the particular constituent, b^2 is its diffusion constant, and v^2 is related to the mean square speed of the atoms.

6. Let $F(\mathbf{x},t)=e^{-at}\psi(\mathbf{x},t), a=im$, then $\psi(\mathbf{x},t)$ satisfies Klein–Gordon equation
$$(\partial_t^2-\nabla^2+m^2)\psi(\mathbf{x},t)=0. \tag{6}$$

7. The first order version of telegrapher equation discussed in section 3 is essentially Dirac equation in two space–time dimensions.

Kac's method offers a new way of looking at these problems, and consequently a new method of solving them. In the next section, we develop a random walk picture of the telegrapher equation.

3. FROM RANDOM MOTION TO TELEGRAPHER EQUATION

It is well-known that the probability density of a Brownian particle (Wiener process) is described by a diffusion equation. Less well known is the fact that the probability density of a "Poissonian particle" is described by a telegrapher equation. We shall presently review the case of the diffusion equation [32], and then discuss the case of the telegrapher equation.

Consider a particle undergoing a random walk (random motion) on a one space and one time lattice, starting at the origin and taking a sequence of steps of equal length Δx. Each step is either forward or backward, with probability $1/2$. What is the probability, $P(N,m)$, that the particle arrives at the site m after N steps? Suppose it takes r steps to the right and s steps to the left, to arrive at m means $r-s=m$, and as the total number of steps is N, we have $r+s=N$.

Hence, $r=(N+m)/2$, and $s=(N-m)/2$. Since the order of the steps does not matter, the number of paths leading to the point m is given by

$$A=\frac{N!}{\frac{(N+m)}{2}!\frac{(N-m)}{2}!}. \tag{7}$$

The probability of each such path is given by $(1/2)^N$. Therefore

$$P(N,m)=A\left(\frac{1}{2}\right)^N=\frac{N!}{\frac{(N+m)}{2}!\frac{(N-m)}{2}!}\left(\frac{1}{2}\right)^N\approx\sqrt{\frac{2}{\pi N}}\exp\left(-\frac{m^2}{2N}\right), \tag{8}$$

where the last expression follows from applying the Stirling's approximation formula $\ln N! = N\ln N$. The probability density of the particle $\rho_B(t,x)$ may be approximated by

$$\rho_B(t,x)\approx\frac{P(N,m)}{2\Delta x}\approx\frac{1}{2\Delta x}\sqrt{\frac{2}{\pi N}}\exp\left(-\frac{m^2}{2N}\right). \tag{9}$$

In the continuum limit, Δx and Δt both approach 0, while m and N approach ∞ such that $x=m\Delta x$ and $t=N\Delta t$ are both held fixed. The only way this can be achieved while Eq. (9) yields a non-trivial distribution is to require that they approach 0 in such a way that $D=(\Delta x)^2/2\Delta t$ remains constant. Consequently, Eq. (9) becomes

$$\rho_B(t,x)=\frac{1}{\sqrt{4\pi Dt}}\exp\left(-\frac{x^2}{4Dt}\right), \tag{10}$$

which is the solution of the diffusion equation

$$\partial_t\rho_B=D\partial_x^2\rho_B, \tag{11}$$

with the initial condition

$$\rho_B(0,x)=\delta(x). \tag{12}$$

(Remark: If the forward probability was taken to be p and the backward probability $1-p$, a non-trivial $\rho_B(t,x)$ would require letting $p\to 1/2$ in the continuum limit, yielding a diffusion equation with a drift term [33].)

The diffusion equation (11) can be obtained if we start from the difference equation satisfied by $P(N,m)$:

$$P(N,m)=P(N-1,m-1)\,p_{m-1,m}+P(N-1,m+1)\,p_{m+1,m} \tag{13}$$

where p_{ij} denotes the probability to jump from site i to site j in one time step, and here j is either $i+1$ or $i-1$, and $p_{m-1,m}=p_{m+1,m}=1/2$. Dividing Eq. (13) by $2\Delta x$, we have

$$2\rho_B(t,x)=\rho_B(t-\Delta t,x-\Delta x)+\rho_B(t-\Delta t,x+\Delta x). \tag{14}$$

Expanding the RHS up to first order in Δt but second order in Δx, and then taking the continuum limit while holding $D=(\Delta x)^2/(2\Delta t)$ yields immediately Eq. (11).

That was for the diffusion equation. By changing the rule for the random walk and taking a different continuum limit, we may obtain a different differential equation. To obtain the telegrapher equation, the rule is that in deciding the next step, the direction of the present motion is taken into account such that in a short time interval Δt the probability of reversing direction is proportional to the time interval, namely,

$$P(\text{reversing}) = a(t)\Delta t, \tag{15}$$

and the probability of continuing in the same direction is given by

$$P(\text{not reversing}) = 1 - a(t)\Delta t. \tag{16}$$

The parameter $a(t)$ is appropriately called the intensity and has the dimension of the inverse of time. As for the continuum limit, the magnitude of the velocity, i.e. $\Delta x/\Delta t$ is taken to be constant as Δx and Δt both approach zero. This has the important consequence that Δt is of order Δx, and not of order $(\Delta x)^2$ as in Brownian motion.

Let $\rho_\pm(a,t,x)$ be the probability density of the particle being at $(i\Delta x, j\Delta t)$ and moving in the positive $(+)$ or negative $(-)$ direction. Then, corresponding to Eq. (14), the densities satisfy the following relation:

$$\rho_\pm(a,t,x) = (1 - a(t)\Delta t)\rho_\pm(a,t-\Delta t, x \mp \Delta x) + a(t)\Delta t\, \rho_\mp(a,t-\Delta t, x \pm \Delta x), \tag{17}$$

which upon expanding to first order in Δt and Δx, and dividing by Δt, and then going to the continuum limit with $\Delta x/\Delta t = v = 1$ gives the coupled equations, which we will call the first order telegrapher equations,

$$\partial_t \rho_\pm = -(a(t) \pm \partial_x)\rho_\pm + a(t)\rho_\mp. \tag{18}$$

They can be rewritten in a matrix form

$$\frac{\partial}{\partial t}\begin{pmatrix} \rho_+ \\ \rho_- \end{pmatrix} = M \begin{pmatrix} \rho_+ \\ \rho_- \end{pmatrix}, \tag{19}$$

where

$$M \equiv \begin{pmatrix} -(a(t)+\partial_x) & a(t) \\ a(t) & -(a(t)-\partial_x) \end{pmatrix}, \tag{20}$$

which satisfies the equation $M^2 + 2a(t)M = \partial_x^2$. For simplicity, we assume a =constant, then by differentiating Eq. (19) again with respect to t, we see that in this case $\rho_\pm$ satisfy the (second order) telegrapher equation

$$(\partial_t^2 + 2a\partial_t - \partial_x^2)\begin{pmatrix} \rho_+ \\ \rho_- \end{pmatrix} = 0, \tag{21}$$

or restoring v into the equation yields

$$\left(\frac{1}{v^2}\partial_t^2 + \frac{2a}{v^2}\partial_t - \partial_x^2\right)\begin{pmatrix} \rho_+ \\ \rho_- \end{pmatrix} = 0. \tag{22}$$

4. THE POISSONIAN MOTION

In this section, we shall examine the Poissonian motion of the previous section in detail. We first review the standard limiting procedure in going from the binomial distribution to the Poisson distribution for the case a =constant, known as the Poisson's limit law [34]. In the binomial distribution, the probability for k successes out of n trials is given by, if the probability of success is p,

$$B_k(n;p) = \binom{n}{k} p^k (1-p)^{n-k}, \qquad 0 \le k \le n. \tag{23}$$

If we let p to vary inversely with n such that

$$p = \alpha/n, \tag{24}$$

where α is an arbitrary dimensionless parameter, then it follows from Eqs.(23) and (24) that

$$\frac{B_{k+1}(n;\alpha/n)}{B_k(n;\alpha/n)} = \frac{\alpha}{k+1}\left(1-\frac{k}{n}\right)\left(1-\frac{\alpha}{n}\right)^{-1}, \qquad 0 \le k \le n. \tag{25}$$

Now holding k fixed and letting $n \to \infty$ yields

$$\lim_{n\to\infty} B_{k+1}(n;\alpha/n) = \frac{\alpha}{k+1} \lim_{n\to\infty} B_k(n;\alpha/n). \tag{26}$$

With $B_0 = e^{-\alpha}$, we obtain the Poisson distribution with parameter α

$$\lim_{n\to\infty} B_k(n;\alpha/n) = e^{-\alpha}\alpha^k/k! = P(k \text{ success}). \tag{27}$$

To make connection with our random walk on the lattice, a trial is taken as a jump and a success is taken as a reversal. Then, $n = t/\Delta t$. In the random walk, $p = a\Delta t = at/n$ (see Eq. (15)), then by Eq. (24) we may substitute α by at in Eq. (27), and obtain, with $N(t)$ denoting the number of reversals made by the particle up to time t, the probability for k reversals,

$$P(N(t) = k) = e^{-at}(at)^k/k!. \tag{28}$$

The parameter a is known as the intensity. This completes the review for a =constant.

Now, let's try to do the same for $a = a(t)$. Let $t = n\Delta t$, $t_i = i\Delta t$, $1 \le i \le n$ and $p_i = a(t_i)\Delta t$. Consider the path of a particle with k reversals which occur at times $t_{i_1}, t_{i_2}, \cdots, t_{i_k}, i_1 < i_2 < \cdots < i_k$, and its complement, the "failures," be at times $t_{j_1}, t_{j_2}, \cdots, t_{j_{n-k}}, j_1 < j_2 < \cdots < j_{n-k}$. Since a path is characterized by the reversal times, we may label it by $\omega_k(t_{i_1}, \cdots, t_{i_k})$, and the probability for *such* a path is given by

$$\begin{aligned} P\Big(\omega_k(t_{i_1},\cdots,t_{i_k})\Big) &= p_{i_1}p_{i_2}\cdots p_{i_k}(1-p_{j_1})(1-p_{j_2})\cdots(1-p_{j_{n-k}}), \qquad 0 \le k \le n \\ &= p_{i_1}p_{i_2}\cdots p_{i_k}\frac{(1-p_1)(1-p_2)\cdots(1-p_n)}{(1-p_{i_1})(1-p_{i_2})\cdots(1-p_{i_k})} \\ &= p_{i_1}p_{i_2}\cdots p_{i_k}\left[\prod_{i=1}^{n}(1-p_i)\right]\Big(1+O(\Delta t)\Big). \end{aligned} \tag{29}$$

Now, observe that

$$\prod_{i=1}^{n}(1-p_i) = \prod_{i=1}^{n}(1-a(t_i)\Delta t) = \exp(-\sum_{i=1}^{n} a(t_i)\Delta t) + O\Big((\Delta t)^2\Big)$$

and

$$p_{i_1}p_{i_2}\cdots p_{i_k} = a(t_{i_1})a(t_{i_2})\cdots a(t_{i_k})(\Delta t)^k,$$

therefore, the $n \to \infty(\Delta t \to 0)$ limit with k fixed gives, with the notation $\tau_j \equiv t_{i_j}$,

$$\prod_{i=1}^{n}(1-p_i) \to \exp\left(-\int_0^t a(\tau)d\tau\right) \quad \text{and} \quad p_{i_1}p_{i_2}\cdots p_{i_k} \to \prod_{i=1}^{k} a(\tau_i)d\tau_i. \tag{30}$$

From Eq. (29) and Eq. (30), the probability, denoted by dP, for a path ω_k (namely, the measure) in the continuum limit is given by

$$dP \equiv P\Big(\omega_k(\tau_1,\tau_2,\cdots,\tau_k)\Big) = \begin{cases} e^{-\int_0^t a(\tau)d\tau}\prod_{i=1}^k a(\tau_i)d\tau_i, & k \geq 1; \\ e^{-\int_0^t a(\tau)d\tau} & k=0\,. \end{cases} \tag{31}$$

Therefore the probability $P(N(t)=k)$ for a path with k reversals, regardless of when the reversals occur, is obtained by summing over all possible reversal times as follows:

$$P\Big(N(t)=k\Big) = \int_{\{N(t)=k\}} dP = e^{-\int_0^\tau d\theta a(\theta)}\int_0^t d\tau_k a(\tau_k)\cdots\int_0^{\tau_2} d\tau_1 a(\tau_1). \tag{32}$$

Due to the fact that the integrand is completely symmetric in $\tau_1,\cdots,\tau_k$, the right hand side of Eq. (32) may be rewritten as

$$\frac{1}{k!}\int_0^t d\tau_k\cdots\int_0^t d\tau_2\int_0^t d\tau_1 a(\tau_1)a(\tau_2)\cdots a(\tau_k) = \frac{1}{k!}\left[\int_0^t d\tau a(\tau)\right]^k. \tag{33}$$

Hence, for $a=a(t)$ we obtain the probability for a path with k reversals

$$P\Big(N(t)=k\Big) = e^{-\int_0^t d\theta a(\theta)}\frac{[\int_0^t d\theta a(\theta)]^k}{k!} \qquad k=0,1,\ldots, \tag{34}$$

which reduces to Eq. (28) for a =constant. It follows that the average number of reversals by time t is given by

$$\langle N(t)\rangle = \sum_{k=0}^{\infty} kP\Big(N(t)=k\Big) = \int_0^t d\theta a(\theta), \tag{35}$$

and for a constant, we have

$$\langle N(t)\rangle = at. \tag{36}$$

Incidently, $N(t)$ also provides a convenient way of keeping track of the direction of the velocity of the Poissonian particle which at time t is given by $(-1)^{N(t)}$. (The process $(-1)^{N(t)}$ is known as asynchronous telegraph signal). Hence, the infinitesimal displacement for an infinitesimal time interval dt is given by $dx(t) = (-1)^{N(t)}v(t)dt$. It follows that the displacement of the particle from the origin at time t is given by

$$x(t) = \int_0^t (-1)^{N(\tau)}v(\tau)\,d\tau = v\int_0^t (-1)^{N(\tau)}\,d\tau \tag{37}$$

for v constant. The integral

$$\int_0^t (-1)^{N(\tau)}\,d\tau \equiv S(t) \tag{38}$$

has the dimension of time and takes a random value between $-t$ and t. Its absolute value is the time it would take a particle moving with constant velocity without reversal to move a distance $|x(t)|$. It is called "randomized time" by M. Kac. In the next section, we will use the measure Eq. (31) in finding a formula for its distribution.

5. APPLICATIONS OF THE MEASURE FOR THE POISSONIAN MOTION

Knowing the measure Eq. (31) would allow us to define, via path integral, the average of a function $\phi(x,\tau_1,\cdots,\tau_k)$ of reversal times $\tau_1,\cdots,\tau_k$ over paths that have k or more reversals as

$$\left\langle \phi(x,\tau_1,\cdots,\tau_k)\right\rangle = \mathcal{N}_k \sum_{n=k}^{\infty} \int_{\{N(t)=n\}} dP\, \phi(x,\tau_1,\cdots,\tau_k) \tag{39}$$

where

$$\mathcal{N}_k = \left[\sum_{n=k}^{\infty} P\Big(N(t)=n\Big)\right]^{-1} = e^{\int_0^t a(\tau)d\tau}\left[e^{\int_0^t a(\tau)d\tau} - \sum_{n=0}^{k-1}\frac{(\int_0^t a(\tau)d\tau)^n}{n!}\right]^{-1}, \tag{40}$$

(note that $\mathcal{N}_0 = 1$) and

$$\int_{\{N(t)=n\}} dP = e^{-\int_0^t a(\tau)d\tau}\int_0^t d\tau_n a(\tau_n)\cdots\int_0^{\tau_2} d\tau_1 a(\tau_1), \tag{41}$$

which, for a =constant, reduces to

$$\int_{\{N(t)=n\}} dP = e^{-at}a^n\int_0^t d\tau_n\cdots\int_0^{\tau_3} d\tau_2\int_0^{\tau_2} d\tau_1. \tag{42}$$

We now apply formula (39) for the case a =constant in the following three examples:

1) $\phi = c$. The average of a constant must be equal to the same constant:

$$\langle c\rangle = e^{-at}\sum_{n=0}^{\infty} a^n\int_0^t d\tau_n\cdots\int_0^{\tau_3} d\tau_2\int_0^{\tau_2} d\tau_1\, c \tag{43}$$

$$= ce^{-at}\sum_{n=0}^{\infty}\frac{a^n t^n}{n!} = c. \tag{44}$$

2) $\phi = \tau_k$. The average of the reversal time at which the kth reversal occurs is given by

$$\begin{aligned}\langle\tau_k\rangle &= \left[e^{at} - \sum_{n=0}^{k-1}\frac{(at)^n}{n!}\right]^{-1}\sum_{n=k}^{\infty} a^n\int_0^t d\tau_n\cdots\int_0^{\tau_{k+1}} d\tau_k\,\tau_k\cdots\int_0^{\tau_2} d\tau_1\\ &= \left[e^{at} - \sum_{n=0}^{k-1}\frac{(at)^n}{n!}\right]^{-1}\frac{k}{a}\sum_{n=k}^{\infty}\frac{(at)^{n+1}}{(n+1)!}\\ &= \frac{k}{a}\left[1 - \left(e^{at} - \sum_{n=0}^{k-1}\frac{(at)^n}{n!}\right)^{-1}\frac{(at)^k}{k!}\right].\end{aligned} \tag{45}$$

Graphs of $\langle\tau_k\rangle$ with $a=1$ as a function of t for $k = 1,2,3,4$ and 5 are given in Fig. 1.

The short and the long time behavior of Eq. (45) are easily obtained to be

$$\langle\tau_k\rangle = \frac{k}{k+1}t, \qquad t\to 0, \tag{46}$$

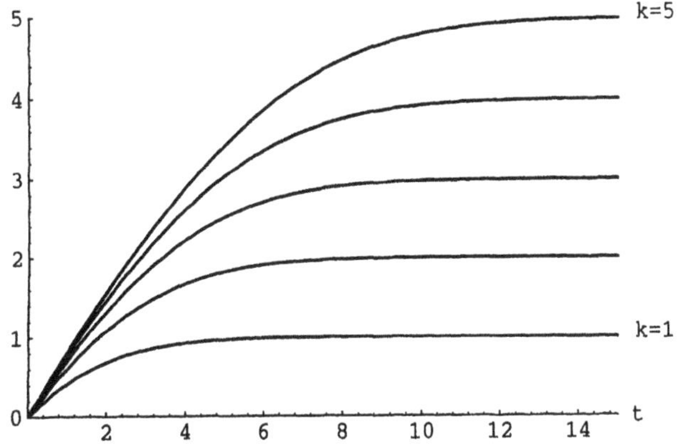

Figure 1. $\langle \tau_k \rangle$ as a function of t for $k = 1,2,3,4$ and 5, and $a = 1$.

and

$$\langle \tau_k \rangle = \frac{k}{a}, \qquad t \to \infty. \tag{47}$$

The long time behavior is what one would expect. Since the average time for a reversal is $1/a$ for a Poisson process of intensity a, and since each reversal is independent of the others, the average time for k reversals is just simply k times that of one reversal. The short time behavior is also what one would expect qualitatively.

3) $\phi = \delta(S(t) - r)$ where the randomized time $S(t)$ was defined by Eq. (38) section 4, with the intensity a constant. Expressed in terms of the reversal times $\tau_1, \cdots, \tau_n$, $S(t)$ reads

$$S(t,\tau_1,\cdots,\tau_n) = \tau_1 - (\tau_2 - \tau_1) + (\tau_3 - \tau_2) - \ldots + (-1)^{n-1}(\tau_n - \tau_{n-1}) + (-1)^n(t - \tau_n). \tag{48}$$

The average of $\delta(S(t) - r)$ is the distribution of $S(t)$ at time t, denoted by $g(a,t,r)$:

$$g(a,t,r) = \Big\langle \delta(S(t) - r) \Big\rangle = \sum_{n=0}^{\infty} \int_{\{N(t)=n\}} dP\, \delta(S(t) - r). \tag{49}$$

Denoting the distribution of $S(t)$ for paths with $N(t) = k$ as $g(a,t,r,N(t) = k)$, we have

$$g(a,t,r) = \sum_{k=0}^{\infty} g(a,t,r,N(t) = k) \tag{50}$$

and

$$g(a,t,r,N(t) = k) = \int_{\{N(t)=k\}} dP\, \delta(S(t) - r). \tag{51}$$

Let's first deal with the $g(a,t,r,N(t) = k)$. For $N(t) = 0$,

$$g(a,t,r,N(t) = 0) = e^{-at}\delta(t - r). \tag{52}$$

To evaluate the integral of Eq. (51) for $N(t) \geq 1$, it is convenient to make the following change of variables, where the Jacobian of the transformation is 1, namely,

$$u_1 = \tau_1, u_2 = \tau_2 - \tau_1, \cdots, u_n = \tau_n - \tau_{n-1}. \tag{53}$$

The u_i are the time intervals between the ith and the $(i+1)^{th}$ reversals and have the property that $0 < \sum_{i=1}^{n} u_i = \tau_n < t$. In the new variables, $S(t)$ becomes

$$S(t) = \begin{cases} -2(u_2+u_4+\cdots+u_{2k})+t, & n=2k; \\ 2(u_1+u_3+\cdots+u_{2k-1})-t, & n=2k-1\,. \end{cases} \tag{54}$$

For the case $n = 2k$, Eq. (51) becomes

$$g(a,t,r,N(t)=2k) = a^{2k}e^{-at}$$
$$\int_0^t du_{2k}\int_0^{t-u_{2k}} du_{2k-1}\cdots\int_0^{t-(u_2\cdots+u_{2k})} du_1\,\delta\Big(-2(u_2+\cdots+u_{2k})+t-r\Big). \tag{55}$$

Now we make another change of variables, whose Jacobian of transformation is again 1:

$$x_k = \sum_{i=1}^{k} u_{2i-1} \quad \text{and} \quad y_k = \sum_{i=1}^{k} u_{2i}. \tag{56}$$

The x_i and y_i are ordered such that $0 < x_1 < x_2 < \cdots < x_k$ and $0 < y_1 < y_2 < \cdots < y_k < t$. As a consequence, Eq. (55) simplifies to

$$\begin{aligned} g(a,t,r,N(t)=2k) &= a^{2k}e^{-at} \\ &\times \int_0^t dy_k\int_0^{t-y_k} dx_k\int_0^{y_k} dy_{k-1}\cdots\int_0^{y_2} dy_1\int_0^{x_k} dx_{k-1}\,\delta(-2y_k+t-r) \\ &= \frac{1}{2}\theta(t-|r|)e^{-at}(t+r)\frac{a^{2k}(t^2-r^2)^{k-1}}{2^{2k-1}(k-1)!k!}. \end{aligned} \tag{57}$$

For the case $n = 2k-1$, by the same analysis, we get

$$g(a,t,r,N(t)=2k-1) = \frac{1}{2}\theta(t-|r|)e^{-at}\frac{a^{2k-1}(t^2-r^2)^{k-1}}{2^{2(k-1)}(k-1)!^2}. \tag{58}$$

Since the series representations for the Modified Bessel functions are

$$I_n(x) = \sum_{k=0}^{\infty}\frac{1}{k!(n+k)!}\left(\frac{x}{2}\right)^{n+2k}, \tag{59}$$

summing Eq. (57) and Eq. (58) over values of $N(t)$, we obtain

$$\sum_{k=1}^{\infty} g(a,t,r,N(t)=2k-1) = \frac{1}{2}ae^{-at}I_0\Big(a\sqrt{t^2-r^2}\Big)\theta(t-|r|) \tag{60}$$

and

$$\sum_{k=1}^{\infty} g(a,t,r,N(t)=2k) = \frac{1}{2}ae^{-at}\frac{t+r}{\sqrt{t^2-r^2}}I_1\Big(a\sqrt{t^2-r^2}\Big)\theta(t-|r|). \tag{61}$$

It follows from Eq. (52), Eq. (60), and Eq. (61) that, for $a =$ constant, the distribution of $S(t)$ at time t is

$$g(a,t,r) = e^{-at}\delta(t-r)+\frac{1}{2}ae^{-at}\Big[I_0\Big(a\sqrt{t^2-r^2}\Big)+\frac{t+r}{\sqrt{t^2-r^2}}I_1\Big(a\sqrt{t^2-r^2}\Big)\Big]\theta(t-|r|). \tag{62}$$

This method of deriving $g(a,t,r)$ is based on Zastawniak [20], but in a less sophisticated mathematical language.

It can be checked that the mean and variance of $S(t)$ are given by

$$\langle r\rangle = \frac{1}{2a}(1-e^{-2at}) \quad\text{and}\quad \sigma_r^2 = \langle r^2\rangle - \langle r\rangle^2 = \frac{t}{a} - \frac{(1-e^{-2at})(3-e^{-2at})}{4a^2}. \tag{63}$$

The distribution $g(a,t,r)$ has the dimension of $1/t$. It will be useful to introduce an associated dimensionless distribution $g(\alpha,\beta)$, the distribution of the ratio $\beta = r/t$, where $\alpha = at$. This is in line with the spirit advocated by C. DeWitt-Morette and P. Cartier [26]. By the relation

$$\int_{-t}^{t} g(a,t,r)dr = 1 = \int_{-1}^{1} g(\alpha,\beta)d\beta, \tag{64}$$

we have

$$\begin{aligned} g(\alpha,\beta) &= tg(a,t,r)|_{at=\alpha,r/t=\beta} = e^{-\alpha}\delta(1-\beta) \\ &+ \frac{\alpha}{2}e^{-\alpha}\left[I_0\left(\alpha\sqrt{1-\beta^2}\right) + \frac{1+\beta}{\sqrt{1-\beta^2}}I_1\left(\alpha\sqrt{1-\beta^2}\right)\right]\theta(1-|\beta|). \end{aligned} \tag{65}$$

The mean and variance of β are

$$\langle\beta\rangle = \frac{1}{2\alpha}(1-e^{-2\alpha}) \quad\text{and}\quad \sigma_\beta^2 = \langle\beta^2\rangle - \langle\beta\rangle^2 = \frac{1}{\alpha} - \frac{(1-e^{-2\alpha})(3-e^{-2\alpha})}{4\alpha^2}. \tag{66}$$

From $g(\alpha,\beta)$, the distribution $\rho(a,t,x)$ of the displacement $x(t) = vS(t)$ of the Poissonian particle, which is assumed to have started by moving to the right, can be obtained from the relation

$$\int_{-vt}^{vt} \rho(a,t,x)dx = 1 = \int_{-1}^{1} g(\alpha,\beta)d\beta \tag{67}$$

to be

$$\rho(a,t,x) = \frac{1}{vt}g\left(\alpha = at, \beta = \frac{x}{vt}\right) \tag{68}$$

$$\begin{aligned} &= e^{-at}\delta(x-vt) \\ &+ \frac{a}{2v}e^{-at}\left[I_0\left(\frac{a}{v}\sqrt{v^2t^2-x^2}\right) + \frac{vt+x}{\sqrt{v^2t^2-x^2}}I_1\left(\frac{a}{v}\sqrt{v^2t^2-x^2}\right)\right]\theta(vt-|x|). \end{aligned} \tag{69}$$

In the next section, we shall gain further appreciation of the Poissonian motion by relating it to the more familiar Brownian motion.

6. CONNECTION BETWEEN POISSONIAN MOTION AND BROWNIAN MOTION

By changing the rules for the random walk and taking a different continuum limit, we obtain the telegrapher equation instead of the diffusion equation. However, the rules for the Brownian motion are, in some sense, special cases of those of the Poissonian motion if the parameters of the latter are allowed to approach infinite values in the continuum limit. First, in the Brownian motion each step is either forward or backward, with probability $1/2$, which is equivalent to letting $a\Delta t = 1/2$ in the Poissonian motion. Hence, a will diverge as $1/(2\Delta t)$ in the continuum limit. Second, holding $(\Delta x)^2/\Delta t = 2D$ in the continuum limit in the Brownian motion is equivalent to letting $v = \Delta x/\Delta t$ diverges as $2D/\Delta x$. Therefore, the lattice version of the telegrapher equation must yield the diffusion equation under these special conditions.

If we naively substitute $a = 1/(2\Delta t)$ and $v = \Delta x/\Delta t$ in (the continuum) telegrapher equation Eq. (22), the first term drops out as v diverges, while the coefficient of the second term becomes

$$\frac{2a}{v^2} = 2\left(\frac{1}{2\Delta t}\right)\left(\frac{\Delta t}{\Delta x}\right)^2 = \frac{\Delta t}{(\Delta x)^2} = \frac{1}{2D}, \tag{70}$$

and the telegrapher equation reduces to

$$\partial_t g - 2D\partial_x^2 g = 0, \tag{71}$$

a diffusion equation, but with a diffusion constant twice too big! What is wrong with these substitutions?

To appreciate the blunder, we should return to the lattice version of the telegrapher equation Eq. (17), and substitute $a\Delta t = 1/2$ there to obtain

$$\rho_\pm(a,t,x) = \frac{1}{2}\rho_\pm(a,t-\Delta t,x\mp\Delta x) + \frac{1}{2}\rho_\mp(a,t-\Delta t,x\pm\Delta x). \tag{72}$$

Then the terms on the RHS are to be expanded with the condition that $(\Delta x)^2/(\Delta t) = 2D$. If we do these steps, then only do we end up with the correct diffusion equation with the correct diffusion constant. What was wrong with the naive substitutions was that in deriving Eq. (18) (and hence Eq. (22)) from Eq. (17), only the first order terms in Δt and Δx were used. Consequently, the correct diffusion equation, which requires keeping the second-order $(\Delta x)^2$ terms in the lattice telegrapher equation, could not be recovered from Eq. (22).

We could have, of course, just as Kac did, let $a \to \infty$ and $v \to \infty$ such that $2a/v^2 = 1/D$ (i.e. without going through the steps shown in Eq. (70)), and obtained the correct diffusion equation for *all* time. However, it is not necessary to go into the non-physical regime in order to make connection with Brownian motion. It can be shown explicitly (for time-independent and finite a and v) that the distribution of the Poissonian motion will tend to that of the Brownian motion at large time.

It is convenient to carry out the analysis using the dimensionless $g(\alpha,\beta)$, as given by Eq. (65). From the series expansions $I_0(x) = 1 + x^2/8 + \cdots$ and $I_1(x)/x = 1/2(1 + x^2/8 + \cdots)$, we have

$$g(\alpha,\beta \to \pm 1) = e^{-\alpha}\delta(1-\beta) + \frac{\alpha}{2}\left(1 + \left\{\begin{array}{c} 2\alpha \\ 0 \end{array}\right\}\right)e^{-\alpha}. \tag{73}$$

It follows that

$$g(\alpha \to \infty, \beta \to \pm 1) \to 0. \tag{74}$$

From the asymptotic expansions

$$I_0(z) = \frac{1}{\sqrt{2\pi z}}\exp\left[z + \frac{1}{8z} + \frac{1}{16z^2} + O\left(\frac{1}{z^3}\right)\right] \tag{75}$$

and

$$I_1(z) = \frac{1}{\sqrt{2\pi z}}\exp\left[z - \frac{3}{8z} - \frac{3}{16z^2} + O\left(\frac{1}{z^3}\right)\right], \tag{76}$$

it can be shown that $g(\alpha,\beta)$ becomes for large α and small β

$$g(\alpha,\beta)=\sqrt{\frac{\alpha}{2\pi}}\exp\left[-\frac{\alpha}{2}\left(\beta-\frac{1}{2\alpha}\right)^2\right]\left[1+O\left(\alpha\beta^4\right)+O\left(\beta^2\right)+O\left(\beta/\alpha\right)\right], \tag{77}$$

whose width (standard deviation) shrinks as $1/\sqrt{\alpha}$. Hence Eq. (77) describes $g(\alpha,\beta)$ well for the range of β for which $g(\alpha,\beta)$ is significantly larger than zero. Since $g(\alpha,\beta)$ is continuous in β for $-1<\beta<1$, and it approaches zero as $|\beta|\to 1$, Eq. (77) is good for all β for large α. Furthermore, since

$$\beta^k\exp\left[-\frac{\alpha}{2}\left(\beta-\frac{1}{2\alpha}\right)^2\right],\quad k=2,4, \tag{78}$$

as a function of β peak at $\beta\sim const./\sqrt{\alpha}$ for large α, the β in the correction terms can be replaced by $1/\sqrt{\alpha}$, and consequently,

$$g(\alpha,\beta)=\sqrt{\frac{\alpha}{2\pi}}\exp\left[-\frac{\alpha}{2}\left(\beta-\frac{1}{2\alpha}\right)^2\right]\left[1+O\left(\frac{1}{\alpha}\right)\right]. \tag{79}$$

For convenience, we denote the Gaussian ($g(\alpha,\beta)$ without the correction term) as

$$g_B(\alpha,\beta)=\sqrt{\frac{\alpha}{2\pi}}\exp\left[-\frac{\alpha}{2}\left(\beta-\frac{1}{2\alpha}\right)^2\right]. \tag{80}$$

(Remark: A naive application of the Central limit theorem to obtain the large α limit of $g(\alpha,\beta)$ yields also $g_B(\alpha,\beta)$.)

Fig. 2 shows how this Brownian approximation improves for all β as α increases.

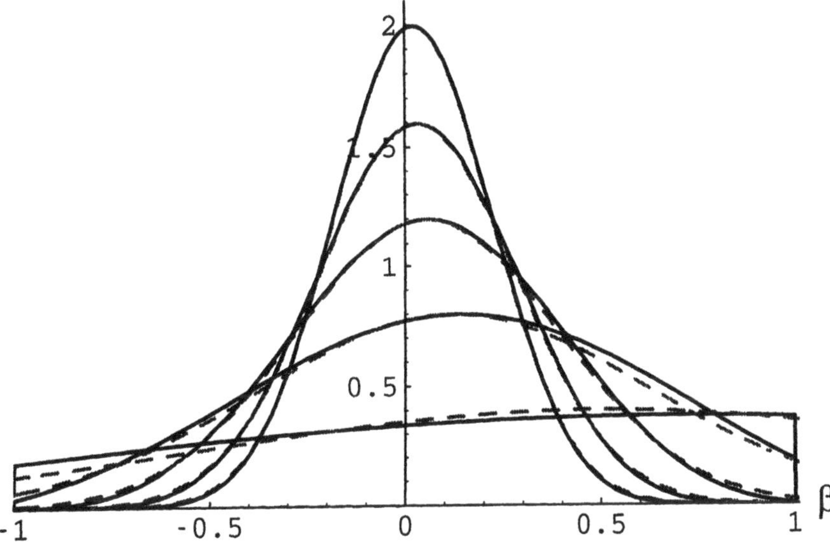

Figure 2. $g_B(\alpha,\beta)$ is a good approximation for all β when α is large. Solid lines refer to $g(\alpha,\beta)$ whereas dashed lines refer to $g_B(\alpha,\beta)$ for $\alpha=1,4,9,16$ and 25.

We can define a measure of the error incurred by the approximation by integrating over β the absolute value of deviation between $g(\alpha,\beta)$ and the Gaussian, namely

$$\text{error}=\int\left|g(\alpha,\beta)-g_B(\alpha,\beta)\right|d\beta. \tag{81}$$

A numerical evaluation with β ranges over four standard deviations (Note: The area under the Gaussian over this range is ≈ 0.99994 using Mathematica [35]) gives the results in Fig. 3, from which we obtain the following approximation for the error:

$$\text{error} \approx \frac{.25}{\alpha}. \tag{82}$$

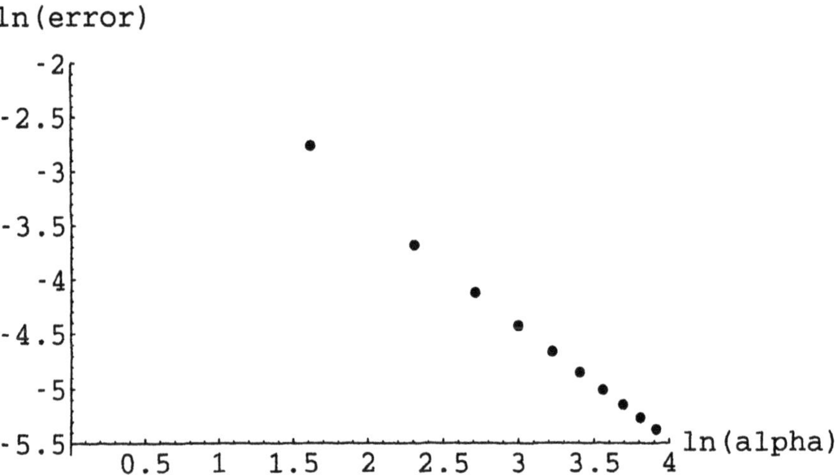

Figure 3. Behavior of the error when approximating $g(\alpha,\beta)$ by $g_B(\alpha,\beta)$.

The corresponding large time limit of $g(a,t,r)$ would be

$$g_B(a,t,r) = \frac{1}{t} g_B(\alpha = at, \beta = r/t) = \sqrt{\frac{a}{2\pi t}} \exp\left[-\frac{a}{2t}\left(r - \frac{1}{2a}\right)^2\right], \quad t \to \infty. \tag{83}$$

To compare with the familiar Brownian motion, we shall examine $\rho(a,t,x)$. From Eq. (79), Eq. (68) and Eq. (10), we have

$$\begin{aligned}\rho(a,t,x) &= \frac{1}{\sqrt{4\pi Dt}} \exp\left[-\frac{(x-\frac{v}{2a})^2}{4Dt}\right]\left[1 + O\left(\frac{1}{at}\right)\right] \\ &= \rho_B\left(t, x - \frac{v}{2a}\right)\left[1 + O\left(\frac{1}{at}\right)\right], \qquad \text{where} \quad D = v^2/(2a).\end{aligned} \tag{84}$$

That is to say, the large time limit of the Poissonian motion with *finite* a and v is the Brownian motion with diffusion constant D whose value is $v^2/(2a)$. The Brownian distribution centers at $v/(2a)$, and its variance is given by $2Dt = v^2 t/a$. The parameters a and v no longer need to take on unphysical values. Fig. 4 shows how $\rho(a,t,x)$ with $v = 1$ and $a = 1$ tends to $\rho_B(t, x - \frac{v}{2a})$ with $D = 1/2$ as t increases. If we had assumed a symmetric starting direction (i.e. half the particles started moving right, and half started moving left) for the Poissonian motion, then instead of Eq. (84), we would have, to order $1/(at)$,

$$\rho(a,t,x) \approx \frac{1}{\sqrt{4\pi Dt}} \exp\left[-\frac{x^2}{4Dt}\right] = \rho_B(t,x). \tag{85}$$

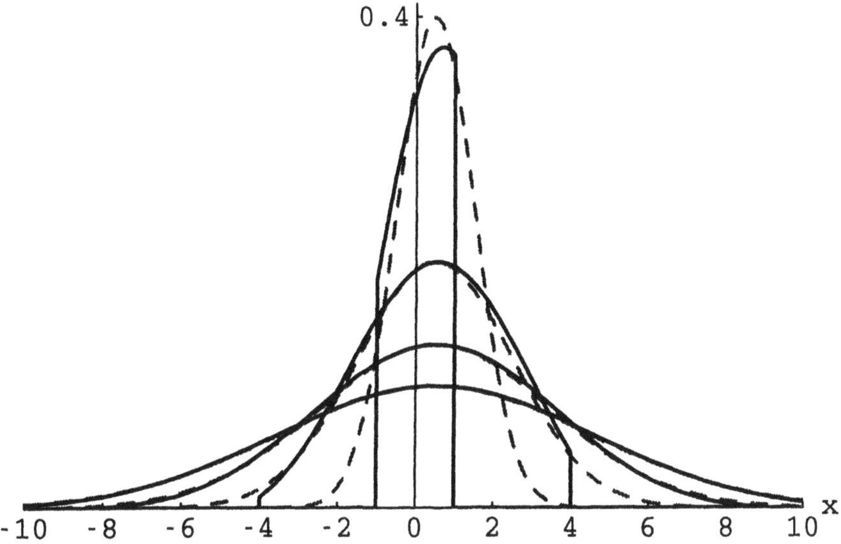

Figure 4. Comparing $\rho(a,t,x)$ with $v=1$ and $a=1$ with the corresponding $\rho_B(t,x-\frac{v}{2a})$ with $D=1/2$ for $t=1,4,9$ and 16.

6.1. Densities and Green Functions

How does the kind of relation between ρ_B and the retarded Green function for the diffusion equation look like in the case of the telegrapher equation?

Recall that

$$\rho_B(t,x)=\frac{1}{\sqrt{4\pi Dt}}\exp\left(\frac{-x^2}{4Dt}\right) \tag{86}$$

is the solution for the diffusion equation for $t>0$, having the property that the particle is initially at $x=0$, namely,

$$\lim_{t\to 0}\rho_B(t,x)=\delta(x). \tag{87}$$

The retarded Green function is given by

$$G_{ret}(t,x)=\theta(t)\rho_B(t,x). \tag{88}$$

For the first order telegrapher equation Eq. (19), there is an analogous way of expressing the Green function in terms of the densities. Let $\rho^+(a,t,x)$ and $\rho^-(a,t,x)$ denote the solutions for which the particle started moving to the right and left, respectively. The initial conditions are thus

$$\rho^+(a,0,x)=\begin{pmatrix}\delta(x)\\ 0\end{pmatrix} \quad \text{and} \quad \rho^-(a,0,x)=\begin{pmatrix}0\\ \delta(x)\end{pmatrix}. \tag{89}$$

Then, the retarded Green function is given by

$$G_{ret}(a,t,x)=\theta(t)\Big(\rho^+(a,t,x) \quad \rho^-(a,t,x)\Big), \tag{90}$$

which the reader can easily verify that it satisfies

$$(\partial_t-M)G_{ret}(a,t,x)=\delta(x)\delta(t), \tag{91}$$

in view of Eq. (19).

The explicit expression for ρ_+^+ (ρ_-^+) is obtained by summing over those contributions from particles which have undergone even (odd) number of reversals by time t. (They are moving to the right (left) at time t.) From Eq. (52), Eq. (60), and Eq. (61), we have, for $a =$ constant and with the notation $d \equiv \sqrt{v^2t^2 - x^2}$,

$$\rho^+(a,t,x) = \begin{pmatrix} \rho_+^+ \\ \rho_-^+ \end{pmatrix} = e^{-at} \begin{pmatrix} \delta(vt-x) + \frac{1}{2}a(vt+x)d^{-1}I_1(ad)\theta(vt-|x|) \\ \frac{1}{2}aI_0(ad)\theta(vt-|x|) \end{pmatrix}. \tag{92}$$

By replacing x with $-x$, and interchange the elements of the columns in Eq. (92), we obtain

$$\rho^-(a,t,x) = \begin{pmatrix} \rho_+^- \\ \rho_-^- \end{pmatrix} = e^{-at} \begin{pmatrix} \frac{1}{2}aI_0(ad)\theta(vt-|x|) \\ \delta(vt+x) + \frac{1}{2}a(vt-x)d^{-1}I_1(ad)\theta(vt-|x|) \end{pmatrix}. \tag{93}$$

The connection between the Poissonian motion and Brownian motion in one space and one time dimensions is summarized in Table 1. This better understanding on the Poisson

Table 1. A summary of the connection between Poissonian motion and Brownian motion

	Poissonian	Brownian
Continuum limit $(\Delta t \to 0, \Delta x \to 0)$	$a\Delta t = 0$ (a finite) $\Delta x/\Delta t = v$ (finite) $(\Delta x)^2/\Delta t = 0$	$a\Delta t = 1/2$ (a diverges) $\Delta x/\Delta t$ diverges $(\Delta x)^2/\Delta t = 2D$
Eq. for density	$\frac{1}{v^2}\partial_t^2\rho + \frac{2a}{v^2}\partial_t\rho - \partial_x^2\rho = 0$	$\partial_t\rho_B - D\partial_x^2\rho_B = 0$
Long time behavior	$\rho \to \rho_B$	
Relation between density and Green function	(First order version) $\theta(t)(g^+\ g^-)$	$\theta(t)\rho_B$

motion will be useful for learning Kac's method of solving the telegrapher equation.

7. KAC'S METHOD (IN NEW FORM)

In this section, we state Kac's method of solving telegrapher equation (in its new form which includes contributions from S. Kaplan, C. DeWitt-Morette, and S. K. Foong), and then provide a partial proof before a complete one in section 9. We then state the solution of the wave equation without dissipation in terms of the initial conditions, and apply Kac's method to a few simple examples, and end with a comparison with the Green function method of solving the telegrapher equation.

7.1. The Method

Kac's path integral solution of the telegrapher equation

$$\frac{1}{v^2}\frac{\partial^2 F}{\partial t^2} + \frac{2a(t)}{v^2}\frac{\partial F}{\partial t} - O_{\mathbf{x}}F = 0, \tag{94}$$

$$F(\mathbf{x},0) = \kappa(\mathbf{x}), \qquad \left.\frac{\partial F}{\partial t}\right|_{t=0} = \xi(\mathbf{x}),$$

where $a(t) = a_1(t)$ — denoted as such in order to distinguish it from $a_2(t)$ to be defined in Eq. (97) — is a non-negative, continuous function defined on $[0,\infty)$, is given by [1, 3]

$$F(\mathbf{x},t) = \langle \phi(\mathbf{x},S(t)) \rangle, \tag{95}$$

where $\phi(\mathbf{x},t)$ possesses the same initial conditions as $F(\mathbf{x},t)$ and satisfies

$$(\partial_t^2 - O_{\mathbf{x}})\phi = 0. \tag{96}$$

Here $S(t)$ is the randomized time as defined by Eq. (38).

In fact, if

$$a(t) = a_2(t) = \begin{cases} \text{finite}, & t = 0\,; \\ \gamma t^q, & q < 0,\ t > 0, \end{cases} \tag{97}$$

where γ is a positive constant, Eq. (95) still holds [19] although $a_2(t)$ diverges as $t \to 0^+$, provided the second boundary condition is modified to

$$\lim_{t\to 0^+} \frac{\partial F}{\partial t} = \begin{cases} \left.\frac{\partial \phi(x,t)}{\partial t}\right|_{t=0}, & q > -1; \\ 0, & q \leq -1\,. \end{cases} \tag{98}$$

With the measure Eq. (31) and $S(t)$ expressed in terms of the reversal times Eq. (48), we obtain a path integral solution of telegrapher equation

$$F(\mathbf{x},t) = \langle \phi(\mathbf{x},S(t)) \rangle = \sum_{n=0}^{\infty} \int_{\{N(t)=n\}} dP\, \phi\Big(\mathbf{x}, S(t,\tau_1,\cdots,\tau_n)\Big). \tag{99}$$

However, knowing the distribution of $S(t)$, the calculation of F is simply given by the following one-dimensional ordinary integral:

$$F(\mathbf{x},t) = \langle \phi(\mathbf{x},S(t)) \rangle = \int_{-\infty}^{\infty} \phi(\mathbf{x},r) g(a,t,r)\, dr = \int_{-t}^{t} \phi(\mathbf{x},r) g(a,t,r)\, dr, \tag{100}$$

where the last equality follows because $g(a,t,r) = 0$ for $|r| \geq t$. This new form for the solution in terms of an one-dimensional integral was emphasized in the papers [10, 11, 12].

Special Case: a =constant. Substituting $g(a,t,r)$ (Eq. (62)) for a constant into Eq. (100), we have

$$F(\mathbf{x},t) = e^{-at}\phi(\mathbf{x},t) + \frac{1}{2} a e^{-at} \int_{-t}^{t} \phi(\mathbf{x},r) \left[I_0\left(a\sqrt{t^2-r^2}\right) + \frac{t+r}{\sqrt{t^2-r^2}} I_1\left(a\sqrt{t^2-r^2}\right) \right] dr. \tag{101}$$

For large time $at >> 1$, because of Eq. (83), Eq. (101) simplifies to

$$F(\mathbf{x},t) = \sqrt{\frac{a}{2\pi t}} \int_{-\infty}^{\infty} \phi(\mathbf{x},r) \exp\left[-\frac{a}{2t} \left(r - \frac{1}{2a} \right)^2 \right] dr + O\Big(1/(at)\Big). \tag{102}$$

Therefore, if one is interested in $F(\mathbf{x},t)$ only for large time, than it may be sufficient to evaluate this simpler integral.

7.2. Proof ($a = \text{constant}$)

For simplicity, we shall first prove Eq. (100) for $a =$ constant. The proof for $a = a(t)$ will be deferred to section 9.

First we prove the initial conditions. The first is

$$F(\mathbf{x},0) = \int_{-\infty}^{\infty} \phi(\mathbf{x},r) g(a,0,r)\, dr = \int_{-\infty}^{\infty} \phi(\mathbf{x},r)\delta(r)\, dr = \phi(\mathbf{x},0). \tag{103}$$

The second is $\partial_t F(\mathbf{x},0)$. From Eq. (100), we have

$$\partial_t F(\mathbf{x},0) = \int_{-\infty}^{\infty} \phi(\mathbf{x},r)\partial_t g(a,0,r)\, dr. \tag{104}$$

The particle is assumed to have started by moving to the right at the origin, hence for a very short time $t = \varepsilon$, $g(\varepsilon,r) = \delta(\varepsilon - r)$. Therefore,

$$\partial_t g|_{t=\varepsilon} = \partial_\varepsilon \delta(\varepsilon - r) = -\partial_r \delta(\varepsilon - r). \tag{105}$$

It follows from (104), by an integration by parts, that

$$\partial_t F(\mathbf{x},0) = \int_{-\infty}^{\infty} \left[\partial_r \phi(\mathbf{x},r)\right] \delta(r)\, dr = \partial_t \phi(\mathbf{x},0). \tag{106}$$

We now show that F given by Eq. (100) satisfies the second order telegrapher equation. Comparing with Eq. (21) and Eq. (37) we see that the distribution of $S(t)$ satisfies Eq. (21) with the replacement $x \to r$, namely

$$\left(\partial_t^2 + 2a\partial_t - \partial_r^2\right) g(a,t,r) = 0. \tag{107}$$

Since $(\partial_t^2 - O_\mathbf{x})\phi = 0$, and $O_\mathbf{x}$ is a linear spatial operator, we have

$$O_\mathbf{x} F(\mathbf{x},t) = \int_{-\infty}^{\infty} \left[\partial_r^2 \phi(\mathbf{x},r)\right] g(a,t,r)\, dr = \int_{-\infty}^{\infty} \phi(\mathbf{x},r)\partial_r^2 g(a,t,r)\, dr. \tag{108}$$

Therefore,

$$\left(\partial_t^2 + 2a\partial_t - O_\mathbf{x}\right) F(\mathbf{x},t) = \int_{-\infty}^{\infty} \phi(x,r)\left(\partial_t^2 + 2a\partial_t - \partial_r^2\right) g(a,t,r)\, dr = 0, \tag{109}$$

since g satisfies the telegrapher equation Eq. (107). QED.

7.3. Some $\phi(\mathbf{x},t)$

Since Kac's method relies on the knowledge of $\phi(\mathbf{x},t)$, it would be appropriate to state some here. For the case $O_\mathbf{x} = \nabla^2$, the solutions $\phi(\mathbf{x},t)$ for unbounded domain in arbitrary space dimensions are given in [36]. For space dimensions $n = 1, 2$, and 3, they are respectively:

1) The d'Alembert solution

$$\phi(x,t) = \frac{1}{2}\left[\kappa(x - vt) + \kappa(x + vt)\right] + \frac{1}{2v}\int_{x-vt}^{x+vt} \xi(\zeta) d\zeta. \tag{110}$$

2) The Parseval solution

$$\phi(x_1,x_2,t) = \frac{1}{2\pi v}\frac{\partial}{\partial t}\int\int_{R(t)} \frac{\kappa(x_1+\zeta_1,x_2+\zeta_2)}{\sqrt{v^2t^2-(\zeta_1^2+\zeta_2^2)}}d\zeta_1 d\zeta_2 + \frac{1}{2\pi v}\int\int_{R(t)} \frac{\xi(x_1+\zeta_1,x_2+\zeta_2)}{\sqrt{v^2t^2-(\zeta_1^2+\zeta_2^2)}}d\zeta_1 d\zeta_2, \qquad (111)$$

where $R(t)$ is the region $(\zeta_1,\zeta_2)|\zeta_1^2+\zeta_2^2 \leq v^2t^2$.

3) The Poisson solution

$$\phi(\mathbf{x},t) = \frac{\partial}{\partial t}\Big(t\omega[\kappa;\mathbf{x},t]\Big) + t\omega[\xi;\mathbf{x},t], \qquad (112)$$

where

$$\omega[h;\mathbf{x},t] = \frac{1}{4\pi}\int_0^{2\pi}\int_0^{\pi} h(x_1+vt\sin\theta\cos\psi, x_2+vt\sin\theta\sin\psi, x_3+vt\cos\theta)\sin\theta d\theta d\psi.$$

7.4. Examples

We give here a few examples showing the use of $g(a,t,r)$ in obtaining $F(x,t)$. The parameter a is assumed constant.

1. $\phi(x,t) = k\delta(x-vt)$, where k is a constant with the dimension of length, i.e. ϕ is a scalar delta-function pulse propagating to the right. The damped pulse is given by

$$F(x,t) = k\int \delta(x-vr)g(a,t,r)dr = (k/v)g(a,t,x/v) = k\rho(a,t,x), \qquad (113)$$

where $\rho(a,t,x)$ is given in Eq. (69). We see that the delta pulse diminishes in height exponentially as it moves forward, and a tail of increasing prominence begins to develop as soon as it leaves the origin. Part of the pulse moves in the backward direction! In other words, a pulse propagating forward in a stretched string in air will have part of it moving backwards when placed in water.

2. We can then write down, from the result in example 1), the corresponding damped version of the impulse delivered at the origin and propagates outwards as a spherical wave $\phi(R,t) = (k/R)\delta(R-vt)$ in three dimensions, where k here is a constant with the dimension of length squared and R is the radial distance from the origin, namely

$$F(R,t) = \frac{k}{R}e^{-at}\delta(R-vt) + \frac{ak\,e^{-at}}{2v\;R}\left[I_0\Big(\frac{a}{v}\sqrt{v^2t^2-R^2}\Big) + \frac{vt+R}{\sqrt{v^2t^2-R^2}}I_1\Big(\frac{a}{v}\sqrt{v^2t^2-R^2}\Big)\right]\theta(vt-R). \qquad (114)$$

3. $\phi(x,t) = \sin(kx-\omega t)$. In this case, the integrals on the right hand side of Eq. (100) can be done [19]. With the substitution $\Omega = \sqrt{\omega^2-a^2}$, we obtain

$$F(x,t) = e^{-at}\left[\sin kx\Big(\cos\Omega t + \frac{a}{\Omega}\sin\Omega t\Big) - \frac{\omega}{\Omega}\cos kx\sin\Omega t\right]. \qquad (115)$$

It is instructive to approximate this in power of a/ω:

$$F(x,t) = \frac{\omega}{\Omega} e^{-at} \left[\sin(kx - \Omega t) + \frac{a}{\omega} \sin kx \sin \Omega t - \frac{a^2}{2\omega^2} \sin kx \cos \Omega t + \cdots \right]. \quad (116)$$

We see that the zero order effect is that the frequency is reduced from ω to Ω, and the velocity is reduced by a factor Ω/ω, and the amplitude is multiplied by a time-dependent exponential damping factor.

4. Similarly, the damped $F(R,t)$ for the spherical wave $\phi(R,t) = \sin(kR - \omega t)/R$ in three dimensions is given by

$$F(R,t) = \frac{e^{-at}}{R} \left[\sin kR \left(\cos \Omega t + \frac{a}{\Omega} \sin \Omega t \right) - \frac{\omega}{\Omega} \cos kR \sin \Omega t \right]. \quad (117)$$

7.5. Comparison with Green Function Method

The Green function method given in classics such as Morse and Feshbach [37] for solving initial value problems of partial differential equation applies in a more general context that includes time-dependent sources. As far as I am aware, the present Kac's method has not been generalized to such general context. Nevertheless, within the context where Kac's method applies, it shows the relationship between the solution in the absence of dissipation and that in the presence of dissipation. The Green function method does not show such a relationship, at least not in an obvious way. We shall cash in on such a relationship to offer an unconventional way of waveform restoration in section 11.

To illustrate the discussion above, some explicit equations are given below. The solution of the telegrapher equation Eq. (94) for $O_{\mathbf{x}} = \nabla^2$, for infinite domain and in the absence of source, is given in the Green function method as

$$F(\mathbf{x},t) = \frac{2a}{v^2} \int d\mathbf{x}_0 \kappa(\mathbf{x}_0) G(\mathbf{x},t;\mathbf{x}_0,0) + \frac{1}{v^2} \int d\mathbf{x}_0 \left[G(\mathbf{x},t;\mathbf{x}_0,0)\xi(\mathbf{x}_0) - \kappa(\mathbf{x}_0)\partial_{t_0} G(\mathbf{x},t;\mathbf{x}_0,t_0)|_{t_0=0} \right]. \quad (118)$$

The infinite domain Green's functions for the telegrapher equation for various dimensions are given in [37], with $r = |\mathbf{x} - \mathbf{x}_0|/v$ and $\tau = t - t_0$, as

$$G(r,\tau) = \frac{v}{2} e^{-a\tau} I_0\left(a\sqrt{\tau^2 - r^2}\right) \theta(\tau - r), \qquad (n = 1), \quad (119)$$

$$G(r,\tau) = \frac{e^{-a\tau}}{2\pi\sqrt{\tau^2 - r^2}} \left[1 + 2\sinh^2\left(\frac{a}{2}\sqrt{\tau^2 - r^2}\right) \right] \theta(\tau - r), \qquad (n = 2), \quad (120)$$

and

$$G(r,\tau) = \frac{1}{4\pi r} e^{-a\tau} \left[\frac{1}{v}\delta(\tau - r) + \frac{ar}{v\sqrt{\tau^2 - r^2}} I_1\left(a\sqrt{\tau^2 - r^2}\right) \right] \theta(\tau - r), \qquad (n = 3). \quad (121)$$

As we can see in Eq. (118), unlike Kac's method, the Green function method emphasizes the dependence of the solution on the initial conditions. It would be interesting at this point to substitute Eq. (119) into Eq. (118) to obtain, with $d \equiv \sqrt{v^2t^2-(x-x_0)^2}$,

$$F(x,t)=\frac{1}{2}e^{-at}\left\{\kappa(x+vt)+\kappa(x-vt)+\frac{1}{v}\int_{x-vt}^{x+vt}dx_0\xi(x_0)I_0\left(\frac{ad}{v}\right)\right.$$
$$\left.+\frac{a}{v}\int_{x-vt}^{x+vt}dx_0\kappa(x_0)\left[I_0\left(\frac{ad}{v}\right)+\frac{vt}{d}I_1\left(\frac{ad}{v}\right)\right]\right\}, \qquad (122)$$

whose appearance is rather different from that given by Kac's method (although they are the same $F(x,t)$, as one can verify).

8. SOME EXPERIMENTAL CONSIDERATIONS

In this section, we shall consider a voltage pulse propagating in a typical copper transmission line of diameter of the order .5 cm. The value of a in this case is of the order $3\times10^5\,\text{sec}^{-1}$. The speed of propagation is of the order c, and by setting $c=1$, time would be measured in meters, and we have the equivalence $1\,\text{sec}=3\times10^8 m$. Hence, $a=10^{-3}m^{-1}$. Suppose the initial pulse is of unit height and length L meters situated between $x=-100$m and $x=0$, and is propagating to the right, then $F(x,t)$ is given by, with $u\equiv\sqrt{t^2-r^2}$,

$$F(x,t)=e^{-at}\left\{1+\frac{1}{2}a\int_x^t dr\left[I_0(au)+\frac{t+r}{u}I_1(au)\right]\right\} \quad \text{for} \quad x+L>t>x \qquad (123)$$

and

$$F(x,t)=\frac{1}{2}ae^{-at}\int_x^{x+L}dr\left[I_0(au)+\frac{t+r}{u}I_1(au)\right] \quad \text{for} \quad t>x+L. \qquad (124)$$

Suppose L is 100m long, and the pulse is observed at the position $x=100$m, then by a numerical evaluation of Eq. (123) and Eq. (124), we obtain $F(x,t)$ as shown in Fig. 5.

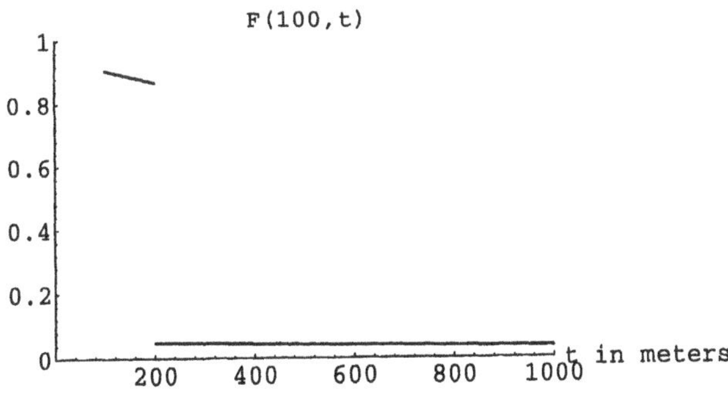

Figure 5. $F(100,t)$ for an initial pulse of length 100m situated between $x=-100$ and $x=0$, and propagating to the right.

Experimental work in this area and waveform restoration, designing and prediction (see section 11, 12, 13) is under planning.

9. (MODIFIED) KAPLAN'S GENERALIZATION

Kac's original solution was for a constant, and Kaplan generalized it to a time-dependent $a(t) = a_1(t)$. We now review Kaplan's generalization, and see how it may also work for $a(t) = a_2(t)$ with minor modifications, as stated by Eq. (98), of the boundary conditions. (Definitions for a_1 and a_2 are given in section 7.1.) Let $u(t) \equiv \langle \phi(S(t)) \rangle$, where the $\mathbf{x}$ dependence is suppressed, we want to prove that $u(t)$ satisfies

$$u''(t) + 2a(t)u'(t) = \langle \phi''(S(t)) \rangle, \tag{125}$$

where the prime $'$ denotes differentiation with respect to the argument. Let $y(t) \equiv u'(t)$ and $f(t) \equiv \langle \phi''(S(t)) \rangle$, Eq. (125) reads

$$y'(t) + 2a(t)y(t) = f(t). \tag{126}$$

With the use of the integrating factor $\exp(2\int_0^t a(\tau)d\tau)$, the solution of Eq. (126) may be expressed as

$$y(t) = y(\varepsilon)\exp\left(-2\int_\varepsilon^t a(\tau)d\tau\right) + \int_\varepsilon^t d\theta f(\theta)\exp\left(-2\int_\theta^t a(\tau)d\tau\right). \tag{127}$$

It follows that

$$u(t) = u(\varepsilon') + u'(\varepsilon)\int_{\varepsilon'}^t d\tau\, e^{-2\int_\varepsilon^\tau a(s)ds} + \int_{\varepsilon'}^t d\tau \int_\varepsilon^\tau d\theta f(\theta) e^{-2\int_\theta^\tau a(s)ds}. \tag{128}$$

Taking the limit $\varepsilon, \varepsilon' \to 0$, and noting $S(\varepsilon) \leq \varepsilon$, and so $\lim_{\varepsilon\to 0} S(\varepsilon) = 0$ independent of paths, we have

$$\lim_{\varepsilon\to 0} u(\varepsilon) = \left\langle \lim_{\varepsilon\to 0} \phi(S(\varepsilon)) \right\rangle = \phi(0), \tag{129}$$

and assuming $\lim_{t\to 0} \phi'(t) = \phi'(0)$, we have

$$\lim_{\varepsilon\to 0} u'(\varepsilon) = \lim_{\varepsilon\to 0}\left\langle \phi'(S(\varepsilon))S'(\varepsilon)\right\rangle = \lim_{\varepsilon\to 0}\left\langle \phi'(S(\varepsilon))(-1)^{N(\varepsilon)}\right\rangle = \phi'(0)\lim_{\varepsilon\to 0}\left\langle (-1)^{N(\varepsilon)}\right\rangle. \tag{130}$$

Depending on the behavior of $N(t)$ or $a(t)$ in the $t \to 0$ limit, the above average may assume different values. With the use of Eq. (34), it may be evaluated as follows:

$$\begin{aligned}
\left\langle (-1)^{N(\tau)} \right\rangle &= \sum_{k=0}^{\infty} (-1)^{N(\tau)=k} P(N(\tau) = k) \\
&= e^{-\int_0^\tau a(s)ds} \sum_{k=0}^{\infty} \frac{(-1)^k}{k!}\left(\int_0^\tau a(s)ds\right)^k \\
&= e^{-2\int_0^\tau a(s)ds} = \lim_{\varepsilon\to 0} e^{-2\int_\varepsilon^\tau a(s)ds} \\
&= \begin{cases} e^{-2\int_0^\tau a(s)ds}, & a = a_1; a = a_2, q > -1\,; \\ 0, & a = a_2, q \leq -1, \gamma > 0, \end{cases}
\end{aligned} \tag{131}$$

and therefore

$$\lim_{\varepsilon\to 0}\left\langle (-1)^{N(\varepsilon)}\right\rangle = \begin{cases} 1, & a = a_1; \quad a = a_2, \quad q > -1\,; \\ 0, & a = a_2, \quad q \leq -1, \quad \gamma > 0\,. \end{cases} \tag{132}$$

(For $a = a_1$, the limiting process is not necessary; we can just set $\varepsilon = 0$.) From Eq. (130) and Eq. (132), we have

$$\lim_{\varepsilon\to 0} u'(\varepsilon) = \begin{cases} \phi'(0), & a = a_1; \quad a = a_2, \quad q > -1; \\ 0, & a = a_2, \quad q \le -1, \quad \gamma > 0\,. \end{cases} \tag{133}$$

It follows that for $a = a_1$, and $a = a_2, \quad q > -1$, Eq. (128) becomes

$$u(t) = \phi(0) + \phi'(0)\int_0^t d\tau\, e^{-2\int_0^\tau a(s)ds} + \int_0^t d\tau \int_0^\tau d\theta \Big\langle \phi''(S(\theta)) \Big\rangle e^{-2\int_\theta^\tau a(s)ds}, \tag{134}$$

and for $a = a_2, \quad q \le -1$, the second term in Eq. (134) vanishes, and we have

$$u(t) = \phi(0) + \int_0^t d\tau \int_0^\tau d\theta \Big\langle \phi''(S(\theta)) \Big\rangle e^{-2\int_\theta^\tau a(s)ds}. \tag{135}$$

Thus, the job of proving Eq. (125) translates to proving Eqs.(134) and (135). Because ϕ is continuous it can be written as a power series $\phi(t) = \sum_{k=0}^{\infty} b_k t^k$ whose convergence is uniform, meaning for any small $\varepsilon > 0$, there is an integer M' such that for $M > M'$ we have $|\phi(t) - \sum_{k=0}^{M} b_k t^k| < \varepsilon$ for all values of t for which ϕ is continuous. The last statement is the Weierstrass approximation theorem.

We begin the proof of Eqs.(134) and (135) by writing $\phi(S(t))$ as a power series in S:

$$\phi(S(t)) = \sum_{k=0}^{\infty} b_k S^k(t) = \phi(0) + \phi'(0)S(t) + \sum_{k=2}^{\infty} b_k S^k(t). \tag{136}$$

Taking the average of the above equation, we have

$$\langle \phi(S(t)) \rangle = \left\langle \sum_{k=0}^{\infty} b_k S^k(t) \right\rangle = \phi(0) + \phi'(0)\langle S(t) \rangle + \sum_{k=2}^{\infty} b_k \Big\langle S^k(t) \Big\rangle. \tag{137}$$

Using Eq. (131), the average $\langle S(t) \rangle$ is given by

$$\langle S(t) \rangle = \int_0^t \Big\langle (-1)^{N(\tau)} \Big\rangle d\tau = \begin{cases} \int_0^t e^{-2\int_0^\tau a(s)ds} d\tau, & a = a_1; \quad a = a_2,\ q > -1; \\ 0, & a = a_2, \quad q \le -1,\ \gamma > 0. \end{cases} \tag{138}$$

For $k \ge 2$, we have

$$\langle S^k(t) \rangle = k! \int_0^t d\tau_k \int_0^{\tau_k} d\tau_{k-1} \cdots \int_0^{\tau_2} d\tau_1 \Big\langle (-1)^{N(\tau_1)+\cdots+N(\tau_k)} \Big\rangle,\ k \ge 0\,. \tag{139}$$

Since $\tau_1 < \tau_2 < \cdots < \tau_k$, it is useful to write the integrand as

$$\begin{aligned} \Big\langle (-1)^{N(\tau_1)+\cdots+N(\tau_k)} \Big\rangle &= \Big\langle (-1)^{N(\tau_1)+\cdots+N(\tau_{k-2})} (-1)^{2N(\tau_{k-1})} (-1)^{N(\tau_k)-N(\tau_{k-1})} \Big\rangle \\ &= \Big\langle (-1)^{N(\tau_1)+\cdots+N(\tau_{k-2})} (-1)^{N(\tau_k)-N(\tau_{k-1})} \Big\rangle. \end{aligned} \tag{140}$$

Since the number of reversals $N(\tau_k) - N(\tau_{k-1})$ during the time interval $\tau_k - \tau_{k-1}$ is independent of the numbers of reversals $N(\tau_1), \cdots, N(\tau_{k-1})$ of earlier times by the definition of the Poisson process, we can factor the above expectation as a product of two expectations, namely

$$\begin{aligned} &\Big\langle (-1)^{N(\tau_1)+\cdots+N(\tau_{k-2})} \Big\rangle \Big\langle (-1)^{N(\tau_k)-N(\tau_{k-1})} \Big\rangle \\ &= \Big\langle (-1)^{N(\tau_1)+\cdots+N(\tau_{k-2})} \Big\rangle \exp -2\int_{\tau_{k+1}}^{\tau_{k+2}} a(s)ds, \end{aligned}$$

which when substituted in Eq. (139) gives an iterative relation for the $\langle S^k(t)\rangle$ which after renaming τ_k as τ and τ_{k-1} as θ

$$\langle S^k(t)\rangle = k(k-1)\int_0^t d\tau \int_0^\tau d\theta \langle S^{k-2}(\theta)\rangle e^{-2\int_\theta^\tau a(s)ds}, \qquad k \geq 2. \tag{141}$$

Noting that

$$\sum_{k=2}^{\infty} b_k k(k-1)\langle S^{k-2}(\theta)\rangle = \langle \phi''(S(\theta))\rangle, \tag{142}$$

we obtain from Eq. (141)

$$\sum_{k=2}^{\infty} b_k \langle S^k(t)\rangle = \int_0^t d\tau \int_0^\tau d\theta \langle \phi''(S(\theta))\rangle e^{-2\int_\theta^\tau a(s)ds}. \tag{143}$$

Collecting Eqs.(137), (138) and (143), we have proved Eqs.(134) and (135), and hence Kaplan's result Eq. (125).

From Eq. (125) to Kaplan's generalization is just a step away. If $\phi(\mathbf{x},t)$ satisfies the wave equation without dissipation, Eq. (96), then the right hand side of Eq. (125) becomes

$$\langle \phi''(\mathbf{x},S(t))\rangle = v^2\langle O_\mathbf{x}\phi(\mathbf{x},S(t))\rangle = v^2 O_\mathbf{x}\langle \phi(\mathbf{x},S(t))\rangle = v^2 O_\mathbf{x} u(\mathbf{x},t), \tag{144}$$

and recalling Eq. (129) and Eq. (133), we obtain the telegrapher equation

$$\begin{aligned} u''(\mathbf{x},t) + 2a(t)u'(\mathbf{x},t) - v^2 O_\mathbf{x} u(\mathbf{x},t) &= 0 \\ u(\mathbf{x},0) &= \phi(\mathbf{x},0) \\ \lim_{t\to 0} u'(\mathbf{x},t) &= \begin{cases} \phi'(\mathbf{x},0), & a=a_1;\ a=a_2,\ q>-1; \\ 0, & a=a_2,\ q\leq -1,\ \gamma>0\,. \end{cases} \end{aligned} \tag{145}$$

Note that u is just $F(\mathbf{x},t)$.

10. CALCULATION SCHEME FOR THE DISTRIBUTIONS

If the distribution $g(a,t,r)$ is known, the solution $F(\mathbf{x},t)$ may be obtained as in Eq. (100). This section deduces from Eq. (125), a calculational scheme for obtaining $g(a,t,r)$.

If we choose $\phi(S(t)) = e^{zS(t)}$, and define $\eta(t,z) = \langle e^{zS(t)}\rangle$ (the moment generating function), then we obtain the following differential equation for η from Eq. (125),

$$\begin{aligned} \eta''(t,z) + 2a(t)\eta'(t,z) - z^2\eta(t,z) &= 0, \\ \eta(0,z) &= 1, \\ \lim_{t\to 0}\eta'(t,z) &= \begin{cases} 1, & a=a_1;\quad a=a_2,\quad q>-1\,; \\ 0, & a=a_2,\quad q\leq -1,\quad \gamma>0\,. \end{cases} \end{aligned} \tag{146}$$

Here the prime $'$ denotes differentiation with respect to the first argument t. This offers a way of calculating the distribution of $S(t)$: Writing

$$\eta(t,z) = \int_{-\infty}^{\infty} e^{zr} g(a,t,r)dr,$$

and then setting $z = -i\rho$, and applying an inverse Fourier transform to η, the distribution is given by the formula

$$g(a,t,r) = \frac{1}{2\pi}\int_{-\infty}^{\infty} e^{i\rho r}\eta(t,-i\rho)d\rho. \tag{147}$$

That is, a scheme for obtaining an explicit expression for $g(a,t,r)$ consists of two steps: solving the differential equation Eq. (146) and evaluating the integral Eq. (147). Both the steps may not be easy. In fact, Eq. (146) may be related to the Schrödinger equation with mass $m = 1/2$ and potential $V = a^2 + a'$, and there are few which are exactly soluble. This suggests the use of any of the approximation methods, for example WKB, developed for quantum mechanics when no exact solution is in sight.

Besides the present scheme and the Zastawniak method (Example 3, Section (5)), there are also other analytical methods of calculating the distribution $g(a,t,r)$ when a is *constant*, for example: Laplace transform[10, 11, 12] method, and checkerboard method[8]. (A discussion on how to extract the distribution from the checkerboard method is given in [12]). However, it appears that only the present scheme is convenient for the case $a(t)$, a function of time.

Using the scheme consisting of Eq. (146) and Eq. (147), we obtain $g(a,t,r)$ for three cases: the intensity a =constant, $a(t) = \gamma/t, t > 0$, and $a(t) = \gamma t$ [16, 19] where γ is a positive constant.

1) a=constant. The scheme gives the same answer as eq.(62).

2) $a(t) = \gamma/t, t > 0$.

$$g(t,r) = \begin{cases} \delta(r), & t = 0\,; \\ B(\frac{1}{2},\gamma)^{-1}t^{-1}\left(1-\frac{r^2}{t^2}\right)^{\gamma-1}, & |r| < t; \\ 0, & |r| > t\,, \end{cases} \tag{148}$$

where B denotes the Beta function.

3) $a(t) = \gamma t$.

$$g(t,r) = \frac{1}{2\pi}\exp(-\gamma t^2)\int_{-\infty}^{\infty} e^{i\rho r}\left[\Phi\left(\frac{1}{2}-\frac{\rho^2}{4\gamma},\frac{1}{2};\gamma t^2\right) - i\rho t\Phi\left(1-\frac{\rho^2}{4\gamma},\frac{3}{2};\gamma t^2\right)\right]d\rho, \tag{149}$$

where Φ is the degenerate (Kummer's) hypergeometric function.

As for the numerical simulation of the random walk to obtain $g(a,t,r)$, we refer the readers to [21, 22].

11. WAVEFORM RESTORATION

For sections 11, 12 and 13, a is assumed constant, and $g(a,t,r)$ will be written simply as $g(t,r)$.

11.1. Restorer No. 1

Since g is known, Eq. (100) may be paraphrased as: Knowing ϕ yields F. It is natural to ask if the converse, knowing F yields ϕ, is also true. It would be — if g had an inverse other than the trivial $a = 0$ case — for we could have constructed the would-have-been wave by inverting Eq. (100). (*After these lectures were delivered, I did find a conjecture for the* g^{-1}

under a certain non-restrictive assumption, see the next sub-section.) However, its symmetric and antisymmetric parts (up to a factor of 2), with respect to r, namely

$$h_1(t,r) \equiv g(t,r) + g(t,-r) = e^{-at}\left\{\delta(t-r) + \theta(t-r)a\left[I_0(au) + \frac{t}{u}I_1(au)\right]\right\} \tag{150}$$

and

$$h_2(t,r) \equiv g(t,r) - g(t,-r) = e^{-at}\left[\delta(t-r) + \theta(t-r)\frac{ar}{u}I_1(au)\right], \tag{151}$$

do possess inverses h_1^{-1} and h_2^{-1} for $r > 0$:

$$h_1^{-1}(t,\tau) = e^{a\tau}\left\{\delta(t-\tau) - \theta(t-\tau)\left[ae^{a\tau} - \frac{2}{\pi}\int_0^a dy \frac{\sin ty}{y}\left[w(y)\cosh(\tau w(y)) + a\sinh(\tau w(y))\right]\right]\right\}, \tag{152}$$

where $w(y) \equiv \sqrt{a^2 - y^2}$, and

$$h_2^{-1}(t,\tau) = e^{a\tau}\left[\delta(t-\tau) - \theta(t-\tau)\frac{a\tau}{\sqrt{t^2-\tau^2}}J_1\left(a\sqrt{t^2-\tau^2}\right)\right], \tag{153}$$

satisfying for each value of i,

$$\int_r^t d\tau h_i^{-1}(t,\tau)h_i(\tau,r) = \delta(t,r), \qquad r > 0. \tag{154}$$

The derivation of these inverses is rather long but interesting, and the readers are encouraged to consult [25] for details. (Incidently, the offer of 1000F at Cargese for evaluating the integral in Eq. (152) has not yet been claimed). Similar techniques are used in obtaining the conjecture Eq. (167). With these inverses, it is then possible to restore the ϕ provided additional requirements are satisfied, as explained below.

Now, denoting the symmetric part and anti-symmetric part of the distortions-free ϕ by $\phi_1(\mathbf{x},t) = [\phi(\mathbf{x},t) + \phi(\mathbf{x},-t)]/2$ and $\phi_2(\mathbf{x},t) = [\phi(\mathbf{x},t) - \phi(\mathbf{x},-t)]/2$ respectively, we observe that $F(\mathbf{x},t)$ may be written as

$$\begin{aligned} F(\mathbf{x},t) &= \int_{-t}^t \phi(\mathbf{x},r)g(t,r)\,dr = \int_{-t}^t [\phi_1(\mathbf{x},r) + \phi_2(\mathbf{x},r)]g(t,r)\,dr \\ &= \int_0^t [\phi_1(\mathbf{x},r) + \phi_2(\mathbf{x},r)]g(t,r)\,dr + \int_0^t [\phi_1(\mathbf{x},r) - \phi_2(\mathbf{x},r)]g(t,-r)\,dr \\ &= F_1(\mathbf{x},t) + F_2(\mathbf{x},t) \end{aligned} \tag{155}$$

where the component parts F_i are defined by

$$F_i(\mathbf{x},t) = \int_0^t dr\, h_i(t,r)\phi_i(\mathbf{x},r), \qquad i = 1,2. \tag{156}$$

These components are not the symmetric part and anti-symmetric part of $F(\mathbf{x},t)$. In fact, F_1 was the original Kac's solution, i.e. it satisfies the *same* equation (94) as F but with the Cauchy data: $F_1(\mathbf{x},0) = \kappa(\mathbf{x})$ and $\partial F_1/\partial t|_{t=0} = 0$. The identity of F_2 is immediately revealed since both F and F_1 satisfy Eq. (94), so must F_2, with $F_2(\mathbf{x},0)=0$ and $\partial F_2/\partial t|_{t=0}=\xi(\mathbf{x})$. Thus, given the Cauchy data $\kappa(\mathbf{x})$ and $\xi(\mathbf{x})$ for F are the same as given the Cauchy data for F_i, and we can choose to propagate F_i. It should also be noted that the Cauchy data for ϕ_i are the same as those for F_i.

By application of Eq. (154), and the fact that $h_i(\tau,r)=0$ for $\tau<r$, Eq. (156) may be inverted to yield ϕ_i

$$\phi_i(\mathbf{x},t)=\int_0^t d\tau\, h_i^{-1}(t,\tau)F_i(\mathbf{x},\tau), \tag{157}$$

and then by summing the ϕ_i yields ϕ:

$$\phi(\mathbf{x},t)=\phi_1(\mathbf{x},t)+\phi_2(\mathbf{x},t). \tag{158}$$

These facts imply the following strategy for the restoration of $\phi(\mathbf{x},t)$ from $F(\mathbf{x},t)$ at $\mathbf{x}$: 1) Propagate F_1 and F_2 separately. 2) From F_i recover ϕ_i by using h_i^{-1}. 3) Obtain ϕ by adding ϕ_i. This restorer will be called Restorer No. 1.

It is appropriate here to make a comparison between a conventional method which assume the knowledge of $F(\mathbf{x},t_1)$ and $\partial F/\partial t(\mathbf{x},t_1)$ over all space at some time t_1. From these, the functional form of F is defined, and hence the Cauchy data for ϕ (by setting $t=0$ in F), from which one then solves for $\phi(\mathbf{x},t)$ for all $\mathbf{x}$ and t. In contrast, the present scheme requires only knowing at *a* point in space the wave F_i received over time, and without the knowledge of them at other points, the wave that *would have been* received at *the* point in the absence of dissipation is then restored.

11.2. Restorer No. 2

Now we come to the conjecture mentioned at the beginning of this section. If $\phi(\mathbf{x},t)=0$ for $t<0$, Eq. (100) becomes

$$F(\mathbf{x},t)=\int_0^\infty \phi(\mathbf{x},r)g(t,r)\,dr. \tag{159}$$

Is there a g^{-1} such that for $r>0$,

$$\int_0^t d\tau g^{-1}(t,\tau)g(\tau,r)=\delta(t-r) \quad ? \tag{160}$$

Let

$$g^{-1}(t,\tau)=e^{a\tau}\left\{\delta(t-\tau)+\theta(t-\tau)\sum_{k=1}^{\infty}c_k(t,\tau)a^k\right\}. \tag{161}$$

Substituting the above ansatz, and the series representation of the modified Bessel functions in g into Eq. (160), the coefficients $c_k^i(t,\tau)$ can be determined, term by term, from the recurrence relations obtained by setting the coefficient of $a^k=0$ in the resulting equation. It is found that

$$c_1(t,\tau)=-\frac{1}{2}, \tag{162}$$

$$c_{2k+1}(t,\tau)=\frac{(-1)^{k+1}u^{2k}}{2^{2k+1}k!(k+1)!}\frac{[t+(2k+1)\tau]}{(t+\tau)}, \qquad k=1,2,3,4, \quad \text{and } 5 \tag{163}$$

and

$$c_{2k+2}(t,\tau)=\frac{-(-1)^{k-1}\tau u^{k-1}}{2^{2k-1}(k-1)!k!},\qquad k=0,1,2,3,4,\quad\text{and}\quad 5,\tag{164}$$

where $u=\sqrt{t^2-\tau^2}$. Conjecturing that this is true for all integers $k\geq 0$, and recalling the series representation for the Bessel function, we have

$$\begin{aligned}\sum_{k=1,3,5\cdots}c_k(t,\tau)a^k&=-\frac{a}{2}-\frac{at}{2(t+\tau)}\left[\frac{2}{au}J_1(au)-1\right]-\frac{a\tau}{2(t+\tau)}\left[2J_1'(au)-1\right]\\&=-\frac{t-\tau}{t+\tau}\frac{J_1(au)}{u}-\frac{a\tau}{t+\tau}J_o(au)\end{aligned}\tag{165}$$

and

$$\sum_{k=2,4,6\cdots}c_k(t,\tau)a^k=-\frac{a\tau}{u}J_1(au).\tag{166}$$

It follows from Eqs. (161), (165), (166) that the conjecture for g^{-1} is, with $u=\sqrt{t^2-\tau^2}$,

$$g^{-1}(t,\tau)=e^{a\tau}\left\{\delta(t-\tau)-\theta(t-\tau)\left[\left(\frac{t-\tau}{t+\tau}+a\tau\right)\frac{J_1(au)}{u}+\frac{a\tau}{t+\tau}J_o(au)\right]\right\}.\tag{167}$$

Hence, given $F(\mathbf{x},t)$ at $\mathbf{x}$, we conjecture that $\phi(\mathbf{x},t)$ can be restored as follows:

$$\phi(\mathbf{x},t)=\int_0^t g^{-1}(t,\tau)F(\mathbf{x},\tau)\,d\tau,\tag{168}$$

with $g^{-1}(t,\tau)$ given by Eq. (167) provided $\phi(\mathbf{x},t)=0$ for $t<0$. This restorer will be called Restorer No.2. To serve as a check on the conjecture, I considered two examples: 1)$\phi(x,t)=\theta(t-x)\sin(x-t)$ and 2)$\phi(x,t)=\theta(x+1-t)-\theta(x-t)$, and took $x=1$ to satisfy the requirement that $\phi(x,t)=0$ for $t<0$. Numerical evaluation of $\phi(x,t)$ using the formula Eq. (168) with g^{-1} given by Eq. (167) and $F(x,t)$ given by Eq. (159) gave excellent numerical results.

12. WAVEFORM DESIGNER

Restorer No. 1 can be turned to function as a waveform designer. We shall again limit our consideration to only one dimension, in which case the desired waveform can be *any* form. Suppose we desire a unit square pulse at an arbitrary but *fixed* point $x_1\geq 0$, namely $F(x_1,t)=\theta(t-x_1)-\theta(t-(x_1+1))$ that switches on at time $t=x_1$ and off at $t=x_1+1$, assuming unit velocity. How should the initial data $\kappa(x)$ and $\xi(x)$ be chosen?

The designer is assumed to have control over $\kappa(x)$ and $\xi(x)$. By choosing $\xi(x)=0$, we will be dealing with F_1 alone (F_2 vanishes.) From the d'Alembert solution Eq. (110), we see that in the absence of dissipation, the waveform

$$\phi_1(x,t)=\frac{1}{2}\Big(\kappa(x-vt)+\kappa(x+vt)\Big)\tag{169}$$

is simply $\kappa(x)$ propagating to the right and to the left without change in shape except for a reduction in amplitude by $1/2$. And in the presence of dissipation described by the parameter

a, $F_1(x_1,t)$ results. This means that by determining the would-have-been wave $\phi_1(x_1,t)$, for t in the range $[x_1,T)$ via Eq. (157), (Although T could be an infinitely long time in principle, it is some large time in practice), $\kappa(x)$ is determined by setting

$$\kappa(x_1-t)=2\theta(t-x_1)\phi_1(x_1,t). \tag{170}$$

For an illustrative example, the readers are referred to [25].

13. PREDICTING TELEGRAPH WAVEFORM

We are interested also in the question of predicting the telegraph waveforms at a desired point from the waveforms received over time at another point.

The question has been considered by J. A. Stratton[29] in the context of electromagnetic waves propagating in a homogeneous, isotropic, and unbounded conducting medium. But he started from the second order telegrapher equation, whereas we shall concentrate here on the first order version. Because of this difference in approach, a different method of predicting the waveform was found [28].

The method employed below is apparently far from the path integral method. This section is motivated by the physical problem at hand, and is included for completeness.

By taking the sum and difference of the two equations in Eq. (18), we have

$$\frac{\partial}{\partial t}\begin{pmatrix}\Psi_1\\ \Psi_2\end{pmatrix}=\begin{pmatrix}0 & \partial_x\\ \partial_x & -2a\end{pmatrix}\begin{pmatrix}\Psi_1\\ \Psi_2\end{pmatrix}. \tag{171}$$

where the notation Ψ_i are used to emphasize that they may represent quantities other than densities. Examples of Ψ_1 are the electric field E and $\rho_-+\rho_+$. Examples of Ψ_2 are the magnetic field B and $\rho_- - \rho_+$.

Let $\psi(x,t)\equiv e^{at}\Psi(x,t)$. Then Eq. (171) becomes

$$i\partial_x\psi=H\psi \tag{172}$$

where

$$H=i\sigma_1\partial_t-a\sigma_2, \tag{173}$$

and the σ_i are the Pauli spin matrices

$$\sigma_1=\begin{pmatrix}0 & 1\\ 1 & 0\end{pmatrix},\quad \sigma_2=\begin{pmatrix}0 & -i\\ i & 0\end{pmatrix},\quad \text{and}\quad \sigma_3=\begin{pmatrix}1 & 0\\ 0 & -1\end{pmatrix}. \tag{174}$$

Hence, the formal solution is given by

$$\psi(x,t)=e^{-i(x-x_0)H}\psi(x_0,t). \tag{175}$$

H has a complete and orthogonal set of eigenvectors $|\omega,\sigma\rangle$ whose representation in $|t\rangle$ basis is

$$\langle t|\omega,\sigma\rangle=\phi_{\omega,\sigma}(t)=\frac{1}{2\sqrt{\pi\sigma\rho}}\begin{pmatrix}-\omega+ia\\ \sigma\rho\end{pmatrix}e^{i\omega t} \tag{176}$$

with eigenvalues $\sigma\rho$ where $\sigma=-1,1$, and $\rho=\sqrt{\omega^2+a^2}$. If we expand $\psi(x_0,t)$ in terms of these eigenfunctions, namely

$$\psi(x_0,t)=\sum_\sigma\int_{-\infty}^{\infty}d\omega\, c_{\omega,\sigma}(x_0)\phi_{\omega,\sigma}(t), \tag{177}$$

where, with $\phi^\dagger$ denoting the adjoint of ϕ,

$$c_{\omega,\sigma}(x_0) = \int_{-\infty}^{\infty} dt\, \phi^\dagger_{\omega,\sigma}(t)\psi(x_0,t), \tag{178}$$

Eq. (175) becomes

$$\psi(x,t) = \sum_\sigma \int_{-\infty}^{\infty} d\omega\, e^{-i(x-x_0)\sigma p} c_{\omega,\sigma}(x_0)\phi_{\omega,\sigma}(t), \tag{179}$$

and inserting the expression (178) for $c_{\omega,\sigma}(x_0)$, the dependence of $\psi(x,t)$ on $\psi(x_0,t_0)$ for $-\infty < t_0 < \infty$ is made explicit:

$$\psi(x,t) = \int_{-\infty}^{\infty} dt_0 K(x-x_0,t-t_0)\psi(x_0,t_0), \tag{180}$$

where the kernel

$$K(x-x_0,t-t_0) = \int_{-\infty}^{\infty} d\omega \sum_{\sigma=\pm 1} e^{-i(x-x_0)\sigma p}\phi_{\omega,\sigma}(t)\phi^\dagger_{\omega,\sigma}(t_0). \tag{181}$$

Recalling that $\psi(x,t) \equiv e^{at}\Psi(x,t)$, Eq. (180) becomes

$$\Psi(x,t) = e^{-at}\int_{-\infty}^{\infty} dt_0 K(x-x_0,t-t_0)e^{at_0}\Psi(x_0,t_0). \tag{182}$$

Eq. (182) is the master formula, whose explicit form is given in Eq. (191). It says that knowing the waveforms Ψ_1 and Ψ_2 at x_0, then Ψ_1 and Ψ_2 at x is known, or in other words, an observer at x_0 can predict the waveform at x. In Stratton's method, Ψ_1 and $\partial\Psi_1/\partial x$ at x_0 are needed in order to predict Ψ_1 at x.

The kernel given by Eq. (181) can be explicitly evaluated as follows: Performing the sum, substituting for the eigenfunctions, and renaming $t-t'$ as t and $x-x_0$ as x, we obtain

$$\begin{aligned}
&K(x,t) \\
&= \frac{1}{4\pi}\int_{-\infty}^{\infty} d\omega \frac{1}{p}\left[\begin{pmatrix} p & -\omega+ia \\ -(\omega+ia) & p \end{pmatrix} e^{i(\omega t - px)} + \begin{pmatrix} p & \omega-ia \\ (\omega+ia) & p \end{pmatrix} e^{i(\omega t + px)}\right] \\
&= \frac{1}{4\pi}\int_{-\infty}^{\infty} d\omega \frac{1}{p}\left[\begin{pmatrix} i\partial_x & i\partial_t+ia \\ i\partial_t-ia & i\partial_x \end{pmatrix} e^{i(\omega t - px)} + \begin{pmatrix} -i\partial_x & -i\partial_t-ia \\ -i\partial_t+ia & -i\partial_x \end{pmatrix} e^{i(\omega t + px)}\right] \\
&= \frac{i}{4\pi}\begin{pmatrix} \partial_x & \partial_t+a \\ \partial_t-a & \partial_x \end{pmatrix}\int_{-\infty}^{\infty} d\omega \frac{e^{-ipx}-e^{ipx}}{p} e^{i\omega t} \\
&= \frac{1}{\pi}\begin{pmatrix} \partial_x & \partial_t+a \\ \partial_t-a & \partial_x \end{pmatrix}\int_0^{\infty} d\omega \frac{\sin(px)}{p}\cos(\omega t).
\end{aligned} \tag{183}$$

The last integral is given by [38]

$$\int_0^{\infty} d\omega \frac{\sin(px)}{p}\cos(\omega t) = \begin{cases} \frac{\pi}{2}J_0(a\sqrt{x^2-t^2}), & x>t>0 \\ 0 & t>x>0 \end{cases}, \tag{184}$$

which can be rewritten as

$$\int_0^{\infty} d\omega \frac{\sin(px)}{p}\cos(\omega t) = \frac{\pi}{2}J(x,t), \text{ for all } x \text{ and } t, \tag{185}$$

where

$$J(x,t) = \varepsilon(x)\theta\Big(|x|-|t|\Big)J_0\Big(a\sqrt{x^2-t^2}\Big) \tag{186}$$

and

$$\varepsilon(x) = \begin{cases} 1, & x>0 \\ -1, & x<0. \end{cases} \tag{187}$$

Hence,

$$K(x,t) = \frac{1}{2}\begin{pmatrix} \partial_x & \partial_t + a \\ \partial_t - a & \partial_x \end{pmatrix} J(x,t). \tag{188}$$

Within the integral of Eq. (180), $\partial_x J(x,t)$ and $\partial_t J(x,t)$ of Eq. (188) can be shown to be effectively given by

$$\partial_x J(x,t) = \frac{1}{2}\left[\delta(|x|-|t|) - \theta(|x|-|t|)\frac{a|x|}{\sqrt{x^2-t^2}}J_1\left(a\sqrt{x^2-t^2}\right)\right] \tag{189}$$

and

$$\partial_t J(x,t) = -\frac{1}{2}\left[\varepsilon(t)\varepsilon(x)\delta(|x|-|t|) - \varepsilon(x)\theta(|x|-|t|)\frac{at}{\sqrt{x^2-t^2}}J_1\left(a\sqrt{x^2-t^2}\right)\right]. \tag{190}$$

Substituting these expressions for $\partial_x J$ and $\partial_t J$, Eq. (182) can be simplified to, with the notation $s \equiv x - x_0$, and $u \equiv \sqrt{s^2 - t_0^2}$,

$$\begin{aligned}\Psi_{1,2}(x,t) = {} & \frac{1}{2}\left[e^{a|s|}\Psi_{1,2}(x_0,t+|s|) + e^{-a|s|}\Psi_{1,2}(x_0,t-|s|)\right] \\ & + \frac{1}{2}\varepsilon(s)\left[e^{a|s|}\Psi_{2,1}(x_0,t+|s|) - e^{-a|s|}\Psi_{2,1}(x_0,t-|s|)\right] \\ & - \frac{a|s|}{2}\int_{-|s|}^{|s|} dt_0 \frac{J_1(au)}{u} e^{at_0}\Psi_{1,2}(x_0,t+t_0) \\ & \pm \frac{a}{2}\varepsilon(s)\int_{-|s|}^{|s|} dt_0 \left[J_0(au) \mp \frac{t_0 J_1(au)}{u}\right] e^{at_0}\Psi_{2,1}(x_0,t+t_0). \end{aligned} \tag{191}$$

We consider here several examples of the applications of Eq. (191).

1) Let

$$\Psi_1(x_0,t) = \delta(t), \quad \text{and} \quad \Psi_2(x_0,t) = 0.$$

The integrals in Eq. (191) can be easily evaluated, and we obtain, with $u \equiv \sqrt{s^2-t^2}$,

$$\Psi_1(x,t) = \frac{1}{2}\left\{\left[e^{a|s|}\delta(t+|s|) + e^{-a|s|}\delta(t-|s|)\right] - \theta(|s|-|t|)a|s|e^{-at}\frac{J_1(au)}{u}\right\} \tag{192}$$

and

$$\Psi_2(x,t) = \frac{1}{2}\left\{\left[e^{as}\delta(t+s) - e^{-as}\delta(t-s)\right] - \theta(|s|-|t|)a\varepsilon(s)e^{-at}\left[J_0(au) - \frac{tJ_1(au)}{u}\right]\right\}. \tag{193}$$

2) Let

$$\Psi_1(x_0,t) = e^{-at}\sin t, \quad \text{and} \quad \Psi_2(x_0,t) = e^{-at}\cos t.$$

We obtain, with $b = \sqrt{(a^2+1)}$,

$$\begin{pmatrix} \Psi_1(x,t) \\ \Psi_2(x,t) \end{pmatrix} = \begin{pmatrix} \cos(bs) - \frac{1}{b}\sin(bs) & \frac{a}{b}\sin(bs) \\ -\frac{a}{b}\sin(bs) & \cos(bs) + \frac{1}{b}\sin(bs) \end{pmatrix} \begin{pmatrix} \Psi_1(x_0,t) \\ \Psi_2(x_0,t) \end{pmatrix}. \quad (194)$$

Suppose $a << 1$, and $s << 1$, and we neglect terms of order a^2 and s^2, then

$$\begin{pmatrix} \Psi_1(x,t) \\ \Psi_2(x,t) \end{pmatrix} = \begin{pmatrix} 1-s & as \\ -as & 1+s \end{pmatrix} \begin{pmatrix} \Psi_1(x_0,t) \\ \Psi_2(x_0,t) \end{pmatrix}. \quad (195)$$

3) Let

$$\Psi_1(0,t) = e^{-at}\theta(t)\sin t, \text{ and } \Psi_2(0,t) = e^{-at}\theta(t)\cos t.$$

By numerical integration of Eq. (191), we obtain the waveforms $\Psi_i(1,t)$ given in Fig. 6 and Fig. 7 below:

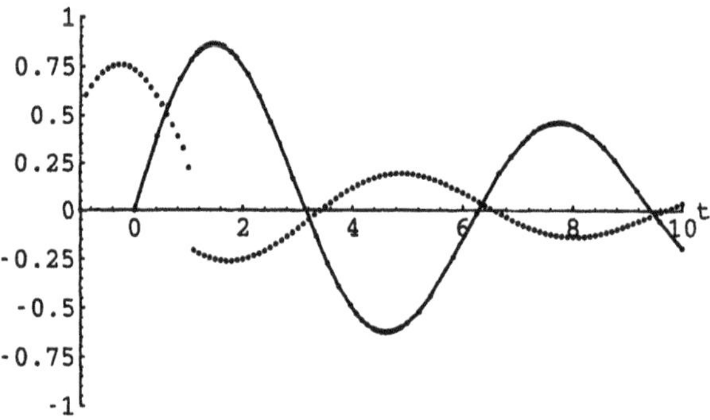

Figure 6. $\Psi_1(1,t)$ (dotted line) for $a = 0.1$. Solid line shows $\Psi_1(0,t)$.

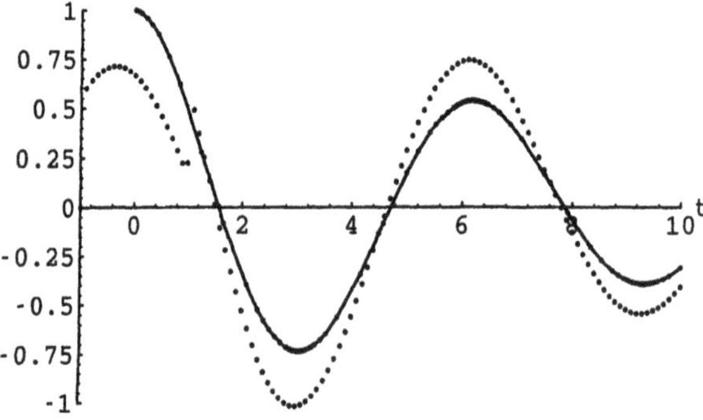

Figure 7. $\Psi_2(1,t)$ (dotted line) for $a = 0.1$. Solid line shows $\Psi_2(0,t)$.

The discontinuities of the dotted line graphs are caused by the arrival of waves from the other side of x_0.

ACKNOWLEDGMENTS

I thank the organizers of this stimulating institute for the kind invitation. My interest on this subject was initiated by Prof. Cécile DeWitt-Morette, and her continued interest in it has helped to motivate the current work. I also thank my other collaborators on the subject Bira van Kolck, and F. Nakamura, and especially S. Kanno for fruitful collaborations, and E. Ishiguro for the interest in the experimental aspects. Over the years, I have benefitted from discussions or comments on the subject from many people, among them are M. Agu, K. Bichteler, S. P. Chia, J. DeLyra, P. Enders, T. Fujiwara, T. Gallivan, R. Hersh T. Ichinose, T. Jacobson, K. Kakazu, K. Kitahara, M. Koyanagi, H. Leydolt, Y. Lobanov, S. Matsumoto, D. Mugnai, H. Nishimori, E. Orsingher, A. Ranfagni, L. Schulman, M. Seino, T. Y. Tou, G. Weiss, D. Wenger, K. M. Yoo, and T. Zastawniak. The interest and comments by the lecturers and participants of the institute especially P. Cartier, C. DeWitt-Morette, B. DeWitt, S. Fetodov, L. Kauffman, and J. Tomazelli were inspiring. I thank Prof. M. Koyanagi for proofreading part of the manuscript and for suggestions on improving the English.

This work is currently supported by a Grant-in-Aid for Scientific Research (C) (No. 08640510) from The Ministry of Education, Science, Sports, and Culture.

REFERENCES

[1] M. Kac, "Some stochastic problems in physics and mathematics," colloquium lectures in the pure and applied sciences, No. 2, Magnolia Petroleum Co. and Socony Mobil Oil Co. (1956), Hectographed. (Translations in Polish and Russian are available in book form.); Rocky Mountain J. of Math. 4, 497–509 (1974)

[2] S. Goldstein, "On diffusion by discontinuous movements, and on the telegraph equation," Quant. J. Mech. Appl. Math. 4, 129–156 (1951)

[3] S. Kaplan, "Differential equations in which the Poisson process plays a role," Bull. Am. Math. Soc. **70**, 264–268(1964)

[4] R. P. Feynman and A. R. Hibbs, *Quantum Mechanics and Path Integrals* (McGraw-Hill, 1965)

[5] R. Hersh, "Random evolutions: A survey of results and problems," Rocky Mountain J. of Math. 4 (1974), 443–477; and "Stochastic solutions of hyperbolic equations," in Partial Differential Equations and Related Topics, eds. A. Dold and B. Eckmann, Lecture Notes in Mathematics #446 (Springer-Verlag, Berlin, 1975), p. 283–300

[6] B. Gaveau, T. Jacobson, M. Kac, and L. S. Schulman, "Relativistic Extension of the Analogy between Quantum Mechanics and Brownian Motion," Phys. Rev. Lett. **53**, 419–422 (1984)

[7] T. Ichinose, "Path Integral for the Dirac Equation in two space-time dimensions," Proc. of the Japan Academy, **58**, Ser. A, No. 7 (1982)

[8] T. Jacobson and L. S. Schulman, "Quantum Stochastics: The passage from a relativistic to a non-relativistic path integral," J. Phys. A 17, 375 (1984)

[9] S. N. Ethier and T. G. Kurtz, Markov Processes Characterization and Convergence (John Wiley and Sons, New York, 1986), p. 468 – 491

[10] C. DeWitt-Morette and S. K. Foong,"Path integral solutions of wave equations with dissipation," Phys. Rev. Lett. 62, 2201–2204 (1989)

[11] C. DeWitt-Morette and S. K. Foong, "Kac's solution of the telegrapher equation, revisited: Part I," in Developments in General Relativity, Astrophysics and Quantum Theory, A Jubilee Volume in Honour of Nathan Rosen, eds. F. I. Cooperstock et. al. (I. O. P. Publishing, Bristol, 1990), p.351–366.

[12] S. K. Foong, "Kac's solution of the telegrapher equation, revisited: Part II," in Developments in General Relativity, Astrophysics and Quantum Theory, A Jubilee Volume in Honour of Nathan Rosen, eds. F. I. Cooperstock et. al. (I. O. P. Publishing, Bristol, 1990), p.351–366.

[13] E. Orsingher, " Probability law, flow function, maximum distribution of wave-governed random motions and their connection with Kirchoff's laws," Stochastic Process. Appl. 34, 49–66 (1990)

[14] B. Gaveau and L. S. Schulman, "Charged polymer in an electric field," Phys. Rev. A 42, 3470–3475 (1990)

[15] T. Nakamura, "A nonstandard representation of Feynman's path integrals," J. Math. Phys. 32(2), 457–463 (1991)

[16] S. K. Foong and U. van Kolck, "Poisson random walk for solving wave equations," Prog. Theor. Phys. **87**, 285–292(1992)

[17] D. Mugnai, A. Ranfagni, R. Ruggeri and A. Agresti, "Path integral solution of the telegrapher equation: An application to tunneling time determination," Phys. Rev. Lett. **68**, 259–262(1992)
[18] D. G. C. McKeon and G. N. Ord, "Time Reversal in Stochastic Processes and the Dirac Equation," Phys. Rev. Lett. **69**, 3–4(1992)
[19] S. K. Foong, "Path integral solution of the telegrapher equation," in *Lectures on Path Integration: Trieste 1991 Adriatico Research Conference, Trieste, Italy, 26 August–6 September 1991*, edited by H. Cerdeira et. al., 427–449 (World Scientific, Singapore, 1993)
[20] T. Zastawniak, "Evaluation of the distribution of the random variable $S(t) = \int_0^t (-1)^{N(u)} du$", in *Lectures on Path Integration: Trieste 1991 Adriatico Research Conference, Trieste, Italy, 26 August–6 September 1991*, edited by H. Cerdeira et. al., 446–448 (World Scientific, Singapore, 1993) (Appendix to Ref. [19])
[21] P. Enders and D. D. Cogan, "The correlated random walk as a general discrete model and as numerical routine," School of Information Systems, University of East Anglia, England, preprint
[22] S. K. Foong and F. E. Nakamura, "Path integral solutions of telegrapher's equation: Numerical approach," in *Path Integral from meV to MeV: Tutzing '92* (Proceedings of the 4th International Conference on Path Integrals from meV to MeV, Tutzing 1992), edited by H. Grabert, et. al., 268–275(World Scientific, 1993)
[23] J. Masoliver, J. M. Porra, G. H. Weiss,"Solutions of the telegrapher's equation in the presence of reflecting and partly reflecting boundaries," Phys. Rev. E, **48**, 939–944 (1993)
[24] S. K. Foong and S. Kanno, "Properties of the telegrapher's random process with or without a trap," Stochastic Processes and their Appl., **53**, p.147–173 (1994)
[25] S. K. Foong,"Overcoming dissipative distortions by a waveform restorer and designer," J. Math. Phys.(Special Issue on Functional Integration), **36**, p.2324–2340 (1995)
[26] C. DeWitt-Morette and P. Cartier, "A new perspective on functional integration" Path Integrals: Dubna '96 (Proceedings of the Dubna Joint Meeting of Int. Seminar "Path Integrals: Theory& Applications" and "5th Int. Conf. "Path Integrals from meV to MeV"), Edited by V. S. Yarunin and M. A. Smondyrev, (JINR, Dubna, 96.5.27–31)
[27] L. H. Kauffman and H. P. Noyes, "Discrete physics and the Dirac equation," To appear in Physics Letters A.
[28] S. Kanno and S. K. Foong, "Predicting telegraph waveform," presentation at the Inaugural Conference of APCTP (Asia Pacific Center for Theoretical Physics) (Seoul, 96.6.4–10)
[29] J. A. Stratton, Electromagnetic Theory (McGraw Hill, 1941)
[30] D. D. Joseph and L. Preziosi, "Heat waves," Rev. Mod. Phys. **61**, 41–73 (1989); D. D. Joseph and L. Preziosi, "Addendum to the paper 'Heat waves'," Rev. Mod. Phys. **62**, 375–391 (1990)
[31] W. Band, *Introduction to Mathematical Physics* (D. Van Nostrand Company, Inc. 1959)
[32] S. Chandrasekhar, "Stochastic problems in physics and astronomy," Rev. Mod. Phys., 1–89 (1943). Reprinted in *Selected papers on Noise and Stochastic Processes*, Ed. N. Wax, p.3–91 (Dover, 1954)
[33] L. S. Schulman, Techniques and Applications of Path Integration (John Wiley, New York, 1981)
[34] K. L. Chung, Elementary Probability Theory with Stochastic Processes, 3rd Ed. (Springer-Verlag, New York, 1979)
[35] Mathematica, Wolfram Research, Inc. (1988)
[36] D. Zwillinger, Handbook of Differential Equations(Academic Press, 1989)
[37] P. M. Morse, and H. Feshbach, *Methods of Theoretical Physics* (McGraw-Hill, 1953)
[38] I. S. Gradshteyn and I. M. Ryzhik, Tables of Integrals, Series and Products (Academic Press, 1980).

MONTE CARLO METHODS IN STATISTICAL MECHANICS: FOUNDATIONS AND NEW ALGORITHMS

A. Sokal

Department of Physics
New York University
4 Washington Place
New York, NY
E-mail: SOKAL@NYU.EDU

These notes are an updated version of lectures given at the Cours de Troisième Cycle de la Physique en Suisse Romande (Lausanne, Switzerland) in June 1989. We thank the Troisième Cycle de la Physique en Suisse Romande and Professor Michel Droz for kindly giving permission to reprint these notes.

NOTE TO THE READER

The following notes are based on my course "Monte Carlo Methods in Statistical Mechanics: Foundations and New Algorithms" given at the Cours de Troisième Cycle de la Physique en Suisse Romande (Lausanne, Switzerland) in June 1989, and on my course "Multi-Grid Monte Carlo for Lattice Field Theories" given at the Winter College on Multilevel Techniques in Computational Physics (Trieste, Italy) in January–February 1991.

The reader is warned that some of this material is out-of-date (this is particularly true as regards reports of numerical work). For lack of time, I have made no attempt to update the text, but I have added footnotes marked "**Note Added 1996**" that correct a few errors and give additional bibliography.

My first two lectures at Cargèse 1996 were based on the material included here. My third lecture described the new finite-size-scaling extrapolation method of [97, 98, 99, 100, 101, 102, 103].

1. INTRODUCTION

The goal of these lectures is to give an introduction to current research on Monte Carlo methods in statistical mechanics and quantum field theory, with an emphasis on:

1) the conceptual foundations of the method, including the possible dangers and misuses, and the correct use of statistical error analysis; and

2) new Monte Carlo algorithms for problems in critical phenomena and quantum field theory, aimed at reducing or eliminating the "critical slowing-down" found in conventional algorithms.

These lectures are aimed at a mixed audience of theoretical, computational and mathematical physicists — some of whom are currently doing or want to do Monte Carlo studies themselves, others of whom want to be able to evaluate the reliability of published Monte Carlo work.

Before embarking on 9 hours of lectures on Monte Carlo methods, let me offer a warning:

Monte Carlo is an extremely bad *method; it should be used only when all alternative methods are worse.*

Why is this so? Firstly, *all* numerical methods are potentially dangerous, compared to analytic methods; there are more ways to make mistakes. Secondly, as numerical methods go, Monte Carlo is one of the least efficient; it should be used only on those intractable problems for which all other numerical methods are even *less* efficient.

Let me be more precise about this latter point. Virtually all Monte Carlo methods have the property that the statistical error behaves as

$$\text{error} \sim \frac{1}{\sqrt{\text{computational budget}}}$$

(or worse); this is an essentially universal consequence of the central limit theorem. It may be possible to improve the *proportionality constant* in this relation by a factor of 10^6 or more — this is one of the principal subjects of these lectures — but the overall $1/\sqrt{n}$ behavior is inescapable. This should be contrasted with traditional deterministic numerical methods whose rate of convergence is typically something like $1/n^4$ or e^{-n} or e^{-2^n}. Therefore, Monte Carlo methods should be used only on those extremely difficult problems in which all alternative numerical methods behave even *worse* than $1/\sqrt{n}$.

Consider, for example, the problem of numerical integration in d dimensions, and let us compare Monte Carlo integration with a traditional deterministic method such as Simpson's rule. As is well known, the error in Simpson's rule with n nodal points behaves asymptotically as $n^{-4/d}$ (for smooth integrands). In low dimension ($d < 8$) this is much better than Monte Carlo integration, but in high dimension ($d > 8$) it is much worse. So it is not surprising that Monte Carlo is the method of choice for performing high-dimensional integrals. It is still a bad method: with an error proportional to $n^{-1/2}$, it is difficult to achieve more than 4 or 5 digits accuracy. But numerical integration in high dimension is very difficult; though Monte Carlo is bad, all other known methods are worse.*

In summary, Monte Carlo methods should be used only when neither analytic methods nor deterministic numerical methods are workable (or efficient). One general domain of application of Monte Carlo methods will be, therefore, to systems with *many degrees of*

*This discussion of numerical integration is grossly oversimplified. Firstly, there are deterministic methods better than Simpson's rule; and there are also sophisticated Monte Carlo methods whose asymptotic behavior (on smooth integrands) behaves as n^{-p} with p strictly greater than 1/2[1, 2]. Secondly, for all these algorithms (except standard Monte Carlo), the asymptotic behavior as $n \to \infty$ may be irrelevant in practice, because it is achieved only at ridiculously large values of n. For example, to carry out Simpson's rule with even 10 nodes per axis (a very coarse mesh) requires $n = 10^d$, which is unachievable for $d \gtrsim 10$.

freedom, far from the perturbative regime. But such systems are precisely the ones of greatest interest in statistical mechanics and quantum field theory!

It is appropriate to close this introduction with a general methodological observation, ably articulated by Wood and Erpenbeck [3]:

> . . . these [Monte Carlo] investigations share some of the features of ordinary experimental work, in that they are susceptible to both statistical and systematic errors. With regard to these matters, we believe that papers should meet much the same standards as are normally required for experimental investigations. We have in mind the inclusion of estimates of statistical error, descriptions of experimental conditions (i.e., parameters of the calculation), relevant details of apparatus (program) design, comparisons with previous investigations, discussion of systematic errors, etc. Only if these are provided will the results be trustworthy guides to improved theoretical understanding.

2. DYNAMIC MONTE CARLO METHODS: GENERAL THEORY

All Monte Carlo work has the same general structure: given some probability measure π on some configuration space S, we wish to generate many random samples from π. How is this to be done?

Monte Carlo methods can be classified as *static* or *dynamic*. Static methods are those that generate a sequence of *statistically independent* samples from the desired probability distribution π. These techniques are widely used in Monte Carlo numerical integration in spaces of not-too-high dimension [2]. But they are unfeasible for most applications in statistical physics and quantum field theory, where π is the Gibbs measure of some rather complicated system (extremely many coupled degrees of freedom).

The idea of *dynamic* Monte Carlo methods is to invent a *stochastic process* with state space S having π as its unique equilibrium distribution. We then simulate this stochastic process on the computer, starting from an arbitrary initial configuration; once the system has reached equilibrium, we measure time averages, which converge (as the run time tends to infinity) to π-averages. In physical terms, we are inventing a *stochastic time evolution* for the given system. Let us emphasize, however, that this time evolution *need not correspond to any real "physical" dynamics*: rather, the dynamics is simply a numerical algorithm, and it is to be chosen, like all numerical algorithms, on the basis of its computational efficiency.

In practice, the stochastic process is always taken to be a *Markov process*. So our first order of business is to review some of the general theory of Markov chains.* For simplicity let us assume that the state space S is discrete (i.e., finite or countably infinite). Much of the theory for general state space can be guessed from the discrete theory by making the obvious replacements (sums by integrals, matrices by kernels), although the proofs tend to be considerably harder.

Loosely speaking, a Markov chain (= discrete-time Markov process) with state space S is a sequence of S-valued random variables $X_0, X_1, X_2, \ldots$ such that successive transitions $X_t \to X_{t+1}$ are statistically independent ("the future depends on the past only through the present"). More precisely, a Markov chain is specified by two ingredients:

*The books of Kemeny and Snell [4] and Iosifescu [5] are excellent references on the theory of Markov chains with *finite* state space. At a somewhat higher mathematical level, the books of Chung [6] and Nummelin [7] deal with the cases of *countable* and *general* state space, respectively.

- The *initial distribution* α. Here α is a probability distribution on S, and the process will be defined so that $\mathbb{P}(X_0 = x) = \alpha_x$.

- The *transition probability matrix* $P = \{p_{xy}\}_{x,y \in S} = \{p(x \to y)\}_{x,y \in S}$. Here P is a matrix satisfying $p_{xy} \geq 0$ for all x,y and $\sum_y p_{xy} = 1$ for all x. The process will be defined so that $\mathbb{P}(X_{t+1} = y | X_t = x) = p_{xy}$.

The Markov chain is then completely specified by the joint probabilities

$$\mathbb{P}(X_0 = x_0, X_1 = x_1, X_2 = x_2, \ldots, X_n = x_n) \equiv \alpha_{x_0} p_{x_0 x_1} p_{x_1 x_2} \cdots p_{x_{n-1} x_n}. \tag{2.1}$$

This product structure expresses the fact that "successive transitions are independent."

Next we define the *n-step transition probabilities*

$$p_{xy}^{(n)} = \mathbb{P}(X_{t+n} = y | X_t = x). \tag{2.2}$$

Clearly $p_{xy}^{(0)} = \delta_{xy}$, $p_{xy}^{(1)} = p_{xy}$, and in general $\{p_{xy}^{(n)}\}$ are the matrix elements of P^n.

A Markov chain is said to be *irreducible* if from each state it is possible to get to each other state: that is, for each pair $x,y \in S$, there exists an $n \geq 0$ for which $p_{xy}^{(n)} > 0$. We shall be considered almost exclusively with irreducible Markov chains.

For each state x, we define the *period of* x (denoted d_x) to be the greatest common divisor of the numbers $n > 0$ for which $p_{xx}^{(n)} > 0$. If $d_x = 1$, the state x is called *aperiodic*. It can be shown that, in an irreducible chain, all states have the same period; so we can speak of the chain having period d. Moreover, the state space can then be partitioned into subsets $S_1, S_2, \ldots, S_d$ around which the chain moves cyclically, i.e., $p_{xy}^{(n)} = 0$ whenever $x \in S_i$, $y \in S_j$ with $j - i \neq n \pmod{d}$. Finally, it can be shown that a chain is *irreducible and aperiodic* if and only if, for each pair x,y, there exists N_{xy} such that $p_{xy}^{(n)} > 0$ for *all* $n \geq N_{xy}$.

We now come to the fundamental topic in the theory of Markov chains, which is the problem of convergence to equilibrium. A probability measure $\pi = \{\pi_x\}_{x \in S}$ is called a *stationary distribution* (or *invariant distribution* or *equilibrium distribution*) for the Markov chain P in case

$$\sum_x \pi_x p_{xy} = \pi_y \qquad \text{for all } y. \tag{2.3}$$

A stationary probability distribution need not exist; but if it does, then a lot more follows:

Theorem 1 *Let P be the transition probability matrix of an* irreducible *Markov chain of period d. If a stationary probability distribution π exists, then it is unique, and $\pi_x > 0$ for all x. Moreover,*

$$\lim_{n \to \infty} p_{xy}^{(nd+r)} = \begin{cases} d\pi_y & \text{if } x \in S_i,\ y \in S_j \text{ with } j - i = r \pmod{d} \\ 0 & \text{if } x \in S_i,\ y \in S_j \text{ with } j - i \neq r \pmod{d} \end{cases} \tag{2.4}$$

for all x,y. In particular, if P is aperiodic, then

$$\lim_{n \to \infty} p_{xy}^{(n)} = \pi_y. \tag{2.5}$$

This theorem shows that the Markov chain converges as $t \longrightarrow \infty$ to the equilibrium distribution π, irrespective of the initial distribution α. Moreover, under the conditions of this theorem much more can be proven — for example, a strong law of large numbers, a central limit theorem, and a law of the iterated logarithm. For statements and proofs of all these theorems, we refer the reader to Chung [6].

We can now see how to set up a dynamic Monte Carlo method for generating samples from the probability distribution π. It suffices to invent a transition probability matrix $P = \{p_{xy}\} = \{p(x \longrightarrow y)\}$ satisfying the following two conditions:

(A) *Irreducibility.** For each pair $x,y \in S$, there exists an $n \geq 0$ for which $p_{xy}^{(n)} > 0$.

(B) *Stationarity of π.* For each $y \in S$,

$$\sum_x \pi_x p_{xy} = \pi_y. \tag{2.6}$$

Then Theorem 1 (together with its more precise counterparts) shows that simulation of the Markov chain P constitutes a legitimate Monte Carlo method for estimating averages with respect to π. We can start the system in any state x, and the system is guaranteed to converge to equilibrium as $t \longrightarrow \infty$ [at least in the averaged sense (2.4)]. Long-time averages of any observable f will converge with probability 1 to π-averages (strong law of large numbers), and will do so with fluctuations of size $\sim n^{-1/2}$ (central limit theorem). In practice we will discard the data from the initial transient, i.e., before the system has come close to equilibrium, but in principle this is not necessary (the bias is of order n^{-1}, hence asymptotically much smaller than the statistical fluctuations).

So far, so good! But while this is a *correct* Monte Carlo algorithm, it may or may not be an *efficient* one. The key difficulty is that the successive states $X_0, X_1, \ldots$ of the Markov chain are *correlated* — perhaps very strongly — so the variance of estimates produced from the dynamic Monte Carlo simulation may be much higher than in static Monte Carlo (independent sampling). To make this precise, let $f = \{f(x)\}_{x \in S}$ be a real-valued function defined on the state space S (i.e., a real-valued observable) that is square-integrable with respect to π. Now consider the *stationary* Markov chain (i.e., start the system in the stationary distribution π, or equivalently, "equilibrate" it for a very long time prior to observing the system). Then $\{f_t\} \equiv \{f(X_t)\}$ is a stationary stochastic process with mean

$$\mu_f \equiv \langle f_t \rangle = \sum_x \pi_x f(x) \tag{2.7}$$

and *unnormalized autocorrelation function*†

$$\begin{aligned} C_{ff}(t) &\equiv \langle f_s f_{s+t} \rangle - \mu_f^2 \\ &= \sum_{x,y} f(x) \left[\pi_x p_{xy}^{(|t|)} - \pi_x \pi_y\right] f(y). \end{aligned} \tag{2.8}$$

*We avoid the term "ergodicity" because of its multiple and conflicting meanings. In the physics literature, and in the mathematics literature on Markov chains with *finite* state space, "ergodic" is typically used as a synonym for "irreducible" [4, Section 2.4] [5, Chapter 4]. However, in the mathematics literature on Markov chains with *general* state space, "ergodic" is used as a synonym for "irreducible, aperiodic and positive Harris recurrent" [7, p. 114] [8, p. 169].

†In the statistics literature, this is called the *autocovariance function*.

The *normalized autocorrelation function* is then

$$\rho_{ff}(t) \equiv C_{ff}(t)/C_{ff}(0). \tag{2.9}$$

Typically $\rho_{ff}(t)$ decays exponentially ($\sim e^{-|t|/\tau}$) for large t; we define the *exponential autocorrelation time*

$$\tau_{\exp,f} = \limsup_{t\to\infty} \frac{t}{-\log|\rho_{ff}(t)|} \tag{2.10}$$

and

$$\tau_{\exp} = \sup_f \tau_{\exp,f}. \tag{2.11}$$

Thus, $\tau_{\exp}$ is the relaxation time of the slowest mode in the system. (If the state space is infinite, $\tau_{\exp}$ might be $+\infty$!)

An equivalent definition, which is useful for rigorous analysis, involves considering the spectrum of the transition probability matrix P considered as an operator on the Hilbert space $l^2(\pi)$.* It is not hard to prove the following facts about P:

(a) The operator P is a contraction. (In particular, its spectrum lies in the closed unit disk.)

(b) 1 is a simple eigenvalue of P, as well as of its adjoint P^*, with eigenvector equal to the constant function **1**.

(c) If the Markov chain is aperiodic, then 1 is the only eigenvalue of P (and of P^*) on the unit circle.

(d) Let R be the spectral radius of P acting on the orthogonal complement of the constant functions:

$$R \equiv \inf\left\{r\colon \operatorname{spec}(P \restriction \mathbf{1}^\perp) \subset \{\lambda\colon |\lambda| \le r\}\right\}. \tag{2.12}$$

Then $R = e^{-1/\tau_{\exp}}$.

Facts (a)–(c) are a generalized Perron–Frobenius theorem [9]; fact (d) is a consequence of a generalized spectral radius formula [10, Propositions 2.3–2.5].

The rate of convergence to equilibrium from an initial nonequilibrium distribution can be bounded above in terms of R (and hence $\tau_{\exp}$). More precisely, let ν is a probability measure on S, and let us define its deviation from equilibrium in the l^2 sense,

$$d_2(\nu;\pi) \equiv \left\|\frac{\nu}{\pi} - 1\right\|_{l^2(\pi)} = \sup_{\|f\|_{l^2(\pi)}\le 1} \left|\int f\,d\nu - \int f\,d\pi\right|. \tag{2.13}$$

Then, clearly,

$$d_2(\alpha P^t;\pi) \le \|P^t \restriction \mathbf{1}^\perp\| d_2(\alpha;\pi). \tag{2.14}$$

And by the spectral radius formula,

$$\|P^t \restriction \mathbf{1}^\perp\| \sim R^t = \exp(-t/\tau_{\exp}) \tag{2.15}$$

asymptotically as $t \to \infty$, with equality for all t if P is self-adjoint (see below).

---

*$l^2(\pi)$ is the space of complex-valued functions on S that are square-integrable with respect to π: $\|f\| \equiv (\sum_x \pi_x |f(x)|^2)^{1/2} < \infty$. The inner product is given by $(f,g) \equiv \sum_x \pi_x f(x)^* g(x)$.

On the other hand, for a given observable f we define the *integrated autocorrelation time*

$$\tau_{\mathrm{inf},f} = \frac{1}{2} \sum_{t=-\infty}^{\infty} \rho_{ff}(t) \tag{2.16}$$
$$= \frac{1}{2} + \sum_{t=1}^{\infty} \rho_{ff}(t)$$

[The factor of $\frac{1}{2}$ is purely a matter of convention; it is inserted so that $\tau_{\mathrm{inf},f} \approx \tau_{\mathrm{exp},f}$ if $\rho_{ff}(t) \sim e^{-|t|/\tau}$ with $\tau \gg 1$.] The integrated autocorrelation time controls the statistical error in Monte Carlo measurements of $\langle f \rangle$. More precisely, the sample mean

$$\bar{f} \equiv \frac{1}{n} \sum_{t=1}^{n} f_t \tag{2.17}$$

has variance

$$\mathrm{var}(\bar{f}) = \frac{1}{n^2} \sum_{r,s=1}^{n} C_{ff}(r-s) \tag{2.18}$$

$$= \frac{1}{n} \sum_{t=-(n-1)}^{n-1} \left(1 - \frac{|t|}{n}\right) C_{ff}(t) \tag{2.19}$$

$$\approx \frac{1}{n} (2\tau_{\mathrm{inf},f}) C_{ff}(0) \quad \text{for } n \gg \tau \tag{2.20}$$

Thus, the variance of $\bar{f}$ is a factor $2\tau_{\mathrm{inf},f}$ larger than it would be if the $\{f_t\}$ were statistically independent. Stated differently, the number of "effectively independent samples" in a run of length n is roughly $n/2\tau_{\mathrm{inf},f}$.

It is sometimes convenient to measure the integrated autocorrelation time in terms of the equivalent pure exponential decay that would produce the same value of $\sum_{t=-\infty}^{\infty} \rho_{ff}(t)$, namely

$$\widetilde{\tau}_{\mathrm{inf},f} \equiv \frac{-1}{\log\left(\frac{2\tau_{\mathrm{inf},f}-1}{2\tau_{\mathrm{inf},f}+1}\right)}. \tag{2.21}$$

This quantity has the nice feature that a sequence of uncorrelated data has $\widetilde{\tau}_{\mathrm{inf},f} = 0$ (but $\tau_{\mathrm{inf},f} = \frac{1}{2}$). Of course, $\widetilde{\tau}_{\mathrm{inf},f}$ is ill-defined if $\tau_{\mathrm{inf},f} < \frac{1}{2}$, as can occasionally happen in cases of anticorrelation.

In summary, the autocorrelation times τ_{exp} and $\tau_{\mathrm{inf},f}$ play different roles in Monte Carlo simulations. τ_{exp} places an upper bound on the number of iterations n_{disc} which should be discarded at the beginning of the run, before the system has attained equilibrium; for example, $n_{\mathrm{disc}} \approx 20\tau_{\mathrm{exp}}$ is usually more than adequate. On the other hand, $\tau_{\mathrm{inf},f}$ determines the statistical errors in Monte Carlo measurements of $\langle f \rangle$, once equilibrium has been attained.

Most commonly it is assumed that τ_{exp} and $\tau_{\mathrm{inf},f}$ are of the same order of magnitude, at least for "reasonable" observables f. But this is *not* true in general. In fact, in statistical-mechanical problems near a critical point, one usually expects the autocorrelation function $\rho_{ff}(t)$ to obey a dynamic scaling law [11] of the form

$$\rho_{ff}(t;\beta) \sim |t|^{-a} F\left((\beta - \beta_c)\,|t|^{b}\right) \tag{2.22}$$

valid in the region

$$|t| \gg 1, \quad |\beta-\beta_c| \ll 1, \quad |\beta-\beta_c|\,|t|^b \text{ bounded.} \tag{2.23}$$

Here $a,b>0$ are dynamic critical exponents and F is a suitable scaling function; β is some "temperature-like" parameter, and β_c is the critical point. Now suppose that F is continuous and strictly positive, with $F(x)$ decaying rapidly (e.g., exponentially) as $|x| \to \infty$. Then it is not hard to see that

$$\tau_{\text{exp},f} \sim |\beta-\beta_c|^{-1/b} \tag{2.24}$$

$$\tau_{\text{inf},f} \sim |\beta-\beta_c|^{-(1-a)/b} \tag{2.25}$$

$$\rho_{ff}(t;\beta=\beta_c) \sim |t|^{-a} \tag{2.26}$$

so that $\tau_{\text{exp},f}$ and $\tau_{\text{inf},f}$ have *different* critical exponents unless $a=0$.* Actually, this should not be surprising: replacing "time" by "space," we see that $\tau_{\text{exp},f}$ is the analogue of a correlation length, while $\tau_{\text{inf},f}$ is the analogue of a susceptibility; and (2.24)–(2.26) are the analogue of the well-known scaling law $\gamma=(2-\eta)\nu$—clearly $\gamma \neq \nu$ in general! So it is crucial to distinguish between the two types of autocorrelation time.

Returning to the general theory, we note that one convenient way of satisfying condition (B) is to satisfy the following *stronger* condition:

$$(\text{B}') \quad \text{For each pair } x,y \in S,\ \pi_x p_{xy} = \pi_y p_{yx}. \tag{2.27}$$

[Summing (B′) over x, we recover (B).] (B′) is called the *detailed-balance condition*; a Markov chain satisfying (B′) is called *reversible*.† (B′) is equivalent to the *self-adjointness* of P as on operator on the space $l^2(\pi)$. In this case, the spectrum of P is *real* and lies in the closed interval $[-1,1]$; we define

$$\lambda_{\text{min}} = \inf \text{spec}(P \restriction \mathbf{1}^{\perp}) \tag{2.28}$$

$$\lambda_{\text{max}} = \sup \text{spec}(P \restriction \mathbf{1}^{\perp}) \tag{2.29}$$

From (2.12) we have

$$\tau_{\text{exp}} = \frac{-1}{\log \max[|\lambda_{\text{min}}|, \lambda_{\text{max}}]}. \tag{2.30}$$

For many purposes, only the spectrum near $+1$ matters, so it is useful to define the *modified exponential autocorrelation time*

$$\tau'_{\text{exp}} = \begin{cases} -1/\log \lambda_{\text{max}} & \text{if } \lambda_{\text{max}} > 0 \\ +\infty & \text{if } \lambda_{\text{max}} \le 0 \end{cases} \tag{2.31}$$

Now let us apply the spectral theorem to the operator P: it follows that the autocorrelation function $\rho_{ff}(t)$ has a spectral representation

$$\rho_{ff}(t) = \int_{\lambda_{\text{min},f}}^{\lambda_{\text{max},f}} \lambda^{|t|}\, d\sigma_{ff}(\lambda) \tag{2.32}$$

*Our discussion of this topic in [12] is incorrect.

†For the physical significance of this term, see Kemeny and Snell [4, section 5.3] or Iosifescu [5, section 4.5].

with a *nonnegative* spectral weight $d\sigma_{ff}(\lambda)$ supported on an interval

$$[\lambda_{\min,f}, \lambda_{\max,f}] \subset [\lambda_{\min}, \lambda_{\max}].$$

Clearly

$$\tau_{\exp,f} = \frac{-1}{\log\max[|\lambda_{\min,f}|, \lambda_{\max,f}]} \tag{2.33}$$

and we can define

$$\tau'_{\exp,f} = \begin{cases} -1/\log\lambda_{\max,f} & \text{if } \lambda_{\max,f} > 0 \\ 0 & \text{if } \lambda_{\max,f} \leq 0 \end{cases} \tag{2.34}$$

(if $\lambda_{\max,f} \geq 0$). On the other hand,

$$\tau_{\mathrm{inf},f} = \frac{1}{2} \int\limits_{\lambda_{\min,f}}^{\lambda_{\max,f}} \frac{1+\lambda}{1-\lambda} d\sigma_{ff}(\lambda) \tag{2.35}$$

It follows that

$$\tau_{\mathrm{inf},f} \leq \frac{1}{2}\left(\frac{1+e^{-1/\tau'_{\exp,f}}}{1-e^{-1/\tau'_{\exp,f}}}\right) \leq \frac{1}{2}\left(\frac{1+e^{-1/\tau'_{\exp}}}{1-e^{-1/\tau'_{\exp}}}\right) \approx \tau'_{\exp}. \tag{2.36}$$

Moreover, since $\lambda \mapsto (1+\lambda)/(1-\lambda)$ is a convex function, Jensen's inequality implies that

$$\tau_{\mathrm{inf},f} \geq \frac{1}{2}\frac{1+\rho_{ff}(1)}{1-\rho_{ff}(1)}. \tag{2.37}$$

If we define the *initial autocorrelation time*

$$\tau_{\mathrm{init},f} = \begin{cases} -1/\log\rho_{ff}(1) & \text{if } \rho_{ff}(1) \geq 0 \\ \text{undefined} & \text{if } \rho_{ff}(1) < 0 \end{cases} \tag{2.38}$$

then these inequalities can be summarized conveniently as

$$\tau_{\mathrm{init},f} \leq \widetilde{\tau}_{\mathrm{inf},f} \leq \tau'_{\exp,f} \leq \tau'_{\exp}. \tag{2.39}$$

Conversely, it is not hard to see that

$$\sup_{f\in l^2(\pi)} \tau_{\mathrm{init},f} = \sup_{f\in l^2(\pi)} \widetilde{\tau}_{\mathrm{inf},f} = \sup_{f\in l^2(\pi)} \tau'_{\exp,f} = \tau'_{\exp}; \tag{2.40}$$

it suffices to choose f so that its spectral weight $d\sigma_{ff}$ is supported in a very small interval near $\lambda_{\max}$.

Finally, let us make a remark about transition probabilities P that are "built up out of" other transition probabilities $P_1, P_2, \ldots, P_n$:

a) If $P_1, P_2, \ldots, P_n$ satisfy the stationarity condition (resp. the detailed-balance condition) for π, then so does any convex combination $P = \sum_{i=1}^n \lambda_i P_i$. Here $\lambda_i \geq 0$ and $\sum_{i=1}^n \lambda_i = 1$.

b) If $P_1, P_2, \ldots, P_n$ satisfy the stationarity condition for π, then so does the product $P = P_1 P_2 \cdots P_n$. (Note, however, that P does *not* in general satisfy the detailed-balance condition, even if the individual P_i do.*)

Algorithmically, the convex combination amounts to choosing *randomly*, with probabilities $\{\lambda_i\}$, from among the "elementary operations" P_i. (It is crucial here that the λ_i are *constants*, independent of the current configuration of the system; only in this case does P leave π stationary in general.) Similarly, the product corresponds to performing *sequentially* the operations $P_1, P_2, \ldots, P_n$.

3. STATISTICAL ANALYSIS OF DYNAMIC MONTE CARLO DATA

Many published Monte Carlo studies contain statements like:

> We ran for a total of 100000 iterations, discarding the first 50000 iterations (for equilibration) and then taking measurements once every 100 iterations.

It is important to emphasize that unless further information is given — namely, the autocorrelation time of the algorithm — *such statements have no value whatsoever.*

Is a run of 100000 iterations long enough? Are 50000 iterations sufficient for equilibration? That depends on how big the autocorrelation time is. The purpose of this lecture is to give some practical advice for choosing the parameters of a dynamic Monte Carlo simulation, and to give an introduction to the statistical theory that puts this advice on a sound mathematical footing.

There are two fundamental — and quite distinct — issues in dynamic Monte Carlo simulation:

- *Initialization bias.* If the Markov chain is started in a distribution α that is not equal to the stationary distribution π, then there is an "initial transient" in which the data do not reflect the desired equilibrium distribution π. This results in a *systematic error* (bias).
- *Autocorrelation in equilibrium.* The Markov chain, once it reaches equilibrium, provides *correlated* samples from π. This correlation causes the *statistical error* (variance) to be a factor $2\tau_{\text{int},f}$ larger than in independent sampling.

Let us discuss these issues in turn.

Initialization bias. Often the Markov chain is started in some chosen configuration x; then $\alpha = \delta_x$. For example, in an Ising model, x might be the configuration with "all spins up"; this is sometimes called an *ordered* or *cold* start. Alternatively, the Markov chain might be started in a random configuration chosen according to some simple probability distribution α. For example, in an Ising model, we might initialize the spins randomly and independently, with equal probabilities of up and down; this is sometimes called a *random* or *hot* start. In all these cases, the initial distribution α is clearly *not* equal to the equilibrium distribution π. Therefore, the system is initially "out of equilibrium." Theorem 1 guarantees that the system approaches equilibrium as $t \to \infty$, but we need to know something about the *rate* of convergence to equilibrium.

---

*Recall that if A and B are self-adjoint operators, then AB is self-adjoint *if and only if* A and B commute.

Using the exponential autocorrelation time τ_{exp}, we can set an *upper bound* on the amount of time we have to wait before equilibrium is "for all practical purposes" attained. For example, if we wait a time $20\tau_{\text{exp}}$, then the deviation from equilibrium (in the l^2 sense) will be at most e^{-20} ($\approx 2 \times 10^{-9}$) times the initial deviation from equilibrium. There are two difficulties with this bound. Firstly, it is usually impossible to apply in practice, since we almost never know τ_{exp} (or a rigorous upper bound for it). Secondly, even if we can apply it, it may be overly conservative; indeed, there exist perfectly reasonable algorithms in which $\tau_{\text{exp}} = +\infty$ (see Sections 7 and 8).

Lacking rigorous knowledge of the autocorrelation time τ_{exp}, we should try to estimate it both *theoretically* and *empirically*. To make a heuristic theoretical estimate of τ_{exp}, we attempt to understand the physical mechanism(s) causing slow convergence to equilibrium; but it is always possible that we have overlooked one or more such mechanisms, and have therefore grossly underestimated τ_{exp}. To make a rough empirical estimate of τ_{exp}, we measure the autocorrelation function $C_{ff}(t)$ for a suitably large set of observables f [see below]; but there is always the danger that our chosen set of observables has failed to include one that has strong enough overlap with the slowest mode, again leading to a gross underestimate of τ_{exp}.

On the other hand, the actual rate of convergence to equilibrium from a given initial distribution α may be much faster than the worst-case estimate given by τ_{exp}. So it is usual to determine *empirically* when "equilibrium" has been achieved, by plotting selected observables as a function of time and noting when the initial transient appears to end. More sophisticated statistical tests for initialization bias can also be employed [13].

In all empirical methods of determining when "equilibrium" has been achieved, a serious danger is the possibility of *metastability*. That is, it could *appear* that equilibrium has been achieved, when in fact the system has only settled down to a long-lived (metastable) region of configuration space that may be very *far* from equilibrium. The only sure-fire protection against metastability is a *proof* of an upper bound on τ_{exp} (or more generally, on the deviation from equilibrium as a function of the elapsed time t). The next-best protection is a convincing heuristic argument that metastability is unlikely (i.e., that τ_{exp} is not too large); but as mentioned before, even if one rules out several potential physical mechanisms for metastability, it is very difficult to be certain that one has not overlooked others. If one cannot rule out metastability on theoretical grounds, then it is helpful at least to have an idea of what the possible metastable regions look like; then one can perform several runs with different initial conditions typical of each of the possible metastable regions, and test whether the answers are consistent. For example, near a first-order phase transition, most Monte Carlo methods suffer from metastability associated with transitions between configurations typical of the distinct pure phases. We can try initial conditions typical of each of these phases (e.g., for many models, a "hot" start and a "cold" start). Consistency between these runs does not *guarantee* that metastability is absent, but it does give increased confidence. Plots of observables as a function of time are also useful indicators of possible metastability. But when all is said and done, no purely empirical estimate of τ from a run of length n can be guaranteed to be even approximately correct. What we *can* say is that if $\tau_{\text{estimated}} \ll n$, then *either* $\tau \approx \tau_{\text{estimated}}$ or else $\tau \gtrsim n$.

Once we know (or guess) the time needed to attain "equilibrium," what do we do with it? The answer is clear: we discard the data from the initial transient, up to some time n_{disc}, and include only the subsequent data in our averages. In principle, this is (asymptotically) unnecessary, because the systematic errors from this initial transient will be of order τ/n, while the statistical errors will be of order $(\tau/n)^{1/2}$. But in practice, the coefficient of τ/n in the systematic error may be fairly large, if the initial distribution is very far from equilibrium. By throwing away the data from the initial transient, we lose nothing, and avoid a potentially large systematic error.

Autocorrelation in equilibrium. As explained in the preceding lecture, the variance of the sample mean $\bar{f}$ in a dynamic Monte Carlo method is a factor $2\tau_{\mathrm{inf},f}$ higher than it would be in independent sampling. Otherwise put, a run of length n contains only $n/2\tau_{\mathrm{inf},f}$ "effectively independent data points."

This has several implications for Monte Carlo work. On the one hand, it means that the the *computational efficiency* of the algorithm is determined principally by its autocorrelation time. More precisely, if one wishes to compare two alternative Monte Carlo algorithms for the same problem, then *the better algorithm is the one that has the smaller autocorrelation time, when time is measured in units of computer (CPU) time.* [In general there may arise tradeoffs between "physical" autocorrelation time (i.e., τ measured in *iterations*) and computational complexity *per iteration*.] So accurate measurements of the autocorrelation time are essential to evaluating the computational efficiency of competing algorithms.

On the other hand, even for a fixed algorithm, knowledge of $\tau_{\mathrm{inf},f}$ is essential for determining run lengths — is a run of 100000 sweeps long enough? — and for setting error bars on estimates of $\langle f\rangle$. Roughly speaking, error bars will be of order $(\tau/n)^{1/2}$; so if we want 1% accuracy, then we need a run of length $\approx 10000\tau$, and so on. Above all, there is a basic self-consistency requirement: the run length n must be $\gg$ than the estimates of τ produced by that same run, otherwise *none* of the results from that run should be believed. Of course, while self-consistency is a *necessary* condition for the trustworthiness of Monte Carlo data, it is not a *sufficient* condition; there is always the danger of metastability.

Already we can draw a conclusion about the relative importance of initialization bias and autocorrelation as difficulties in dynamic Monte Carlo work. Let us *assume* that the time for initial convergence to equilibrium is comparable to (or at least not too much larger than) the equilibrium autocorrelation time $\tau_{\mathrm{inf},f}$ (for the observables f of interest) — this is often but not always the case. Then initialization bias is a relatively trivial problem compared to autocorrelation in equilibrium. To eliminate initialization bias, it suffices to discard $\approx 20\tau$ of the data at the beginning of the run; but to achieve a reasonably small statistical error, it is necessary to make a run of length $\approx 1000\tau$ or more. So the data that must be discarded at the beginning, n_{disc}, is a negligible fraction of the total run length n. This estimate also shows that the exact value of n_{disc} is not particularly delicate: anything between $\approx 20\tau$ and $\approx n/5$ will eliminate essentially all initialization bias while paying less than a 10% price in the final error bars.

In this remainder of this lecture I would like to discuss in more detail the statistical analysis of dynamic Monte Carlo data (assumed to be already "in equilibrium"), with emphasis on how to estimate the autocorrelation time $\tau_{\mathrm{inf},f}$ and how to compute valid error bars. What is involved here is a branch of mathematical statistics called *time-series analysis*. An excellent exposition can be found in the books of of Priestley [14] and Anderson [15].

Let $\{f_t\}$ be a real-valued stationary stochastic process with mean

$$\mu \equiv \langle f_t\rangle, \tag{3.1}$$

unnormalized autocorrelation function

$$C(t) \equiv \langle f_s f_{s+t}\rangle - \mu^2, \tag{3.2}$$

normalized autocorrelation function

$$\rho(t) \equiv C(t)/C(0), \tag{3.3}$$

and integrated autocorrelation time

$$\tau_{\text{inf}} = \frac{1}{2} \sum_{t=-\infty}^{\infty} \rho(t). \tag{3.4}$$

Our goal is to estimate μ, $C(t)$, $\rho(t)$ and τ_{inf} based on a finite (but large) sample $f_1, \ldots, f_n$ from this stochastic process.

The "natural" estimator of μ is the sample mean

$$\overline{f} \equiv \frac{1}{n} \sum_{i=1}^{n} f_i. \tag{3.5}$$

This estimator is unbiased (i.e., $\langle \overline{f} \rangle = \mu$) and has variance

$$\text{var}(\overline{f}) = \frac{1}{n} \sum_{t=-(n-1)}^{n-1} \left(1 - \frac{|t|}{n}\right) C(t) \tag{3.6}$$

$$\approx \frac{1}{n} (2\tau_{\text{inf}})\, C(0) \qquad \text{for } n \gg \tau \tag{3.7}$$

Thus, even if we are interested only in the static quantity μ, it is necessary to estimate the dynamic quantity τ_{inf} in order to determine valid error bars for μ.

The "natural" estimator of $C(t)$ is

$$\widehat{C}(t) \equiv \frac{1}{n-|t|} \sum_{i=1}^{n-|t|} (f_i - \mu)(f_{i+|t|} - \mu) \tag{3.8}$$

if the mean μ is known, and

$$\widehat{\widehat{C}}(t) \equiv \frac{1}{n-|t|} \sum_{i=1}^{n-|t|} (f_i - \overline{f})(f_{i+|t|} - \overline{f}) \tag{3.9}$$

if the mean μ is unknown. We emphasize the conceptual distinction between the autocorrelation function $C(t)$, which for each t is a *number*, and the estimator $\widehat{C}(t)$ or $\widehat{\widehat{C}}(t)$, which for each t is a *random variable*. As will become clear, this distinction is also of *practical* importance. $\widehat{C}(t)$ is an unbiased estimator of $C(t)$, and $\widehat{\widehat{C}}(t)$ is almost unbiased (the bias is of order $1/n$) [15, p. 463]. Their variances and covariances are [15, pp. 464–471] [14, pp. 324–328]

$$\text{var}(\widehat{C}(t)) = \frac{1}{n} \sum_{m=-\infty}^{\infty} \left[C(m)^2 + C(m+t)C(m-t) + \kappa(t,m,m+t)\right] + o\left(\frac{1}{n}\right) \tag{3.10}$$

$$\text{cov}(\widehat{C}(t), \widehat{C}(u)) = \frac{1}{n} \sum_{m=-\infty}^{\infty} [C(m)C(m+u-t) + C(m+u)C(m-t) + \kappa(t,m,m+u)] + o\left(\frac{1}{n}\right) \tag{3.11}$$

where $t, u \geq 0$ and κ is the connected 4-point autocorrelation function

$$\kappa(r,s,t) \equiv \langle (f_i - \mu)(f_{i+r} - \mu)(f_{i+s} - \mu)(f_{i+t} - \mu) \rangle \\ - C(r)C(t-s) - C(s)C(t-r) - C(t)C(s-r). \tag{3.12}$$

To leading order in $1/n$, the behavior of $\widehat{\widehat{C}}$ is identical to that of $\widehat{C}$.

The "natural" estimator of $\rho(t)$ is

$$\widehat{\rho}(t) \equiv \widehat{C}(t)/\widehat{C}(0) \tag{3.13}$$

if the mean μ is known, and

$$\widehat{\widehat{\rho}}(t) \equiv \widehat{\widehat{C}}(t)/\widehat{\widehat{C}}(0) \tag{3.14}$$

if the mean μ is unknown. The variances and covariances of $\widehat{\rho}(t)$ and $\widehat{\widehat{\rho}}(t)$ can be computed (for large n) from (3.11); we omit the detailed formulae.

The "natural" estimator of τ_{inf} would seem to be

$$\widehat{\tau}_{\text{inf}} \overset{?}{\equiv} \frac{1}{2} \sum_{t=-(n-1)}^{n-1} \widehat{\rho}(t) \tag{3.15}$$

(or the analogous thing with $\widehat{\widehat{\rho}}$), *but this is wrong!* The estimator defined in (3.15) has a variance that does not go to zero as the sample size n goes to infinity [14, pp. 420–431], so it is clearly a very bad estimator of τ_{inf}. Roughly speaking, this is because the sample autocorrelations $\widehat{\rho}(t)$ for $|t| \gg \tau$ contain much "noise" but little "signal"; and there are so many of them (order n) that the noise adds up to a total variance of order 1. (For a more detailed discussion, see [14, pp. 432–437].) The solution is to cut off the sum in (3.15) using a "window" $\lambda(t)$ which is ≈ 1 for $|t| \lesssim \tau$ but ≈ 0 for $|t| \gg \tau$:

$$\widehat{\tau}_{\text{inf}} \equiv \frac{1}{2} \sum_{t=-(n-1)}^{n-1} \lambda(t) \widehat{\rho}(t). \tag{3.16}$$

This retains most of the "signal" but discards most of the "noise." A good choice is the rectangular window

$$\lambda(t) = \begin{cases} 1 & \text{if } |t| \leq M \\ 0 & \text{if } |t| > M \end{cases} \tag{3.17}$$

where M is a suitably chosen cutoff. This cutoff introduces a bias

$$\text{bias}(\widehat{\tau}_{\text{inf}}) = -\frac{1}{2} \sum_{|t|>M} \rho(t) + o\left(\frac{1}{n}\right). \tag{3.18}$$

On the other hand, the variance of $\widehat{\tau}_{\text{inf}}$ can be computed from (3.11); after some algebra, one obtains

$$\text{var}(\widehat{\tau}_{\text{inf}}) \approx \frac{2(2M+1)}{n} \tau_{\text{inf}}^2, \tag{3.19}$$

where we have made the approximation $\tau \ll M \ll n$.* The choice of M is thus a tradeoff between bias and variance: the bias can be made small by taking M large enough so that $\rho(t)$

*We have also assumed that the only strong peaks in the Fourier transform of $C(t)$ are at zero frequency. This assumption is valid if $C(t) \geq 0$, but could fail if there are strong *anti*correlations.

is negligible for $|t| > M$ (e.g., $M =$ a few times τ usually suffices), while the variance is kept small by taking M to be no larger than necessary consistent with this constraint. We have found the following "automatic windowing" algorithm [16] to be convenient: choose M to be the smallest integer such that $M \geq c\,\widehat{\tau}_{\text{inf}}(M)$. If $\rho(t)$ were roughly a pure exponential, then it would suffice to take $c \approx 4$ (since $e^{-4} < 2\%$). However, in many cases $\rho(t)$ is expected to have an asymptotic or pre-asymptotic decay slower than exponential, so it is usually prudent to take c at least 6, and possibly as large as 10.

We have found this automatic windowing procedure to work well in practice, *provided* that a sufficient quantity of data is available ($n \gtrsim 1000\tau$). However, at present we have very little understanding of the conditions under which this windowing algorithm may produce biased estimates of τ_{inf} or of its own error bars. Further theoretical and experimental study of the windowing algorithm — e.g., experiments on various exactly-known stochastic processes, with various run lengths — would be highly desirable.

4. CONVENTIONAL MONTE CARLO ALGORITHMS FOR SPIN MODELS

In this lecture we describe the construction of dynamic Monte Carlo algorithms for models in statistical mechanics and quantum field theory. Recall our goal: given a probability measure π on the state space S, we wish to construct a transition matrix $P = \{p_{xy}\}$ satisfying:

(A) *Irreducibility.* For each pair $x, y \in S$, there exists an $n \geq 0$ for which $p_{xy}^{(n)} > 0$.

(B) *Stationarity of* π. For each $y \in S$,

$$\sum_x \pi_x p_{xy} = \pi_y. \tag{4.1}$$

A sufficient condition for (B), which is often more convenient to verify, is:

(B′) *Detailed balance for* π. For each pair $x, y \in S$, $\pi_x p_{xy} = \pi_y p_{yx}$.

A very general method for constructing transition matrices satisfying detailed balance for a given distribution π was introduced in 1953 by Metropolis *et al.* [17], with a slight extension two decades later by Hastings [18]. The idea is the following: Let $P^{(0)} = \{p_{xy}^{(0)}\}$ be an *arbitrary* irreducible transition matrix on S. We call $P^{(0)}$ the *proposal matrix*; we shall use it to generate *proposed* moves $x \longrightarrow y$ that will then be accepted or rejected with probabilities a_{xy} and $1 - a_{xy}$, respectively. If a proposed move is rejected, then we make a "null transition" $x \longrightarrow x$. Therefore, the transition matrix $P = \{p_{xy}\}$ of the full algorithm is

$$\begin{aligned} p_{xy} &= p_{xy}^{(0)}\, a_{xy} \qquad \text{for } x \neq y \\ p_{xx} &= p_{xx}^{(0)} + \sum_{y \neq x} p_{xy}^{(0)}\,(1 - a_{xy}) \end{aligned} \tag{4.2}$$

where of course we must have $0 \leq a_{xy} \leq 1$ for all x, y. It is easy to see that P satisfies detailed balance for π if and only if

$$\frac{a_{xy}}{a_{yx}} = \frac{\pi_y\, p_{yx}^{(0)}}{\pi_x\, p_{xy}^{(0)}} \tag{4.3}$$

for all pairs $x \neq y$. But this is easily arranged: just set

$$a_{xy} = F\left(\frac{\pi_y p_{yx}^{(0)}}{\pi_x p_{xy}^{(0)}}\right), \tag{4.4}$$

where $F\colon [0,+\infty] \to [0,1]$ is any function satisfying

$$\frac{F(z)}{F(1/z)} = z \qquad \text{for all } z. \tag{4.5}$$

The choice suggested by Metropolis *et al.* is

$$F(z) = \min(z,1); \tag{4.6}$$

this is the *maximal* function satisfying (4.5). Another choice sometimes used is

$$F(z) = \frac{z}{1+z}. \tag{4.7}$$

Of course, it is still necessary to check that P is irreducible; this is usually done on a case-by-case basis.

Note that if the proposal matrix $P^{(0)}$ happens to *already* satisfy detailed balance for π, then we have $\pi_y p_{yx}^{(0)} / \pi_x p_{xy}^{(0)} = 1$, so that $a_{xy} = 1$ (if we use the Metropolis choice of F) and $P = P^{(0)}$. On the other hand, no matter what $P^{(0)}$ is, we obtain a matrix P that satisfies detailed balance for π. So the Metropolis–Hastings procedure can be thought of as a prescription for minimally modifying a given transition matrix $P^{(0)}$ so that it satisfies detailed balance for π.

Many textbooks and articles describe the Metropolis–Hastings procedure only in the special case in which the proposal matrix $P^{(0)}$ is *symmetric*, namely $p_{xy}^{(0)} = p_{yx}^{(0)}$. In this case (4.4) reduces to

$$a_{xy} = F\left(\frac{\pi_y}{\pi_x}\right). \tag{4.8}$$

In statistical mechanics we have

$$\pi_x = \frac{e^{-\beta E_x}}{\sum\limits_y e^{-\beta E_y}} = \frac{e^{-\beta E_x}}{Z}, \tag{4.9}$$

and hence

$$\frac{\pi_y}{\pi_x} = e^{-\beta(E_y - E_x)}. \tag{4.10}$$

Note that the partition function Z has disappeared from this expression; this is crucial, as Z is almost never explicitly computable! Using the Metropolis acceptance probability $F(z) = \min(z,1)$, we obtain the following rules for acceptance or rejection:

- If $\Delta E \equiv E_y - E_x \leq 0$, then we accept the proposal *always* (i.e., with probability 1).
- If $\Delta E > 0$, then we accept the proposal with probability $e^{-\beta \Delta E}$ (< 1). That is, we choose a random number r uniformly distributed on $[0,1]$, and we accept the proposal if $r \leq e^{-\beta \Delta E}$.

But there is nothing special about $P^{(0)}$ being symmetric; *any* proposal matrix $P^{(0)}$ is perfectly legitimate, and the Metropolis–Hastings procedure is defined quite generally by (4.4).

Let us emphasize once more that the Metropolis–Hastings procedure is a *general technique*; it produces an infinite family of different algorithms depending on the choice of the proposal matrix $P^{(0)}$. In the literature the term "Metropolis algorithm" is often used to denote the algorithm resulting from some *particular* commonly-used choice of $P^{(0)}$, but it is important not to be misled.

To see the Metropolis–Hastings procedure in action, let us consider a typical statistical-mechanical model, the *Ising model*: On each site i of some finite d-dimensional lattice, we place a random variable σ_i taking the values ± 1. The Hamiltonian is

$$H(\sigma) = -\sum_{\langle ij \rangle} \sigma_i \sigma_j, \tag{4.11}$$

where the sum runs over all nearest-neighbor pairs. The corresponding Gibbs measure is

$$\pi(\sigma) = Z^{-1} \exp[-\beta H(\sigma)], \tag{4.12}$$

where Z is a normalization constant (the partition function). Two different proposal matrices are in common use:

1) *Single-spin-flip (Glauber) dynamics.* Fix some site i. The proposal is to flip σ_i, hence

$$p_i^{(0)}(\{\sigma\} \to \{\sigma'\}) = \begin{cases} 1 & \text{if } \sigma_i' = -\sigma_i \text{ and } \sigma_j' = \sigma_j \text{ for all } j \neq i \\ 0 & \text{otherwise} \end{cases} \tag{4.13}$$

Here $P_i^{(0)}$ is symmetric, so the acceptance probability is

$$a_i(\{\sigma\} \to \{\sigma'\}) = \min(e^{-\beta \Delta E}, 1), \tag{4.14}$$

where

$$\Delta E \equiv E(\{\sigma'\}) - E(\{\sigma\}) = 2\sigma_i \sum_{j \text{ n.n. of } i} \sigma_j. \tag{4.15}$$

So ΔE is easily computed by comparing the status of σ_i and its neighbors.

This defines a transition matrix P_i in which only the spin at site i is touched. The full "single-spin-flip Metropolis algorithm" involves sweeping through the entire lattice in either a random or periodic fashion, i.e., either

$$P = \frac{1}{V} \sum_i P_i \qquad \text{(random site updating)} \tag{4.16}$$

or

$$P = P_{i_1} P_{i_2} \cdots P_{i_V} \qquad \text{(sequential site updating)} \tag{4.17}$$

(here V is the volume). In the former case, the transition matrix P satisfies detailed balance for π. In the latter case, P does *not* in general satisfy detailed balance for π, but it does satisfy stationarity for π, which is all that really matters.

It is easy to check that P is irreducible, except in the case of sequential site updating with $\beta = 0$.*

*__Note Added 1996:__ This statement is **wrong**!!! For the Metropolis acceptance probability $F(z) = \min(z, 1)$, one may construct examples of nonergodicity for sequential site updating at *any* β [104, 105]: the idea is to construct configurations for which *every* proposed update (working sequentially) has $\Delta E = 0$ and is thus accepted. Of course, the ergodicity can be restored by taking any $F(z) < 1$, such as the choice $F(z) = z/(1+z)$.

2) *Pair-interchange (Kawasaki) dynamics.* Fix a pair of sites i,j. The proposal is to interchange σ_i and σ_j, hence

$$p^{(0)}_{(ij)}(\{\sigma\} \to \{\sigma'\}) = \begin{cases} 1 & \text{if } \sigma'_i = \sigma_j,\ \sigma'_j = \sigma_i \text{ and } \sigma'_k = \sigma_k \text{ for all } k \neq i,j \\ 0 & \text{otherwise} \end{cases} \tag{4.18}$$

The rest of the formulae are analogous to those for single-spin-flip dynamics. The overall algorithm is again constructed by a random or periodic sweep over a suitable set of pairs i,j, usually taken to be nearest-neighbor pairs. It should be noted that this algorithm is *not* irreducible, as it conserves the total magnetization $\mathcal{M} = \sum_i \sigma_i$. But it is irreducible on subspaces of fixed $\mathcal{M}$ (except for sequential updating with $\beta = 0$).

A very different approach to constructing transition matrices satisfying detailed balance for π is the *heat-bath method*. This is best illustrated in a specific example. Consider again the Ising model (4.12), and focus on a single site i. Then the conditional probability distribution of σ_i, given all the other spins $\{\sigma_j\}_{j\neq i}$, is

$$P^{\pi}(\sigma_i | \{\sigma_j\}_{j\neq i}) = \text{const}\left(\{\sigma_j\}_{j\neq i}\right) \times \exp\left[\beta\sigma_i \sum_{j \text{ n.n. of } i} \sigma_j\right]. \tag{4.19}$$

(Note that this conditional distribution is precisely that of a single Ising spin σ_i in an "effective magnetic field" produced by the fixed neighboring spins σ_j.) The heat-bath algorithm updates σ_i by choosing a new spin value σ'_i, *independent of the old value* σ_i, from the conditional distribution (4.19); all the other spins $\{\sigma_j\}_{j\neq i}$ remain unchanged.* As in the Metropolis algorithm, this operation is carried out over the whole lattice, either randomly or sequentially.

Analogous algorithms can be developed for more complicated models, e.g., $P(\varphi)$ models, σ-models and lattice gauge theories. In each case, we focus on a single field variable (holding all the other variables fixed), and give it a new value, *independent of the old value*, chosen from the appropriate conditional distribution. Of course, the feasibility of this algorithm depends on our ability to construct an efficient subroutine for generating the required single-site (or single-link) random variables. But even if this algorithm is not always the most efficient one in practice, it serves as a clear standard of comparison, which is useful in the development of more sophisticated algorithms.

A more general version of the heat-bath idea is called *partial resampling*: here we focus on a set of variables rather than only one, and the new values need not be independent of the old values. That is, we divide the variables of the system, call them $\{\varphi\}$, into two subsets, call them $\{\psi\}$ and $\{\theta\}$. For fixed values of the $\{\theta\}$ variables, π induces a conditional probability distribution of $\{\psi\}$ given $\{\theta\}$, call it $P^{\pi}(\{\psi\}|\{\theta\})$. Then any algorithm for updating $\{\psi\}$ with $\{\theta\}$ *fixed* that leaves invariant all of the distributions $P^{\pi}(\cdot|\{\theta\})$ will also leave invariant π. One possibility is to use an *independent resampling* of $\{\psi\}$: we throw away the old values $\{\psi\}$, and take $\{\psi'\}$ to be a new random variable chosen from the probability distribution $P^{\pi}(\cdot|\{\theta\})$, independent of the old values. Independent resampling might also called *block heat-bath updating*. On the other hand, if $\{\psi\}$ is a large set of variables, independent resampling is probably unfeasible, but we are free to use *any* updating that leaves invariant the appropriate conditional distributions. Of course, in this generality "partial resampling"

*The alert reader will note that *in the Ising case* the heat-bath algorithm is equivalent to the single-spin-flip Metropolis algorithm with the choice $F(z) = z/(1+z)$ of acceptance function. But this correspondence does not hold for more complicated models.

includes *all* dynamic Monte Carlo algorithms — we could just take $\{\psi\}$ to be the entire system — but it is in many cases conceptually useful to focus on some subset of variables. The partial-resampling idea will be at the heart of the multi-grid Monte Carlo method (Section 5) and the embedding algorithms (Section 6).

We have now defined a rather large class of dynamic Monte Carlo algorithms: the single-spin-flip Metropolis algorithm, the single-site heat-bath algorithm, and so on. How well do these algorithms perform? Away from phase transitions, they perform rather well. However, near a phase transition, the autocorrelation time grows rapidly. In particular, near a critical point (second-order phase transition), the autocorrelation time typically diverges as

$$\tau \sim \min(L, \xi)^z, \tag{4.20}$$

where L is the linear size of the system, ξ is the correlation length of an infinite-volume system at the same temperature, and z is a *dynamic critical exponent*. This phenomenon is called *critical slowing-down*; it severely hampers the study of critical phenomena by Monte Carlo methods. Most of the remainder of these lectures will be devoted, therefore, to describing recent progress in inventing new Monte Carlo algorithms with radically reduced critical slowing-down.

The critical slowing-down of the conventional algorithms arises fundamentally from the fact that their updates are *local*: in a single step of the algorithm, "information" is transmitted from a given site only to its nearest neighbors. Crudely one might guess that this "information" executes a random walk around the lattice. In order for the system to evolve to an "essentially new" configuration, the "information" has to travel a distance of order ξ, the (static) correlation length. One would guess, therefore, that $\tau \sim \xi^2$ near criticality, i.e., that the dynamic critical exponent z equals 2. This guess is correct for the Gaussian model (free field).* For other models, we have a situation analogous to theory of static critical phenomena: the dynamic critical exponent is a nontrivial number that characterizes a rather large class of algorithms (a so-called "dynamic universality class"). In any case, for most models of interest, the dynamic critical exponent for local algorithms is close to 2 (usually somewhat higher) [21]. Accurate measurements of dynamic critical exponents are, however, very difficult — even more difficult than measurements of static critical exponents — and require enormous quantities of Monte Carlo data: run lengths of $\approx 10000\tau$, when τ is itself getting large!

We can now make a rough estimate of the computer time needed to study the Ising model near its critical point, or quantum chromodynamics near the continuum limit. Each sweep of the lattice takes a time of order L^d, where d is the spatial (or space–"time") dimensionality of the model. And we need $\approx 2\tau$ sweeps in order to get *one* "effectively independent" sample. So this means a computer time of order $L^d \xi^z \gtrsim \xi^{d+z}$.† For high-precision statistics one might want 10^6 "independent" samples. The reader is invited to plug in $\xi = 100$, $d = 3$ (or $d = 4$ if

*Indeed, for the Gaussian model this random-walk picture can be made rigorous: see [19] combined with [20, Section 8].

†Clearly one must take $L \gtrsim \xi$ in order to avoid severe finite-size effects. Typically one approaches the critical point with $L \approx c\xi$, where $c \approx 2-4$, and then uses finite-size scaling [22, 23] to extrapolate to the infinite-volume limit. **Note Added 1996:** Recently, radical advances have been made in applying finite-size scaling to Monte Carlo simulations (see [97, 98] and especially [99, 100, 101, 102, 103]); the preceding two sentences can now be seen to be far too pessimistic. For reliable extrapolation to the infinite-volume limit, L and ξ must both be $\gg 1$, but the ratio L/ξ can in some cases be as small as 10^{-3} or even smaller (depending on the model and on the quality of the data). However, reliable extrapolation requires careful attention to systematic errors arising from correction-to-scaling terms. In practice, reliable infinite-volume values (with both statistical and systematic errors of order a few percent) can be obtained at $\xi \sim 10^5$ from lattices of size $L \sim 10^2$, at least in some models [101, 102, 103].

you're an elementary-particle physicist) and get depressed. It should be emphasized that the factor ξ^d is inherent in *all* Monte Carlo algorithms for spin models and field theories (but not for self-avoiding walks, see Section 7). The factor ξ^z could, however, conceivably be reduced or eliminated by a more clever algorithm.

What is to be done? Our knowledge of the physics of critical slowing-down tells us that the slow modes are the long-wavelength modes, if the updating is purely local. The natural solution is therefore to speed up those modes by some sort of *collective-mode* (nonlocal) updating. It is necessary, then, to *identify physically* the appropriate collective modes, and to devise an *efficient computational algorithm* for speeding up those modes. These two goals are unfortunately in conflict; it is very difficult to devise collective-mode algorithms that are not so nonlocal that their computational cost outweighs the reduction in critical slowing-down. Specific implementations of the collective-mode idea are thus highly model-dependent. At least three such algorithms have been invented so far:

- Fourier acceleration [24]
- Multi-grid Monte Carlo (MGMC) [25, 20, 26]
- The Swendsen–Wang algorithm [27] and its generalizations

Fourier acceleration and MGMC are very similar in spirit (though quite different technically). Their performance is thus probably qualitatively similar, in the sense that they probably work well for the same models and work badly for the same models. In the next lecture we give an introduction to the MGMC method; in the following lecture we discuss algorithms of Swendsen–Wang type.

5. MULTI-GRID ALGORITHMS

The phenomenon of critical slowing-down is not confined to Monte Carlo simulations: very similar difficulties were encountered long ago by numerical analysts concerned with the numerical solution of partial differential equations. An ingenious solution, now called the multi-grid (MG) method, was proposed in 1964 by the Soviet numerical analyst Fedorenko [28]: the idea is to consider, in addition to the original ("fine-grid") problem, a sequence of auxiliary "coarse-grid" problems that approximate the behavior of the original problem for excitations at successively longer length scales (a sort of "coarse-graining" procedure). The local updates of the traditional algorithms are then supplemented by coarse-grid updates. To a present-day physicist, this philosophy is remarkably reminiscent of the renormalization group — so it is all the more remarkable that it was invented two years before the work of Kadanoff [29] and seven years before the work of Wilson [30]! After a decade of dormancy, multi-grid was revived in the mid-1970's [31], and was shown to be an extremely efficient computational method. In the 1980's, multi-grid methods have become an active area of research in numerical analysis, and have been applied to a wide variety of problems in *classical* physics [32, 33]. Very recently [25] it was shown how a stochastic generalization of the multi-grid method — multi-grid Monte Carlo (MGMC) — can be applied to problems in *statistical*, and hence also Euclidean *quantum*, physics.

In this lecture we begin by giving a brief introduction to the deterministic multi-grid method; we then explain the stochastic analogue.* But it is worth indicating now the basic

*For an excellent introduction to the deterministic multi-grid method, see Briggs [34]; more advanced topics are covered in the book of Hackbusch [32]. Both MG and MGMC are discussed in detail in [20].

idea behind this generalization. There is a strong analogy between solving lattice systems of equations (such as the discrete Laplace equation) and making Monte Carlo simulations of lattice random fields. Indeed, given a Hamiltonian $H(\varphi)$, the deterministic problem is that of *minimizing* $H(\varphi)$, while the stochastic problem is that of *generating random samples* from the Boltzmann–Gibbs probability distribution $e^{-\beta H(\varphi)}$. The statistical-mechanical problem reduces to the deterministic one in the zero-temperature limit $\beta \longrightarrow +\infty$.

Many (but not all) of the deterministic iterative algorithms for minimizing $H(\varphi)$ can be generalized to stochastic iterative algorithms — that is, dynamic Monte Carlo methods — for generating random samples from $e^{-\beta H(\varphi)}$. For example, the Gauss–Seidel algorithm for minimizing H and the heat-bath algorithm for random sampling from $e^{-\beta H}$ are very closely related. Both algorithms sweep successively through the lattice, working on one site x at a time. The Gauss–Seidel algorithm updates φ_x so as to *minimize* the Hamiltonian $H(\varphi) = H(\varphi_x, \{\varphi_y\}_{y\neq x})$ when all the other fields $\{\varphi_y\}_{y\neq x}$ are held fixed at their current values. The heat-bath algorithm gives φ_x a new *random* value chosen from the probability distribution $\exp[-H(\varphi_x, \{\varphi_y\}_{y\neq x})]$, with all the fields $\{\varphi_y\}_{y\neq x}$ again held fixed. As $\beta \longrightarrow +\infty$ the heat-bath algorithm approaches Gauss–Seidel. A similar correspondence holds between MG and MGMC.

Before entering into details, let us emphasize that although the multi-grid method and the block-spin renormalization group (RG) are based on very similar *philosophies* — dealing with a single length scale at a time — they are in fact *very different*. In particular, the conditional coarse-grid Hamiltonian employed in the MGMC method is *not* the same as the renormalized Hamiltonian given by a block-spin RG transformation. The RG transformation computes the *marginal*, not the conditional, distribution of the block means — that is, it *integrates* over the complementary degrees of freedom, while the MGMC method *fixes* these degrees of freedom at their current (random) values. The conditional Hamiltonian employed in MGMC is given by an explicit finite expression, while the marginal (RG) Hamiltonian cannot be computed in closed form. The failure to appreciate these distinctions has unfortunately led to much confusion in the literature.*

5.1. Deterministic Multi-Grid

In this section we give a pedagogical introduction to multi-grid methods in the simplest case, namely the solution of deterministic linear or nonlinear systems of equations.

Consider, for purposes of exposition, the lattice Poisson equation $-\Delta\varphi = f$ in a region $\Omega \subset \mathbf{Z}^d$ with zero Dirichlet data. Thus, the equation is

$$(-\Delta\varphi)_x \equiv 2d\varphi_x - \sum_{x':\,|x-x'|=1} \varphi_{x'} = f_x \tag{5.1}$$

for $x \in \Omega$, with $\varphi_x \equiv 0$ for $x \notin \Omega$. This problem is equivalent to minimizing the quadratic Hamiltonian

$$H(\varphi) = \frac{1}{2}\sum_{\langle xy\rangle}(\varphi_x - \varphi_y)^2 - \sum_x f_x\varphi_x. \tag{5.2}$$

More generally, we may wish to solve a linear system

$$A\varphi = f, \tag{5.3}$$

*For further discussion, see [20, Section 10.1].

where for simplicity we shall assume A to be symmetric and positive-definite. This problem is equivalent to minimizing

$$H(\varphi) = \frac{1}{2}\langle\varphi, A\varphi\rangle - \langle f, \varphi\rangle. \tag{5.4}$$

Later we shall consider also non-quadratic Hamiltonians.

Our goal is to devise a rapidly convergent iterative method for solving numerically the linear system (5.3). We shall restrict attention to first-order stationary linear iterations of the general form

$$\varphi^{(n+1)} = M\varphi^{(n)} + Nf, \tag{5.5}$$

where $\varphi^{(0)}$ is an arbitrary initial guess for the solution. Obviously, we must demand at the very least that the true solution $\varphi \equiv A^{-1}f$ be a fixed point of (5.5); imposing this condition for all f, we conclude that

$$N = (I - M)A^{-1}. \tag{5.6}$$

The iteration (5.5) is thus completely specified by its *iteration matrix* M. Moreover, (5.5)-(5.6) imply that the error $e^{(n)} \equiv \varphi^{(n)} - \varphi$ satisfies

$$e^{(n+1)} = Me^{(n)}. \tag{5.7}$$

That is, the iteration matrix is the amplification matrix for the error. It follows easily that the iteration (5.5) is convergent for all initial vectors $\varphi^{(0)}$ if and only if the spectral radius $\rho(M) \equiv \lim_{n\to\infty} \|M^n\|^{1/n}$ is < 1; and in this case the convergence is exponential with asymptotic rate at least $\rho(M)$, i.e.,

$$\|\varphi^{(n)} - \varphi\| \le K n^p \rho(M)^n \tag{5.8}$$

for some $K, p < \infty$ (K depends on $\varphi^{(0)}$).

Now let us return to the specific system (5.1). One simple iterative algorithm arises by solving (5.1) repeatedly for φ_x:

$$\varphi_x^{(n+1)} = \frac{1}{2d}\left[\sum_{x':\,|x-x'|=1} \varphi_{x'}^{(n)} + f_x\right]. \tag{5.9}$$

(5.9) is called the *Jacobi iteration*. It is convenient to consider also a slight generalization of (5.9): let $0 < \omega \le 1$, and define

$$\varphi_x^{(n+1)} = (1-\omega)\varphi_x^{(n)} + \frac{\omega}{2d}\left[\sum_{x':\,|x-x'|=1} \varphi_{x'}^{(n)} + f_x\right]. \tag{5.10}$$

(5.10) is called the *damped Jacobi iteration* with damping parameter ω; for $\omega = 1$ it reduces to the ordinary Jacobi iteration.

It can be shown [35] that the spectral radius $\rho(M_{DJ,\omega})$ of the damped Jacobi iteration matrix is less than 1, so that the iteration (5.10) converges exponentially to the solution φ. This would appear to be a happy situation. Unfortunately, however, the convergence factor $\rho(M_{DJ,\omega})$ is usually very close to 1, so that many iterations are required in order to reduce the error $\|\varphi^{(n)} - \varphi\|$ to a small fraction of its initial value. Insight into this phenomenon can be gained by considering the simple *model problem* in which the domain Ω is a square $\{1,\ldots,L\} \times \{1,\ldots,L\}$. In this case we can solve exactly for the eigenvectors and eigenvalues of $M_{DJ,\omega}$: they are

$$\varphi_x^{(p)} = \sin p_1 x_1 \sin p_2 x_2 \tag{5.11}$$

$$\lambda_p = (1-\omega) + \frac{\omega}{2}(\cos p_1 + \cos p_2) \tag{5.12}$$

where $p_1, p_2 = \frac{\pi}{L+1}, \frac{2\pi}{L+1}, \ldots, \frac{L\pi}{L+1}$. The spectral radius of $M_{DJ,\omega}$ is the eigenvalue of largest magnitude, namely

$$\begin{aligned} \rho(M_{DJ,\omega}) = \lambda_{\frac{\pi}{L+1},\frac{\pi}{L+1}} &= 1 - \omega\left[1 - \cos\frac{\pi}{L+1}\right] \\ &= 1 - O(L^{-2}). \end{aligned} \tag{5.13}$$

It follows that $O(L^2)$ iterations are needed for the damped Jacobi iteration to converge adequately. This represents an enormous computational labor when L is large.

It is easy to see what is going on here: *the slow modes* ($\lambda_p \approx 1$) *are the long-wavelength modes* ($p_1, p_2 \ll 1$). [If $\omega \approx 1$, then some modes with wavenumber $p = (p_1, p_2) \approx (\pi, \pi)$ have eigenvalue $\lambda_p \approx -1$ and so also are slow. This phenomenon can be avoided by taking ω significantly less than 1; for simplicity we shall henceforth take $\omega = \frac{1}{2}$, which makes $\lambda_p \geq 0$ for all p.] It is also easy to see physically *why* the long-wavelength modes are slow. The key fact is that the (damped) Jacobi iteration is *local*: in a single step of the algorithm, "information" is transmitted only to nearest neighbors. One might guess that this "information" executes a random walk around the lattice; and for the true solution to be reached, "information" must propagate from the boundaries to the interior (and back and forth until "equilibrium" is attained). This takes a time of order L^2, in agreement with (5.13). In fact, this random-walk picture can be made rigorous [19].

This is an example of a *critical phenomenon*, in precisely the same sense that the term is used in statistical mechanics. The Laplace operator $A = -\Delta$ is critical, inasmuch as its Green function A^{-1} has long-range correlations (power-law decay in dimension $d > 2$, or growth in $d \leq 2$). This means that the solution of Poisson's equation in one region of the lattice depends strongly on the solution in distant regions of the lattice; "information" must propagate globally in order for "equilibrium" to be reached. Put another way, excitations at many length scales are significant, from one lattice spacing at the smallest to the entire lattice at the largest. The situation would be very different if we were to consider instead the Helmholtz–Yukawa equation $(-\Delta + m^2)\varphi = f$ with $m > 0$: its Green function has exponential decay with characteristic length m^{-1}, so that regions of the lattice separated by distances $\gg m^{-1}$ are essentially decoupled. In this case, "information" need only propagate a distance of order $\min(m^{-1}, L)$ in order for "equilibrium" to be reached. This takes a time of order $\min(m^{-2}, L^2)$, an estimate which can be confirmed rigorously by computing the obvious generalization of (5.12)–(5.13). On the other hand, as $m \to 0$ we recover the Laplace operator with its attendant difficulties: $m = 0$ is a *critical point*. We have here an example of *critical slowing-down* in classical physics.

The general structure of a remedy should now be obvious to physicists reared on the renormalization group: don't try to deal with all length scales at once, but define instead a *sequence* of problems in which each length scale, beginning with the smallest and working towards the largest, can be dealt with separately. An algorithm of precisely this form was proposed in 1964 by the Soviet numerical analyst Fedorenko [28], and is now called the *multi-grid method.*

Note first that the *only* slow modes in the damped Jacobi iteration are the long-wavelength modes (provided that ω is not near 1): as long as, say, $\max(p_1, p_2) \geq \frac{\pi}{2}$, we have $0 \leq \lambda_p \leq \frac{3}{4}$ (for $\omega = \frac{1}{2}$), independent of L. It follows that the short-wavelength components of the error $e^{(n)} \equiv \varphi^{(n)} - \varphi$ can be effectively killed by a few (say, five or ten) damped Jacobi iterations.

The remaining error has primarily long-wavelength components, and so is slowly varying in x-space. But a slowly varying function can be well represented on a coarser grid: if, for example, we were told $e_x^{(n)}$ only at *even* values of x, we could nevertheless reconstruct with high accuracy the function $e_x^{(n)}$ at *all* x by, say, linear interpolation. This suggests an improved algorithm for solving (5.1): perform a few damped Jacobi iterations on the original grid, until the (unknown) error is smooth in x-space; then set up an auxiliary coarse-grid problem whose solution will be approximately this error (this problem will turn out to be a Poisson equation on the coarser grid); perform a few damped Jacobi iterations on the coarser grid; and then transfer (interpolate) the result back to the original (fine) grid and add it in to the current approximate solution.

There are two advantages to performing the damped Jacobi iterations on the coarse grid. Firstly, the iterations take less work, because there are fewer lattice points on the coarse grid (2^{-d} times as many for a factor-of-2 coarsening in d dimensions). Secondly, with respect to the coarse grid the long-wavelength modes no longer have such long wavelength: *their wavelength has been halved* (i.e., their wavenumber has been doubled). This suggests that those modes with, say, $\max(p_1, p_2) \geq \frac{\pi}{4}$ can be effectively killed by a few damped Jacobi iterations on the coarse grid. And then we can transfer the remaining (smooth) error to a yet coarser grid, and so on recursively. These are the essential ideas of the multi-grid method.

Let us now give a precise definition of the multi-grid algorithm. For simplicity we shall restrict attention to problems defined in variational form*: thus, the goal is to minimize a real-valued function ("Hamiltonian") $H(\varphi)$, where φ runs over some N-dimensional real vector space U. We shall treat quadratic and non-quadratic Hamiltonians on an equal footing. In order to specify the algorithm we must specify the following ingredients:

1) A sequence of *coarse-grid spaces* $U_M \equiv U, U_{M-1}, U_{M-2}, \ldots, U_0$. Here $\dim U_l \equiv N_l$ and $N = N_M > N_{M-1} > N_{M-2} > \cdots > N_0$.

2) *Prolongation* (or "interpolation") *operators* $p_{l,l-1}: U_{l-1} \to U_l$ for $1 \leq l \leq M$.

3) *Basic* (or "smoothing") *iterations* $\mathcal{S}_l: U_l \times \mathcal{H}_l \to U_l$ for $0 \leq l \leq M$. Here $\mathcal{H}_l$ is a space of "possible Hamiltonians" defined on U_l; we discuss this in more detail below. The role of $\mathcal{S}_l$ is to take an approximate minimizer φ'_l of the Hamiltonian H_l and compute a new (hopefully better) approximate minimizer $\varphi''_l = \mathcal{S}_l(\varphi'_l, H_l)$. [For the present we can imagine that $\mathcal{S}_l$ consists of a few iterations of damped Jacobi for the Hamiltonian H_l.] Most generally, we shall use two smoothing iterations, $\mathcal{S}_l^{pre}$ and $\mathcal{S}_l^{post}$; they may be the same, but need not be.

4) *Cycle control parameters* (integers) $\gamma_l \geq 1$ for $1 \leq l \leq M$, which control the number of times that the coarse grids are visited.

We discuss these ingredients in more detail below.

The multi-grid algorithm is then defined recursively as follows:

procedure $mgm(l, \varphi, H_l)$

comment This algorithm takes an approximate minimizer φ of the Hamiltonian H_l, and overwrites it with a better approximate minimizer.

---

*In fact, the multi-grid method can be applied to the solution of linear or nonlinear systems of equations, whether or not these equations come from a variational principle. See, for example, [32] and [20, Section 2].

$\varphi \leftarrow S_l^{pre}(\varphi, H_l)$
if $l > 0$ **then**
compute $H_{l-1}(\cdot) \equiv H_l(\varphi + p_{l,l-1}\cdot)$
$\psi \leftarrow 0$
for $j = 1$ **until** γ_l **do** $mgm(l-1, \psi, H_{l-1})$
$\varphi \leftarrow \varphi + p_{l,l-1}\psi$
endif
$\varphi \leftarrow S_l^{post}(\varphi, H_l)$
end

Here is what is going on: We wish to solve the minimize the Hamiltonian H_l, and are given as input an approximate minimizer φ. The algorithm consists of three steps:

1) *Pre-smoothing.* We apply the basic iteration (e.g., a few sweeps of damped Jacobi) to the given approximate minimizer. This produces a better approximate minimizer in which the high-frequency (short-wavelength) components of the error have been reduced significantly. Therefore, the error, although still large, is *smooth* in x-space (whence the name "smoothing iteration").

2) *Coarse-grid correction.* We want to move rapidly towards the minimizer φ^* of H_l, using coarse-grid updates. Because of the pre-smoothing, the error $\varphi - \varphi^*$ is a smooth function in x-space, so it should be well approximated by fields in the range of the prolongation operator $p_{l,l-1}$. We will therefore carry out a coarse-grid update in which φ is replaced by $\varphi + p_{l,l-1}\psi$, where ψ lies in the coarse-grid subspace U_{l-1}. A sensible goal is to attempt to choose ψ so as to minimize H_l; that is, we attempt to minimize

$$H_{l-1}(\psi) \equiv H_l(\varphi + p_{l,l-1}\psi). \tag{5.14}$$

To carry out this approximate minimization, we use a few (γ_l) iterations of the best algorithm we know — namely, multi-grid itself! And we start at the best approximate minimizer we know, namely $\psi = 0$! The goal of this coarse-grid correction step is to reduce significantly the low-frequency components of the error in φ (hopefully without creating large new high-frequency error components).

3) *Post-smoothing.* We apply, for good measure, a few more sweeps of the basic smoother. (This would protect against any high-frequency error components which may inadvertently have been created by the coarse-grid correction step.)

The foregoing constitutes, of course, a single step of the multi-grid algorithm. In practice this step would be repeated several times, as in any other iteration, until the error has been reduced to an acceptably small value. The advantage of multi-grid over the traditional (e.g., damped Jacobi) iterative methods is that, with a suitable choice of the ingredients $p_{l,l-1}$, S_l and so on, only a few (maybe five or ten) iterations are needed to reduce the error to a small value, *independent of the lattice size L.* This contrasts favorably with the behavior (5.13) of the damped Jacobi method, in which $O(L^2)$ iterations are needed.

The multi-grid algorithm is thus a general framework; the user has considerable freedom in choosing the specific ingredients, which must be adapted to the specific problem. We now discuss briefly each of these ingredients; more details can be found in Chapter 3 of the book of Hackbusch [32].

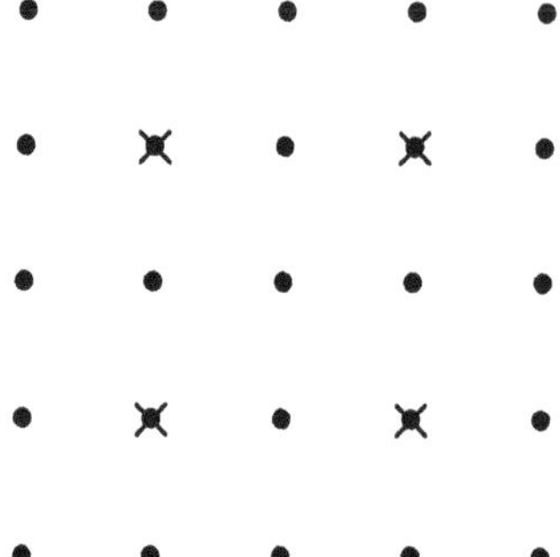

Figure 1. Standard coarsening (factor-of-2) in dimension $d = 2$. Dots are fine-grid sites and crosses are coarse-grid sites.

Coarse grids. Most commonly one uses a uniform factor-of-2 coarsening between each grid Ω_l and the next coarser grid Ω_{l-1}. The coarse-grid points could be either a subset of the fine-grid points (Fig. 1) or a subset of the dual lattice (Fig. 2). These schemes have obvious generalizations to higher-dimensional cubic lattices. In dimension $d = 2$, another possibility is a uniform factor-of-$\sqrt{2}$ coarsening (Fig. 3); note that the coarse grid is again a square lattice, rotated by 45°. Figs. 1–3 are often referred to as "standard coarsening," "staggered coarsening," and "red-black (or checkerboard) coarsening", respectively. Coarsenings by a larger factor (e.g., 3) could also be considered, but are generally disadvantageous. Note that each of the above schemes works also for periodic boundary conditions provided that the linear size L_l of the grid Ω_l is *even*. For this reason it is most convenient to take the linear size $L \equiv L_M$ of the original (finest) grid $\Omega \equiv \Omega_M$ to be a power of 2, or at least a power of 2 times a small integer. Other definitions of coarse grids (e.g., anisotropic coarsening) are occasionally appropriate.

Prolongation operators. For a coarse grid as in Fig. 2, a natural choice of prolongation

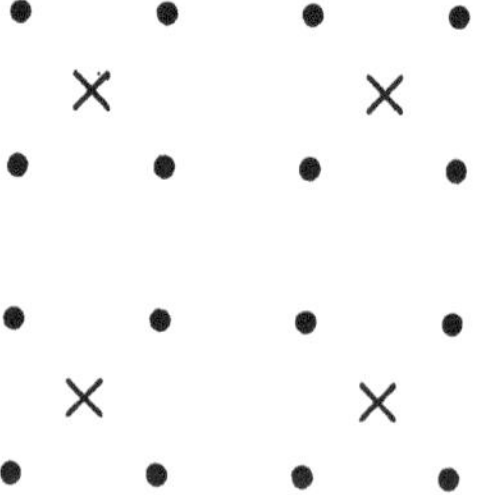

Figure 2. Staggered coarsening (factor-of-2) in dimension $d = 2$.

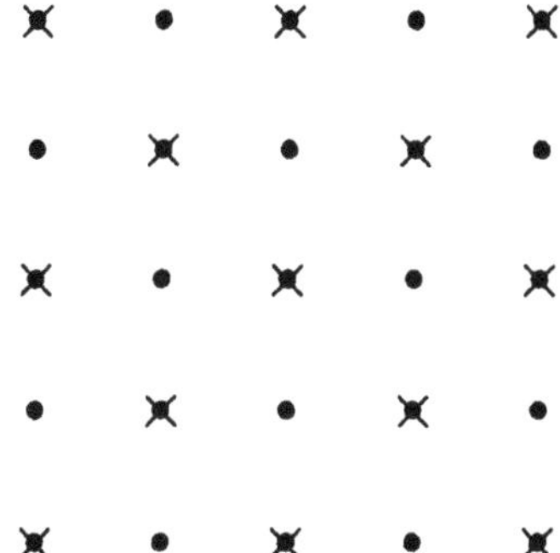

Figure 3. Red-black coarsening (factor-of-$\sqrt{2}$) in dimension $d = 2$.

operator is *piecewise-constant injection*:

$$(p_{l,l-1}\varphi_{l-1})_{x_1 \pm \frac{1}{2}, x_2 \pm \frac{1}{2}} = (\varphi_{l-1})_{x_1, x_2} \qquad \text{for all } x \in \Omega_{l-1} \tag{5.15}$$

(illustrated here for $d = 2$). It can be represented in an obvious shorthand notation by the stencil

$$\begin{bmatrix} 1 & 1 \\ 1 & 1 \end{bmatrix}. \tag{5.16}$$

For a coarse grid as in Fig. 1, a natural choice is *piecewise-linear interpolation*, one example of which is the *nine-point prolongation*

$$\begin{bmatrix} \frac{1}{4} & \frac{1}{2} & \frac{1}{4} \\ \frac{1}{2} & 1 & \frac{1}{2} \\ \frac{1}{4} & \frac{1}{2} & \frac{1}{4} \end{bmatrix}. \tag{5.17}$$

Higher-order interpolations (e.g., quadratic or cubic) can also be considered. All these prolongation operators can easily be generalized to higher dimensions.

We have ignored here some important subtleties concerning the treatment of the boundaries in defining the prolongation operator. Fortunately we shall not have to worry much about this problem, since most applications in statistical mechanics and quantum field theory use periodic boundary conditions.

Coarse-grid Hamiltonians. What does the coarse-grid Hamiltonian H_{l-1} look like? If the fine-grid Hamiltonian H_l is quadratic,

$$H_l(\varphi) = \frac{1}{2}\langle \varphi, A_l \varphi \rangle - \langle f_l, \varphi \rangle, \tag{5.18}$$

then so is the coarse-grid Hamiltonian H_{l-1}:

$$H_{l-1}(\psi) \equiv H_l(\varphi + p_{l,l-1}\psi) \tag{5.19}$$

$$= \frac{1}{2}\langle \psi, A_{l-1}\psi \rangle - \langle d, \psi \rangle + \text{const}, \tag{5.20}$$

where

$$A_{l-1} \equiv p^*_{l,l-1} A_l p_{l,l-1} \tag{5.21}$$

$$d \equiv p^*_{l,l-1}(f - A_l\varphi) \tag{5.22}$$

The coarse-grid problem is thus also a linear equation whose right-hand side is just the "coarse-graining" of the residual $r \equiv f - A_l\varphi$; this coarse-graining is performed using the adjoint of the interpolation operator $p_{l,l-1}$. The exact form of the coarse-grid operator A_{l-1} depends on the fine-grid operator A_l and on the choice of interpolation operator $p_{l,l-1}$. For example, if A_l is the nearest-neighbor Laplacian and $p_{l,l-1}$ is piecewise-constant injection, then it is easily checked that A_{l-1} is also a nearest-neighbor Laplacian (multiplied by an extra 2^{d-1}). On the other hand, if $p_{l,l-1}$ is piecewise-linear interpolation, then A_{l-1} will have nearest-neighbor *and next-nearest-neighbor* terms (but nothing worse than that).

Clearly, the point is to choose classes of Hamiltonians $\mathcal{H}_l$ with the property that if $H_l \in \mathcal{H}_l$ and $\varphi \in U_l$, then the coarse-grid Hamiltonian H_{l-1} defined by (5.14) necessarily lies in $\mathcal{H}_{l-1}$. In particular, it is convenient (though not in principle necessary) to choose all the Hamiltonians to have the same "functional form"; this functional form must be one which is stable under the coarsening operation (5.14). For example, suppose that the Hamiltonian H_l is a φ^4 theory with nearest-neighbor gradient term and possibly site-dependent coefficients:

$$H_l(\varphi) = \frac{\alpha}{2} \sum_{|x-x'|=1} (\varphi_x - \varphi_{x'})^2 + \sum_x V_x(\varphi_x), \tag{5.23}$$

where

$$V_x(\varphi_x) = \lambda\varphi_x^4 + \kappa_x\varphi_x^3 + A_x\varphi_x^2 + h_x\varphi_x. \tag{5.24}$$

Suppose, further, that the prolongation operator $p_{l,l-1}$ is piecewise-constant injection (5.15). Then the coarse-grid Hamiltonian $H_{l-1}(\psi) \equiv H_l(\varphi + p_{l,l-1}\psi)$ can easily be computed: it is

$$H_{l-1}(\psi) = \frac{\alpha'}{2} \sum_{|y-y'|=1} (\psi_y - \psi_{y'})^2 + \sum_y V'_y(\psi_y) + \text{const}, \tag{5.25}$$

where

$$V'_y(\psi_y) = \lambda'\psi_y^4 + \kappa'_y\psi_y^3 + A'_y\psi_y^2 + h'_y\psi_y \tag{5.26}$$

and

$$\alpha' = 2^{d-1}\alpha \tag{5.27}$$

$$\lambda' = 2^d\lambda \tag{5.28}$$

$$\kappa'_y = \sum_{x\in B_y} (4\lambda\varphi_x + \kappa_x) \tag{5.29}$$

$$A'_y = \sum_{x\in B_y} (6\lambda\varphi_x^2 + 3\kappa_x\varphi_x + A_x) \tag{5.30}$$

$$h'_y = \sum_{x\in B_y} (4\lambda\varphi_x^3 + 3\kappa_x\varphi_x^2 + 2A_x\varphi_x + h_x) \tag{5.31}$$

Here B_y is the block consisting of those 2^d sites of grid Ω_l which are affected by interpolation from the coarse-grid site $y \in \Omega_{l-1}$ (see Figure 2). Note that the coarse-grid Hamiltonian H_{l-1} has the same functional form as the "fine-grid" Hamiltonian H_l: it is specified by the coefficients α', λ', κ'_y, A'_y and h'_y. The step "*compute H_{l-1}*" therefore means to compute these coefficients. Note also the importance of allowing in (5.23) for φ^3 and φ terms and for site-dependent coefficients: even if these are not present in the original Hamiltonian $H \equiv H_M$, they will be generated on coarser grids. Finally, we emphasize that the coarse-grid Hamiltonian H_{l-1} depends implicitly on the current value of the fine-lattice field $\varphi \in U_l$; although our notation suppresses this dependence, it should be kept in mind.

Basic (smoothing) iterations. We have already discussed the damped Jacobi iteration as one possible smoother. Note that in this method only the "old" values $\varphi^{(n)}$ are used on the right-hand side of (5.9)/(5.10), even though for some of the terms the "new" value $\varphi^{(n+1)}$ may already have been computed. An alternative algorithm is to use at each stage on the right-hand side the "newest" available value. This algorithm is called the *Gauss–Seidel iteration.** Note that the Gauss–Seidel algorithm, unlike the Jacobi algorithm, depends on the *ordering* of the grid points. For example, if a 2-dimensional grid is swept in *lexicographic order* $(1,1)$, $(2,1)$, $\ldots$, $(L,1)$, $(1,2)$, $(2,2)$, $\ldots$, $(L,2)$, $\ldots$, $(1,L)$, $(2,L)$, $\ldots$, (L,L), then the Gauss–Seidel iteration becomes

$$\varphi_{x_1,x_2}^{(n+1)} = \frac{1}{4}[\varphi_{x_1+1,x_2}^{(n)} + \varphi_{x_1-1,x_2}^{(n+1)} + \varphi_{x_1,x_2+1}^{(n)} + \varphi_{x_1,x_2-1}^{(n+1)} + f_{x_1,x_2}]. \tag{5.32}$$

Another convenient ordering is the *red-black* (or *checkerboard*) *ordering*, in which the "red" sublattice $\Omega^r = \{x \in \Omega: x_1 + \cdots + x_d \text{ is even}\}$ is swept first, followed by the "black" sublattice $\Omega^b = \{x \in \Omega: x_1 + \cdots + x_d \text{ is odd}\}$. Note that the ordering of the grid points *within* each sublattice is irrelevant [for the usual nearest-neighbor Laplacian (5.1)], since the matrix A does not couple sites of the same color. This means that red-black Gauss–Seidel is particularly well suited to vector or parallel computation. Note that the red-black ordering makes sense with periodic boundary conditions only if the linear size L_l of the grid Ω_l is *even*.

It turns out that Gauss–Seidel is a better smoother than damped Jacobi (even if the latter is given its optimal ω). Moreover, Gauss–Seidel is easier to program and requires only half the storage space (no need for separate storage of "old" and "new" values). The only reason we introduced damped Jacobi at all is that it is easier to understand and to analyze.

Many other smoothing iterations can be considered, and can be advantageous in anisotropic or otherwise singular problems [32, Section 3.3 and Chapters 10–11]. But we shall stick to ordinary Gauss–Seidel, usually with red-black ordering.

Thus, S_l^{pre} and S_l^{post} will consist, respectively, of m_1 and m_2 iterations of the Gauss–Seidel algorithm. The balance between pre-smoothing and post-smoothing is usually not very crucial; only the total $m_1 + m_2$ seems to matter much. Indeed, one (but not both!) of m_1 or m_2 could be zero, i.e., either the pre-smoothing or the post-smoothing could be omitted entirely. Increasing m_1 and m_2 improves the convergence rate of the multi-grid iteration, but at the expense of increased computational labor per iteration. The optimal tradeoff seems to be achieved in most cases with $m_1 + m_2$ between about 2 and 4. The coarsest grid Ω_0 is a special case: it usually has so few grid points (perhaps only one!) that S_0 can be an exact solver.

The variational point of view gives special insight into the Gauss–Seidel algorithm, and shows how to generalize it to non-quadratic Hamiltonians. When updating site x, the new value φ_x' is chosen so as to *minimize* the Hamiltonian $H(\varphi) = H(\varphi_x, \{\varphi_y\}_{y \neq x})$ when all the other fields $\{\varphi_y\}_{y \neq x}$ are held fixed at their current values. The natural generalization of Gauss–Seidel to non-quadratic Hamiltonians is to adopt this variational definition: φ_x' should be the *absolute minimizer* of $H(\varphi) = H(\varphi_x, \{\varphi_y\}_{y \neq x})$. [If the absolute minimizer is non-unique, then one such

*It is amusing to note that "Gauss did not use a cyclic order of relaxation, and ... Seidel specifically recommended against using it" [36, p. 44*n*]. See also [37].

minimizer is chosen by some arbitrary rule.] This algorithm is called nonlinear Gauss–Seidel with exact minimization (NLGSEM) [38]. This definition of the algorithm presupposes, of course, that it is feasible to carry out the requisite exact one-dimensional minimizations. For example, for a φ^4 theory it would be necessary to compute the absolute minimum of a quartic polynomial in one variable. In practice these one-dimensional minimizations might themselves be carried out iteratively, e.g., by some variant of Newton's method.

Cycle control parameters and computational labor. Usually the parameters γ_l are all taken to be equal, i.e., $\gamma_l = \gamma \geq 1$ for $1 \leq l \leq M$. Then one iteration of the multi-grid algorithm at level M comprises one visit to grid M, γ visits to grid $M-1$, γ^2 visits to grid $M-2$, and so on. Thus, γ determines the degree of emphasis placed on the coarse-grid updates. ($\gamma = 0$ would correspond to the pure Gauss–Seidel iteration on the finest grid alone.)

We can now estimate the computational labor required for one iteration of the multi-grid algorithm. Each visit to a given grid involves $m_1 + m_2$ Gauss–Seidel sweeps on that grid, plus some computation of the coarse-grid Hamiltonian and the prolongation. The work involved is proportional to the number of lattice points on that grid. Let W_l be the work required for these operations on grid l. Then, for grids defined by a factor-of-2 coarsening in d dimensions, we have

$$W_l \approx 2^{-d(M-l)} W_M, \tag{5.33}$$

so that the total work for one multi-grid iteration is

$$\begin{aligned} \text{work}(MG) &= \sum_{l=M}^{0} \gamma^{M-l} W_l \\ &\approx W_M \sum_{l=M}^{0} (\gamma 2^{-d})^{M-l} \\ &\leq W_M (1-\gamma 2^{-d})^{-1} \qquad \text{if } \gamma < 2^d. \end{aligned} \tag{5.34}$$

Thus, provided that $\gamma < 2^d$, the work required for one entire multi-grid iteration is no more than $(1-\gamma 2^{-d})^{-1}$ times the work required for $m_1 + m_2$ Gauss–Seidel iterations (plus a little auxiliary computation) on the finest grid alone — *irrespective of the total number of levels.* The most common choices are $\gamma = 1$ (which is called the V-cycle) and $\gamma = 2$ (the W-cycle).

Convergence proofs. For certain classes of Hamiltonians H — primarily quadratic ones — and suitable choices of the coarse grids, prolongations, smoothing iterations and cycle control parameters, it can be proven rigorously* that the multi-grid iteration matrices M_l satisfy a *uniform* bound

$$\|M_l\| \leq C < 1, \tag{5.35}$$

valid *irrespective of the total number of levels.* Thus, a fixed number of multi-grid iterations (maybe five or ten) are sufficient to reduce the error to a small value, *independent of the lattice size L.* In other words, *critical slowing-down has been completely eliminated.*

The rigorous convergence proofs are somewhat arcane, so we cannot describe them here in any detail, but certain general features are worth noting. The convergence proofs are most straightforward when linear or higher-order interpolation and restriction are used, and $\gamma > 1$ (e.g., the W-cycle). When either low-order interpolation (e.g., piecewise-constant) or $\gamma = 1$

*For a detailed exposition of multi-grid convergence proofs, see [32, Chapters 6–8, 10, 11], [40] and the references cited therein. The additional work needed to handle the piecewise-constant interpolation can be found in [41].

(the V-cycle) is used, the convergence proofs become much more delicate. Indeed, if *both* piecewise-constant interpolation and a V-cycle are used, then the uniform bound (5.35) has *not* yet been proven, and it is most likely *false*! To some extent these features may be artifacts of the current methods of proof, but we suspect that they do also reflect real properties of the multi-grid method, and so the convergence proofs may serve as guidance for practice. For example, in our work we have used piecewise-constant interpolation (so as to preserve the simple nearest-neighbor coupling on the coarse grids), and thus for safety we stick to the W-cycle. There is in any case much room for further research, both theoretical and experimental.

To recapitulate, the extraordinary efficiency of the multi-grid method arises from the combination of two key features:

1) The convergence estimate (5.35). This means that only $O(1)$ iterations are needed, independent of the lattice size L.

2) The work estimate (5.34). This means that each iteration requires only a computational labor of order L^d (the fine-grid lattice volume).

It follows that the complete solution of the minimization problem, to any specified accuracy ε, requires a computational labor of order L^d.

Unigrid point of view. Let us look again at the multi-grid algorithm from the variational standpoint. One natural class of iterative algorithms for minimizing H are the so-called *directional methods*: let $p_0, p_1, \ldots$ be a sequence of "direction vectors" in U, and define $\varphi^{(n+1)}$ to be that vector of the form $\varphi^{(n)} + \lambda p_n$ which minimizes H. The algorithm thus travels "downhill" from $\varphi^{(n)}$ along the line $\varphi^{(n)} + \lambda p_n$ until reaching the minimum of H, then switches to direction p_{n+1} starting from this new point $\varphi^{(n+1)}$, and so on. For a suitable choice of the direction vectors $p_0, p_1, \ldots$, this method can be proven to converge to the global minimum of H [38, pp. 513–520].

Now, some iterative algorithms for minimizing $H(\varphi)$ can be recognized as special cases of the directional method. For example, the Gauss–Seidel iteration is a directional method in which the direction vectors are chosen to be unit vectors $e_1, e_2, \ldots, e_N$ (i.e., vectors which take the value 1 at a single grid point and zero at all others), where $N = \dim U$. [One step of the Gauss–Seidel iteration corresponds to N steps of the directional method.] Similarly, it is not hard to see [39] that the multi-grid iteration with the variational choices of restriction and coarse-grid operators, and with Gauss–Seidel smoothing at each level, is itself a directional method: some of the direction vectors are the unit vectors $e_1^{(M)}, e_2^{(M)}, \ldots, e_{N_M}^{(M)}$ of the fine-grid space, but other direction vectors are the images in the fine-grid space of the unit vectors of the coarse-grid spaces, i.e., they are $p_{M,l} e_1^{(l)}, p_{M,l} e_2^{(l)}, \ldots, p_{M,l} e_{N_l}^{(l)}$. The exact order in which these direction vectors are interleaved depends on the parameters m_1, m_2 and γ which define the cycling structure of the multi-grid algorithm. For example, if $m_1 = 1$, $m_2 = 0$ and $\gamma = 1$, the order of the direction vectors is $\{M\}$, $\{M-1\}, \ldots, \{0\}$, where $\{l\}$ denotes the sequence $p_{M,l} e_1^{(l)}, p_{M,l} e_2^{(l)}, \ldots, p_{M,l} e_{N_l}^{(l)}$. If $m_1 = 0$, $m_2 = 1$ and $\gamma = 1$, the order is $\{0\}, \{1\}, \ldots, \{M\}$. The reader is invited to work out other cases.

Thus, the multi-grid algorithm (for problems defined in variational form) is a directional method in which the direction vectors include both "single-site modes" $\{M\}$ and also "collective modes" $\{M-1\}, \{M-2\}, \ldots, \{0\}$ on all length scales. For example, if $p_{l,l-1}$ is piecewise-constant injection, then the direction vectors are characteristic functions χ_B (i.e., functions which are 1 on the block $B \subset \Omega$ and zero outside B), where the sets B are successively single sites, cubes of side 2, cubes of side 4, and so on. Similarly, if $p_{l,l-1}$ is linear interpolation, then the direction vectors are triangular waves of various widths.

The multi-grid algorithm has thus an alternative interpretation as a collective-mode algorithm working solely in the fine-grid space U. We emphasize that this "unigrid" viewpoint [39] is mathematically fully equivalent to the recursive definition given earlier. But it gives, we think, an important additional insight into what the multi-grid algorithm is really doing.

For example, for the simple model problem (Poisson equation in a square), we know that the "correct" collective modes are sine waves, in the sense that these modes diagonalize the Laplacian, so that in this basis the Jacobi or Gauss–Seidel algorithm would give the exact solution in a *single* iteration ($M_{\text{Jacobi}} = M_{\text{GS}} = 0$). On the other hand, the multi-grid method uses square-wave (or triangular-wave) updates, which are not exactly the "correct" collective modes. Nevertheless, the multi-grid convergence proofs [32, 40, 41] assure us that they are "close enough": the norm of the multi-grid iteration matrix M_l is bounded away from 1, uniformly in the lattice size, so that an accurate solution is reached in a *very few* MG iterations (in particular, critical slowing-down is completely eliminated). This viewpoint also explains why MG convergence is more delicate for piecewise-constant interpolation than for piecewise-linear: the point is that a sine wave (or other slowly varying function) can be approximated to arbitrary accuracy (in energy norm) by piecewise-linear functions but *not* by piecewise-constant functions.

We remark that McCormick and Ruge [39] have advocated the "unigrid" idea not just as an alternate point of view on the multi-grid algorithm, but as an alternate *computational procedure*. To be sure, the unigrid method is somewhat simpler to program, and this could have pedagogical advantages. But one of the key properties of the multi-grid method, namely the $O(L^d)$ computational labor per iteration, is sacrificed in the unigrid scheme. Instead of (5.33)–(5.34) one has

$$W_l \approx W_M \tag{5.36}$$

and hence

$$\begin{aligned} \text{work}(UG) &\approx W_M \sum_{l=M}^{0} \gamma^{M-l} \\ &\sim \begin{cases} MW_M & \text{if } \gamma = 1 \\ \gamma^M W_M & \text{if } \gamma > 1 \end{cases} \end{aligned} \tag{5.37}$$

Since $M \approx \log_2 L$ and $W_M \sim L^d$, we obtain

$$\text{work}(UG) \sim \begin{cases} L^d \log L & \text{if } \gamma = 1 \\ L^{d+\log_2 \gamma} & \text{if } \gamma > 1 \end{cases} \tag{5.38}$$

For a V-cycle the additional factor of $\log L$ is perhaps not terribly harmful, but for a W-cycle the additional factor of L is a severe drawback (though not as severe as the $O(L^2)$ critical slowing-down of the traditional algorithms). Thus, we do *not* advocate the use of unigrid as a computational method if there is a viable multi-grid alternative. Unigrid could, however, be of interest in cases where true multi-grid is unfeasible, as may occur for non-Abelian lattice gauge theories.

Multi-grid algorithms can also be devised for some models in which state space is a nonlinear manifold, such as nonlinear σ-models and lattice gauge theories [20, Sections 3–5].

The simplest case is the XY model: both the fine-grid and coarse-grid field variables are *angles*, and the interpolation operator is piecewise-constant (with angles added modulo 2π). Thus, a coarse-grid variable ψ_y specifies the angle by which the 2^d spins in the block B_y are to be simultaneously rotated. A similar strategy can be employed for nonlinear σ-models taking values in a group G (the so-called "principal chiral models"): the coarse-grid variable ψ_y simultaneously left-multiplies the 2^d spins in the block B_y. For nonlinear σ-models taking values in a nonlinear manifold M on which a group G acts [e.g., the n-vector model with $M = S_{n-1}$ and $G = SO(n)$], the coarse-grid-correction moves are still simultaneous rotation; this means that while the fine-grid fields lie in M, the coarse-grid fields all lie in G. Similar ideas can be applied to lattice gauge theories; the key requirement is to respect the geometric (parallel-transport) properties of the theory. Unfortunately, the resulting algorithms appear to be practical only in the abelian case. (In the non-abelian case, the coarse-grid Hamiltonian becomes too complicated.) Much more work needs to be done on devising good interpolation operators for non-abelian lattice gauge theories.

5.2. Multi-Grid Monte Carlo

Classical equilibrium statistical mechanics is a natural generalization of classical statics (for problems posed in variational form): in the latter we seek to minimize a Hamiltonian $H(\varphi)$, while in the former we seek to generate random samples from the Boltzmann–Gibbs probability distribution $e^{-\beta H(\varphi)}$. The statistical-mechanical problem reduces to the deterministic one in the zero-temperature limit $\beta \longrightarrow +\infty$.

Likewise, many (but not all) of the deterministic iterative algorithms for minimizing $H(\varphi)$ can be generalized to stochastic iterative algorithms — that is, dynamic Monte Carlo methods — for generating random samples from $e^{-\beta H(\varphi)}$. For example, the stochastic generalization of the Gauss–Seidel algorithm (or more generally, nonlinear Gauss–Seidel with exact minimization) is the single-site heat-bath algorithm; and the stochastic generalization of multi-grid is multi-grid Monte Carlo.

Let us explain these correspondences in more detail. In the Gauss–Seidel algorithm, the grid points are swept in some order, and at each stage the Hamiltonian is minimized as a function of a single variable φ_x, with all other variables $\{\varphi_y\}_{y\neq x}$ being held fixed. The single-site heat-bath algorithm has the same general structure, but the new value φ'_x is chosen randomly from the conditional distribution of $e^{-\beta H(\varphi)}$ given $\{\varphi_y\}_{y\neq x}$, i.e., from the one-dimensional probability distribution

$$P(\varphi'_x)\, d\varphi'_x = \text{const} \times \exp\left[-\beta H(\varphi'_x, \{\varphi_y\}_{y\neq x})\right] d\varphi'_x \tag{5.39}$$

(where the normalizing constant depends on $\{\varphi_y\}_{y\neq x}$). It is not difficult to see that this operation leaves invariant the Gibbs distribution $e^{-\beta H(\varphi)}$. As $\beta \longrightarrow +\infty$ it reduces to the Gauss–Seidel algorithm.

It is useful to visualize geometrically the action of the Gauss–Seidel and heat-bath algorithms within the space U of all possible field configurations. Starting at the current field configuration φ, the Gauss–Seidel and heat-bath algorithms propose to move the system along the line in U consisting of configurations of the form $\varphi' = \varphi + t\delta_x$ $(-\infty < t < \infty)$, where δ_x denotes the configuration which is 1 at site x and zero elsewhere. In the Gauss–Seidel algorithm, t is chosen so as to minimize the Hamiltonian restricted to the given line; while in the heat-bath algorithm, t is chosen randomly from the the conditional distribution of $e^{-\beta H(\varphi)}$ restricted to the given line, namely the one-dimensional distribution with probability density $P_{\text{cond}}(t) \sim \exp[-H_{\text{cond}}(t)] \equiv \exp[-H(\varphi + t\delta_x)]$.

The method of *partial resampling* generalizes the heat-bath algorithm in two ways:

1) The "fibers" used by the algorithm need not be lines, but can be higher-dimensional linear or even nonlinear manifolds.

2) The new configuration φ' need not be chosen *independently* of the old configuration φ (as in the heat-bath algorithm); rather, it can be selected by any updating procedure which leaves invariant the conditional probability distribution of $e^{-\beta H(\varphi)}$ restricted to the fiber.

The multi-grid Monte Carlo (MGMC) algorithm is a partial-resampling algorithm in which the "fibers" are the sets of field configurations that can be obtained one from another by a coarse-grid-correction step, i.e., the sets of fields $\varphi + p_{l,l-1}\psi$ with φ fixed and ψ varying over U_{l-1}. These fibers form a family of parallel affine subspaces in U_l, of dimension $N_{l-1} = \dim U_{l-1}$.

The ingredients of the MGMC algorithm are identical to those of the deterministic MG algorithm, with one exception: the deterministic smoothing iteration $\mathcal{S}_l$ is replaced by a stochastic smoothing iteration (for example, single-site heat-bath). That is, $\mathcal{S}_l(\cdot, H_l)$ is a stochastic updating procedure $\varphi_l \to \varphi_l'$ that leaves invariant the Gibbs distribution $e^{-\beta H_l}$:

$$\int d\varphi_l\, e^{-\beta H_l(\varphi_l)}\, P_{\mathcal{S}_l(\cdot,H_l)}(\varphi_l \to \varphi_l') = d\varphi_l'\, e^{-\beta H_l(\varphi_l')}. \tag{5.40}$$

The MGMC algorithm is then defined as follows:

procedure *mgmc*(l, φ, H_l)

comment This algorithm updates the field φ in such a way as to leave invariant the probability distribution $e^{-\beta H_l}$.

$\varphi \leftarrow \mathcal{S}_l^{pre}(\varphi, H_l)$

if $l > 0$ **then**

compute $H_{l-1}(\cdot) \equiv H_l(\varphi + p_{l,l-1}\cdot)$

$\psi \leftarrow 0$

for $j = 1$ **until** γ_l **do** *mgmc*$(l-1, \psi, H_{l-1})$

$\varphi \leftarrow \varphi + p_{l,l-1}\psi$

endif

$\varphi \leftarrow \mathcal{S}_l^{post}(\varphi, H_l)$

end

The alert reader will note that this algorithm is *identical* to the deterministic MG algorithm presented earlier; only the meaning of $\mathcal{S}_l$ is different.

The validity of the MGMC algorithm is proven inductively, starting at level 0 and working upwards. That is, if *mgmc*$(l-1, \cdot, H_{l-1})$ is a stochastic updating procedure that leaves invariant the probability distribution $e^{-\beta H_{l-1}}$, then *mgmc*$(l, \cdot, H_l)$ leaves invariant $e^{-\beta H_l}$. Note that the coarse-grid-correction step of the MGMC algorithm differs from the heat-bath algorithm in that the new configuration φ' is *not* chosen independently of the old configuration φ; to do so would be impractical, since the fiber has such high dimension. Rather, φ' (or what is equivalent, ψ) is chosen by a *valid updating procedure* — namely, MGMC itself!

The MGMC algorithm has also an alternate interpretation — the *unigrid* viewpoint — in which the fibers are one-dimensional and the resamplings are independent. More precisely, the fibers are lines of the form $\varphi' = \varphi + t\chi_B$ $(-\infty < t < \infty)$, where χ_B denotes the function which is 1 for sites belonging to the block B and zero elsewhere. The sets B are taken successively to be single sites, cubes of side 2, cubes of side 4, and so on. (If linear interpolation were used, then the "direction vectors" χ_B would be replaced by triangular waves of various widths.) Just as the deterministic unigrid algorithm chooses t so as to minimize the "conditional Hamiltonian" $H_{\text{cond}}(t) \equiv H(\varphi + t\chi_B)$, so the stochastic unigrid algorithm chooses t randomly from the one-dimensional distribution with probability density $P_{\text{cond}}(t) \sim \exp[-H_{\text{cond}}(t)]$. Conceptually this algorithm is no more complicated than the single-site heat-bath algorithm. But physically it is of course very different, as the direction vectors χ_B represent *collective modes* on all length scales.

We emphasize that the stochastic unigrid algorithm is *mathematically and physically equivalent* to the multi-grid Monte Carlo algorithm described above. But it is useful, we believe, to be able to look at MGMC from either of the two points of view: independent resamplings in one-dimensional fibers, or non-independent resamplings (defined recursively) in higher-dimensional (coarse-grid) fibers. On the other hand, the two algorithms are not *computationally* equivalent. One MGMC sweep requires a CPU time of order volume (provided that $\gamma < 2^d$), while the time for a unigrid sweep grows faster than the volume [cf. the work estimates (5.34) and (5.38)]. Therefore, we advocate unigrid only as a conceptual device, not as a computational algorithm.

How well does MGMC perform? The answer is highly model-dependent:

• For the *Gaussian model*, it can be proven rigorously [25, 20, 41] that τ is *bounded* as criticality is approached (empirically $\tau \approx 1-2$); therefore, critical slowing-down is *completely eliminated*. The proof is a simple Fock-space argument, combined with the convergence proof for deterministic MG; this will be discussed in Section 5.3.

• For the φ^4 *model*, numerical experiments [25] show that τ diverges with the *same* dynamic critical exponent as in the heat-bath algorithm; the gain in efficiency thus approaches a *constant* factor $F(\lambda)$ near the critical point. This behavior can be understood [20, Section 9.1] as due to the double-well nature of the φ^4 potential, which makes MGMC ineffective on large blocks. Thus, the correct collective modes at long length scales are nonlinear excitations *not* well modelled by $\varphi \to \varphi + t\chi_B$. (See Section 6 for an algorithm that appears to model these excitations well, at least for λ not too small.)

• For the $d = 2$ *XY model*, our numerical data [26] show a more complicated behavior: As the critical temperature is approached from above, τ diverges with a dynamic critical exponent $z = 1.4 \pm 0.3$ for the MGMC algorithm (in either V-cycle or W-cycle), compared to $z = 2.1 \pm 0.3$ for the heat-bath algorithm. Thus, critical slowing-down is significantly reduced but is still very far from being eliminated. On the other hand, below the critical temperature, τ is very small ($\approx 1-2$), uniformly in L and β (at least for the W-cycle); critical slowing-down appears to be completely eliminated. This very different behavior in the two phases can be understood physically: in the low-temperature phase the main excitations are spin waves, which are well handled by MGMC (as in the Gaussian model); but near the critical temperature the important excitations are widely separated vortex–antivortex pairs, which are apparently not easily created by the MGMC updates.

• For the $O(4)$ nonlinear σ-model in two dimensions, which is asymptotically free, preliminary data [42] show a very strong reduction, but not the total elimination, of critical slowing-down. For a W-cycle we find that $z \approx 0.6$ (I emphasize that these data are *very* preliminary!). Previously, Goodman and I had argued heuristically [20, Section 9.3] that for asymptotically free theories with a continuous symmetry group, MGMC (with a W-cycle) should completely eliminate critical slowing-down except for a possible logarithm. But our reasoning may now need to be re-examined!*

5.3. Stochastic Linear Iterations for the Gaussian Model

In this section we analyze an important class of Markov chains, the stochastic linear iterations for Gaussian models [20, Section 8].† This class includes, among others, the single-site heat-bath algorithm (with deterministic sweep of the sites), the stochastic SOR algorithm [44, 45] and the multi-grid Monte Carlo algorithm — all, of course, in the Gaussian case only. We show that the behavior of the stochastic algorithm is completely determined by the behavior of the corresponding deterministic algorithm for solving linear equations. In particular, we show that the exponential autocorrelation time τ_{exp} of the stochastic linear iteration is *equal* to the relaxation time of the corresponding linear iteration.

Consider any quadratic Hamiltonian

$$H(\varphi) = \frac{1}{2}(\varphi, A\varphi) - (f, \varphi), \tag{5.41}$$

where A is a symmetric positive-definite matrix. The corresponding Gaussian measure

$$d\pi(\varphi) = \text{const} \times e^{-\frac{1}{2}(\varphi, A\varphi) + (f, \varphi)}\, d\varphi \tag{5.42}$$

has mean $A^{-1}f$ and covariance matrix A^{-1}. Next consider any first-order stationary linear stochastic iteration of the form

$$\varphi^{(n+1)} = M\varphi^{(n)} + Nf + Q\xi^{(n)}, \tag{5.43}$$

where M, N and Q are fixed matrices and the $\xi^{(n)}$ are independent Gaussian random vectors with mean zero and covariance matrix C. The iteration (5.43) has a unique stationary distribution if and only if the spectral radius $\rho(M) \equiv \lim_{n \to \infty} \|M^n\|^{1/n}$ is < 1; and in this case the stationary distribution is the desired Gaussian measure (5.42) for all f if and only if

$$N = (I - M)A^{-1} \tag{5.44a}$$

$$QCQ^T = A^{-1} - MA^{-1}M^T \tag{5.44b}$$

(here ^{T} denotes transpose).

The reader will note the close analogy with the deterministic linear problem (5.3)–(5.6). Indeed, (5.44a) is identical with (5.6); and if we take the "zero-temperature limit" in which H is replaced by βH with $\beta \to +\infty$, then the Gaussian measure (5.42) approaches a delta function concentrated at the unique minimum of H (namely, the solution of the linear equation $A\varphi = f$), and the "noise" term disappears ($Q \to 0$), so that the stochastic iteration (5.43) turns into the deterministic iteration (5.5). That is:

(a) The linear deterministic problem is the zero-temperature limit of the Gaussian stochastic problem; and the first-order stationary linear deterministic iteration is the zero-temperature

---

***Note Added 1996:** For work on multi-grid Monte Carlo covering the period 1992–96, see [106, 107, 108, 103] and the references cited therein.

†Some of this material appears also in [43].

limit of the first-order stationary linear stochastic iteration. Therefore, any stochastic linear iteration for generating samples from the Gaussian measure (5.42) gives rise to a deterministic linear iteration for solving the linear equation (5.3), simply by setting $Q = 0$.

(b) Conversely, the stochastic problem and iteration are the nonzero-temperature generalizations of the deterministic ones. In principle this means that a deterministic linear iteration for solving (5.3) can be generalized to a stochastic linear iteration for generating samples from (5.42), if and only if the matrix $A^{-1} - MA^{-1}M^T$ is positive-semidefinite: just choose a matrix Q satisfying (5.44b). In practice, however, such an algorithm is computationally tractable only if the matrix Q has additional nice properties such as sparsity (or triangularity with a sparse inverse).

Examples. 1. *Single-site heat bath (with deterministic sweep of the sites) = stochastic Gauss–Seidel.* On each visit to site i, φ_i is replaced by a new value φ_i' chosen independently from the conditional distribution of (5.42) with $\{\varphi_j\}_{j\neq i}$ fixed at their current values: that is, φ_i' is Gaussian with mean $(f_i - \sum_{j\neq i} a_{ij}\varphi_j)/a_{ii}$ and variance $1/a_{ii}$. When updating φ_i at sweep $n+1$, the variables φ_j with $j < i$ have already been visited on this sweep, hence have their "new" values $\varphi_j^{(n+1)}$, while the variables φ_j with $j > i$ have not yet been visited on this sweep, and so have their "old" values $\varphi_j^{(n)}$. It follows that

$$\varphi_i^{(n+1)} = a_{ii}^{-1}\left(f_i - \sum_{j<i} a_{ij}\varphi_j^{(n+1)} - \sum_{j>i} a_{ij}\varphi_j^{(n)}\right) + a_{ii}^{-1/2}\xi_i^{(n)}, \tag{5.45}$$

where ξ has covariance matrix I. A little algebra brings this into the matrix form (5.43) with

$$M = -(D+L)^{-1}L^T \tag{5.46a}$$

$$N = (D+L)^{-1} \tag{5.46b}$$

$$Q = (D+L)^{-1}D^{1/2} \tag{5.46c}$$

where D and L are the diagonal and lower-triangular parts of the matrix A, respectively. It is straightforward to verify that (5.44a,b) are satisfied.* The single-site heat-bath algorithm is clearly the stochastic generalization of the Gauss–Seidel algorithm.

2. *Stochastic SOR.* For models which are Gaussian (or more generally, "multi-Gaussian"), Adler [44] and Whitmer [45] have shown that the successive over-relaxation (SOR) iteration admits a stochastic generalization, namely

$$\varphi_i^{(n+1)} = (1-\omega)\varphi_i^{(n)} + \omega a_{ii}^{-1}\left(f_i - \sum_{j<i} a_{ij}\varphi_j^{(n+1)} - \sum_{j>i} a_{ij}\varphi_j^{(n)}\right) + \left(\frac{\omega(2-\omega)}{a_{ii}}\right)^{1/2}\xi_i^{(n)}, \tag{5.47}$$

where $0 < \omega < 2$. For $\omega = 1$ this reduces to the single-site heat-bath algorithm. This is easily seen to be of the form (5.43) with

$$M = -(D+\omega L)^{-1}\left[(\omega-1)D + \omega L^T\right] \tag{5.48a}$$

*We remark that this verification never uses the fact that D is diagonal or that L is lower triangular. It is sufficient to have $A = D + L + L^T$ with D symmetric positive-definite and $D+L$ nonsingular. However, for the method to be practical, it is important that $D^{1/2}$ and $(D+L)^{-1}$ be "easy" to compute when applied to a vector.

$$N = \omega(D+\omega L)^{-1} \tag{5.48b}$$

$$Q = [\omega(2-\omega)]^{1/2}(D+\omega L)^{-1}D^{1/2} \tag{5.48c}$$

where D and L are as before. It is straightforward to verify that (5.44a,b) are satisfied.*

3. *Multi-Grid Monte Carlo (MGMC).* The multi-grid Monte Carlo algorithm *mgmc* (defined in Section 5.2) is identical to the corresponding deterministic multi-grid algorithm *mgm* (defined in Section 5.1) except that S_l is a stochastic rather than deterministic updating. Consider, for example, the case in which S_l is a stochastic *linear* updating (e.g., single-site heat-bath). Then the MGMC is also a stochastic linear updating of the form (5.43): in fact, M equals M_{MG}, the iteration matrix of the corresponding deterministic MG method, and N equals N_{MG}; the matrix Q is rather complicated, but since the MGMC algorithm is correct, Q must satisfy (5.44b). [The easiest way to see that $M = M_{\mathrm{MG}}$ is to imagine what would happen if all the random numbers $\xi^{(n)}$ were zero. Then the stochastic linear updating would reduce to the corresponding deterministic updating, and hence the same would be true for the MGMC updating as a whole.]

4. *Langevin equation with small time step.* As far as I know, there does not exist any useful stochastic generalization of the Jacobi iteration. However, let us discretize the Langevin equation

$$\frac{d\varphi}{dt} = -\frac{1}{2}C(A\varphi - f) + \xi, \tag{5.49}$$

where ξ is Gaussian white noise with covariance matrix C, using a small time step δ. The result is an iteration of the form (5.43) with

$$M = I - \frac{\delta}{2}CA \tag{5.50a}$$

$$N = \frac{\delta}{2}C \tag{5.50b}$$

$$Q = \delta^{1/2}I \tag{5.50c}$$

This satisfies (5.44a) exactly, but satisfies (5.44b) only up to an error of order δ. If $C = D^{-1}$, these M,N correspond to a damped Jacobi iteration with $\omega = \delta/2 \ll 1$.

It is straightforward to analyze the dynamic behavior of the stochastic linear iteration (5.43). Using (5.43) and (5.44) to express $\varphi^{(n)}$ in terms of the independent random variables $\varphi^{(0)}, \xi^{(0)}, \xi^{(1)}, \ldots, \xi^{(n-1)}$, we find after a bit of manipulation that

$$\langle \varphi^{(n)} \rangle = M^n \langle \varphi^{(0)} \rangle + (I - M^n)A^{-1}f \tag{5.51}$$

and

$$\begin{aligned} \operatorname{cov}(\varphi^{(s)}, \varphi^{(t)}) &= M^s \operatorname{cov}(\varphi^{(0)}, \varphi^{(0)})(M^T)^t \\ &\quad + \begin{cases} [A^{-1} - M^s A^{-1}(M^T)^s](M^T)^{t-s} & \text{if } s \le t \\ M^{s-t}[A^{-1} - M^t A^{-1}(M^T)^t] & \text{if } s \ge t \end{cases} \end{aligned} \tag{5.52}$$

Now let us either start the stochastic process in equilibrium

$$\langle \varphi^{(0)} \rangle = A^{-1}f \tag{5.53a}$$

$$\operatorname{cov}(\varphi^{(0)}, \varphi^{(0)}) = A^{-1} \tag{5.53b}$$

or else let it relax to equilibrium by taking $s,t \longrightarrow +\infty$ with $s-t$ fixed. Either way, we conclude that in equilibrium (5.43) defines a Gaussian stationary stochastic process with mean $A^{-1}f$ and autocovariance matrix

$$\operatorname{cov}(\varphi^{(s)},\varphi^{(t)}) = \begin{cases} A^{-1}(M^T)^{t-s} & \text{if } s \le t \\ M^{s-t}A^{-1} & \text{if } s \ge t \end{cases} \tag{5.54}$$

Moreover, since the stochastic process is Gaussian, all higher-order time-dependent correlation functions are determined in terms of the mean and autocovariance. Thus, the matrix M determines the autocorrelation functions of the Monte Carlo algorithm.

Another way to state these relationships is to recall [46, 47] that the Hilbert space $L^2(\pi)$ is isomorphic to the bosonic Fock space $\mathcal{F}(U)$ built on the "energy Hilbert space" (U,A): the "n-particle states" are the homogeneous Wick polynomials of degree n in the shifted field $\widetilde{\varphi} = \varphi - A^{-1}f$. (If U is one-dimensional, these are just the Hermite polynomials.) Then the transition probability $P(\varphi^{(n)} \longrightarrow \varphi^{(n+1)})$ induces on the Fock space an operator

$$P = \Gamma(M^T) \equiv I \oplus M^T \oplus (M^T \otimes M^T) \oplus \cdots \tag{5.55}$$

that is the second quantization of the operator M^T on the energy Hilbert space (see [20, Section 8] for details). It follows from (5.55) that

$$\|\Gamma(M)^n \upharpoonright \mathbf{1}^{\perp}\|_{L^2(\pi)} = \|M^n\|_{(U,A)} \tag{5.56}$$

and hence that

$$\rho(\Gamma(M) \upharpoonright \mathbf{1}^{\perp}) = \rho(M). \tag{5.57}$$

Moreover, P is self-adjoint on $L^2(\pi)$ [i.e., satisfies detailed balance] if and only if M is self-adjoint with respect to the energy inner product, i.e.,

$$MA = AM^T; \tag{5.58}$$

and in this case

$$\rho(\Gamma(M) \upharpoonright \mathbf{1}^{\perp}) = \|\Gamma(M) \upharpoonright \mathbf{1}^{\perp}\|_{L^2(\pi)} = \rho(M) = \|M\|_{(U,A)}. \tag{5.59}$$

In summary, we have shown that the dynamic behavior of any stochastic linear iteration is completely determined by the behavior of the corresponding deterministic linear iteration. In particular, the exponential autocorrelation time τ_{exp} (slowest decay rate of any autocorrelation function) is given by

$$\exp(-1/\tau_{\text{exp}}) = \rho(M), \tag{5.60}$$

and this decay rate is achieved by at least one observable which is linear in the field φ. In other words, the (worst-case) convergence rate of the Monte Carlo algorithm is precisely *equal* to the (worst-case) convergence rate of the corresponding deterministic iteration.

In particular, for Gaussian MGMC, the convergence proofs for deterministic multi-grid [32, 40, 41] combined with the arguments of the present section prove rigorously that *critical slowing-down is completely eliminated* (at least for a W-cycle). That is, the autocorrelation time τ of the MGMC method is *bounded* as criticality is approached (empirically $\tau \approx 1-2$).

6. SWENDSEN–WANG ALGORITHMS

A very different type of collective-mode algorithm was proposed two years ago* by Swendsen and Wang [27] for Potts spin models. Since then, there has been an explosion of work trying to understand why this algorithm works so well (and why it does not work even better), and trying to improve or generalize it. The basic idea behind all algorithms of Swendsen–Wang type is to augment the given model by means of *auxiliary variables*, and then to simulate this augmented model. In this lecture we describe the Swendsen–Wang (SW) algorithm and review some of the proposed variants and generalizations.

Let us first recall that the q-state Potts model [48, 49] is a generalization of the Ising model in which each spin σ_i can take q distinct values rather than just two ($\sigma_i = 1,2,\ldots,q$); here q is an integer ≥ 2. Neighboring spins prefer to be in the same state, and pay an energy price if they are not. The Hamiltonian is therefore

$$H(\sigma) = \sum_{\langle ij \rangle} J_{ij}\,(1 - \delta_{\sigma_i,\sigma_j}) \tag{6.1}$$

with $J_{ij} \geq 0$ for all i,j ("ferromagnetism"), and the partition function is

$$\begin{aligned} Z &= \sum_{\{\sigma\}} \exp[-H(\sigma)] \\ &= \sum_{\{\sigma\}} \exp\left[\sum_{\langle ij \rangle} J_{ij}\,(\delta_{\sigma_i,\sigma_j} - 1)\right] \\ &= \sum_{\{\sigma\}} \prod_{\langle ij \rangle} [(1 - p_{ij}) + p_{ij}\delta_{\sigma_i,\sigma_j}] \end{aligned} \tag{6.2}$$

where we have defined $p_{ij} = 1 - \exp(-J_{ij})$. The Gibbs measure $\mu_{\text{Potts}}(\sigma)$ is, of course,

$$\begin{aligned} \mu_{\text{Potts}}(\sigma) &= Z^{-1} \exp\left[\sum_{\langle ij \rangle} J_{ij}\,(\delta_{\sigma_i,\sigma_j} - 1)\right] \\ &= Z^{-1} \prod_{\langle ij \rangle} [(1 - p_{ij}) + p_{ij}\delta_{\sigma_i,\sigma_j}] \end{aligned} \tag{6.3}$$

We now use the deep identity

$$a + b = \sum_{n=0}^{1} [a\delta_{n,0} + b\delta_{n,1}] \tag{6.4}$$

on each bond $\langle ij \rangle$; that is, we introduce on each bond $\langle ij \rangle$ an auxiliary variable n_{ij} taking the values 0 and 1, and obtain

$$Z = \sum_{\{\sigma\}} \sum_{\{n\}} \prod_{\langle ij \rangle} [(1 - p_{ij})\,\delta_{n_{ij},0} + p_{ij}\delta_{n_{ij},1}\delta_{\sigma_i,\sigma_j}]. \tag{6.5}$$

***Note Added 1996:** Now nearly *ten* years ago! The great interest in algorithms of Swendsen–Wang type (frequently called *cluster algorithms*) has not abated: see the references cited in the "Notes Added" below, plus many others. See also [109, 110] for reviews that are slightly more up-to-date than the present notes (1990 rather than 1989).

Let us now take seriously the $\{n\}$ as dynamical variables: we can think of n_{ij} as an *occupation variable* for the bond $\langle ij \rangle$ (1 = occupied, 0 = empty). We therefore define the *Fortuin–Kasteleyn–Swendsen–Wang (FKSW) model* to be a joint model having q-state Potts spins σ_i at the sites and occupation variables n_{ij} on the bonds, with joint probability distribution

$$\mu_{\text{FKSW}}(\sigma,n) = Z^{-1} \prod_{\langle ij \rangle} [(1-p_{ij})\,\delta_{n_{ij},0} + p_{ij}\delta_{n_{ij},1}\,\delta_{\sigma_i,\sigma_j}]. \tag{6.6}$$

Finally, let us see what happens if we sum over the $\{\sigma\}$ at fixed $\{n\}$. Each occupied bond $\langle ij \rangle$ imposes a constraint that the spins σ_i and σ_j must be in the same state, but otherwise the spins are unconstrained. We therefore group the sites into connected clusters (two sites are in the same cluster if they can be joined by a path of occupied bonds); then all the spins within a cluster must be in the same state (all q values are equally probable), and distinct clusters are independent. It follows that

$$Z = \sum_{\{n\}} \left(\prod_{\langle ij \rangle : n_{ij}=1} p_{ij} \right) \left(\prod_{\langle ij \rangle : n_{ij}=0} (1-p_{ij}) \right) q^{C(n)}, \tag{6.7}$$

where $C(n)$ is the number of connected clusters (including one-site clusters) in the graph whose edges are the bonds having $n_{ij} = 1$. The corresponding probability distribution,

$$\mu_{\text{RC}}(n) = Z^{-1} \left(\prod_{\langle ij \rangle : n_{ij}=1} p_{ij} \right) \left(\prod_{\langle ij \rangle : n_{ij}=0} (1-p_{ij}) \right) q^{C(n)}, \tag{6.8}$$

is called the *random-cluster model with parameter q*. This is a generalized bond-percolation model, with non-local correlations coming from the factor $q^{C(n)}$; for $q = 1$ it reduces to ordinary bond percolation. Note, by the way, that in the random-cluster model (unlike the Potts and FKSW models), q is merely a parameter; it can take any positive real value, not just $2, 3, \ldots$. So the random-cluster model defines, in some sense, an analytic continuation of the Potts model to non-integer q; ordinary bond percolation corresponds to the "one-state Potts model."

We have already verified the following facts about the FKSW model:

a) $Z_{\text{Potts}} = Z_{\text{FKSW}} = Z_{\text{RC}}$.

b) The marginal distribution of μ_{FKSW} on the Potts variables $\{\sigma\}$ (integrating out the $\{n\}$) is precisely the Potts model $\mu_{\text{Potts}}(\sigma)$.

c) The marginal distribution of μ_{FKSW} on the bond occupation variables $\{n\}$ (integrating out the $\{\sigma\}$) is precisely the random-cluster model $\mu_{\text{RC}}(n)$.

The conditional distributions of μ_{FKSW} are also simple:

d) The conditional distribution of the $\{n\}$ given the $\{\sigma\}$ is as follows: independently for each bond $\langle ij \rangle$, one sets $n_{ij} = 0$ in case $\sigma_i \neq \sigma_j$, and sets $n_{ij} = 0, 1$ with probability $1 - p_{ij}, p_{ij}$, respectively, in case $\sigma_i = \sigma_j$.

e) The conditional distribution of the $\{\sigma\}$ given the $\{n\}$ is as follows: independently for each connected cluster, one sets all the spins σ_i in the cluster to the same value, chosen equiprobably from $\{1, 2, \ldots, q\}$.

These facts can be used for both analytic and numerical purposes. For example, by using facts (b), (c) and (e) we can prove an identity relating expectations in the Potts model to connection probabilities in the random-cluster model:

$$\begin{aligned}
\langle \delta_{\sigma_i,\sigma_j} \rangle_{Potts,q} &= \langle \delta_{\sigma_i,\sigma_j} \rangle_{FKSW,q} && \text{[by (b)]} \\
&= \langle E(\delta_{\sigma_i,\sigma_j} | \{n\}) \rangle_{FKSW,q} \\
&= \left\langle \frac{(q-1)\gamma_{ij}+1}{q} \right\rangle_{FKSW,q} && \text{[by (e)]} \\
&= \left\langle \frac{(q-1)\gamma_{ij}+1}{q} \right\rangle_{RC,q} && \text{[by (c)]}
\end{aligned} \tag{6.9}$$

Here

$$\gamma_{ij} \equiv \gamma_{ij}(n) \equiv \begin{cases} 1 & \text{if } i \text{ is connected to } j \\ 0 & \text{if } i \text{ is not connected to } j \end{cases} \tag{6.10}$$

and $E(\cdot | \{n\})$ denotes conditional expectation given $\{n\}$.* For the Ising model with the usual convention $\sigma = \pm 1$, (6.9) can be written more simply as

$$\langle \sigma_i \sigma_j \rangle_{\text{Ising}} = \langle \gamma_{ij} \rangle_{RC,q=2}. \tag{6.11}$$

Similar identities can be proven for higher-order correlation functions, and can be employed to prove Griffiths-type correlation inequalities for the Potts model [51, 52].

On the other hand, Swendsen and Wang [27] exploited facts (b)–(e) to devise a radically new type of Monte Carlo algorithm. The Swendsen–Wang algorithm (SW) simulates the joint model (6.6) by alternately applying the conditional distributions (d) and (e) — that is, by alternately generating new bond occupation variables (independent of the old ones) given the spins, and new spin variables (independent of the old ones) given the bonds. Each of these operations can be carried out in a computer time of order volume: for generating the bond variables this is trivial, and for generating the spin variable it relies on efficient (linear-time) algorithms for computing the connected clusters.† It is trivial that the SW algorithm leaves invariant the Gibbs measure (6.6), since any product of conditional probability operators has this property. It is also easy to see that the algorithm is ergodic, in the sense that every configuration $\{\sigma, n\}$ having nonzero μ_{FKSW}-measure is accessible from every other. So the SW algorithm is at least a *correct* algorithm for simulating the FKSW model. It is also an algorithm for simulating the Potts and random-cluster models, since expectations in these two models are equal to the corresponding expectations in the FKSW model.

Historical remark. The random-cluster model was introduced in 1969 by Fortuin and Kasteleyn [58]; they derived the identity $Z_{\text{Potts}} = Z_{\text{RC}}$, along with the correlation-function identity (6.9) and some generalizations. These relations were rediscovered several times

*For an excellent introduction to conditional expectations, see [50].

†Determining the connected components of an undirected graph is a classic problem of computer science. The depth-first-search and breadth-first-search algorithms[53] have a running time of order V, while the Fischer–Galler–Hoshen–Kopelman algorithm (in one of its variants)[54] has a worst-case running time of order $V \log V$, and an observed mean running time of order V in percolation-type problems[55]. Both of these algorithms are non-vectorizable. Shiloach and Vishkin[56] have invented a SIMD parallel algorithm, and we have very recently vectorized it for the Cyber 205, obtaining a speedup of a factor of 11 over scalar mode. We are currently carrying out a comparative test of these three algorithms, as a function of lattice size and bond density [57]. In view of the extraordinary performance of the SW algorithm (see below) and the fact that virtually all its CPU time is spent finding connected components, we feel that the desirability of finding improved algorithms for this problem is self-evident.

Table 1. Susceptibility χ and autocorrelation time $\tau_{\mathrm{inf},\mathcal{E}}$ ($\mathcal{E}$ = energy $\approx$ slowest mode) for two-dimensional Ising model at criticality, using Swendsen–Wang algorithm. Standard error is shown in parentheses.

$d = 2$ Ising Model				
L	χ		$\tau_{\mathrm{inf},\mathcal{E}}$	
64	1575	(10)	5.25	(0.30)
128	5352	(53)	7.05	(0.67)
256	17921	(109)	6.83	(0.40)
512	59504	(632)	7.99	(0.81)

during the subsequent two decades [59]. Surprisingly, however, no one seems to have noticed the *joint* probability distribution μ_{FKSW} that underlay all these identities; this was discovered implicitly by Swendsen and Wang [27], and was made explicit by Edwards and Sokal [60].

It is certainly plausible that the SW algorithm might have less critical slowing-down than the conventional (single-spin-update) algorithms: the reason is that a local move in one set of variables can have highly nonlocal effects in the other. For example, setting $n_b = 0$ on a single bond may disconnect a cluster, causing a big subset of the spins in that cluster to be flipped simultaneously. In some sense, therefore, the SW algorithm is a collective-mode algorithm in which the collective modes are *chosen by the system* rather than imposed from the outside as in multi-grid. (The miracle is that this is done in a way that preserves the correct Gibbs measure.)

How well does the SW algorithm perform? In at least some cases, the performance is nothing short of extraordinary. Table 1 shows some preliminary data [61] on a two-dimensional Ising model at the bulk critical temperature. These data are consistent with the estimate $\tau_{SW} \sim L^{\approx 0.35}$ [27].* By contrast, the conventional single-spin-flip algorithms for the two-dimensional Ising model have $\tau_{\mathrm{conv}} \sim L^{\approx 2.1}$ [21]. So the advantage of Swendsen–Wang over conventional algorithms (for this model) grows asymptotically like $L^{\approx 1.75}$. To be sure, one iteration of the Swendsen–Wang algorithm may be a factor of ~ 10 more costly in CPU time than one iteration of a conventional algorithm (the exact factor depends on the efficiency of the cluster-finding subroutine). But the SW algorithm wins already for modest values of L.

For other Potts models, the performance of the SW algorithm is less spectacular than for the two-dimensional Ising model, but it is still very impressive. In Table 2 we give the current best estimates of the dynamic critical exponent z_{SW} for q-state Potts models in d dimensions, as a function of q and d.†

All these exponents are much lower than the $z \gtrsim 2$ observed in the single-spin-flip algorithms.

Although the SW algorithm performs impressively well, we understand very little about *why* these exponents take the values they do. Some cases are easy. If $q = 1$, then all

*But *precisely because* τ rises so slowly with L, good estimates of the dynamic critical exponent will require the use of *extremely* large lattices. Even with lattices up to $L = 512$, we are unable to distinguish convincingly between $z \approx 0.35$ and $z \approx 0$. **Note Added 1996:** For more recent and much more precise data, see [111, 112]. These data are consistent with $\tau_{SW} \sim L^{\approx 0.24}$, but they are also consistent with $\tau_{SW} \sim \log^2 L$ [112]. It is *extremely* difficult to distinguish a small power from a logarithm.

†**Note Added 1996:** For more recent data, see [111, 112] for $d = 2, q = 2$; [113] for $d = 2, q = 3$; and [112, 114] for $d = 2, q = 4$. The situation for $d = 3, q = 2$ is extremely unclear: exponent estimates by different workers (using different methods) disagree wildly.

Table 2. Current best estimates of the dynamic critical exponent z for the Swendsen–Wang algorithm. Estimates are taken from [27] for $d = 2,3$, $q = 2$; [62] for $d = 2$, $q = 3,4$; and [63] for $d = 4$, $q = 2$. Error bar is a 95% confidence interval.

	Estimates of z_{SW}			
	$q=1$	$q=2$	$q=3$	$q=4$
$d=1$	0	0	0	0
$d=2$	0	≈ 0.35	0.55 ± 0.03	≈ 1 (exact??)
$d=3$	0	≈ 0.75	—	—
$d=4$	0	1 (exact?)	—	—

spins are in the same state (the *only* state!), and all bonds are thrown independently, so the autocorrelation time is zero. (Here the SW algorithm just reduces to the standard *static* algorithm for independent bond percolation.) If $d = 1$ (more generally, if the lattice is a *tree*), the SW dynamics is exactly soluble: the behavior of each bond is independent of each other bond, and $\tau_{\exp} \longrightarrow -1/\log(1-1/q) < \infty$ as $\beta \longrightarrow +\infty$. But the remainder of our understanding is very murky. Two principal insights have been obtained so far:

a) A calculation yielding $z_{SW} = 1$ in a mean-field (Curie–Weiss) Ising model [64]. This suggests (but of course does not prove) that $z_{SW} = 1$ for Ising models ($q = 2$) in dimension $d \geq 4$.

b) A rigorous proof that $z_{SW} \geq \alpha/\nu$ [62]. This bound, while valid for all d and q, is extremely far from sharp for the Ising models in dimensions 3 and higher. But it is reasonably good for the 3- and 4-state Potts models in two dimensions, and in the latter case it may even be sharp.*

But much remains to be understood!

The Potts model with q *large* behaves very differently. Instead of a critical point, the model undergoes a *first-order* phase transition: in two dimensions, this occurs when $q > 4$, while in three or more dimensions, it is believed to occur already when $q \geq 3$ [49]. At a first-order transition, *both* the conventional algorithms and the Swendsen–Wang algorithm have an extremely severe slowing-down (*much* more severe than the slowing-down at a critical point): right at the transition temperature, we expect $\tau \sim \exp(cL^{d-1})$. This is because sets of configurations typical of the ordered and disordered phases are separated by free-energy barriers of order L^{d-1}, i.e., by sets of intermediate configurations that contain interfaces of surface area $\sim L^{d-1}$ and therefore have an equilibrium probability $\sim \exp(-cL^{d-1})$.†

Wolff [65] has proposed a interesting modification of the SW algorithm, in which one builds only a *single* cluster (starting at a randomly chosen site) and flips it. Clearly, one step of the single-cluster SW algorithm makes less change in the system than one step of the standard

*__Note Added 1996:__ See [112, 113, 114] for more recent data on the possible sharpness of the Li–Sokal bound for two-dimensional models with $q = 2,3,4$. These data appear to be consistent with sharpness *modulo a logarithm*, i.e., $\tau_{SW}/C_H \sim \log L$.

†**Note Added 1996:** Great progress has been made in recent years in Monte Carlo methods for systems undergoing a first-order phase transition. The key idea is to simulate a suitably chosen *non*-Boltzmann probability distribution ("umbrella," "multicanonical," ...) and then apply reweighting methods. The exponential slowing-down $\tau \sim \exp(cL^{d-1})$ resulting from barrier penetration is replaced by a power-law slowing-down $\tau \sim L^p$ resulting from diffusion, *provided that* the distribution to be simulated is suitably chosen. See [115, 116, 117, 118, 119] for reviews.

SW algorithm, but it also takes much less work. If one enumerates the cluster using depth-first-search or breadth-first-search, then the CPU time is proportional to the size of the cluster; and by the Fortuin–Kasteleyn identity (6.9), the mean cluster size is equal to the susceptibility:

$$\sum_j \langle \gamma_{ij} \rangle = \sum_j \left\langle \frac{q\delta_{\sigma_i,\sigma_j} - 1}{q-1} \right\rangle \equiv \chi. \tag{6.12}$$

So the relevant quantity in the single-cluster SW algorithm is the dynamic critical exponent measured *in CPU units*:

$$z_{1-\text{cluster,CPU}} = z_{1-\text{cluster}} - \left(d - \frac{\gamma}{\nu}\right). \tag{6.13}$$

The value of the single-cluster algorithm is that the probability of choosing a cluster is proportional to its size (since we pick a random *site*), so the work is concentrated preferentially on larger clusters — and we think that is these clusters which are most important near the critical point. So it would not be surprising if $z_{1-\text{cluster,CPU}}$ were smaller than z_{SW}. Preliminary measurements indicate that $z_{1-\text{cluster,CPU}}$ is about the same as z_{SW} for the two-dimensional Ising model, but is significantly smaller for the three- and four-dimensional Ising models [66]. But a convincing theoretical understanding of this behavior is lacking.

Several other generalizations of the SW algorithm for Potts models have been proposed.* One is a multi-grid extension of the SW algorithm: the idea is to carry out only a partial FKSW transformation, but then to apply this concept recursively [67]. This algorithm *may* have a dynamic critical exponent that is smaller than that of standard SW (but the claims that $z = 0$ are in my opinion unconvincing).† A second generalization, which works only in *two* dimensions, augments the SW algorithm by transformations to the dual lattice [68]. This algorithm appears to achieve a modest improvement in critical slowing-down in the scaling region $|\beta - \beta_c| \sim L^{-1/\nu}$.

Finally, the SW algorithm can be generalized in a straightforward manner to Potts lattice gauge theories (more precisely, lattice gauge theories with a *finite abelian* gauge group G and *Potts (δ-function) action*). Preliminary results for the three-dimensional Z_2 gauge theory yield a dynamic critical exponent roughly equal to that of the ordinary SW algorithm for the three-dimensional Ising model (to which the gauge theory is dual) [69].

In the past year there has been a flurry of papers trying to generalize the SW algorithm to non-Potts models. Interesting proposals for a direct generalization of the SW algorithm were made by Edwards and Sokal [60] and by Niedermayer [70]. But the most promising ideas at present seem to be the *embedding algorithms* proposed by Brower and Tamayo [71] for one-component φ^4 models and by Wolff [65, 72] for multi-component $O(n)$-invariant models.

The idea of the embedding algorithms is to find Ising-like variables underlying a general spin variable, and then to update the resulting Ising model using the ordinary SW algorithm

***Note Added 1996:** In addition to the generalizations discussed below, see [120, 112] for SW-type algorithms for the Ashkin–Teller model; see [121] for a clever SW-type algorithm for the fully frustrated Ising model; and see [122, 123, 124, 100] for a very interesting SW-type algorithm for *antiferromagnetic* Potts models, based on the "embedding" idea to be described below.

†**Note Added 1996:** For at least some versions of the multi-grid SW algorithm, it can be proven [125] that the bound $z_{MGSW} \geq \alpha/\nu$ holds. Thus, the critical slowing-down is *not* completely eliminated in any model in which the specific heat is divergent.

(or the single-cluster variant). For one-component spins, this embedding is the obvious decomposition into magnitude and sign. Consider, therefore, a one-component model with Hamiltonian

$$H(\varphi) = -\beta \sum_{\langle xy \rangle} \varphi_x \cdot \varphi_y + \sum_x V(\varphi_x), \tag{6.14}$$

where $\beta \geq 0$ and $V(\varphi) = V(-\varphi)$. We write

$$\varphi_x = \varepsilon_x |\varphi_x|, \tag{6.15}$$

where the signs $\{\varepsilon\}$ are Ising variables. For *fixed* values of the magnitudes $\{|\varphi|\}$, the conditional probability distribution of the $\{\varepsilon\}$ is given by an Ising model with *ferromagnetic* (though space- dependent) couplings $J_{xy} \equiv \beta|\varphi_x||\varphi_y|$. Therefore, the $\{\varepsilon\}$ model can be updated by the Swendsen–Wang algorithm. (Heat-bath or MGMC sweeps must also be performed, in order to update the magnitudes.) For the two-dimensional φ^4 model, Brower and Tamayo [71] find a dynamic critical behavior identical to that of the SW algorithm for the two-dimensional Ising model — just as one would expect based on the idea that the "important" collective modes in the φ^4 model are spin flips.

Wolff's embedding algorithm for n-component models ($n \geq 2$) is equally simple. Consider an $O(n)$-invariant model with Hamiltonian

$$H(\sigma) = -\beta \sum_{\langle xy \rangle} \sigma_x \cdot \sigma_y + \sum_x V(|\sigma_x|), \tag{6.16}$$

with $\beta \geq 0$. Now fix a unit vector $\mathbf{r} \in \mathbb{R}^n$; then any spin vector $\sigma_x \in \mathbb{R}^n$ can be written uniquely (except for a set of measure zero) in the form

$$\sigma_x = \sigma_x^\perp + \varepsilon_x |\sigma_x \cdot \mathbf{r}| \mathbf{r}, \tag{6.17}$$

where

$$\sigma_x^\perp = \sigma_x - (\sigma_x \cdot \mathbf{r})\mathbf{r} \tag{6.18}$$

is the component of σ_x perpendicular to $\mathbf{r}$, and

$$\varepsilon_x = \operatorname{sgn}(\sigma_x \cdot \mathbf{r}) \tag{6.19}$$

takes the values ± 1. Therefore, for *fixed* values of the $\{\sigma^\perp\}$ and $\{|\sigma \cdot \mathbf{r}|\}$, the probability distribution of the $\{\varepsilon\}$ is given by an Ising model with *ferromagnetic* couplings $J_{xy} \equiv \beta|\sigma_x \cdot \mathbf{r}||\sigma_y \cdot \mathbf{r}|$. The algorithm is then: Choose at random a unit vector $\mathbf{r}$; fix the $\{\sigma^\perp\}$ and $\{|\sigma \cdot \mathbf{r}|\}$ at their current values, and update the $\{\varepsilon\}$ by the standard Swendsen–Wang algorithm (or the single-cluster variant). Flipping ε_x corresponds to reflecting σ_x in the hyperplane perpendicular to $\mathbf{r}$.

At first thought it may seem strange (and somehow "unphysical") to try to find Ising-like (i.e., discrete) variables in a model with a *continuous* symmetry group. However, upon reflection (pardon the pun) one sees what is going on: if the spin configuration is slowly varying (e.g., a long-wavelength spin wave), then the clusters tend to break along the surfaces where J_{xy} is small, hence where $\sigma \cdot \mathbf{r} \approx 0$. Then flipping ε_x on some clusters corresponds to a "soft" change near the cluster boundaries but a "hard" change in the cluster interiors, i.e., a long-wavelength collective mode (Figure 4). So it is conceivable that the algorithm could have very small (or even zero!) critical slowing-down in models where the important large-scale collective modes are spin waves.

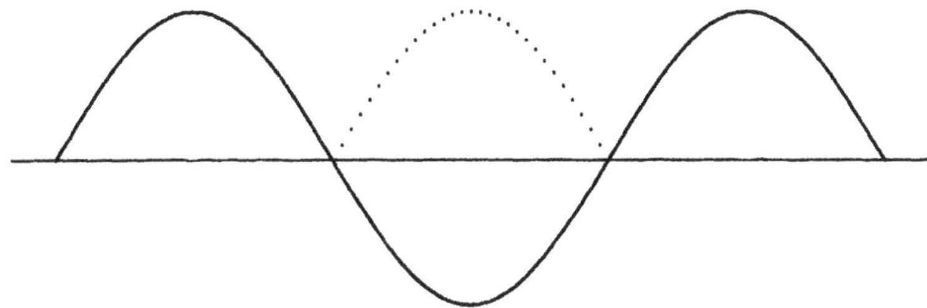

Figure 4. Action of the Wolff embedding algorithm on a long-wavelength spin wave. For simplicity, both spin space ($\boldsymbol{\sigma}$) and physical space (x) are depicted as one-dimensional.

Even more strikingly, consider a two-dimensional XY-model configuration consisting of a widely separated vortex–antivortex pair: in the continuum limit this is given by

$$\theta(z) = \operatorname{Im}\log(z-a) - \operatorname{Im}\log(z+a) \tag{6.20}$$

where $z \equiv x_1 + ix_2$ and $\boldsymbol{\sigma} \equiv (\cos\theta, \sin\theta)$. Then the surface $\boldsymbol{\sigma}\cdot\mathbf{r} = 0$ is an ellipse passing *directly through* the vortex and antivortex*. Reflecting the spins inside this ellipse produces a configuration $\{\boldsymbol{\sigma}_{\text{new}}\}$ that is a continuous map of the doubly punctured plane into the semicircle $|\boldsymbol{\sigma}| = 1, \boldsymbol{\sigma}\cdot\mathbf{r} \geq 0$. Such a map necessarily has zero winding number around the points $\pm a$. So the Wolff update has destroyed the vorticity!†

Therefore, the key collective modes in the two-dimensional XY and $O(n)$ models — spin waves and (for the XY case) vortices — are well "encoded" by the Ising variables $\{\varepsilon\}$. So it is quite plausible that critical slowing-down could be eliminated or almost eliminated. In fact, tests of this algorithm on the two-dimensional XY ($n = 2$), classical Heisenberg ($n = 3$) and $O(4)$ models are consistent with the complete absence of critical slowing-down [72, 73].

In view of the extraordinary success of the Wolff algorithm for spin models, it is tempting to try to extend it to lattice gauge theories with continuous gauge group [for example, $U(1)$, $SU(N)$ or $SO(N)$]. But I have nothing to report at present!‡

7. ALGORITHMS FOR THE SELF-AVOIDING WALK§

An *N–step self-avoiding walk* ω on a lattice $\mathcal{L}$ is a sequence of *distinct* points ω_0, $\omega_1, \ldots, \omega_N$ in $\mathcal{L}$ such that each point is a nearest neighbor of its predecessor. Let c_N [resp. $c_N(x)$] be the number of N-step SAWs starting at the origin and ending anywhere [resp. ending

*With $\mathbf{r} = (\cos\phi, \sin\phi)$, the equation of this ellipse is $x_1^2 + (x_2 + a\tan\phi)^2 = a^2\sec^2\phi$.

†This picture of the action of the Wolff algorithm on vortex–antivortex pairs was developed in discussions with Richard Brower and Robert Swendsen.

‡**Note Added 1996:** See [126] for a general theory of Wolff-type embedding algorithms as applied to nonlinear σ-models (spin models taking values in a Riemannian manifold M). Roughly speaking, we argue that a Wolff-type embedding algorithm can work well — in the sense of having a dynamic critical exponent z significantly less than 2 — *only if* M is a discrete quotient of a product of spheres: for example, a sphere S^{N-1} or a real projective space $RP^{N-1} \equiv S^{N-1}/Z_2$. In particular, we do *not* expect Wolff-type embedding algorithms to work well for σ-models taking values in $SU(N)$ for $N \geq 3$.

§**Note Added 1996:** An extensive and up-to-date review of Monte Carlo algorithms for the self-avoiding walk can be found in [127]. A briefer version is [128].

at x]. Let $\langle \omega_N^2 \rangle$ be the mean-square end-to-end distance of an N-step SAW. These quantities are believed to have the asymptotic behavior

$$c_N \sim \mu^N N^{\gamma-1} \tag{7.1}$$

$$c_N(x) \sim \mu^N N^{\alpha_{\rm sing}-2} \qquad (x \text{ fixed } \neq 0) \tag{7.2}$$

$$\langle \omega_N^2 \rangle \sim N^{2\nu} \tag{7.3}$$

as $N \longrightarrow \infty$; here γ, $\alpha_{\rm sing}$ and ν are critical exponents, while μ (the connective constant of the lattice) is the analogue of a critical temperature. The SAW has direct application in polymer physics [74], and is indirectly relevant to ferromagnetism and quantum field theory by virtue of its equivalence with the $n \longrightarrow 0$ limit of the n-vector model [75].

The SAW has some advantages over spin systems for Monte Carlo work: Firstly, one can work directly with SAWs on an infinite lattice; there are no systematic errors due to finite-volume corrections. Secondly, there is no L^d (or ξ^d) factor in the computational work, so one can go closer to criticality. Thus, the SAW is an exceptionally advantageous "laboratory" for the numerical study of critical phenomena.

Different aspects of the SAW can be probed in three different ensembles*:

- Free-endpoint grand canonical ensemble (variable N, variable x)
- Fixed-endpoint grand canonical ensemble (variable N, fixed x)
- Canonical ensemble (fixed N, variable x)

In the remainder of this section we survey some typical Monte Carlo algorithms for these ensembles.

Free-endpoint grand canonical ensemble. Here the configuration space S is the set of all SAWs, of arbitrary length, starting at the origin and ending anywhere. The grand partition function is

$$\Xi(\beta) = \sum_{N=0}^{\infty} \beta^N c_N \tag{7.4}$$

and the Gibbs measure is

$$\pi(\omega) = \Xi(\beta)^{-1} \times \beta^{|\omega|}. \tag{7.5}$$

The "monomer activity" β is a user-chosen parameter satisfying $0 \leq \beta < \beta_c \equiv \mu^{-1}$. As β approaches the critical activity β_c, the average walk length $\langle N \rangle$ tends to infinity. The connective constant μ and the critical exponent γ can be estimated from the Monte Carlo data, using the method of maximum likelihood [76].

A dynamic Monte Carlo algorithm for this ensemble was proposed by Berretti and Sokal [76]. The algorithm's elementary moves are as follows: either one attempts to append a new step to the walk, with equal probability in each of the q possible directions (here q is the coordination number of the lattice); or else one deletes the last step from the walk. In the former case, one must check that the proposed new walk is self-avoiding; if it isn't, then the attempted move is rejected and the old configuration is counted again in the sample ("null transition"). If an attempt is made to delete a step from an already-empty walk, then a null

*The proper terminology for these ensembles is unclear to me. Perhaps the grand canonical and canonical ensembles ought to be called "canonical" and "microcanonical," respectively, reserving the term "grand canonical" for ensembles of *many* SAWs. **Note Added 1996:** I now prefer calling these the "variable-length" and "fixed-length" ensembles, respectively, and avoiding the "–canonical" terms entirely; this avoids ambiguity.

transition is also made. The relative probabilities of $\Delta N = +1$ and $\Delta N = -1$ attempts are chosen to be

$$\mathbb{P}(\Delta N = +1 \text{ attempt}) = \frac{q\beta}{1+q\beta} \tag{7.6}$$

$$\mathbb{P}(\Delta N = -1 \text{ attempt}) = \frac{1}{1+q\beta} \tag{7.7}$$

Therefore, the transition probability matrix is

$$p(\omega \to \omega') = \begin{cases} \frac{\beta}{1+q\beta}\chi_{\mathrm{SAW}}(\omega') & \text{if } \omega \prec \omega' \\ \frac{1}{1+q\beta} & \text{if } \omega' \prec \omega \text{ or } \omega = \omega' = \emptyset \\ \frac{\beta}{1+q\beta}A(\omega) & \text{if } \omega = \omega' = \emptyset \end{cases} \tag{7.8}$$

where

$$\chi_{\mathrm{SAW}}(\omega') = \begin{cases} 1 & \text{if } \omega' \text{ is a SAW} \\ 0 & \text{if } \omega' \text{ is not a SAW} \end{cases} \tag{7.9}$$

Here $\omega \prec \omega'$ denotes that the walk ω' is a one-step extension of ω; and $A(\omega)$ is the number of *non-self-avoiding* walks ω' with $\omega \prec \omega'$. It is easily verified that this transition matrix satisfies detailed balance for the Gibbs distribution π, i.e.,

$$\pi(\omega)\,p(\omega \to \omega') = \pi(\omega')\,p(\omega' \to \omega) \tag{7.10}$$

for all $\omega, \omega' \in S$. It is also easily verified that the algorithm is ergodic (= irreducible): to get from a walk ω to another walk ω', it suffices to use $\Delta N = -1$ moves to transform ω into the empty walk $\emptyset$, and then use $\Delta N = +1$ moves to build up the walk ω'.

Let us now analyze the critical slowing-down of the Berretti–Sokal algorithm. We can argue heuristically that

$$\tau \sim \langle N \rangle^2. \tag{7.11}$$

To see this, consider the quantity $N(t) = |\omega|(t)$, the number of steps in the walk at time t. This quantity executes, crudely speaking, a random walk (with drift) on the nonnegative integers; the average time to go from some point N to the point 0 (i.e., the empty walk) is of order N^2. Now, each time the empty walk is reached, all memory of the past is erased; future walks are then independent of past ones. Thus, the autocorrelation time ought to be of order $\langle N^2 \rangle$, or equivalently $\langle N \rangle^2$.

This heuristic argument can be turned into a rigorous proof of a *lower bound* $\tau \geq \text{const} \times \langle N \rangle^2$ [10]. However, as an argument for an *upper bound* of the same form, it is not entirely convincing, as it assumes without proof that the *slowest* mode is the one represented by $N(t)$. With considerably more work, it is possible to prove an upper bound on τ that is only slightly weaker than the heuristic prediction: $\tau \leq \text{const} \times \langle N \rangle^{1+\gamma}$ [10, 77].* (Note that the critical exponent γ is believed to equal $43/32$ in dimension $d = 2$, ≈ 1.16 in $d = 3$, and 1 in $d \geq 4$.) In fact, we suspect [78] that the true behavior is $\tau \sim \langle N \rangle^p$ for some exponent p *strictly between* 2 and $1 + \gamma$. A deeper understanding of the dynamic critical behavior of the Berretti–Sokal algorithm would be of definite value.

It is worth comparing the computational work required for SAW versus Ising simulations: $\langle N \rangle^{\approx 2} \sim \xi^{\approx 2/\nu} = \xi^{\approx 3.4}$ for the $d = 3$ SAW, versus $\xi^{d+z} = \xi^{\approx 5.0}$ (resp.$\xi^{\approx 3.8}$) for the $d = 3$ Ising

---

*All these proofs are discussed in Section 8.

model using the Metropolis (resp. Swendsen–Wang) algorithm. This vindicates our assertion that the SAW is an advantageous model for Monte Carlo studies of critical phenomena.

Fixed-endpoint grand canonical ensemble. The configuration space S is the set of all SAWs, of arbitrary length, starting at the origin and ending at the fixed site x ($\neq 0$). The ensemble is as in the free-endpoint case, with c_N replaced by $Nc_N(x)$. The connective constant μ and the critical exponent α_{sing} can be estimated from the Monte Carlo data, using the method of maximum likelihood.

A dynamic Monte Carlo algorithm for this ensemble was proposed by Berg and Foerster [79] and Aragão de Carvalho, Caracciolo and Fröhlich (BFACF) [75, 80]. The elementary moves are local deformations of the chain, with $\Delta N = 0, \pm 2$. The critical slowing-down in the BFACF algorithm is quite subtle. On the one hand, Sokal and Thomas [81] have proven the surprising result that $\tau_{\text{exp}} = +\infty$ for *all* $\beta \neq 0$ (see Section 8). On the other hand, numerical experiments [12, 82] show that $\tau_{\text{inf},N} \sim \langle N \rangle^{3.0 \pm 0.4}$ (in $d = 2$). Clearly, the BFACF dynamics is not well understood at present: further work, both theoretical and numerical, is needed.

In addition, Caracciolo, Pelissetto and Sokal [82] are studying a "hybrid" algorithm that combines local (BFACF) moves with non-local (cut-and-paste) moves. The critical slowing-down, measured in CPU units, appears to be reduced slightly compared to the pure BFACF algorithm: $\tau \sim \langle N \rangle^{\approx 2.3}$ in $d = 2$.

Canonical ensemble. Algorithms for this ensemble, based on local deformations of the chain, have been used by polymer physicists for more than 25 years [83, 84]. So the recent proof [85] that all such algorithms are nonergodic (= not irreducible) comes as a slight embarrassment. Fortunately, there does exist a *non-local* fixed-N algorithm which is ergodic: the "pivot" algorithm, invented by Lal [86] and independently reinvented by MacDonald *et al.* [87] and by Madras [16]. The elementary move is as follows: choose at random a pivot point k along the walk ($1 \leq k \leq N-1$); choose at random a non-identity element of the symmetry group of the lattice (rotation or reflection); then apply the symmetry-group element to $\omega_{k+1}, \ldots, \omega_N$ using ω_k as a pivot. The resulting walk is accepted if it is self-avoiding; otherwise it is rejected and the walk ω is counted once more in the sample. It can be proven [16] that this algorithm is ergodic and preserves the equal-weight probability distribution.

At first thought the pivot algorithm sounds terrible (at least it did to me): for N large, nearly all the proposed moves will get rejected. This is in fact true: the acceptance fraction behaves N^{-p} as $N \to \infty$, where $p \approx 0.19$ in $d = 2$ [16]. On the other hand, the pivot moves are very radical: after very few (5 or 10) accepted moves the SAW will have reached an "essentially new" conformation. One conjectures, therefore, that $\tau \sim N^p$. Actually it is necessary to be a bit more careful: for *global* observables f (such as the end-to-end distance ω_N^2) one expects $\tau_{\text{inf},f} \sim N^p$; but *local* observables (such as the angle between the 17th and 18th bonds of the walk) are expected to evolve a factor of N more slowly: $\tau_{\text{inf},f} \sim N^{1+p}$. Thus, the *slowest* mode is expected to behave as $\tau_{\text{exp}} \sim N^{1+p}$. For the pivot algorithm applied to *ordinary* random walk one can calculate the dynamical behavior exactly [16]: for *global* observables f the autocorrelation function behaves roughly like

$$\rho_{ff}(t) \sim \sum_{i=1}^{n} \left(1 - \frac{i}{n}\right)^t, \tag{7.12}$$

from which it follows that

$$\tau_{\text{inf},f} \sim \log N \tag{7.13}$$

$$\tau_{\text{exp},f} \sim N \tag{7.14}$$

— in agreement with our heuristic argument modulo logarithms. For the SAW, it is found numerically [16] that $\tau_{\mathrm{inf},f} \sim N^{\approx 0.20}$ in $d = 2$, also in close agreement with the heuristic argument.

A careful analysis of the computational complexity of the pivot algorithm [36] shows that one "effectively independent" sample (at least as regards *global* observables) can be produced in a computer time of order N. This is a factor N more efficient than the Berretti–Sokal algorithm, a fact which opens up exciting prospects for high-precision Monte Carlo studies of critical phenomena in the SAW. Thus, with a modest computational effort (300 hours on a Cyber 170-730), Madras and I found $\nu = 0.7496 \pm 0.0007$ (95% confidence limits) for 2-dimensional SAWs of lengths $200 \leq N \leq 10000$ [16]. We hope to carry out soon a convincing numerical test of hyperscaling in the three-dimensional SAW.*

8. RIGOROUS ANALYSIS OF DYNAMIC MONTE CARLO ALGORITHMS

In this final lecture, we discuss techniques for proving rigorous lower and upper bounds on the autocorrelation times of dynamic Monte Carlo algorithms. This topic is of primary interest, of course, to mathematical physicists: it constitutes the first steps toward a rigorous theory of dynamic critical phenomena, along lines parallel to the well-established rigorous theory of static critical phenomena. But these proofs are, I believe, also of some importance for practical Monte Carlo work, as they give insight into the physical basis of critical slowing-down and may point towards improved algorithms with reduced critical slowing-down.

There is a big difference between the techniques used for proving lower and upper bounds, and it is easy to understand this physically. To prove a *lower* bound on the autocorrelation time, it suffices to find *one* physical reason why the dynamics should be slow, i.e., to find *one* "slow mode." This physical insight can often be converted directly into a rigorous proof, using the variational method described below. On the other hand, to prove an *upper* bound on the autocorrelation time, it is necessary to understand *all* conceivable physical reasons for slowness, and to prove that *none* of them cause too great a slowing-down. This is extremely difficult to do, and has been carried to completion in very few cases.

We shall be obliged to restrict attention to *reversible* Markov chains [i.e., those satisfying the detailed-balance condition (2.27)], as these give rise to self-adjoint operators. Non-reversible Markov chains, corresponding to non-self-adjoint operators, are much more difficult to analyze rigorously.

The principal method for proving lower bounds on $\tau_{\exp}$ (and in fact on $\tau_{\mathrm{inf},f}$) is the *variational* (or *Rayleigh–Ritz*) *method.* Let us recall some of the theory of reversible Markov chains from Section 2: Let π be a probability measure, and let $P = \{p_{xy}\}$ be a transition probability matrix that satisfies detailed balance with respect to π. Now P acts naturally on functions (observables) according to

$$(Pf)(x) \equiv \sum_y p_{xy} f(y). \tag{8.1}$$

*__Note Added 1996:__ This study has now been carried out [129], and yields $\nu = 0.5877 \pm 0.0006$ (subjective 68% confidence limits), based on SAWs of length up to 80000 steps. Proper treatment of corrections to scaling is crucial in obtaining this estimate. We also show that the interpenetration ratio Ψ approaches its limiting value $\Psi^* = 0.2471 \pm 0.0003$ *from above*, contrary to the prediction of the two-parameter renormalization-group theory. We have critically reexamined this theory and shown where the error lies [130, 129].

In particular, when acting on the space $l^2(\pi)$ of π-square-integrable functions, the operator P is a *self-adjoint contraction*. Its spectrum therefore lies in the interval $[-1,1]$. Moreover, P has an eigenvalue 1 with eigenvector equal to the constant function 1. Let Π be the orthogonal projector in $l^2(\pi)$ onto the constant functions. Then, for each real-valued observable $f \in l^2(\pi)$, we have

$$\begin{aligned} C_{ff}(t) &= (f, (P^{|t|} - \Pi)f) \\ &= (f, (I - \Pi)P^{|t|}(I - \Pi)f) \end{aligned} \tag{8.2}$$

where $(g,h) \equiv \langle g^* h \rangle_\pi$ denotes the inner product in $l^2(\pi)$. By the spectral theorem, this can be written as

$$C_{ff}(t) = \int_{-1}^{1} \lambda^{|t|}\, d\nu_{ff}(\lambda), \tag{8.3}$$

where $d\nu_{ff}$ is a *positive* measure. It follows that

$$\begin{aligned} \tau_{\mathrm{inf},f} &= \frac{1}{2} \frac{\int_{-1}^{1} \frac{1+\lambda}{1-\lambda}\, d\nu_{ff}(\lambda)}{\int_{-1}^{1} d\nu_{ff}(\lambda)} \\ &\geq \frac{1}{2} \frac{1+\rho_{ff}(1)}{1-\rho_{ff}(1)} \end{aligned} \tag{8.4}$$

by Jensen's inequality (since the function $\lambda \mapsto (1+\lambda)/(1-\lambda)$ is convex).*

Our method will be to compute explicitly a lower bound on the normalized autocorrelation function at time lag 1,

$$\rho_{ff}(1) = \frac{C_{ff}(1)}{C_{ff}(0)}, \tag{8.5}$$

for a suitably chosen trial observable f. Equivalently, we compute an upper bound on the Rayleigh quotient

$$\frac{(f, (I-P)f)}{(f, (I-\Pi)f)} = \frac{C_{ff}(0) - C_{ff}(1)}{C_{ff}(0)} = 1 - \rho_{ff}(1). \tag{8.6}$$

The crux of the matter is to pick a trial observable f that has a strong enough overlap with the "slowest mode." A useful formula is

$$(f, (I-P)f) = \frac{1}{2} \sum_{x,y} \pi_x p_{xy} \, |f(x) - f(y)|^2. \tag{8.7}$$

That is, the numerator of the Rayleigh quotient is half the mean-square change in f in a single time step.

Example 1. *Single-site heat-bath algorithm.* Let us consider, for simplicity, a translation-invariant spin model in a periodic box of volume V. Let $P_i = \{P_i(\varphi \to \varphi')\}$ be the transition probability matrix associated with applying the heat-bath operation at site i.† Then the operator P_i has the following properties:

*It also follows from (8.3) that $\rho_{ff}(t) \geq \rho_{ff}(1)^{|t|}$ for *even* values of t. Moreover, this holds for odd values of t if $d\nu_{ff}$ is supported on $\lambda \geq 0$ (though not necessarily otherwise); this is the case for the heat-bath and Swendsen–Wang algorithms, in which $P \geq 0$. Therefore, the decay as $t \to \infty$ of $\rho_{ff}(t)$ is also bounded below in terms of $\rho_{ff}(1)$.

†Probabilistically, P_i is the conditional expectation $E^\pi(\cdot|\{\varphi_j\}_{j \neq i})$. Analytically, P_i is the orthogonal projection in $l^2(\pi)$ onto the linear subspace consisting of functions depending only on $\{\varphi_j\}_{j \neq i}$.

(a) P_i is an orthogonal projection. (In particular, $P_i^2 = P_i$ and $0 \leq P_i \leq I$.)

(b) $P_i \mathbf{1} = \mathbf{1}$. (In particular, $\Pi P_i = P_i \Pi = \Pi$.)

(c) $P_i f = f$ if f depends only on $\{\varphi_j\}_{j \neq i}$. (In fact a stronger property holds: $P_i(fg) = f P_i g$ if f depends only on $\{\varphi_j\}_{j \neq i}$. But we shall not need this.)

For simplicity we consider only random site updating. Then the transition matrix of the heat-bath algorithm is

$$P = \frac{1}{V} \sum_i P_i. \tag{8.8}$$

We shall use the notation of the Ising model, but the same proof applies to much more general models.

As explained in Section 4, the critical slowing-down of the local algorithms (such as single-site heat-bath) arises from the fact that large regions in which the net magnetization is positive or negative tend to move slowly. In particular, the total magnetization of the lattice fluctuates slowly. So it is natural to expect that that the total magnetization $\mathcal{M} = \sum_i \sigma_i$ will be a "slow mode." Indeed, let us compute an upper bound on the Rayleigh quotient (8.6) with $f = \mathcal{M}$. The denominator is

$$(\mathcal{M}, (I - \Pi)\mathcal{M}) = \langle \mathcal{M}^2 \rangle - \langle \mathcal{M} \rangle^2 = V\chi \tag{8.9}$$

(since $\langle \mathcal{M} \rangle = 0$ by symmetry). The numerator is

$$\begin{aligned}
(\mathcal{M}, (I - P)\mathcal{M}) &= \frac{1}{V} \sum_{i,j,k} (\sigma_i, (I - P_j)\sigma_k) \\
&= \frac{1}{V} \sum_i (\sigma_i, (I - P_i)\sigma_i) \\
&\leq \frac{1}{V} \sum_i (\sigma_i, \sigma_i) \\
&= \frac{1}{V} \sum_i \langle \sigma_i^2 \rangle \\
&= 1.
\end{aligned} \tag{8.10}$$

Here we first used properties (a) and (c) to deduce that $(\sigma_i, (I - P_j)\sigma_k) = 0$ unless $i = j = k$, and then used property (a) to bound $(\sigma_i, (I - P_i)\sigma_i)$. It follows from (8.9) and (8.10) that

$$1 - \rho_{\mathcal{M},\mathcal{M}}(1) \leq \frac{1}{V\chi} \tag{8.11}$$

and hence that

$$\tau_{\mathrm{inf},\mathcal{M}} \geq V\chi - \frac{1}{2}. \tag{8.12}$$

Note, however, that here time is measured in *hits of a single site*. If we measure time in *hits per site*, then we conclude that

$$\tau_{\mathrm{inf},\mathcal{M}}^{(HPS)} \geq \chi - \frac{1}{2V} \approx \chi. \tag{8.13}$$

This proves a lower bound [88, 89, 90] on the dynamic critical exponent z, namely $z \geq \gamma/\nu$. (Here γ and ν are the static critical exponents for the susceptibility and correlation length, respectively. Note that by the usual static scaling law, $\gamma/\nu = 2-\eta$; and for most models η is very close to zero. So this argument almost makes rigorous (as a *lower bound*) the heuristic argument that $z \approx 2$.)

Virtually the same argument applies, in fact, to *any* reversible single-site algorithm (e.g., Metropolis). The only difference is that the property $0 \le P_i \le I$ is replaced by $-I \le P_i \le I$, so the bound on the numerator is a factor of 2 larger. Hence

$$\tau_{\mathrm{inf},\mathcal{M}}^{(HPS)} \ge \frac{\chi}{2} - \frac{1}{2V} \approx \frac{\chi}{2}. \tag{8.14}$$

Example 2. *Swendsen–Wang algorithm.* Recall that the SW algorithm simulates the joint model (6.6) by alternately applying the conditional distributions of $\{n\}$ given $\{\sigma\}$, and $\{\sigma\}$ given $\{n\}$ — that is, by alternately generating new bond occupation variables (independent of the old ones) given the spins, and new spin variables (independent of the old ones) given the bonds. The transition matrix $P_{\mathrm{SW}} = P_{\mathrm{SW}}(\{\sigma,n\} \to \{\sigma',n'\})$ is therefore a product

$$P_{\mathrm{SW}} = P_{\mathrm{bond}} P_{\mathrm{spin}}, \tag{8.15}$$

where P_{bond} is the conditional expectation operator $E(\cdot|\{\sigma\})$, and P_{spin} is $E(\cdot|\{n\})$.

We shall use the variational method, with f chosen to be the bond density

$$f = \mathcal{N} = \sum_{\langle ij \rangle} n_{ij}. \tag{8.16}$$

To lighten the notation, let us write $\delta_{\sigma_b} \equiv \delta_{\sigma_i,\sigma_j}$ for a bond $b = \langle ij \rangle$. We then have

$$E(n_b|\{\sigma\}) = p_b\, \delta_{\sigma_b} \tag{8.17}$$

$$E(n_b n_{b'}|\{\sigma\}) = p_b\, p_{b'}\, \delta_{\sigma_b}\, \delta_{\sigma_{b'}} \qquad \text{for } b \neq b' \tag{8.18}$$

It follows from (8.17) that

$$\begin{aligned} \langle n_b(t=0)\, n_{b'}(t=1)\rangle &= \langle n_b E(E(n_{b'}|\{\sigma\})|\{n\})\rangle \\ &= \langle n_b E(n_{b'}|\{\sigma\})\rangle \\ &= \langle E(n_b|\{\sigma\}) E(n_{b'}|\{\sigma\})\rangle \\ &= p_b\, p_{b'}\, \langle \delta_{\sigma_b}\, \delta_{\sigma_{b'}}\rangle. \end{aligned} \tag{8.19}$$

The corresponding truncated (connected) correlation function is clearly

$$\langle n_b(t=0); n_{b'}(t=1)\rangle = p_b\, p_{b'}\, \langle \delta_{\sigma_b}; \delta_{\sigma_{b'}}\rangle, \tag{8.20}$$

where $\langle A;B\rangle \equiv \langle AB\rangle - \langle A\rangle\langle B\rangle$. We have thus expressed a dynamic correlation function of the SW algorithm in terms of a static correlation function of the Potts model.

Now let us compute the same quantity at time lag 0: by (8.18), for $b \neq b'$ we have

$$\langle n_b(t=0)\, n_{b'}(t=0)\rangle = p_b\, p_{b'}\, \langle \delta_{\sigma_b}\, \delta_{\sigma_{b'}}\rangle \tag{8.21}$$

and hence

$$\langle n_b(t=0); n_{b'}(t=0)\rangle = p_b\, p_{b'}\, \langle \delta_{\sigma_b}; \delta_{\sigma_{b'}}\rangle. \tag{8.22}$$

On the other hand, for $b = b'$ we clearly have

$$\begin{aligned}\langle n_b(t=0); n_{b'}(t=0)\rangle &= \langle n_b\rangle - \langle n_b\rangle^2 \\ &= p_b\,\langle\delta_{\sigma_b}\rangle - p_b^2\,\langle\delta_{\sigma_b}\rangle^2.\end{aligned} \tag{8.23}$$

Consider now the usual case in which

$$p_b = \begin{cases} p & \text{for } b \in B \\ 0 & \text{otherwise}\end{cases} \tag{8.24}$$

for some family of bonds B. Combining (8.20), (8.22) and (8.23) and summing over b, b', we obtain

$$\langle \mathcal{N}(t=0); \mathcal{N}(t=1)\rangle = p^2\langle \mathcal{E};\mathcal{E}\rangle \tag{8.25}$$

$$\langle \mathcal{N}(t=0); \mathcal{N}(t=0)\rangle = p^2\langle \mathcal{E};\mathcal{E}\rangle - p(1-p)\langle \mathcal{E}\rangle \tag{8.26}$$

where $\mathcal{E} \equiv -\sum_{b\in B}\delta_{\sigma_b} \leq 0$ is the energy. Hence the normalized autocorrelation function at time lag 1 is exactly

$$\rho_{\mathcal{N}\mathcal{N}}(1) \equiv \frac{\langle \mathcal{N}(t=0); \mathcal{N}(t=1)\rangle}{\langle \mathcal{N}(t=0); \mathcal{N}(t=0)\rangle} = 1 - \frac{-(1-p)E}{pC_H - (1-p)E}, \tag{8.27}$$

where $E \equiv V^{-1}\langle\mathcal{E}\rangle$ is the mean energy and $C_H \equiv V^{-1}\langle\mathcal{E};\mathcal{E}\rangle$ is the specific heat (V is the volume). At criticality, $p \to p_{\text{crit}} > 0$ and $E \to E_{\text{crit}} < 0$, so

$$\rho_{\mathcal{N}\mathcal{N}}(1) \geq 1 - \frac{\text{const}}{C_H}. \tag{8.28}$$

We now remark that although $P_{\text{SW}} = P_{\text{bond}}P_{\text{spin}}$ is not self-adjoint, the modified transition matrix $P'_{\text{SW}} \equiv P_{\text{spin}}P_{\text{bond}}P_{\text{spin}}$ *is* self-adjoint (and positive-semidefinite), and $\mathcal{N}$ has the same autocorrelation function for both:

$$(\mathcal{N}, (P_{\text{bond}}P_{\text{spin}})^t\,\mathcal{N}) = (\mathcal{N}, (P_{\text{spin}}P_{\text{bond}}P_{\text{spin}})^t\,\mathcal{N}) \tag{8.29}$$

It follows that

$$\rho_{\mathcal{N}\mathcal{N}}(t) \geq \rho_{\mathcal{N}\mathcal{N}}(1)^{|t|} \geq (1 - \text{const}/C_H)^{|t|} \tag{8.30}$$

and hence

$$\tau_{\text{inf},\mathcal{N}} \geq \text{const} \times C_H. \tag{8.31}$$

This proves a lower bound [62] on the dynamic critical exponent z_{SW}, namely $z_{\text{SW}} \geq \alpha/\nu$. (Here α and ν are the static critical exponents for the susceptibility and correlation length, respectively.) For the two-dimensional Potts models with $q = 2,3,4$, it is known exactly (but non-rigorously!) that $\alpha/\nu = 0, \frac{2}{5}, 1$, with multiplicative logarithmic corrections for $q = 4$ [91]. The bound on z_{SW} may be conceivably be sharp (or sharp up to a logarithm) for $q = 4$ [62].

Example 3. *Berretti–Sokal algorithm for SAWs.* Let us consider the observable

$$f(\omega) = |\omega| (= N), \tag{8.32}$$

the total number of bonds in the walk. We have argued heuristically that $\tau \sim \langle N \rangle^2$; we can now make this argument rigorous as a *lower bound*. Indeed, it suffices to use (8.7): since the maximum change of $|\omega|$ in a single time step is ± 1, it follows that

$$(f,(I-P)f) \leq \frac{1}{2}. \tag{8.33}$$

On the other hand, the denominator of the Rayleigh quotient is

$$(f,(I-\Pi)f) = \langle N^2 \rangle - \langle N \rangle^2. \tag{8.34}$$

Assuming the usual scaling behavior (7.1) for the c_N, we have

$$\langle N \rangle \approx \frac{\gamma}{1-\beta\mu} \tag{8.35}$$

$$\langle N^2 \rangle \approx \frac{\gamma(\gamma+1)}{1-\beta\mu} \tag{8.36}$$

asymptotically as $\beta \uparrow \beta_c = \mu^{-1}$, and hence

$$\tau_{\text{inf},N} \geq \text{const} \times \langle N \rangle^2. \tag{8.37}$$

A very different approach to proving lower bounds on the autocorrelation time τ_{exp} is the *minimum hitting-time argument* [81]. Consider a Markov chain with transition matrix P satisfying detailed balance for some probability measure π. If A and B are subsets of the state space S, let T_{AB} be the minimum time for getting from A to B with nonzero probability, i.e.,

$$T_{\text{AB}} \equiv \min\{n: p_{xy}^{(n)} > 0 \text{ for some } x \in A, y \in B\} \tag{8.38}$$

Then the theorem asserts that if T_{AB} is large and this is not "justified" by the rarity of A and/or B in the equilibrium distribution π, then the autocorrelation time τ_{exp} must be large. More precisely:

Theorem 2 *Consider a Markov chain with transition matrix P satisfying detailed balance for the probability measure π. Let T_{AB} be defined as in (8.38). Then*

$$\tau_{exp} \geq \sup_{A,B \in S} \frac{2(T_{AB}-1)}{-\log(\pi(A)\,\pi(B))} \tag{8.39}$$

Proof. Let $A, B \subset S$, and let $n < T_{\text{AB}}$. Then, by definition of T_{AB},

$$(\chi_A, P^n \chi_B)_{l^2(\pi)} \equiv \sum_{\substack{x\in A \\ y \in B}} \pi_x p_{xy}^{(n)} = 0. \tag{8.40}$$

On the other hand, $P\mathbf{1} = P^*\mathbf{1} = \mathbf{1}$. It follows that

$$(\chi_A - \pi(A)\mathbf{1}, P^n(\chi_B - \pi(B)\mathbf{1}))_{l^2(\pi)} = -\pi(A)\pi(B). \tag{8.41}$$

Now since P is a *self-adjoint* operator, we have

$$\|P^n \restriction \mathbf{1}^{\perp}\| = \|P \restriction \mathbf{1}^{\perp}\|^n = R^n, \tag{8.42}$$

where $R = e^{-1/\tau_{\exp}}$ is the spectral radius (= norm) of $P \restriction \mathbf{1}^{\perp}$. Hence, by the Schwarz inequality,

$$\begin{aligned}
&\left|(\chi_A - \pi(A)\mathbf{1}, P^n(\chi_B - \mathbf{1}))_{l^2(\pi)}\right| \\
&\le R^n \,\|\chi_A - \pi(A)\mathbf{1}\|_{l^2(\pi)} \,\|\chi_B - \pi(B)\mathbf{1}\|_{l^2(\pi)} \\
&= R^n \,\pi(A)^{1/2}\,(1-\pi(A))^{1/2}\,\pi(B)^{1/2}\,(1-\pi(B))^{1/2} \\
&\le R^n \,\pi(A)^{1/2}\,\pi(B)^{1/2}
\end{aligned} \tag{8.43}$$

Combining (8.41) with (8.43) and taking $n = T_{\mathrm{AB}} - 1$, we arrive after a little algebra at (8.39).

Remark. Using Chebyshev polynomials, a stronger bound can be proven: roughly speaking, $\tau_{\exp}$ is bounded below by the *square* of the RHS of (8.39). For details, see [81, Theorem 3.1].

Let us apply this theorem to the BFACF algorithm for variable-length fixed-endpoint SAWs. Let ω^* be a fixed short walk from 0 to x, and let ω^n be a quasi-rectangular walk from 0 to x of linear size $\approx n$. Then $\pi(\omega^*) \sim 1$ and $\pi(\omega^n) \sim \beta^{\approx 4n}$, so that $-\log(\pi(\omega^*)\pi(\omega^n)) \sim n$. On the other hand — and this is the key point — the minimum time required to get from ω^n to ω^* (or vice versa) in the BFACF algorithm is of order n^2, since the *surface area* spanned by $\omega^n \cup \omega^*$ can change by at most one unit in a local deformation. Applying the theorem with $A = \{\omega^n\}$ and $B = \{\omega^*\}$, we obtain

$$\tau_{\exp} \ge \sup_n \frac{\sim n^2}{\sim n} = +\infty. \tag{8.44}$$

As noted at the beginning of this lecture, it is *much* more difficult to prove *upper* bounds on the autocorrelation time — or even to prove that $\tau_{\exp} < \infty$ — and few nontrivial results have been obtained so far.

The only *general* method (to my knowledge) for proving upper bounds on the autocorrelation time is Cheeger's inequality [77, 92, 93], the basic idea of which is to search for "bottlenecks" in the state space.* Consider first the rate of probability flow, in the stationary Markov chain, from a set A to its complement A^c, normalized by the invariant probabilities of A and A^c:

$$k(A) \equiv \frac{\sum\limits_{x\in A, y\in A^c} \pi_x p_{xy}}{\pi(A)\,\pi(A^c)} = \frac{(\chi_A, P\chi_{A^c})_{l^2(\pi)}}{\pi(A)\,\pi(A^c)}. \tag{8.45}$$

Now look for the *worst* such decomposition into A and A^c:

$$k \equiv \inf_{A: 0<\pi(A)<1} k(A). \tag{8.46}$$

***Note Added 1996:** There is now an extensive literature (largely by probabilists and theoretical computer scientists) on upper bounds on the autocorrelation time for Markov chains, based on inequalities of Cheeger, Poincaré, Nash and log-Sobolev types. For reviews, see e.g., [131, 132, 133, 134, 135, 136].

If, for some set A, the flow from A to A^c is very small compared to the invariant probabilities of A and A^c, then intuitively the Markov chain must have very slow convergence to equilibrium (the sets A and A^c are "metastable"). For reversible Markov chains a trivial variational argument makes this intuition rigorous: just take $f = \chi_A$. What is much more exciting is the converse: if there does *not* exist a set A for which the flow from A to A^c is unduly small, then the Markov chain must have *rapid* convergence to equilibrium, in the sense that the modified autocorrelation time τ'_{exp} is small. The precise statement is the following [77, Theorem 2.1]:

Theorem 3 *Let P be a transition probability matrix satisfying detailed balance for π, and let k be defined as above. Then*

$$\frac{-1}{\log(1-k)} \leq \tau'_{\text{exp}} \leq \frac{-1}{\log(1-\frac{k^2}{8})}. \tag{8.47}$$

The proof is not difficult, but enough is enough — the interested reader can look it up in the original paper.

Since this is a *two-sided* bound, we see that τ'_{exp} is finite *if and only if* $k > 0$. So in principle Cheeger's inequality can be used to prove exponential convergence to equilibrium whenever it holds. But in practice it is almost impossible to control the infimum (8.46) over *all* sets A. The most tractable case seems to be when the state space is a *tree*: then A can always be chosen so that it is connected to A^c by a single bond. Using this fact, Lawler and Sokal [77] used Cheeger's inequality to prove

$$\tau'_{\text{exp}} \leq \text{const} \times \langle N \rangle^{2\gamma} \tag{8.48}$$

for the Berretti–Sokal algorithm for SAWs.

A very different proof of an upper bound on τ'_{exp} in the Berretti–Sokal algorithm was given by Sokal and Thomas [10]. Their method is to study in detail the exponential moments of the hitting times from an arbitrary walk ω to the empty walk. Using a sequence of identities, they are able to write an algebraic inequality for such a moment in terms of itself; this inequality says roughly that the moment is either small or huge (where "huge" includes the possibility of $+\infty$) but cannot lie in-between (there is a "forbidden interval"). Then, by a continuity argument, they are able to rule out the possibility that the moment is huge. So it must be small. The final result is

$$\tau'_{\text{exp}} \leq \text{const} \times \langle N \rangle^{1+\gamma}, \tag{8.49}$$

slightly better than the Lawler–Sokal bound.*

For spin models and lattice field theories, almost nothing is known about upper bounds on the autocorrelation time, except at high temperature. For the single-site heat-bath dynamics, it is easy to show that $\tau_{\text{exp}} < \infty$ (uniformly in the volume) above the Dobrushin uniqueness temperature; indeed, this is precisely what the standard proof of the Dobrushin uniqueness theorem [94, 95] does. One expects the same result to hold for *all* temperatures above critical, but this remains an open problem, despite recent progress by Aizenman and Holley [96].†

*__Note Added 1996:__ See also [137], which recovers the bound (8.49) by a much simpler argument using the Poincaré inequality [131].

†__Note Added 1996:__ There has been great progress on this problem in recent years: see [138, 139, 140, 141, 142, 143] and references cited therein.

ACKNOWLEDGMENTS

Many of the ideas reported here have grown out of joint work with my colleagues Alberto Berretti, Frank Brown, Sergio Caracciolo, Robert Edwards, Jonathan Goodman, Tony Guttmann, Greg Lawler, Xiao-Jian Li, Neal Madras, Andrea Pelissetto, Larry Thomas and Dan Zwanziger. I thank them for many pleasant and fruitful collaborations. In addition, I would like to thank Mal Kalos for introducing me to Monte Carlo work and for patiently answering my first naive questions.

The research of the author was supported in part by the U.S. National Science Foundation grants DMS–8705599 and DMS–8911273, the U.S. Department of Energy contract DE-FG02-90ER40581, the John Von Neumann National Supercomputer Center* grant NAC–705, and the Pittsburgh Supercomputing Center grant PHY890035P. Acknowledgment is also made to the donors of the Petroleum Research Fund, administered by the American Chemical Society, for partial support of this research.

REFERENCES

[1] N. Bakhvalov, *Méthodes Numériques* (MIR, Moscou, 1976), Chapter 5.
[2] J. M. Hammersley and D. C. Handscomb, *Monte Carlo Methods* (Methuen, London, 1964), Chapter 5.
[3] W. W. Wood and J. J. Erpenbeck, Ann. Rev. Phys. Chem. **27**, 319 (1976).
[4] J. G. Kemeny and J. L. Snell, *Finite Markov Chains* (Springer, New York, 1976).
[5] M. Iosifescu, *Finite Markov Processes and Their Applications* (Wiley, Chichester, 1980).
[6] K. L. Chung, *Markov Chains with Stationary Transition Probabilities*, 2nd ed. (Springer, New York, 1967).
[7] E. Nummelin, *General Irreducible Markov Chains and Non-Negative Operators* (Cambridge Univ. Press, Cambridge, 1984).
[8] D. Revuz, *Markov Chains* (North-Holland, Amsterdam, 1975).
[9] Z. Šidak, Czechoslovak Math. J. **14**, 438 (1964).
[10] A. D. Sokal and L. E. Thomas, J. Stat. Phys. **54**, 797 (1989).
[11] P. C. Hohenberg and B. I. Halperin, Rev. Mod. Phys. **49**, 435 (1977).
[12] S. Caracciolo and A. D. Sokal, J. Phys. **A19**, L797 (1986).
[13] D. Goldsman, Ph. D. thesis, School of Operations Research and Industrial Engineering, Cornell University (1984); L. Schruben, Oper. Res. **30**, 569 (1982) and **31**, 1090 (1983).
[14] M. B. Priestley, *Spectral Analysis and Time Series*, 2 vols. (Academic, London, 1981), Chapters 5-7.
[15] T. W. Anderson, *The Statistical Analysis of Time Series* (Wiley, New York, 1971).
[16] N. Madras and A. D. Sokal, J. Stat. Phys. **50**, 109 (1988).
[17] N. Metropolis, A. W. Rosenbluth, M. N. Rosenbluth, A. H. Teller and E. Teller, J. Chem. Phys. **21**, 1087 (1953).
[18] W. K. Hastings, Biometrika **57**, 97 (1970).
[19] J. Goodman and N. Madras, Lin. Alg. Appl. **216**, 61 (1995).
[20] J. Goodman and A. D. Sokal, Phys. Rev. **D40**, 2035 (1989).
[21] G. F. Mazenko and O. T. Valls, Phys. Rev. **B24**, 1419 (1981); C. Kalle, J. Phys. **A17**, L801 (1984); J. K. Williams, J. Phys. **A18**, 49 (1985); R. B. Pearson, J. L. Richardson and D. Toussaint, Phys. Rev. **B31**, 4472 (1985); S. Wansleben and D. P. Landau, J. Appl. Phys. **61**, 3968 (1987).
[22] M. N. Barber, in *Phase Transitions and Critical Phenomena*, vol. 8, ed. C. Domb and J. L. Lebowitz (Academic Press, London, 1983).
[23] K. Binder, J. Comput. Phys. **59**, 1 (1985).
[24] G. Parisi, in *Progress in Gauge Field Theory* (1983 Cargèse lectures), ed. G. 't Hooft *et al.* (Plenum, New York, 1984); G. G. Batrouni, G. R. Katz, A. S. Kronfeld, G. P. Lepage, B. Svetitsky and K. G. Wilson, Phys. Rev. **D32**, 2736 (1985); C. Davies, G. Batrouni, G. Katz, A. Kronfeld, P. Lepage, P. Rossi, B. Svetitsky and K. Wilson, J. Stat. Phys. **43**, 1073 (1986); J. B. Kogut, Nucl. Phys. **B275** [FS17], 1 (1986); S. Duane, R. Kenway, B. J. Pendleton and D. Roweth, Phys. Lett. **176B**, 143 (1986); E. Dagotto and J. B. Kogut, Phys. Rev. Lett. **58**, 299 (1987) and Nucl. Phys. **B290** [FS20], 451 (1987).
[25] J. Goodman and A. D. Sokal, Phys. Rev. Lett. **56**, 1015 (1986).
[26] R. G. Edwards, J. Goodman and A. D. Sokal, Nucl. Phys. **B354**, 289 (1991).

*Deceased.

[27] R. H. Swendsen and J.-S. Wang, Phys. Rev. Lett. **58**, 86 (1987).
[28] R. P. Fedorenko, Zh. Vychisl. i Mat. Fiz. **4**, 559 (1964) [USSR Comput. Math. and Math. Phys. **4**, 227 (1964)].
[29] L. P. Kadanoff, Physics **2**, 263 (1966).
[30] K. G. Wilson, Phys. Rev. **B4**, 3174, 3184 (1971).
[31] A. Brandt, in *Proceedings of the Third International Conference on Numerical Methods in Fluid Mechanics* (Paris, July 1972), ed. H. Cabannes and R. Temam (Lecture Notes in Physics #18) (Springer, Berlin, 1973); R. A. Nicolaides, J. Comput. Phys. **19**, 418 (1975) and Math. Comp. **31**, 892 (1977); A. Brandt, Math. Comp. **31**, 333 (1977); W. Hackbusch, in *Numerical Treatment of Differential Equations* (Oberwolfach, July 1976), ed. R. Bulirsch, R. D. Griegorieff and J. Schröder (Lecture Notes in Mathematics #631) (Springer, Berlin, 1978).
[32] W. Hackbusch, *Multi-Grid Methods and Applications* (Springer, Berlin, 1985).
[33] W. Hackbusch and U. Trottenberg, editors, *Multigrid Methods* (Lecture Notes in Mathematics #960) (Springer, Berlin, 1982); Proceedings of the First Copper Mountain Conference on Multigrid Methods, ed. S. McCormick and U. Trottenberg, Appl. Math. Comput. **13**, no. 3-4, 213-470 (1983); D. J. Paddon and H. Holstein, editors, *Multigrid Methods for Integral and Differential Equations* (Clarendon Press, Oxford, 1985); Proceedings of the Second Copper Mountain Conference on Multigrid Methods, ed. S. McCormick, Appl. Math. Comput. **19**, no. 1-4, 1-372 (1986); W. Hackbusch and U. Trottenberg, editors, *Multigrid Methods II* (Lecture Notes in Mathematics #1228) (Springer, Berlin, 1986); S. F. McCormick, editor, *Multigrid Methods* (SIAM, Philadelphia, 1987); *Proceedings of the Third Copper Mountain Conference on Multigrid Methods*, ed. S. McCormick (Dekker, New York, 1988).
[34] W. L. Briggs, *A Multigrid Tutorial* (SIAM, Philadelphia, 1987).
[35] R. S. Varga, *Matrix Iterative Analysis* (Prentice-Hall, Englewood Cliffs, N. J., 1962).
[36] H. R. Schwarz, H. Rutishauer and E. Stiefel, *Numerical Analysis of Symmetric Matrices* (Prentice-Hall, Englewood Cliffs, N. J., 1973).
[37] A. M. Ostrowski, Rendiconti Mat. e Appl. **13**, 140 (1954) at p. 141*n* [A. Ostrowski, *Collected Mathematical Papers*, vol. 1 (Birkhaüser, Basel, 1983), p. 169*n*].
[38] J. M. Ortega and W. C. Rheinboldt, *Iterative Solution of Nonlinear Equations in Several Variables* (Academic Press, New York–London, 1970).
[39] S. F. McCormick and J. Ruge, Math. Comp. **41**, 43 (1983).
[40] J. Mandel, S. McCormick and R. Bank, in *Multigrid Methods*, ed. S. F. McCormick (SIAM, Philadelphia, 1987), Chapter 5.
[41] J. Goodman and A. D. Sokal, unpublished.
[42] R. G. Edwards, S. J. Ferreira, J. Goodman and A. D. Sokal, Nucl. Phys. **B380**, 621 (1992).
[43] S. L. Adler, Phys. Rev. Lett. **60**, 1243 (1988).
[44] S. L. Adler, Phys. Rev. **D23**, 2091 (1981).
[45] C. Whitmer, Phys. Rev. **D29**, 306 (1984).
[46] E. Nelson, in *Constructive Quantum Field Theory*, ed. G. Velo and A. Wightman (Lecture Notes in Physics #25) (Springer, Berlin, 1973).
[47] B. Simon, *The $P(\varphi)_2$ Euclidean (Quantum) Field Theory* (Princeton Univ. Press, Princeton, N. J., 1974), Chapter I.
[48] R. B. Potts, Proc. Camb. Phil. Soc. **48**, 106 (1952).
[49] F. Y. Wu, Rev. Mod. Phys. **54**, 235 (1982); **55**, 315 (E) (1983).
[50] K. Krickeberg, *Probability Theory* (Addison-Wesley, Reading, Mass., 1965), Sections 4.2 and 4.5.
[51] R. H. Schonmann, J. Stat. Phys. **52**, 61 (1988).
[52] A. D. Sokal, unpublished.
[53] E. M. Reingold, J. Nievergelt and N. Deo, *Combinatorial Algorithms: Theory and Practice* (Prentice-Hall, Englewood Cliffs, N. J., 1977), Chapter 8; A. Gibbons, *Algorithmic Graph Theory* (Cambridge University Press, 1985), Chapter 1; S. Even, *Graph Algorithms* (Computer Science Press, Potomac, Maryland, 1979), Chapter 3.
[54] B. A. Galler and M. J. Fischer, Commun. ACM **7**, 301, 506 (1964); D. E. Knuth, *The Art of Computer Programming*, vol. 1, 2nd ed., (Addison-Wesley, Reading, Massachusetts, 1973), pp. 353–355, 360, 572; J. Hoshen and R. Kopelman, Phys. Rev. **B14**, 3438 (1976).
[55] K. Binder and D. Stauffer, in *Applications of the Monte Carlo Method in Statistical Physics*, ed. K. Binder (Springer-Verlag, Berlin, 1984), section 8.4; D. Stauffer, *Introduction to Percolation Theory* (Taylor & Francis, London, 1985).
[56] Y. Shiloach and U. Vishkin, J. Algorithms **3**, 57 (1982).
[57] R. G. Edwards, X.-J. Li and A. D. Sokal, Sequential and vectorized algorithms for computing the connected components of an undirected graph, unpublished (and uncompleted) notes.
[58] P. W. Kasteleyn and C. M. Fortuin, J. Phys. Soc. Japan **26** (Suppl.), 11 (1969); C. M. Fortuin and P. W. Kasteleyn, Physica **57**, 536 (1972); C. M. Fortuin, Physica **58**, 393 (1972); C. M. Fortuin, Physica **59**, 545 (1972).
[59] C.-K. Hu, Phys. Rev. **B29**, 5103 (1984); T. A. Larsson, J. Phys. **A19**, 2383 (1986) and **A20**, 2239 (1987).

[60] R. G. Edwards and A. D. Sokal, Phys. Rev. **D38**, 2009 (1988).
[61] A. D. Sokal, in *Computer Simulation Studies in Condensed Matter Physics: Recent Developments* (Athens, Georgia, 15–26 February 1988), ed. D. P. Landau, K. K. Mon and H.-B. Schüttler (Springer-Verlag, Berlin–Heidelberg, 1988).
[62] X.-J. Li and A. D. Sokal, Phys. Rev. Lett. **63**, 827 (1989).
[63] W. Klein, T. Ray and P. Tamayo, Phys. Rev. Lett. **62**, 163 (1989).
[64] T. S. Ray, P. Tamayo and W. Klein, Phys. Rev. **A39**, 5949 (1989).
[65] U. Wolff, Phys. Rev. Lett. **62**, 361 (1989).
[66] U. Wolff, Phys. Lett. **B228**, 379 (1989); P. Tamayo, R. C. Brower and W. Klein, J. Stat. Phys. **58**, 1083 (1990) and **60**, 889 (1990).
[67] D. Kandel, E. Domany, D. Ron, A. Brandt and E. Loh, Phys. Rev. Lett. **60**, 1591 (1988); D. Kandel, E. Domany and A. Brandt, Phys. Rev. **B40**, 330 (1989).
[68] R. G. Edwards and A. D. Sokal, Swendsen–Wang algorithm augmented by duality transformations for the two-dimensional Potts model, in preparation.
[69] R. Ben-Av, D. Kandel, E. Katznelson, P. G. Lauwers and S. Solomon, J. Stat. Phys. **58**, 125 (1990); R. C. Brower and S. Huang, Phys. Rev. **B41**, 708 (1990).
[70] F. Niedermayer, Phys. Rev. Lett. **61**, 2026 (1988).
[71] R. C. Brower and P. Tamayo, Phys. Rev. Lett. **62**, 1087 (1989).
[72] U. Wolff, Nucl. Phys. **B322**, 759 (1989); Phys. Lett. **B222**, 473 (1989); Nucl. Phys. **B334**, 581 (1990).
[73] R. G. Edwards and A. D. Sokal, Phys. Rev. **D40**, 1374 (1989).
[74] P. G. DeGennes, *Scaling Concepts in Polymer Physics* (Cornell Univ. Press, Ithaca NY, 1979).
[75] C. Aragão de Carvalho, S. Caracciolo and J. Fröhlich, Nucl. Phys. **B215** [FS7], 209 (1983).
[76] A. Berretti and A. D. Sokal, J. Stat. Phys. **40**, 483 (1985).
[77] G. Lawler and A. D. Sokal, Trans. Amer. Math. Soc. **309**, 557 (1988).
[78] G. Lawler and A. D. Sokal, in preparation.
[79] B. Berg and D. Foerster, Phys. Lett. **106B**, 323 (1981).
[80] C. Aragão de Carvalho and S. Caracciolo, J. Physique **44**, 323 (1983).
[81] A. D. Sokal and L. E. Thomas, J. Stat. Phys. **51**, 907 (1988).
[82] S. Caracciolo, A. Pelissetto and A. D. Sokal, J. Stat. Phys. **60**, 1 (1990).
[83] M. Delbrück, in *Mathematical Problems in the Biological Sciences* (Proc. Symp. Appl. Math., vol. 14), ed. R. E. Bellman (American Math. Soc., Providence RI, 1962).
[84] P. H. Verdier and W. H. Stockmayer, J. Chem. Phys. **36**, 227 (1962).
[85] N. Madras and A. D. Sokal, J. Stat. Phys. **47**, 573 (1987).
[86] M. Lal, Molec. Phys. **17**, 57 (1969).
[87] B. MacDonald *et al.*, J. Phys. **A18**, 2627 (1985).
[88] K. Kawasaki, Phys. Rev. **148**, 375 (1966).
[89] B. I. Halperin, Phys. Rev. **B8**, 4437 (1973).
[90] R. Alicki, M. Fannes and A. Verbeure, J. Stat. Phys. **41**, 263 (1985).
[91] M. P. M. den Nijs, J. Phys. **A12**, 1857 (1979); J. L. Black and V. J. Emery, Phys. Rev. **B23**, 429 (1981); B. Nienhuis, J. Stat. Phys. **34**, 731 (1984).
[92] A. Sinclair and M. Jerrum, Information and Computation **82**, 93 (1989).
[93] R. Brooks, P. Doyle and R. Kohn, in preparation.
[94] L. N. Vasershtein, Prob. Inform. Transm. **5**, 64 (1969).
[95] O. E. Lanford III, in *Statistical Mechanics and Mathematical Problems* (Lecture Notes in Physics #20), ed. A. Lenard (Springer, Berlin, 1973).
[96] M. Aizenman and R. Holley, in *Percolation Theory and Ergodic Theory of Infinite Particle Systems*, ed. H. Kesten (Springer, New York, 1987).

ADDED BIBLIOGRAPHY (DECEMBER 1996):

[97] M. Lüscher, P. Weisz and U. Wolff, Nucl. Phys. **B359**, 221 (1991).
[98] J.-K. Kim, Phys. Rev. Lett. **70**, 1735 (1993); Nucl. Phys. B (Proc. Suppl.) **34**, 702 (1994); Phys. Rev. **D50**, 4663 (1994); Europhys. Lett. **28**, 211 (1994); Phys. Lett. **B345**, 469 (1995).
[99] S. Caracciolo, R. G. Edwards, S. J. Ferreira, A. Pelissetto and A. D. Sokal, Phys. Rev. Lett. **74**, 2969 (1995).
[100] S. J. Ferreira and A. D. Sokal, Phys. Rev. **B51**, 6720 (1995).
[101] S. Caracciolo, R. G. Edwards, A. Pelissetto and A. D. Sokal, Phys. Rev. Lett. **75**, 1891 (1995).
[102] G. Mana, A. Pelissetto and A. D. Sokal, Phys. Rev. **D54**, R1252 (1996).
[103] G. Mana, A. Pelissetto and A. D. Sokal, `hep-lat/9610021`, submitted to Phys. Rev. D.

[104] C.-R. Hwang and S.-J. Sheu, in *Stochastic Models, Statistical Methods, and Algorithms in Image Analysis* (Lecture Notes in Statistics #74), ed. P. Barone, A. Frigessi and M. Piccioni (Springer, Berlin, 1992).
[105] B. Gidas, private communication (1991).
[106] T. Mendes and A. D. Sokal, Phys. Rev. **D53**, 3438 (1996).
[107] G. Mana, T. Mendes, A. Pelissetto and A. D. Sokal, Nucl. Phys. B (Proc. Suppl.) **47**, 796 (1996).
[108] T. Mendes and A. Pelissetto and A. D. Sokal, Nucl. Phys. **B477**, 203 (1996).
[109] A. D. Sokal, Nucl. Phys. B (Proc. Suppl.) **20**, 55 (1991).
[110] J.-S. Wang and R. H. Swendsen, Physica **A167**, 565 (1990).
[111] C. F. Baillie and P. D. Coddington, Phys. Rev. Lett. **68**, 962 (1992); and private communication.
[112] J. Salas and A. D. Sokal, J. Stat. Phys. **85**, 297 (1996).
[113] J. Salas and A. D. Sokal, Dynamic critical behavior of the Swendsen–Wang algorithm: The two-dimensional 3-state Potts model revisited, `hep-lat/9605018`, J. Stat. Phys. **87**, no. 1/2 (April 1997).
[114] J. Salas and A. D. Sokal, Logarithmic corrections and finite-size scaling in the two-dimensional 4-State Potts model, `hep-lat/9607030`, submitted to J. Stat. Phys.
[115] G. M. Torrie and J. P. Valleau, Chem. Phys. Lett. **28**, 578 (1974); J. Comput. Phys. **23**, 187 (1977); J. Chem. Phys. **66**, 1402 (1977).
[116] B. A. Berg and T. Neuhaus, Phys. Lett. **B267**, 249 (1991); Phys. Rev. Lett. **68**, 9 (1992).
[117] B. A. Berg, Int. J. Mod. Phys. **C4**, 249 (1993).
[118] A. D. Kennedy, Nucl. Phys. B (Proc. Suppl.) **30**, 96 (1993).
[119] W. Janke, in *Computer Simulations in Condensed Matter Physics VII*, eds. D. P. Landau, K. K. Mon and H.-B. Schüttler (Springer-Verlag, Berlin, 1994).
[120] S. Wiseman and E. Domany, Phys. Rev. E **48**, 4080 (1993).
[121] D. Kandel, R. Ben-Av and E. Domany, Phys. Rev. Lett. **65**, 941 (1990) and Phys. Rev. **B45**, 4700 (1992).
[122] J.-S. Wang, R. H. Swendsen and R. Kotecký, Phys. Rev. Lett. **63**, 109 (1989).
[123] J.-S. Wang, R. H. Swendsen and R. Kotecký, Phys. Rev. **B42**, 2465 (1990).
[124] M. Lubin and A. D. Sokal, Phys. Rev. Lett. **71**, 1778 (1993).
[125] X.-J. Li and A. D. Sokal, Phys. Rev. Lett. **67**, 1482 (1991).
[126] S. Caracciolo, R. G. Edwards, A. Pelissetto and A. D. Sokal, Nucl. Phys. **B403**, 475 (1993).
[127] A. D. Sokal, in *Monte Carlo and Molecular Dynamics Simulations in Polymer Science*, ed. K. Binder (Oxford University Press, New York, 1995).
[128] A. D. Sokal, Nucl. Phys. B (Proc. Suppl.) **47**, 172 (1996).
[129] B. Li, N. Madras and A. D. Sokal, J. Stat. Phys. **80**, 661 (1995).
[130] A. D. Sokal, Europhys. Lett. **27**, 661 (1994).
[131] P. Diaconis and D. Stroock, Ann. Appl. Prob. **1**, 36 (1991).
[132] A. J. Sinclair, *Randomised Algorithms for Counting and Generating Combinatorial Structures* (Birkhäuser, Boston, 1993).
[133] P. Diaconis and L. Saloff-Coste, Ann. Appl. Prob. **3**, 696 (1993).
[134] P. Diaconis and L. Saloff-Coste, Ann. Appl. Prob. **6**, 695 (1996).
[135] P. Diaconis and L. Saloff-Coste, J. Theoret. Prob. **9**, 459 (1996).
[136] M. Jerrum and A. Sinclair, in *Approximation Algorithms for NP-Hard Problems*, ed. D. S. Hochbaum (PWS Publishing, Boston, 1996).
[137] D. Randall and A. J. Sinclair, in *Proceedings of the 5th Annual ACM-SIAM Symposium on Discrete Algorithms* (ACM Press, New York, 1994).
[138] D. W. Stroock and B. Zegarlinski, J. Funct. Anal. **104**, 299 (1992).
[139] D. W. Stroock and B. Zegarlinski, Commun. Math. Phys. **144**, 303 (1992).
[140] D. W. Stroock and B. Zegarlinski, Commun. Math. Phys. **149**, 175 (1992).
[141] S. L. Lu and H. T. Yau, Commun. Math. Phys. **156**, 399 (1993).
[142] F. Martinelli and E. Olivieri, Commun. Math. Phys. **161**, 447, 487 (1994).
[143] D. W. Stroock and B. Zegarlinski, J. Stat. Phys. **81**, 1007 (1995).

7

PATH INTEGRAL SIMULATION OF LONG-TIME DYNAMICS IN QUANTUM DISSIPATIVE SYSTEMS

Nancy Makri

Department of Chemistry
University of Illinois,
Urbana, Illinois 61801

INTRODUCTION

Feynman's path integral approach to time-dependent quantum mechanics[1] has found wide application in many areas of physics. Its most celebrated successes include situations where the effects of a dissipative environment on the system of interest can be adequately represented via a bath of harmonic oscillators. Due to its Gaussian character, the multidimensional bath can be integrated out,[2] giving rise to reduced-dimension descriptions of the dynamics which in simple cases are amenable to a host of analytic approximations. Another major field where path integral ideas have proven extremely useful is quantum statistical mechanics.[3] Expressing equilibrium averages of many-body systems in path integral form allows, after appropriate discretization, numerical evaluation via stochastic integration schemes.[4] By contrast, the use of numerical methods to compute real-time path integral expressions of many-particle systems has not been met with success. The reason behind the failure of numerical schemes lies in the oscillatory nature of the quantum mechanical propagator which renders stochastic integration methods inappropriate.[5]

Recent attempts to bridge the gap between analytic approximations to the dynamics of simple dissipative systems on one hand and accurate numerical simulations of equilibrium properties of many-body systems on the other hand have focused on the development of accurate numerical approaches to the dynamics of quantum mechanical particles in harmonic dissipative environments. The present article describes a systematic approximation to the path integral for quantum dissipative systems which lends itself naturally to iterative numerical algorithms, thereby circumventing the phase cancellation problem of conventional real-time approaches. The resulting methodology allows accurate fully quantum mechanical simulation of processes in harmonic dissipative environments over long times. Selected applications to various problems in physics, chemistry and biology are also reviewed.

Functional Integration, edited by DeWitt-Morette
Plenum Press, New York, 1997

The theoretical framework, along with the numerical methodology, are presented in the next section. Various applications to problems involving quantum phenomena in the condensed phase are presented in the third section. Finally, the last section summarizes with brief concluding remarks.

THEORY

This article deals exclusively with system-bath Lagrangians of the type

$$L = \frac{1}{2} m_0 \dot{s}^2 - V_0(s) + \sum_j \left(\frac{1}{2} m_j \dot{x}_j^2 - \frac{1}{2} m_j \omega_j^2 \left(x_j - \frac{c_j s}{m_j \omega_j^2} \right)^2 \right). \tag{1}$$

Here s is the coordinate of the particle of interest, which is bilinearly coupled to a large number of bath degrees of freedom denoted by x_j. Eq. (1) contains "counterterms" quadratic in the particle coordinate[6] which ensure that essential features of the system (such as the barrier height in the case of a double minimum potential) remain invariant with respect to variation of the system-bath coupling strength. It has been shown that all information about the bath pertinent to the dynamics of the system can be captured in the spectral density

$$J(\omega) = \frac{\pi}{2} \sum_j \frac{c_j^2}{m_j \omega_j} \delta(\omega - \omega_j). \tag{2}$$

Dissipative environments are characterized by continuous spectral densities. A commonly used model involves power-law spectral densities of the type

$$J(\omega) = \gamma \omega^r e^{-\omega/\omega_c}. \tag{3}$$

At low frequencies Eq. (3) approaches the Debye spectrum of an r-dimensional solid and high-frequency components are cut off exponentially via the parameter ω_c. The parameter γ/m_0 characterizes the strength of dissipation.

Within the framework the classical mechanics, Eq. (1) is equivalent to the generalized Langevin equation[7], which describes the motion of the observable system:

$$m_0 \ddot{s}(t) + m_0 \int_{-\infty}^{t} dt' \chi(t-t') \dot{s}(t') + V_0'(s(t)) = \zeta(t) \tag{4}$$

It is seen that apart from Newtonian forces, the particle of interest experiences frictional damping as well as stochastic forces $\zeta(t)$ which satisfy the fluctuation-dissipation theorem.

In a similar spirit, quantum mechanical properties of the system can be extracted from the reduced density operator $\tilde{\rho}(t)$ whose coordinate representation takes the form

$$\tilde{\rho}(s'',s';t) \equiv \mathrm{Tr}_b \langle s'' | e^{-iHt/\hbar} \rho(0) e^{iHt/\hbar} | s' \rangle \ . \tag{5}$$

Here $\rho(0)$ is the initial density operator for the composite system, H is the Hamiltonian that corresponds to the Lagrangian of Eq. (1) and Tr_b denotes the quantum mechanical trace with respect to the bath. If the interaction between system and bath is switched on at $t = 0$, i.e., if the initial density operator has the form

$$\rho(0) = \tilde{\rho}(0)\rho_b(0)\,, \tag{6}$$

then the evolution of the reduced density matrix of the system can be written as the following path integral[2]:

$$\tilde{\rho}(s'',s';t) = \int_{-\infty}^{\infty} ds_0^+ \int_{-\infty}^{\infty} ds_0^- \int \mathcal{D}s^+ \int \mathcal{D}s^- \exp\left(\frac{i}{\hbar}\left(\Phi_0[s^+] - \Phi_0[s^-]\right)\right) \times \langle s_0^+ | \tilde{\rho}(0) | s_0^- \rangle F[s^+, s^-] \tag{7}$$

where s^+ and s^- are forward and backward system paths with endpoints $s^+(0) = s_0^+$, $s^+(t) = s''$, $s^-(0) = s_0^-$ and $s^-(t) = s'$ that correspond to the path integral representation of the positive- and negative-time evolution operators, respectively. We work in the quasi-adiabatic representation[8] (where the counterterms are grouped with the bath in the partitioning of the time-evolution operator) which provides a significant advantage in the numerical implementation discussed later in this section. In this representation Φ_0 is the action functional for the bare system,

$$\Phi_0[s] = \int_0^t \left(\frac{1}{2} m_0 \dot{s}(t')^2 - V_0(s(t'))\right) dt' \tag{8}$$

and F is a modified Feynman-Vernon influence functional[2]. If the bath is initially at thermal equilibrium at the temperature $1/k_B\beta$, where k_B is the Boltzmann constant, the latter is given by the expression

$$F[s^+, s^-] = \exp\left(-\frac{1}{\hbar}\int_0^t dt' \int_0^{t'} dt'' \left(s^+(t') - s^-(t')\right)\left(\alpha(t'-t'')s^+(t'') - \alpha^*(t'-t'')s^-(t'')\right)\right) \times \exp\left(-\frac{i}{\hbar}\int_0^t dt' \sum_j \frac{c_j^2}{2m_j\omega_j^2}\left(s^+(t')^2 - s^-(t')^2\right)\right) \tag{9}$$

The last expression contains interactions nonlocal in time through the bath response function

$$\alpha(t)=\frac{1}{\hbar}\sum_j c_j^2\langle x_j(0)x_j(t)\rangle_\beta=\pi^{-1}\int_0^\infty d\omega\, J(\omega)\left(\coth(\tfrac{1}{2}\hbar\omega\beta)\cos(\omega t)-i\sin(\omega t)\right) \tag{10}$$

where $x_j(t)$ is the Heisenberg operator for the corresponding position variable. These two-time interactions constitute the quantum mechanical analog of memory in the classical generalized Langevin equation. The terms appearing in the single integral of Eq. (9) arise from the adiabatic partitioning of the Hamiltonian.

The features of the bath response function lead to a systematic approximation of the path integral for quantum dissipative systems. In the special case of strictly Ohmic dissipation [cf. Eq. (3) with $r = 1$ and $\omega_c \to \infty$] the dynamics of a classical particle in a dissipative bath is described by the memory-free Langevin equation, i.e., the memory kernel in Eq. (4) becomes proportional to a delta function and the integral over histories of the trajectory disappears. Indeed, this behavior is recovered from the path integral in the case of an Ohmic bath at infinite temperature: in this case the real part of the kernel $\alpha(t)$ behaves as the delta function $\delta(t)$, while its imaginary part is proportional to $\delta'(t)$, implying that the double integral in Eq. (9) reduces to a single integral and the dynamics becomes Markovian[9].

Away from this special case the response function has a width and the dynamics is not Markovian. However, the Fourier-type relationship between this function and the spectral density implies that the influence functional kernel $\alpha(t)$ decays to zero at long time if the bath has a broad spectrum. Such behavior arises from destructive phase interference among an infinity of degrees of freedom and is characteristic of dissipative environments. The shape of the response function is illustrated in Fig. 1 for a

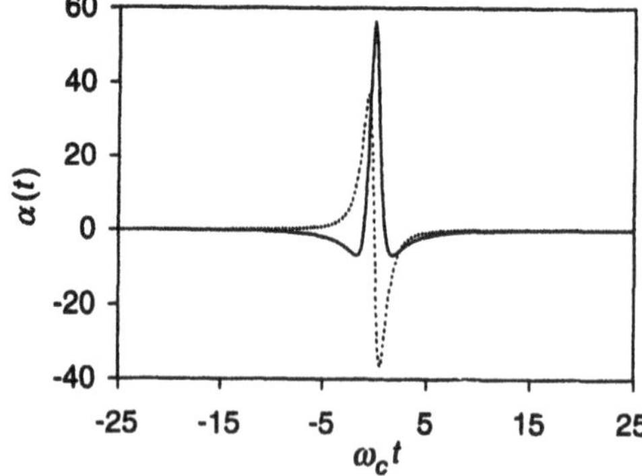

Figure 1. Real (solid line) and imaginary (dashed line) parts of the bath response function, Eq. (10), for an Ohmic spectral density, Eq. (3), with $r=1$ and $\hbar\omega_c\beta = 5$ in arbitrary units.

low-temperature bath characterized by the spectral density of Eq. (3) with $r = 1$ and finite cutoff frequency.

Taking advantage of this feature of dissipative baths, a correlation or memory time t_m can be introduced such that $\alpha(t) \approx 0$ for $t >> t_m$. This allows truncation of the inner integral in Eq. (9) and the influence functional takes the form

$$F[s^+,s^-] \approx \exp\left(-\frac{1}{\hbar}\int_0^t dt' \int_{t_{\text{in}}}^{t'} dt'' \left(s^+(t')-s^-(t')\right)\left(\alpha(t'-t'')s^+(t'')-\alpha^*(t'-t'')s^-(t'')\right)\right)$$
$$\times \exp\left(-\frac{i}{\hbar}\int_0^t dt' \sum_j \frac{c_j^2}{2m_j\omega_j^2}\left(s^+(t')^2 - s^-(t')^2\right)\right) \tag{11}$$

where $t_{\text{in}} = \max\{0,t' - t_m\}$, which becomes an equality as t_m approaches the total propagation time t. This forms the basis for a local path integral expression.

To this end, we define a functional $R[s^\pm(t)_{\mathbb{L}_{k,k'}}]$, where $s^\pm(t)_{\mathbb{L}_{k,k'}}$ are system forward and backward paths defined on the interval $\mathbb{L}_{k,k'} \equiv [kt_m, k't_m)$, where k and k' are integers. This functional is propagated forward in time according to the *local* functional integral

$$R[s^\pm(t)_{\mathbb{L}_{k,k+1}}] = \int \overline{\mathcal{D}}s^\pm(t)_{\mathbb{L}_{k-1,k}}$$
$$\times \exp\left(\frac{i}{\hbar}\left(\Phi_0[s^+(t)_{\mathbb{L}_{k,k+1}}] - \Phi_0[s^-(t)_{\mathbb{L}_{k,k+1}}]\right)\right) R[s^\pm(t)_{\mathbb{L}_{k-1,k}}]$$
$$\times \exp\left\{-\frac{1}{\hbar}\int_{kt_m}^{(k+1)t_m} dt' \int_{t'-t_m}^{t'} dt'' \left(s^+(t')_{\mathbb{L}_{k,k+1}} - s^-(t')_{\mathbb{L}_{k,k+1}}\right)\right.$$
$$\left. \times \left(\alpha(t'-t'')s^+(t'')_{\mathbb{L}_{k-1,k+1}} - \alpha^*(t'-t'')s^-(t'')_{\mathbb{L}_{k-1,k+1}}\right)\right\}$$
$$\times \exp\left\{-\frac{i}{\hbar}\int_{kt_m}^{(k+1)t_m} dt' \sum_j \frac{c_j^2}{2m_j\omega_j^2}\left(s^+(t')^2_{\mathbb{L}_{k,k+1}} - s^-(t')^2_{\mathbb{L}_{k,k+1}}\right)\right\} \tag{12}$$

with initial condition

$$R[s^\pm(t)_{\mathbb{L}_{0,1}}] = \langle s^+(0)|\tilde{\rho}(0)|s^-(0)\rangle \exp\left(\frac{i}{\hbar}\left(\Phi_0[s^+(t)_{\mathbb{L}_{0,1}}] - \Phi_0[s^-(t)_{\mathbb{L}_{0,1}}]\right)\right)$$
$$\times \exp\left\{-\frac{1}{\hbar}\int_0^{t_m} dt' \int_0^{t'} dt'' \left(s^+(t')_{\mathbb{L}_{0,1}} - s^-(t')_{\mathbb{L}_{0,1}}\right)\right.$$
$$\left. \times \left(\alpha(t'-t'')s^+(t'')_{\mathbb{L}_{0,1}} - \alpha^*(t'-t'')s^-(t'')_{\mathbb{L}_{0,1}}\right)\right\}$$
$$\times \exp\left\{-\frac{i}{\hbar}\int_0^{t_m} dt' \sum_j \frac{c_j^2}{2m_j\omega_j^2}\left(s^+(t')^2_{\mathbb{L}_{0,1}} - s^-(t')^2_{\mathbb{L}_{0,1}}\right)\right\} \tag{13}$$

In Eq. (12) the symbol $\int \overline{\mathfrak{D}} s(t)_{\mathrm{L}_{k-1,k}}$ denotes integration over all paths $s(t)$ in the interval from $(k-1)t_m$ to kt_m with endpoint equal to the starting point of the path $s(t)_{\mathrm{L}_{k,k+1}}$. This way paths adjacent in time combine to form a single continuous path that spans a longer time interval. No constraint is imposed on the initial point of the integration paths. It is straightforward to show that the last endpoint of the reduced density functional after N iterations reproduces Eq. (11) at the time Nt_m, i.e.,

$$R[s^+(Nt_m)=s'',\ s^-(Nt_m)=s'] \approx \tilde{\rho}(s'',s';Nt_m)\,. \tag{14}$$

Eq. (12) decomposes the global path integral into a series of local functional integrals involving paths that span shorter time intervals. The accuracy of this approximation improves systematically as the memory time t_m is increased.

Numerical implementation of the above ideas requires discretization of the time. Defining a time step Δt and employing a symmetric splitting of the time evolution operator based on an adiabatic partitioning of the system-bath Hamiltonian,[8]

$$\exp(-iH\Delta t/\hbar) \approx \exp(-iH_{\mathrm{SO}}\Delta t/2\hbar)\exp(-iH_0\Delta t/\hbar)\exp(-iH_{\mathrm{SO}}\Delta t/2\hbar) \tag{15}$$

where H_0 is a reference Hamiltonian along the adiabatic path defined by the line $x_j = c_j s/m_j\omega_j^2$,

$$H_0 = \frac{p_s^2}{2m_0} + V_0(s)\,. \tag{16}$$

and $H_{\mathrm{SO}} \equiv H - H_0$ is a quadratic shifted oscillator Hamiltonian, leads to the following quasi-adiabatic propagator path integral (QUAPI) expression for the reduced density matrix at the total time $t = N\Delta t$,[10]

$$\begin{aligned}\tilde{\rho}(s'',s';t) \approx & \int ds_0^+ \int ds_1^+ \cdots \int ds_{N-1}^+ \int ds_0^- \int ds_1^- \cdots \int ds_{N-1}^- \langle s''|e^{-iH_0\Delta t/\hbar}|s_{N-1}^+\rangle \cdots \\ & \times \langle s_1^+|e^{-iH_0\Delta t/\hbar}|s_0^+\rangle\langle s_0^+|\tilde{\rho}(0)|s_0^-\rangle\langle s_0^-|e^{iH_0\Delta t/\hbar}|s_1^-\rangle \cdots \\ & \times \langle s_{N-1}^-|e^{iH_0\Delta t/\hbar}|s'\rangle F(s_0^+,s_1^+,\ldots,s_{N-1}^+,s'',s_0^-,s_1^-,\ldots s_{N-1}^-,s';\Delta t)\end{aligned} \tag{17}$$

where the initial condition (2.6) was assumed. Here $\{s_0^+,s_1^+,\ldots\}$ and $\{s_0^-,s_1^-,\ldots\}$ denote discretizations of the forward and backward path employed in the path integral representation of the forward and reverse time evolution operators, respectively. The symmetric splitting of the time evolution operator, Eq. (15), is exact in the limit $\Delta t \to 0$. If the propagators involving the one-dimensional reference Hamiltonian are evaluated exactly Eq. (17) becomes exact in the limit of vanishing time step.

In the scheme described below the propagator involving the reference Hamiltonian H_0 is calculated exactly via a basis set expansion.[8] The numerical construction of the adiabatic system propagator guarantees that the path integral expression, Eq. (17), is exact for any value of the time step in the absence of system-bath coupling. With nonzero coupling the quasi-adiabatic propagator splitting is still exact in the limit of a high frequency bath. In the opposite limit of a sluggish bath Eq. (17) is also exact, as finite values of the time step offer adequate discretization of the long-period bath oscillators in that case and the specifics of the propagator splitting become insignificant.

The discretized influence functional has the structure

$$F = \exp\left(-\frac{1}{\hbar}\sum_{k=0}^{N}\sum_{k'=1}^{k}\left(s_k^+ - s_k^-\right)\left(\eta_{kk'}s_{k'}^+ - \eta_{kk'}^* s_{k'}^-\right)\right) \tag{18}$$

where the coefficients $\eta_{kk'}$ constitute the discrete analog of the bath response function, Eq. (10), and have been given explicitly in Ref. 10.

The QUAPI-discretized influence functional can be obtained from Eq. (9) if the forward and backward paths are replaced by piecewise continuous paths consisting of constant s segments, $s^{\pm}(t) \to s_k^{\pm}$ for $(k-½)\Delta t \leq t < (k+½)\Delta t$. This particular stepwise change of the time-dependent force, which is different from that resulting from a simple trapezoid rule-type discretization of the action [2], arises from the exact treatment of the system dynamics. Note that the counterterms cancel terms arising from the displaced harmonic oscillators and therefore do not appear in Eq. (18).

The role of the influence functional in the QUAPI is to include multidimensional nonadiabatic corrections to the exact dynamics along the adiabatic path. These corrections become more accurate as the time slicing of the path integral becomes finer. Because the quasi-adiabatic propagator is by construction exact for any magnitude of the time increment in the two opposite limits of fast or slow coupled oscillators, the QUAPI converges rapidly (i.e., with fairly large time steps) over wide ranges of bath oscillator frequencies.

Using the ideas presented above it is straightforward to convert the QUAPI-discretized path integral into a series of lower-dimension integrals. This is done by recognizing that the magnitude of the coefficients $\eta_{kk'}$ decays rapidly if the difference $|k-k'| \equiv \Delta k$ becomes large.[11] Defining an integer $\Delta k_{\max}$ such that the effective medium memory is t_m = $\Delta k_{\max}\Delta t$ and discarding influence functional interactions between points separated by a time difference exceeding t_m leads to a discretized path integral expression where the nonlocal interactions are of short range compared to the full path integral, Eq. (17).[11–13] The structure of the influence functional couplings after this memory truncation is illustrated in Fig. 2.

Furthermore, it is clear that different paths enter the path integral expression of the reduced density matrix with different weights which depend on the characteristics of the reference system as well as on the bath-induced damping. Within a given desired preci-

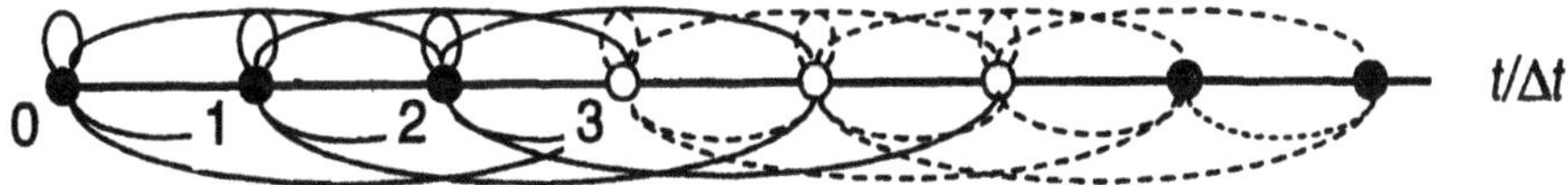

Figure 2. Schematic representation of interactions (curved lines) in the discretized path integral expression for the reduced density matrix. The circles indicate points separated by Δt. Solid and hollow circles correspond to points defining the vector **R** of path segments at times 0 and $3\Delta t$, respectively. The solid lines indicate the interactions included in the propagator matrix. Adapted from Ref. 15.

sion, paths of weight smaller than an appropriate threshold can be ignored. A path that spans the full propagation time $n\Delta k_{max}\Delta t$ can have appreciable weight only if all of its constituent segments that span the intervals $0 \leq t \leq (\Delta k_{max} - 1)\Delta t$, $\Delta k_{max}\Delta t \leq t \leq (2\Delta k_{max} - 1)\Delta t$, ... have weights that are larger than some threshold θ. In general for $\Delta k_{max} >> 1$ the number of path segments with appreciable weight corresponds to a small fraction of the total number of conceivable forward and backward discretized path segments that span the memory length $\Delta k_{max}\Delta t$.[14] The path segments of appreciable weight can form the elements of a vector (array) **R**, which constitutes the discrete analog of the reduced density functional R defined earlier. This vector can be propagated forward according to the following matrix-vector multiplication

$$\mathbf{R}((m+1)\Delta k_{max}\Delta t) = \mathbf{T}\cdot\mathbf{R}(m\Delta k_{max}\Delta t)\,. \tag{19}$$

In the last equation $\mathbf{R}(m\Delta k_{max}\Delta t)$ is a vector of discretized path segments defined on the time interval $m\Delta k_{max}\Delta t \leq t < (m+1)\Delta k_{max}\Delta t$ and **T** is a propagator matrix whose elements T_{ji} consist of all interactions that couple the ith path segment to its jth neighbor in the direction of increasing time.[15]

The interactions accounted for in each step of this procedure are indicated in Fig. 2. The elements of the propagator matrix that connect path segments $(s_k^\pm,\ldots,s_{k+\Delta k_{max}-1}^\pm)_i$ and $(s_{k+\Delta k_{max}}^\pm,\ldots,s_{k+2\Delta k_{max}-1}^\pm)_j$ are given by the relations

$$\begin{aligned} T_{ji} &\equiv T\Big((s_k^\pm,s_{k+1}^\pm,\ldots,s_{k+\Delta k_{max}-1}^\pm)_i,(s_{k+\Delta k_{max}}^\pm,\ldots,s_{k+2\Delta k_{max}-1}^\pm)_j\Big) \\ &= \prod_{n=k}^{k+\Delta k_{max}-1} K(s_n^\pm,s_{n+1}^\pm)I_0(s_n^\pm)I_1(s_n^\pm,s_{n+1}^\pm)I_2(s_n^\pm,s_{n+2}^\pm)\cdots I_{\Delta k_{max}}(s_n^\pm,s_{n+\Delta k_{max}}^\pm) \end{aligned} \tag{20}$$

where $I_{\Delta k}$ denote influence functional interactions between points separated in time by $\Delta k\Delta t$,

$$\begin{aligned} I_0(s_k^\pm) &= \exp\left(-\frac{1}{\hbar}(s_k^+ - s_k^-)(\eta_{kk}s_k^+ - \eta_{kk}^* s_k^-)\right), \\ I_{\Delta k}(s_k^\pm,s_{k+\Delta k}^\pm) &= \exp\left(-\frac{1}{\hbar}(s_{k+\Delta k}^+ - s_{k+\Delta k}^-)(\eta_{k+\Delta k,k}s_k^+ - \eta_{k+\Delta k,k}^* s_k^-)\right), \quad \Delta k > 0 \end{aligned} \tag{21}$$

and K is the bare system propagator matrix:

$$K(s_k^\pm,s_{k+1}^\pm) = \left\langle s_{k+1}^+\right| e^{-iH_0\Delta t/\hbar}\left|s_k^+\right\rangle\left\langle s_k^-\right| e^{iH_0\Delta t/\hbar}\left|s_{k+1}^-\right\rangle. \tag{22}$$

Notice that these expressions involve the coefficients $\eta_{kk'}$ which, unless k or k' are endpoints, depend on the absolute difference $|k-k'|$. For this reason the propagator **T** is independent of time for all but the first and last propagation step. Finally, multiplication of the vector **R** by a special endpoint propagator yields the reduced density matrix at each iteration step.[15]

A final issue of practical significance when dealing with continuous potentials concerns the choice of quadrature scheme employed in the discretization of the system coordinate. System-specific discrete variable representations[16] have been shown[17] to offer maximum efficiency by allowing accurate discretization with sparse grids. Typically, the number of grid points required for accurate representation of the reduced density functional is only slightly larger than the number of thermally populated eigenstates of the reference Hamiltonian at the given temperature.

APPLICATIONS

Reaction Rates

Many chemical reactions can be described as barrier crossing events between two locally stable configurations along a reaction coordinate. For processes occurring in the condensed phase, the reaction coordinate interacts with a dissipative medium of phonon or solvent degrees of freedom. The simplest model involves a double well potential coupled to a harmonic heat bath. Further, the double well is symmetric if the heat of the reaction is equal to zero. This model exhibits a host of intriguing behaviors which arise from the combination of frictional damping, dissipative tunneling and phenomena associated with bound motion in the reactant or product potential well. Perhaps the most interesting feature of dissipative barrier crossing is the non-monotonic variation of the reaction rate with solvent friction, known as Kramers turnover[18]. This behavior and its quantum mechanical generalizations have been the focus of intense theoretical work in recent years (for a recent review see Ref. 19).

Topaler and Makri have reported accurate quantum path integral calculations for the canonical reaction rate constant in a symmetric double well potential coupled to an Ohmic bath with parameters typical of proton transfer reactions.[20]. These calculations, which employed the reactive flux correlation function formalism,[21] confirmed theoretical predictions for the temperature and friction dependence of the rate in the thermally activated regime, i.e. in the portion of parameter space where the rate increases exponentially with temperature. In that regime the reaction rate exhibits a Kramers turnover with tunneling corrections (see Fig. 3). This situation is described very well by analytical theories. Specifically, the path integral results are in excellent agreement with the predictions of quantum Grote-Hynes and quantum transition state approximations[22,23] at strong friction where the reaction is dominated by spatial diffusion, while quantum turnover theories[24,25] apply successfully to the entire friction dependence. Below a characteristic crossover temperature tunneling effects prevail and the rate becomes only weakly dependent on temperature. In that regime analytical or numerical approximations are only qualitatively correct, as can be seen from Fig. 4. The most intriguing feature displayed in this figure is the positive enhancement of the quantum rate constant compared to the prediction of the centroid-based quantum transition state theory[23] in the deep tunneling regime. As classical transition state theory always provides an upper bound to the rate, this is a dynamical effect of purely quantum mechanical nature which arises from vibrational coherence in the potential wells[26]. This coherence effect is illustrated in Fig. 5, which shows the time dependence

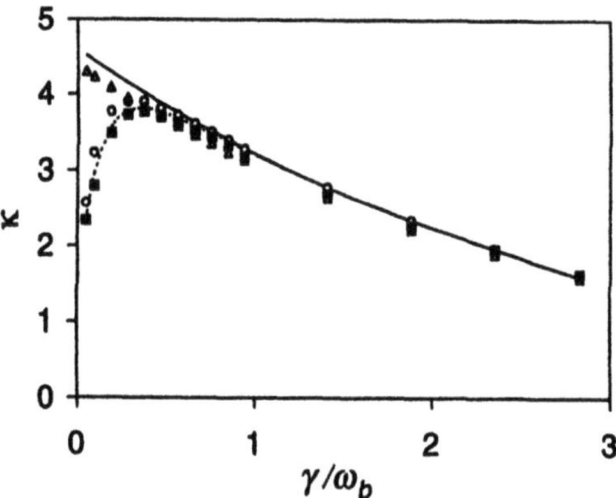

Figure 3. Quantum transmission coefficient for a symmetric double well potential as a function of damping strength at a temperature above crossover. ω_b is the frequency at the top of the potential barrier. Numerical path integral results are shown as solid squares. The solid line and triangles show results obtained by the approximations of references 22 and 23, while the circles and dashed line correspond to the quantum turnover theories discussed in references 24 and 25.

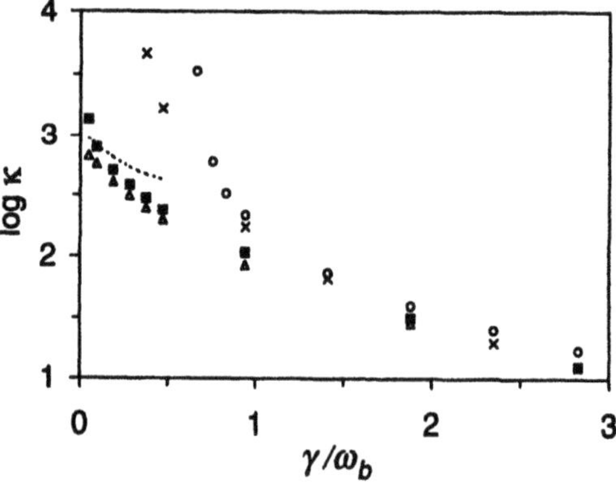

Figure 4. Logarithm of the quantum transmission coefficient for a symmetric double well potential as a function of damping strength at a temperature around crossover. ω_b is the frequency at the top of the potential barrier. Values of γ/ω_b greater than approximately 0.6 correspond to activated dynamics, while tunneling effects become dominant at smaller values of the friction. Numerical path integral results are shown as solid squares. The crosses and dashed line indicate results obtained by approximate evaluation of semiclassical transition state theories while the circles show the results of the approximation discussed in Ref. 22. The triangles correspond to quantum transition state theory results.

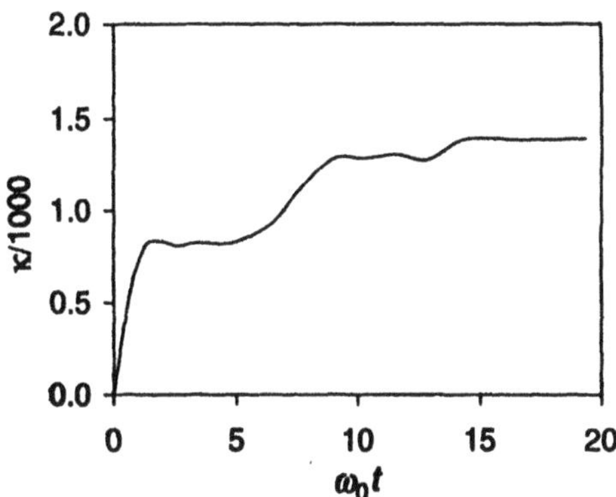

Figure 5. The integrated quantum reactive flux crossing the barrier of a symmetric double well potential as a function of time for the case of low temperature (below crossover) and weak friction. The long-time plateau of this curve yields the quantum transmission coefficient. The vibrational frequency at the potential minimum is indicated by ω_0.

of the quantum reactive flux that crosses the potential barrier. The observed increase of the reactive flux beyond the first step is responsible for the significant enhancement of the rate above its quantum transition state theory value, which corresponds roughly to the first plateau value of the flux correlation function.

Control of Dissipative Tunneling

The interaction of electromagnetic radiation with tunneling systems can lead to a variety of intriguing behaviors and has been a subject of numerous studies. For simplicity, often only the lowest tunneling doublet is considered explicitly, giving rise to a two-level system (TLS) which (within the classical treatment of the radiation field) is driven by a time-dependent force. Coupling of the driven TLS to a dissipative environment results in even more complex dynamical effects. Because the dissipative TLS provides an adequate model of such diverse phenomena as electron transfer reactions, impurity tunneling in crystals and charge oscillations in semiconductor double quantum well structures, its understanding is vital in several areas of chemistry and physics. The addition of coherent radiation introduces new parameters, offering the intriguing possibility of manipulating such processes in a desirable manner. Aside from theoretical and perhaps technological interest in exploring the different behaviors that can arise from the interplay among tunneling, coherent driving and dissipation, the driven TLS in contact with a heat bath constitutes the simplest "control" problem whose study may shed light onto the possibility of overcoming intramolecular energy redistribution to lead a polyatomic system toward a specific product state.

Most previous work on driven two-level systems has been concerned with the issue of localization. In the absence of dissipation, the addition of a simple oscillatory driving term can under certain conditions quench tunneling in a symmetric TLS, maintaining indefinitely spatial localization of a left- (or right-) localized TLS state.[27,28] This phenomenon is related to a degeneracy in the driven TLS quasienergy spectrum when the field intensity and frequency satisfy a certain "optimal localization" condition. Interaction with a dissipative bath offers the possibility of delocalization via destruction of phase coherence, generally leading the TLS to a time-dependent steady state. Recent studies[29–31] have shown that even though dissipation eventually destroys localization, the rate of delocalization is generally small under low temperature and weak friction conditions.

Perhaps surprisingly, localization effects do not disappear gradually as the temperature and dissipation strength are increased. Recent work[32,33] revealed a novel "universal" regime in dissipative two-level systems driven by a high-frequency periodic term of the type $V_0 \cos \omega_0 t$, which represents the classical limit of a monochromatic continuous-wave electromagnetic field. By resorting to the quantized representation of the radiation field and employing semiclassical arguments and results from nonadiabatic rate theory, Makri and Wei argued[32] that the decay rate of an initially localized state at high temperature is largely insensitive to the characteristics of the environment when the dissipation parameter is sufficiently large to suppress phase interference, decreasing as the inverse of the field amplitude. In this regime the lifetime of localized states can be extended significantly with the aid of strong monochromatic fields.

Accurate path integral calculations[32,33] have confirmed this effect and explored the deviations from universal delocalization behavior. Fig. 6 shows the average TLS position as a function of time for an Ohmic TLS driven by an optimal field at the temperature $k_B T = 10\hbar\Omega$, where $2\hbar\Omega$ is the tunneling splitting of the bare TLS, at three values of the dimensionless dissipation parameter $\alpha = 2\gamma/\pi\hbar$. In the absence of dissipation ($\alpha = 0$) it is seen that the average TLS position maintains its initial value of unity at all times that are multiples of the field period. With finite friction the localized state displays essentially in-

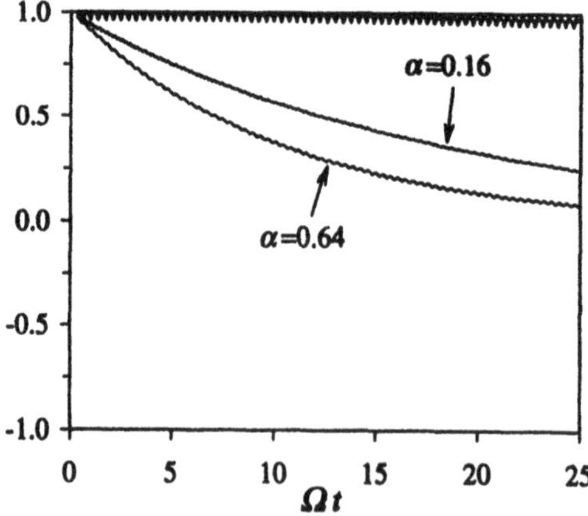

Figure 6. Path integral results for the position expectation value as a function of time at a high temperature $\hbar\Omega\beta = 0.1$ for a TLS coupled to an Ohmic bath with cutoff frequency $\omega_c = 20\ \Omega$ and $\alpha = 0, 0.16$ and 0.64. The field parameters $V_0 = 20\ \hbar\Omega$, $\omega_0 = 16.714\ \Omega$ satisfy the optimal localization condition. The oscillatory line that does not deviate significantly from unity corresponds to the dissipationless case. Adapted from Ref. 33.

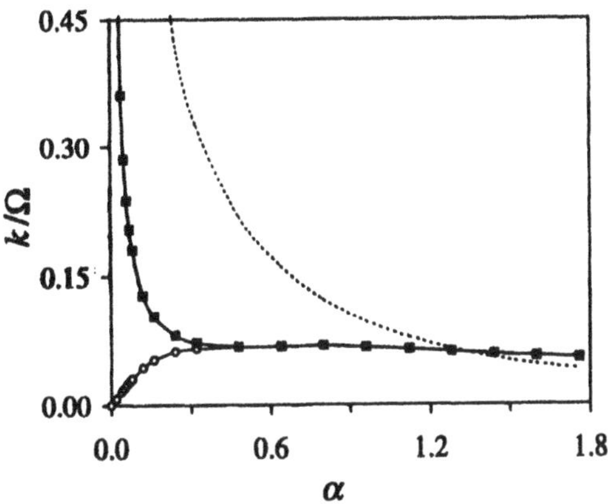

Figure 7. Path integral results for the decay rate of a localized state in a driven TLS coupled to a bath of Ohmic spectral density with $\omega_c = 20\ \Omega$ as a function of the dimensionless dissipation parameter $\hbar\Omega\beta = 0.1$. The field amplitude is $V_0 = 30\ \hbar\Omega$. Solid squares: generic field of frequency $\omega_0 = 15\ \Omega$. Open circles: localizing field of $\omega_0 = 10.87\ \Omega$. The dotted line shows the delocalization rate of the force-free TLS. The semiclassical (Ref. 32) prediction of the plateau value of the rate is $k/\Omega = 0.67$.

coherent dynamics with a small-amplitude oscillatory component superposed on the overall exponential decay.

The TLS delocalization rate is shown in Fig. 7 as a function of the dissipation parameter for an Ohmic TLS in the absence of driving and with two fields of the same amplitude whose frequencies correspond to generic and optimal conditions. It is seen that both driving fields enhance significantly the lifetime of a localized state. Most importantly, the delocalization rate displays the same plateau that spans a wide range of friction strength in both systems driven by the same intensity field, irrespective of the precise values of the driving frequencies and in spite of the fact that the dissipationsless dynamics are vastly different in these two cases. The friction-independent value of the rate in the plateau regime is in excellent agreement with the theoretical prediction based on a semiclassical treatment of the quantized field model.[32] With weak or moderate dissipation the driven TLS decay rate is generally small compared to its field-free value. The path integral results confirmed that the delocalization rate in the plateau regime is also largely independent of temperature.

As shown in Fig. 7, the decay constant deviates from its medium-independent value at small values of the dissipation parameter, exhibiting behavior sensitive to the parameters of the field as well as those of the environment. As phase factors prevail in the weak friction regime, constructive or destructive phase interference leads to increase or decrease of the rate with respect to its plateau value. The non-monotonic variation of these phase factors leads to an interesting "antiresonance" pattern[33] where the TLS lifetime exhibits several maxima which become more pronounced as the field frequency is increased up to about $V_0/\hbar$. This effect is illustrated in Fig. 8. At zero dissipation even small departure from the rigorous localization condition causes destruction of localization and the lifetime

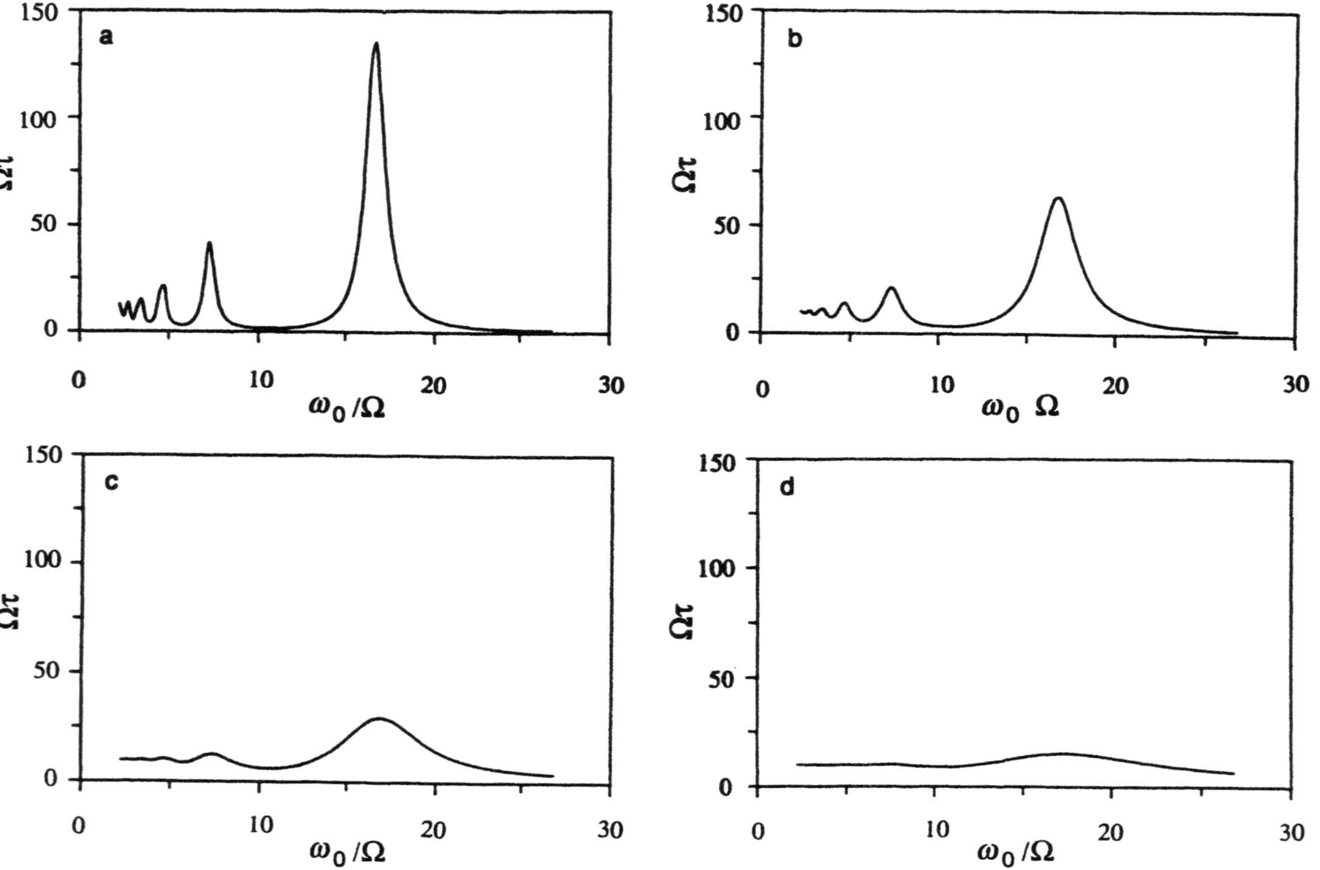

Figure 8. The lifetime of a localized state as a function of driving frequency for the field amplitude $V_0 = 20\ \hbar\Omega$ obtained by iterative evaluation of the path integral. The TLS is coupled to a high-temperature Ohmic bath with $\omega_c = 20\ \Omega$. (a) $\alpha = 0.025$. (b) $\alpha = 0.05$. (c) $\alpha = 0.1$. (d) $\alpha = 0.2$. Adapted from Ref. 33.

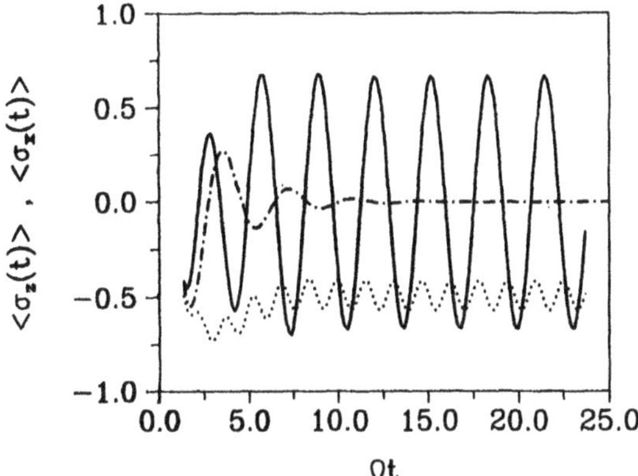

Figure 9. The average TLS position (solid line) and the population difference of the TLS eigenstates (dotted line) plotted as functions of time. The field parameters are $V_0 = 0.5\ \hbar\Omega$ and $\omega_0 = 2\ \Omega$. The temperature is $k_BT = 0.278\ \hbar\Omega$. The bath cutoff frequency is $\omega_c = 7.5\ \Omega$ and the dissipation parameter is $\alpha = 0.16$. The chain-dotted line indicates the field-free dynamics. Adapted from Ref. 31.

peaks become delta functions. With finite friction these peaks acquire a width. As a consequence, the lifetime of localized states can be extended even with significant deviation from optimal conditions. At small driving frequencies neighboring peaks overlap, resulting in small fluctuations of the TLS delocalization rate about the plateau value. At high driving frequency the lifetime decreases monotonically to the force-free dissipative TLS value. Finally, when the dissipation becomes sufficiently strong, state-specific effects induced by phase interference are wiped out and the lifetime becomes only very weakly dependent on driving frequency.

Another interesting effect is observed in two-level systems driven by monochromatic fields nearly resonant with the tunneling frequency at weak dissipation. Numerical path integral calculations[34] demonstrated the possibility of inducing long-lived coherent tunneling motion, thus preventing thermal equilibration. The time evolution of an initially localized state under such conditions is displayed in Fig. 9. As seen in Fig. 10, the steady-state amplitude of the induced tunneling oscillations depends non-monotonically on the field intensity and friction strength, thus exhibiting a quantum stochastic resonance.

Dynamics of Primary Charge Separation in Photosynthetic Reaction Centers

The first few picoseconds of photosynthesis in bacterial reaction centers involve transfer of an electron from an excited special chlorophyll pair P^* to a distant bacteriopheophytin on the L branch of the polypeptide. In spite of the large distance between these two sites (about 17 Å center-to-center), this event takes place with a time constant of

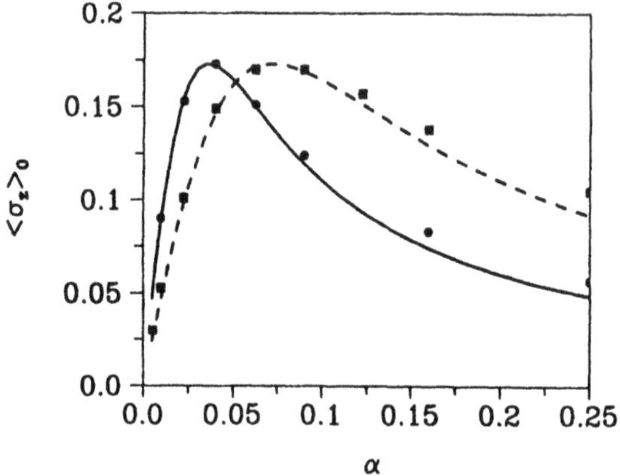

Figure 10. Steady-state oscillation amplitude plotted as a function of the dissipation parameter for $\omega_0 = 2\ \Omega$. Circles and squares correspond to results with $V_0 = 0.5\ \hbar\Omega$ and $V_0 = \hbar\Omega$, respectively. The lines indicate analytical approximations discussed in Ref. 34. The temperature is $k_B T = 4\ \hbar\Omega$ and the bath cutoff frequency $\omega_c = 7.5\ \Omega$. Adapted from Ref. 31.

about 3 ps and with quantum yield that approaches unity. In addition, a weak inverted temperature dependence is observed. Understanding the specifics of this highly efficient electron transfer process has been the subject of numerous experimental and theoretical studies, while its mechanism has been surrounded by significant controversy (for recent reviews see references 34 and 35).

Because of the large separation between donor and acceptor units, the corresponding electronic coupling is estimated to be very small. There is now convincing evidence that a chlorophyll monomer located between the special pair and the bacteriopheophytin is involved by providing a bridge state. As it is not easy to extract the energetics of this state, its role in the electron transfer process is not clear. The simplest scheme involves a two-step sequential mechanism in which the reduced chlorophyll monomer participates as a true intermediate. Another popular scenario is the superexchange mechanism in which the bridge state may enhance the transfer rate without actually accommodating the electron. Time-resolved experiments on wild-type reaction centers do not detect a reduced chlorophyll transient (with a 10–15% detection threshold) and thus do not resolve the mechanism. By contrast, recent experiments on various mutants of *Rhodobacter sphaeroides* have reported significant long-lived population of the bridge state,[35] indicative of a two-step transfer. However, it has been argued[36,37] that the presence of a long-lived reduced chlorophyll intermediate is not sufficient to rule out the superexchange hypothesis and that a large bridge population may result even if that state has energy much higher than that of the electron donor.

Makri and coworkers recently presented accurate path integral calculations of the charge transfer dynamics in a three-state model for the charge transfer in wild-type and modified photosynthetic reaction centers[38,39]. By following the electronic state populations with various combinations of the unknown energy parameters over the initial 17 picoseconds of the process, that work concluded that the presence of a significant long-lived population of the reduced chlorophyll monomer is not compatible with a superexchange mechanism and that the primary charge separation evolves via a two-step mechanism through a low-lying bridge state. Additional simulations have shown the implications of this mechanism to be compatible with a variety of other experimental observations, including the temperature variation of the rate and kinetic effects introduced by specific chemical modifications.

CONCLUDING REMARKS

Exploiting the finite correlation time of dissipative media allows approximation of the path integral for a system coupled to a harmonic bath by a sequence of local functional integrals for an appropriate reduced density functional where the domain of integration extends over paths spanning a finite time length. The exact dynamics is recovered as the included memory length is increased. Extension to time-dependent Lagrangians is straightforward.

In the discrete time representation of the path integral, these ideas give rise to an iterative scheme which involves multiplication of a propagator matrix that connects discretized path segments with a vector of weight-selected path segments which represents the discrete analog of the reduced density functional. Large time steps and therefore high efficiency are achieved by carrying out the calculation in the quasi-adiabatic propagator representation of the path integral, which utilizes the one-dimensional adiabatic potential as the reference in terms of which the time evolution operator is split. System-specific discrete variable representations offer optimal quadratures.

This scheme is applicable to a wide variety of problems involving quantum mechanical effects in dissipative environments. A sample of the situations that can be addressed this way was presented in section III which reviewed recent path integral calculations of reaction rates, exploration of the effects of periodic driving on tunneling dynamics and simulation of the early stages of photosynthetic charge separation. Future applications of this methodology will no doubt uncover further interesting phenomena and aid the design and interpretation of novel experiments.

ACKNOWLEDGMENTS

This work has been supported by a Sloan Research Fellowship and a Packard Fellowship for Science and Engineering.

REFERENCES

1. R. P. Feynman, Space-time approach to non-relativistic quantum mechanics, *Rev. Mod. Phys.*, 20:367 (1948).
2. R. P. Feynman and F. L. Vernon, Jr., The theory of a general quantum system interacting with a linear dissipative system, *Ann. Phys.* 24:118 (1963).
3. R. P. Feynman, *Statistical Mechanics*, Addison-Wesley, Redwood City (1972).

4. N. Metropolis, A. W. Rosenbluth, M. N. Rosenbluth, H. Teller, and E. Teller, *J. Chem. Phys.* 21:1087 (1953).
5. N. Makri, Feynman path integration in quantum dynamics, *Comp. Phys. Comm.* 63:389 (1991).
6. A. J. Leggett, S. Chakravarty, A. T. Dorsey, M. P. A. Fisher, A. Garg, and M. Zwerger, Dynamics of the dissipative two-state system, *Rev. Mod. Phys.* 59:1 (1987).
7. R Kubo, M. Toda, and N. Hashitsume, *Statistical Physics*, 2nd ed., Springer-Verlag, Heidelberg (1991).
8. N. Makri, Improved Feynman propagators on a grid and non-adiabatic corrections within the path integral framework, *Chem. Phys. Lett.* 193:435 (1992).
9. A. O. Caldeira and A. J. Leggett, Path integral approach to quantum Brownian motion, *Physica A*, 121:587 (1983).
10. N. Makri, Numerical path integral techniques for long-time quantum dynamics of dissipative systems, *J. Math. Phys.* 36:2430 (1995).
11. D. E. Makarov and N. Makri, Path integrals for dissipative systems by tensor multiplication: condensed phase quantum dynamics for arbitrarily long time, *Chem. Phys. Lett.* 221:482 (1994).
12. N. Makri and D. E. Makarov, Tensor multiplication for iterative quantum time evolution of reduced density matrices. I. Theory, *J. Chem. Phys.* 102:4600 (1995).
13. N. Makri and D. E. Makarov, Tensor multiplication for iterative quantum time evolution of reduced density matrices. II. Numerical methodology, *J. Chem. Phys.* 102:4611 (1995).
14. E. Sim and N. Makri, Tensor propagator with weight-selected paths for quantum dissipative dynamics with long-memory kernels, *Chem. Phys. Lett.* 249:224 (1996).
15. E. Sim and N. Makri, Filtered propagator functional for iterative dynamics of quantum dissipative systems, *Comp. Phys. Commun.* (in press).
16. Z. Bacic and J. C. Light, Theoretical methods for rovibrational states of floppy molecules, *Annu. Rev. Phys. Chem.* 40:469 (1989).
17. M. Topaler and N. Makri, System-specific discrete variable representations for path integral calculations with quasi-adiabatic propagators, *Chem. Phys. Lett.* 210:448 (1993).
18. H. A. Kramers, Brownian motion an a field of force and the diffusion model of chemical reactions, *Physica (Utrecht)* 7:284 (1940).
19. G. R. Fleming and P. Hänggi, Activated barrier crossing, World Scientific, Singapore (1993).
20. M. Topaler and N. Makri, Quantum rates for a double well coupled to a dissipative bath: accurate path integral results and comparisons with approximate theories, *J. Chem. Phys,* 101:7500 (1994).
21. W. H. Miller, S. D. Schwartz, and J. W. Tromp, Quantum mechanical rate constants for bimolecular reactions, *J. Chem. Phys.* 79:4889 (1983).
22. P. G. Wolynes, Quantum theory of activated events in condensed phases, *Phys. Rev. Lett.* 47:968 (1981).
23. G. A. Voth, Path integral centroid methods in quantum statistical mechanics and dynamics, *Adv. Chem. Phys.* 93:135 (1996).
24. P. Hänggi, E. Pollak, and H. Grabert, Report No. 215, 1989.
25. I. Rips and E. Pollak, Quantum Kramers model: solution of the turnover problem, *Phys. Rev. A* 41:5366 (1990).
26. J. N. Onuchic and P. G. Wolynes, Classical and quantum pictures of reaction dynamics in condensed matter: resonances, dephasing, and all that, *J. Phys. Chem.* 92:6495 (1988).
27. F. Grossmann, T. Dittrich, P. Jung, and P. Hänggi, Coherent destruction of tunneling, *Phys. Rev. Lett.* 67:516 (1991).
28. F. Grossmann and P. Hänggi, Localization in a driven two-level dynamics, *Europhysics Letters* 18:571 (1992).
29. M Grifoni, M. Sassetti, J. Stockburger, and U. Weiss, Nonlinear response of a periodically driven damped two-state system, *Phys. Rev. E* 48:3497 (1993).
30. T. Dittrich, B. Oeschlagel, and P. Hänggi, Driven tunneling with dissipation, *Europhys. Lett.* 22:5 (1993).
31. D. E. Makarov and N. Makri, Control of dissipative tunneling dynamics by continuous wave electromagnetic fields: localization and large-amplitude coherent motion, *Phys. Rev. E* 52:5863 (1995).
32. N. Makri and Liqiang Wei, Universal delocalization rate in driven dissipative two-level systems at high temperature, *Phys. Rev. E* (in press).
33. N. Makri, Stabilization of localized states in dissipative tunneling systems interacting with monochromatic fields, *J. Chem. Phys.* (in press).
34. D. E. Makarov and N. Makri, Stochastic resonance and nonlinear response in double quantum well structures, *Phys. Rev. B* 52:R2257 (1995).
35. S. Schmidt, T. Arlt, P. Hamm, H. Hüber, T. Nägele, J. Wachtveitl, M. Meyer, H. Scheer and W. Zinth, Energetics of the primary electron transfer reaction revealed by ultrafast spectroscopy on modified bacterial reaction centers, *Chem. Phys. Lett.* 223:116 (1994).

36. R. Egger, C. H. Mak, and U. Weiss, Rate concept and retarted master equations for dissipative tight-binding models, *Phys. Rev. E* 50:R655 (1994).
37. R. Egger and C. H. Mak (unpublished).
38. N. Makri, E. Sim, D. E. Makarov, and M. Topaler, Long-time quantum simulation of the primary charge separation in bacterial photosynthesis, *Proc. Natl. Acad. Sci. U.S.A.* 93:3926 (1996).
39. E. Sim and N. Makri, Path integral simulation of charge transfer dynamics in photosynthetic reaction centers, *J. Phys. Chem.* (submitted).

8

DÉVELOPPEMENTS RÉCENTS SUR LES GROUPES DE TRESSES APPLICATIONS À LA TOPOLOGIE ET À L'ALGÈBRE

P. Cartier

École Normale Supérieure
Département de Mathématique
et Informatique
45 rue d'Ulm
F-75230 PARIS CEDEX 05

A mes amis soviétiques
retrouvés grâce à la
"perestroïka"

BRAID GROUPS IN TOPOLOGY AND ALGEBRA

ABSTRACT

Braid groups were invented by Emil Artin in 1925. The most convenient definition of the braid group B_n (n is the number of strings) is as the fundamental group of the configuration space Φ_n parametrizing the subsets with n points in a given pane. From a physical point of view, the space Φ_n is the configuration space of n indistinguishable particles moving in two dimensions. This explains the recent interest for the braid groups in many two-dimensional physical models, in the form of parastatistics and anyons.

From a more mathematical point of view, Markov discovered in 1936 how to construct invariants of knots from linear representations of the braid groups. For a long time, very few representations of the braid groups were known. One of the greatest recent discoveries has been the invention of the Jones polynomial for knots (in 1984). The corresponding representations of B_n were introduced in connection with the von Neumann algebras in operator theory, or in another form, via solutions of the Yang–Baxter equation.

In this paper, after a recollection of known facts about the combinatorics of knots, we develop consequently a categorical point of view. For a long time, categories — in the form described in the 40's by Eilenberg and MacLane — were collections of preexisting mathematical objects connected via certain kind of transformations. In the newest developments, categories appear to be more abstract, defined via generators and relations, and encapsulating a great deal of sophisticated combinatorics.

We report on the main results of Kohno, Drinfeld and Reshetikhin, which provide a link between the main two approaches to knot invariants:

— through the monodromy of the so-called Knizhnik–Zamolodchikov differential equations

— or using representations of quantum groups.

After the writing of this paper — in 1989 — the most dramatic developments were the definition of the Vasiliev invariants, and their explicit description by Kontsevitch, Bar-Natan and myself in 1993.

INTRODUCTION

Les nœuds, ouverts ou fermés, les rubans, les tresses, les cables,$\cdots$ se prêtent à des jeux sans fin. Les applications sérieuses ne manquent pas non plus : cablages électriques, modèles de mécanique des fluides faisant revivre la vieille théorie des tourbillons de Descartes$\cdots$. Dès le siècle dernier, Tait et Thompson, après Cayley, se sont intéressés à ces structures combinatoires. Mais il a fallu attendre le début de ce siècle, avec les travaux de Dehn sur la Topologie, pour obtenir les premiers résultats généraux et rigoureux.

C'est en 1925 qu'Émile Artin [A1] fonde la théorie des tresses. Conformément aux tendances de l'Algèbre allemande contemporaine, il cherche — et réussit avec éclat — comment introduire des groupes adéquats dans le sujet. Aujourd'hui, la définition la plus simple du groupe B_n des tresses à n brins est la suivante : considérons une variété connexe M de dimension d ; *l'espace de configuration* est l'espace des parties finies d'ordre n dans M ; on le note $\Phi_n(M)$. On peut le considérer comme l'espace-quotient $M_*^n \,/\, \mathbf{S}_n$, où M_*^n est l'ensemble des suites (numérotées) de n points distincts $x_1,\cdots,x_n$ de M , et où le groupe symétrique $\mathbf{S}_n$ agit par permutation des n points. Le groupe de tresses généralisées est le groupe fondamental $B_n(M)$ de l'espace $\Phi_n(M)$. Lorsque l'on a $d \geq 3$, ce groupe est le produit semi-direct de $\mathbf{S}_n$ et de $\pi_1(M)^n$; cela provient de ce que M_*^n diffère de M^n par des sous-variétés de codimension ≥ 3 , donc a même groupe fondamental que M^n . Lorsque $d = 1$, considérons seulement le cas $M = \mathbf{R}$; alors $\Phi_n(M)$ est l'ouvert de $\mathbf{R}^n$ formé des points $(x_1,\cdots,x_n)$ tels que $x_1 < \cdots < x_n$, et l'on a $B_n(M) = 0$. Le seul cas intéressant est celui de la dimension d égale à 2. *Le groupe d'Artin est* $B_n = B_n(\mathbf{R}^2)$.

La définition précédente a été introduite par Neuwirth et Fox [A6] en 1962 ! En 1925, Artin avait donné les propriétés de base des groupes B_n au moyen d'arguments de nature combinatoire, en partie heuristiques. Il reviendra de manière plus rigoureuse sur le sujet dans ses articles [A1] de 1947. Rétrospectivement, les arguments d'Artin deuxième manière sont des manipulations sur la suite exacte d'homotopie des fibrés. Encore aujourd'hui, ces sujets de Topologie se divisent *grosso modo* en deux catégories :

— des arguments combinatoires ;

— des théorèmes difficiles (lemme de Dehn, théorème de Markov, théorèmes de Cerf et Kirby) qui permettent en général de montrer qu'une équivalence d'un certain type entre objets géométriques se ramène à une chaîne finie d'équivalences élémentaires.
La difficulté de ces derniers théorèmes est l'un des nombreux avertissements que notre description mathématique du continu n'est peut-être pas adéquate.

L'étude des espaces de configuration est un sujet actif en Topologie ; elle est menée en particulier au Japon et au Viet-Nam (autour de Huynh Mui). Mais les groupes de tresses B_n restaient une curiosité isolée avant 1970. Le réveil a eu plusieurs causes :

a) Arnold s'est intéressé à l'étude des *singularités* et de leur déploiement. Un exemple typique est l'étude des polynômes de degré n à racines simples, soient $P(z) = z^n + a_1 z^{n-1} + \cdots + a_n$; l'hypothèse de simplicité se traduit par la non-nullité du discriminant de P , vu comme polynôme $D(a_1, \cdots, a_n)$ en ses coefficients. Arnold a calculé la cohomologie du complémentaire dans $\mathbf{C}^n$ de l'hypersurface d'équation $D = 0$, espace dont le groupe fondamental est B_n. Un peu plus tard, Brieskorn a généralisé cette étude à certains cas de complémentaires d'une famille finie d'hyperplans (le cas d'Arnold correspond à la famille des hyperplans d'équation $z_i = z_j$ dans $\mathbf{C}^n$ muni des coordonnées $z_1, \cdots, z_n$). Dans ces exemples, *les groupes de tresses* (et certaines de leurs généralisations) *apparaissent comme groupes de monodromie* ; il appartenait à Zamolodchikov [F9] et Kohno [F3] d'étudier les équations différentielles admettant des singularités polaires liées à ces configurations. Celles-ci fournissent des exemples assez simples où tester les théories générales de l'homotopie rationnelle et de la monodromie (voir mes deux exposés précédents [E5] et [F1]). Par ailleurs, l'étude des théories des champs quantiques conformes a récemment conduit Tsuchiya et Kanie [F11] à l'étude systématique de fonctions de plusieurs variables complexes ayant ce type de ramification.

b) Vers 1968, Iwahori a introduit sous le nom d'*algèbres de Hecke* des déformations $\mathcal{H}_n(q)$ de l'algèbre $\mathbf{C}\mathbf{S}_n$ du groupe symétrique (et d'autres groupes similaires). De telles algèbres jouent un rôle important dans l'étude des représentations linéaires des groupes finis simples et des groupes algébriques réductifs sur un corps p-adique (Iwahori, Matsumoto, Tits). Les représentations irréductibles des algèbres $\mathcal{H}_n(q)$ ont été déterminées explicitement par Hoefsmit en 1974 dans [C17] ; les résultats s'insèrent dans le courant actuel de la Combinatoire où l'on étudie les *q-analogues des nombres classiques* (par exemple le polynôme de Gauss $\left[{n \atop i}\right]_q$ qui pour $q = 1$ redonne le coefficient binomial $\binom{n}{i}$). On n'avait guère prêté attention au fait que les représentations de $\mathcal{H}_n(q)$ fournissent de nouvelles représentations du groupe de tresses B_n , jusqu'à la découverte fracassante, en 1984, du *polynôme de Jones associé à un nœud.* Ce polynôme a été aussitôt interprété comme une trace associée aux représentations de B_n fournies par l'algèbre $\mathcal{H}_n(q)$ par Jones lui-même [C3, C5] et par Ocneanu [C8].

c) *En Mécanique Statistique*, l'étude de certains modèles, comme le modèle de Potts, conduit à des algèbres apparentées aux algèbres $\mathcal{H}_n(q)$ et appelées *algèbres de Temperley-Lieb* ; elles peuvent servir de point de départ à la construction du polynôme de Jones. Mais, plus important, on a peu à peu découvert l'importance de la *relation de Yang-Baxter*, dont l'une des formes

$$\hat{R}_{12}\hat{R}_{23}\hat{R}_{12} = \hat{R}_{23}\hat{R}_{12}\hat{R}_{23} \qquad \text{(YB)}$$

est très proche de l'une des relations qui servent de définition au groupe B_n. Toute solution de cette équation fournit automatiquement une représentation du groupe B_n, et de plus les modèles de Mécanique Stat'ebres de Temperley-Lieb ; elles peuvent servir de point de départ à la construction du polynôme de Jones. Mais, plus important, on a peu à peu découvert l'importance de la *relation de Yang-Baxter*, dont l'une des formes

$$\hat{R}_{12}\hat{R}_{23}\hat{R}_{12} = \hat{R}_{23}\hat{R}_{12}\hat{R}_{23} \qquad \text{(YB)}$$

est très proche de l'une des relations qui servent de définition au groupe B_n. Toute solution de cette équation fournit automatiquement une représentatide Yang-Baxter.

d) Drinfeld [D2] a introduit en 1986 sous le nom de "groupes quantiques" une nouvelle classe d'algèbres de Hopf. À chaque groupe de Lie semi-simple complexe G est attaché un groupe quantique G_q, et chaque représentation linéaire de G se "déforme" en une représentation de G_q , qui fournit *ipso facto* une solution de l'équation de Yang-Baxter [Rosso D7]. Il est remarquable qu'en fait *un groupe quantique soit entièrement déterminé par la catégorie de ses représentations* ; il s'agit là d'une généralisation de la dualité de Tannaka-Krein pour les groupes compacts. Mais on doit à Drinfeld, Reshetikhin et Turaev (et indépendamment à Freyd, Yetter et leurs collaborateurs) la profonde remarque que *la combinatoire des tresses et celle des nœuds se traduit dans le langage des catégories* en exploitant les relations de cohérence introduites par Mac Lane [B1]. Il s'agit là de l'idée, souvent défendue par Grothendieck en particulier, que les catégories sont des objets géométriques bien adaptés à l'étude de l'homotopie.

Tous les thèmes précédents semblent étroitement liés les uns aux autres, comme nous essaierons de le montrer dans cet exposé. La théorie est cependant encore très mouvante, et une synthèse définitive semble prématurée.

Pour terminer cette introduction, nous dirons quelques mots de deux autres thèmes dont chacun mériterait un exposé en propre.

La théorie quantique des champs a fait d'énormes progrès récents dans l'étude des *modèles à* 2 *ou* 3 *dimensions* d'espace-temps : théorie des champs conformes, théorie des cordes. Il se peut que la clé des phénomènes de supraconductivité à "haute température" soit à trouver là. Comme J. Fröhlich l'a remarqué dans [H2] et [H3], les arguments classiques qui permettent de lier le spin à la statistique (principe d'exclusion de Pauli) perdent leur validité en dimension 2. La raison en est que le double cône d'influence

d'un point cesse d'avoir *un complémentaire connexe* et qu'on peut alors intrinsèquement distinguer "gauche" et "droite". Au lieu de faire intervenir seulement les deux représentations ρ de degré 1 du groupe $\mathbf{S}_n$, distinguant le cas des *bosons* ($\rho(s) = 1$) de celui des *formions* ($\rho(s) =$ signature de s), on doit considérer les représentations du groupe B_n agissant dans l'espace de Hilbert correspondant à n particules. On rejoint là l'étude faite par Tsuchiya et Kanie [F11] de la monodromie des "fonctions à n points" de la théorie, qui jouent le rôle de moments d'une mesure sur un espace fonctionnel.

En théorie des nombres, une question ouverte — et fondamentale — est de prouver que *tout groupe fini G est le groupe de Galois d'une extension finie L du corps* **Q** *des nombres rationnels.* Cela revient à étudier la structure du groupe de Galois $\mathrm{Gal}(\bar{\mathbf{Q}}/\mathbf{Q})$. Fixons un entier $n \geq 1$; le groupe de tresses B_n se réalise comme un groupe d'automorphismes du groupe libre F_n ; géométriquement, cela revient à considérer F_n comme le groupe fondamental de $\mathbf{C} \backslash S$, où S est une partie finie à n éléments de $\mathbf{C}$. Si l'on suppose que S se compose de nombres rationnels, le groupe de Galois de la plus grande extension du corps de fractions rationnelles $\bar{\mathbf{Q}}(t)$ non ramifiée en dehors de S est le complété profini $\hat{F}_n$ de F_n. On peut plonger B_n dans le groupe des automorphismes de $\hat{F}_n$; soit $\hat{B}_n$ son adhérence dans $Aut(\hat{F}_n)$. Le groupe de Galois $\mathrm{Gal}(\bar{\mathbf{Q}}/\mathbf{Q})$ agit aussi dans $\hat{F}_n$, et l'on peut montrer qu'on définit ainsi un *homomorphisme injectif* ψ_S *de* $Gal(\bar{\mathbf{Q}}/\mathbf{Q})$ *dans le groupe de tresses complété* $\hat{B}_n$. C'est là l'esquisse du programme de Grothendieck [I1] pour l'étude du groupe de Galois. Nous renvoyons à la bibliographie pour les travaux de Ihara, Oda, Deligne... sur ce sujet.

Remerciements : Ils vont aux organisateurs et aux participants du Colloque d'Ascona (octobre 1989) sur "Géométrie et Théorie quantique des champs". En particulier, je remercie Semenov-Tian-Shansky, Reshetikhin, Wasserman pour des entretiens approfondis et aussi Maillet qui m'a grandement aidé dans la collecte des documents. Je remercie aussi Jones et Freyd pour leurs commentaires vigilants sur cette introduction, et M. Gallois pour une relecture attentive.

1. RAPPELS SUR LES NŒUDS ET LES TRESSES

1.1. On appelle *nœud* dans l'espace euclidien $\mathbf{R}^3$ toute partie fermée de $\mathbf{R}^3$ homéomorphe au cercle $\mathbf{S}^1$; un *entrelac* est la réunion d'un nombre fini de nœuds deux à deux disjoints. Il est commode de rajouter un point à l'infini à $\mathbf{R}^3$, et de raisonner dans l'espace $\mathbf{S}^3$ obtenu, homéomorphe à une sphère à 3 dimensions. Nous orienterons l'espace $\mathbf{S}^3$, ainsi que les nœuds et entrelacs considérés.

On dira que *deux entrelacs L et L′* (en particulier deux nœuds) *sont équivalents s'il existe un homéomorphisme de* $\mathbf{S}^3$ *transformant L en L′ et respectant les orientations de L, L′ et* $\mathbf{S}^3$. Le groupe des homéomorphismes orientés de $\mathbf{S}^3$ sera muni de la convergence uniforme ; il est connexe par arcs. Par suite, si deux entrelacs L et L' sont équivalents, il existe une *isotopie* transformant L en L', c'est-à-dire une application continue* $h : I \times \mathbf{S}^3 \to \mathbf{S}^3$ avec les propriétés suivantes :

a) on a $h(0,x) = x$ pour x dans $\mathbf{S}^3$;

b) pour tout t dans I, l'application $h_t : x \mapsto h(t,x)$ est un homéomorphisme de $\mathbf{S}^3$;

c) L' est l'image de L par l'application h_1.

On parle d'*isotopie ambiante* pour les entrelacs, car tout l'espace $\mathbf{S}^3$ se déforme de manière à entraîner L vers L'. Une déformation de L en L', au moyen d'une famille continue d'entrelacs, donnerait une équivalence trop faible.

Il existe des nœuds sauvages, découverts par Fox et Artin ; pour les éviter, nous ne considérerons que des nœuds qui sont équivalents à des lignes polygonales, formées de segments de droites. Les entrelacs sont restreints de même. On peut montrer que si deux nœuds polygonaux L et L' sont équivalents, on peut passer de L à L' par des opérations élémentaires du type décrit ci-dessous, et leurs inverses (en nombre fini)

*On note I l'intervalle $[0,1]$ de $\mathbf{R}$.

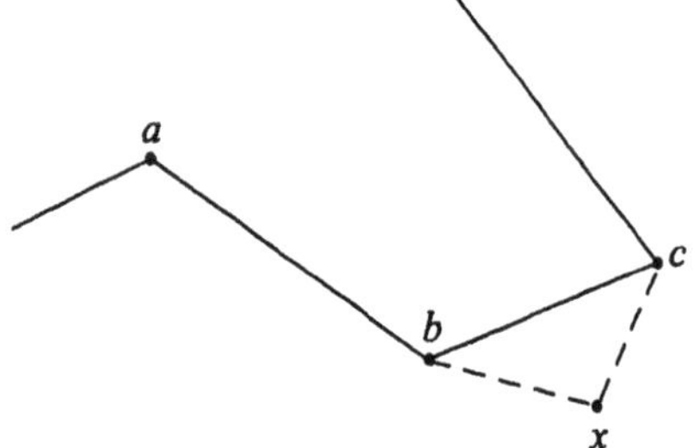

La modification remplace le segment bc par la réunion de bx et xc .

1.2. Pour décrire les entrelacs, on utilise une *projection plane.* Quitte à remplacer un entrelac L par un entrelac équivalent, on peut supposer que la projection orthogonale de L sur un plan P est une réunion de courbes différentiables, avec pour seules singularités des points de croisement doubles avec tangentes distinctes. A chaque croisement, on doit indiquer laquelle des deux branches est au-dessus de l'autre ; compte tenu des orientations, on a les deux dispositions suivantes :

On a indiqué l'indice du croisement égal à $+1$ ou -1 .

A une telle projection D , on associe un *graphe planaire* T , qui ne prend pas en compte les orientations: le complémentaire de D dans le plan P a un nombre fini de composantes connexes qu'on peut colorier avec deux couleurs α et β de telle manière que deux régions qui bordent un même arc de courbe aient des couleurs distinctes. Chacune de ces régions de couleur α définit un sommet de T , et les points de croisement définissent des arêtes de T joignant les deux régions α adjacentes ; chaque arête est munie d'un signe donné par les règles suivantes :

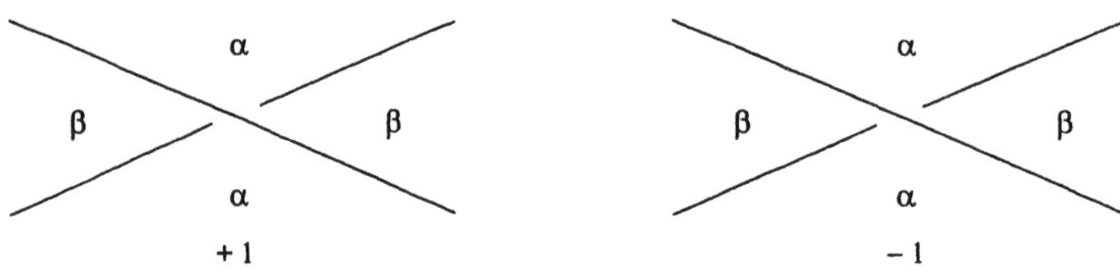

Voici un exemple de graphe associé à une projection de nœud

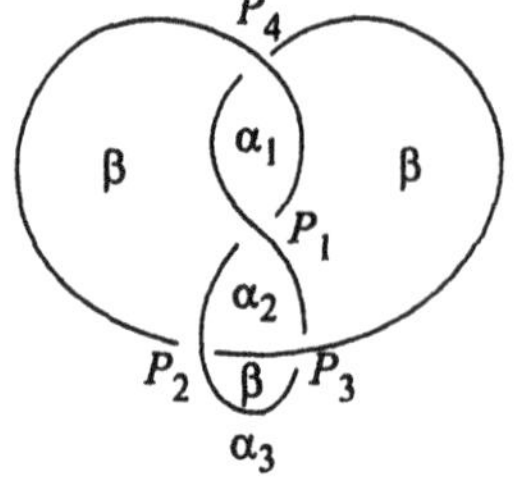

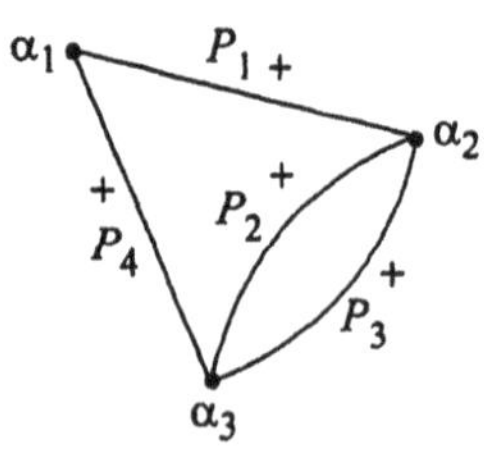

En général, le même entrelac a plusieurs projections non équivalentes par une isotopie du plan. On peut prouver que l'on peut passer d'une projection D_1 à une projection D_2 du même nœud, à isotopie près, par une suite finie de *transformations de Reidemeister*, qui se répartissent en trois types

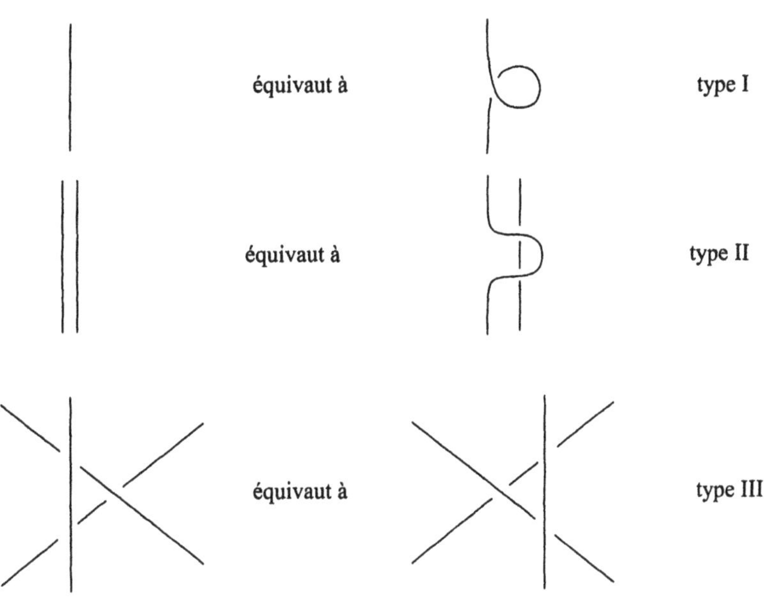

Faire attention aux croisements ; on peut mettre les orientations *ad libitum*.

1.3. Soit L un nœud orienté. Il existe une surface orientée S plongée dans $\mathbf{S}^3$, dont le bord soit L (*surface de Seifert*). Si L' est un nœud orienté disjoint de L , on peut par une isotopie ambiante s'arranger pour que S et L' se coupent transversalement en un nombre fini de points $p_1, \cdots, p_r$. Chacun de ces points p_i définit un nombre ε_i égal à ± 1 et déterminé par les orientations. Le nombre $\varepsilon_1 + \cdots + \varepsilon_r$ s'appelle *le nombre d'entrelacements* de L et L' , noté $i(L,L')$.

De manière plus précise, les groupes d'homologie à coefficients entiers de $\mathbf{S}^3 \backslash L$ sont donnés par

$$H_0(\mathbf{S}^3 \backslash L) = \mathbf{Z} \quad , \quad H_1(\mathbf{S}^3 \backslash L) = \mathbf{Z} \quad , \quad H_i(\mathbf{S}^3 \backslash L) = 0 \qquad \text{pour } i \geq 2 \, .$$

Le générateur γ de $H_1(\mathbf{S}^3 \backslash L)$ peut être choisi de sorte que la classe d'homologie définie par L' dans $\mathbf{S}^3 \backslash L$ soit égale à $i(L,L') . \gamma$.

Comme $H_1(\mathbf{S}^3 \backslash L)$ est isomorphe à $\mathbf{Z}$, il existe une variété M de dimension 3 qui est un revêtement de $\mathbf{S}^3 \backslash L$, galoisien de groupe $\mathbf{Z}$. Le *module d'Alexander* de L , noté $A(L)$, est le groupe d'homologie $H_1(M)$; comme le groupe $\mathbf{Z}$ agit sur M , donc sur son homologie, on peut considérer $A(L)$ comme un module sur l'anneau $\Lambda = \mathbf{Z}[u, u^{-1}]$ des polynômes de Laurent en u , la multiplication par u sur $A(L)$ correspondant à l'action du générateur de $\mathbf{Z}$ agissant sur M . La structure du Λ-module $A(L)$ s'exprime par une suite exacte

$$0 \longrightarrow \Lambda^{2h} \xrightarrow{\alpha} \Lambda^{2h} \longrightarrow A(L) \longrightarrow 0 \, ;$$

l'homomorphisme α est décrit par une matrice de la forme $A(L) = V - u^tV$, où V est une matrice carrée de type $2h \times 2h$ à coefficients entiers, telle que

$$V - {}^tV = \begin{pmatrix} 0 & I_h \\ -I_h & 0 \end{pmatrix} \tag{1.1}$$

(I_h : matrice unité d'ordre h). Le déterminant $\Delta_L(u)$ de la matrice $A(L)$ s'appelle le *polynôme d'Alexander* du nœud L . A multiplication près par une puissance de u, il ne dépend que du nœud L . On le normalise sous la forme

$$\Delta_L(u) = a_0 + a_1 u + \cdots + a_{2m} u^{2m} \tag{1.2}$$

avec $a_0 \neq 0$; on a alors la relation de symétrie $a_i = a_{2m-i}$.

1.4. Compte tenu de l'identification de $\mathbf{R}^2$ et de $\mathbf{C}$, la définition générale donnée dans l'introduction se reformule ainsi : on note $\mathbf{C}_*^n$ l'ensemble des vecteurs à n coordonnées complexes toutes distinctes ; alors B_n est le groupe fondamental de $\mathbf{C}_*^n/\mathbf{S}_n$, le groupe $\mathbf{S}_n$ opérant par permutation des coordonnées. Le groupe fondamental de l'espace $\mathbf{C}_*^n$ s'appelle le *groupe de tresses pures* ; il se note P_n . Alors P_n est un sous-groupe invariant du *groupe de tresses* B_n et le groupe B_n/P_n est canoniquement isomorphe à $\mathbf{S}_n$.

Représentant les éléments de B_n par des lacets dans l'espace $\mathbf{C}_*^n/\mathbf{S}_n$, on obtient la description géométrique suivante. Considérons deux plans parallèles Π_0 et Π_1 dans $\mathbf{R}^3$, et la bande Δ qu'ils limitent. Fixons n points distincts $P_1^0, \cdots, P_n^0$ dans Π_0 et n points distincts $P_1^1, \cdots, P_n^1$ dans Π_1 . Une tresse à n brins est une réunion γ de n arcs simples $\gamma_1, \cdots, \gamma_n$ contenus dans Δ avec les propriétés suivantes :

a) les arcs $\gamma_1, \cdots, \gamma_n$ sont deux à deux disjoints ;

b) chacun de ces arcs coupe tout plan Π , parallèle à Π_0 et Π_1 et contenu dans Δ , en un point et un seul ;

c) chaque arc a l'une de ses extrémités en l'un des points P_i^0 ; il se termine alors en $P_{\pi(i)}^1$, où π est la permutation dans $\mathbf{S}_n$ associée à l'élément de B_n défini par γ .

Pour définir l'équivalence, on introduit le groupe Γ des homéomorphismes de Δ qui fixent chacun des points $P_1^0, \cdots, P_n^0, P_1^1, \cdots, P_n^1$ et transforment en lui-même tout plan Π parallèle à Π_0 et Π_1 ; on note Γ_0 la composante connexe par arcs de l'identité dans Γ . Deux tresses γ et γ' sont équivalentes par Γ si et seulement si elles définissent la même permutation $\pi \in \mathbf{S}_n$. On dira qu'elles sont *isotopes* si elles sont équivalentes par le groupe Γ_0 ; les éléments de B_n sont alors les classes d'isotopie de tresses à n brins.

Choisissons un plan P perpendiculaire à Π_0 et Π_1 , et soit D la bande $P \cap \Delta$ dans P . On projette chaque tresse γ perpendiculairement sur P . Quitte à modifier γ par une isotopie, on peut supposer que la projection de γ se compose de n arcs différentiables qui se coupent transversalement en un nombre fini de points, et qui coupent transversalement chaque droite parallèle à $L_i = \Pi_i \cap P$ en un point unique. On adopte la convention précédente pour indiquer les croisements par-dessus ou par-dessous

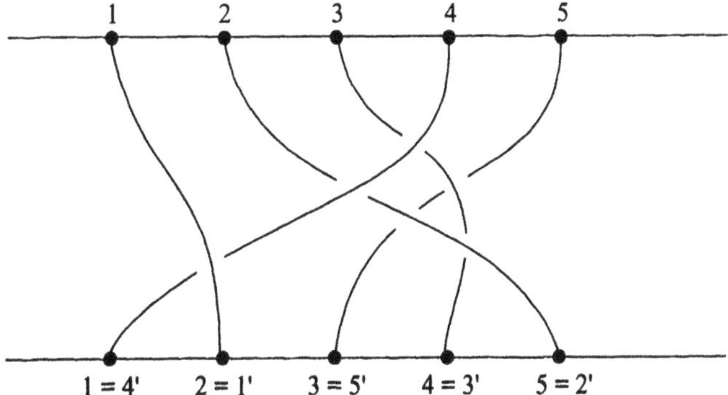

La permutation associée est $i \mapsto i'$, soit ici $\pi = \begin{pmatrix} 1 & 2 & 3 & 4 & 5 \\ 2 & 5 & 4 & 1 & 3 \end{pmatrix}$. On laisse au lecteur le soin de définir l'isotopie convenable sur les projections planes de tresses.

1.5. Structure du groupe des tresses. Le groupe B_n admet $n-1$ générateurs $\sigma_1, \cdots, \sigma_{n-1}$, que l'on définit facilement au moyen des projections planes

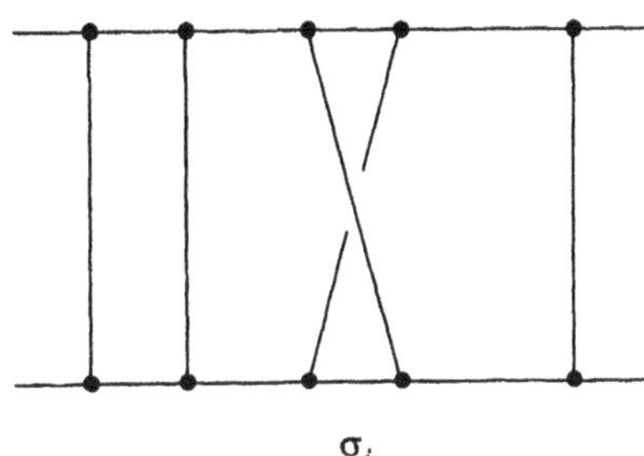

σ_i

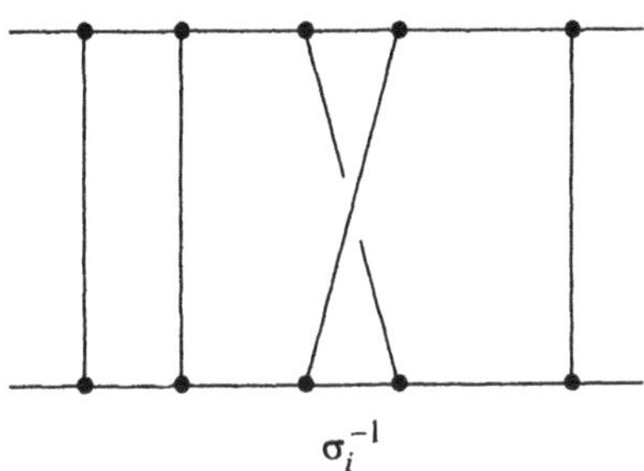

σ_i^{-1}

Le groupe B_n admet la présentation donnée par les relations suivantes* :

$$\sigma_i \sigma_{i+1} \sigma_i = \sigma_{i+1} \sigma_i \sigma_{i+1} \tag{1.3}$$

$$\sigma_i \sigma_j = \sigma_j \sigma_i \tag{1.4}$$

De manière analogue, le groupe symétrique $\mathbf{S}_n$ admet une présentation donnée par les générateurs s_i (transposition de i et $i+1$) pour $1 \leq i \leq n-1$, et les relations

$$s_i s_{i+1} s_i = s_{i+1} s_i s_{i+1} \tag{1.5}$$

$$s_i s_j = s_j s_i \,. \tag{1.6}$$

$$s_i^2 = 1 \tag{1.7}$$

Dans la projection de B_n sur S_n , le générateur σ_i donne s_i . Le noyau P_n est donc le plus petit sous-groupe invariant de B_n qui contienne les σ_i^2 .

On définit un groupe B_∞ ayant une suite infinie de générateurs $\sigma_1, \sigma_2, \cdots$ avec les relations (1.3) et (1.4). Le groupe B_n s'identifie au sous-groupe de B_∞ engendré par $\sigma_1, \cdots, \sigma_{n-1}$. Le plongement de B_n dans B_{n+1} ainsi obtenu se représente géométriquement par l'opération suivante : prendre une tresse à n brins et ajouter un $(n+1)^e$ brin à droite, assez loin pour

*Les permutations se composent comme les fonctions ; ainsi on a $(\pi\pi')(i) = \pi(\pi'(i))$. Pour la composition des tresses, on est donc conduit à les orienter du haut vers le bas, et tt' représente t en dessous de t'.

ne pas être enlacé avec les autres brins. De manière analogue, on a une chaîne croissante $\mathbf{S}_1 \subset \mathbf{S}_2 \subset \cdots \subset \mathbf{S}_n \subset \mathbf{S}_{n+1} \subset \cdots$ fournie par les groupes de permutations, avec la réunion $\mathbf{S}_\infty$.

La projection de $\mathbf{C}_*^{n+1}$ sur $\mathbf{C}_*^n$ qui transforme $(z_1, \cdots, z_n, z_{n+1})$ en $(z_1, \cdots, z_n)$ est une fibration, dont la fibre F est obtenue en ôtant n points dans $\mathbf{C}$. Le groupe fondamental de F est un groupe libre F_n à n générateurs $a_1, \cdots, a_n$, et dans une fibration, le groupe fondamental de la base opère sur celui de la fibre. On en déduit ceci :

Le groupe de tresses B_n s'identifie au groupe des automorphismes φ de F_n possédant les propriétés suivantes :

a) le produit $a_1 \cdots a_n$ est invariant par φ ;

b) il existe une permutation $\pi \in \mathbf{S}_n$ telle que $\varphi(a_i)$ soit conjugué dans F_n à $a_{\pi(i)}$.

Dans cette identification, le générateur σ_i est décrit ainsi :

$$\sigma_i(a_j) = \begin{cases} a_i a_{i+1} a_i^{-1} & \text{si } j = i \\ a_i & \text{si } j = i+1 \\ a_j & \text{si } j = i, i+1 \,. \end{cases} \tag{1.8}$$

En particulier, le groupe de tresses pures P_n est le groupe des automorphismes de F_n qui laissent invariants $a_1 \cdots a_n$ et la classe de conjugaison de chaque a_i . On peut alors identifier *P_{n+1} au produit semi-direct de P_n et de F_n muni de l'action précédente de P_n* ; dans cette identification, on a

$$a_i = (\sigma_{i+1} \cdots \sigma_n)^{-1} \sigma_i^2 (\sigma_{i+1} \cdots \sigma_n) \,. \tag{1.9}$$

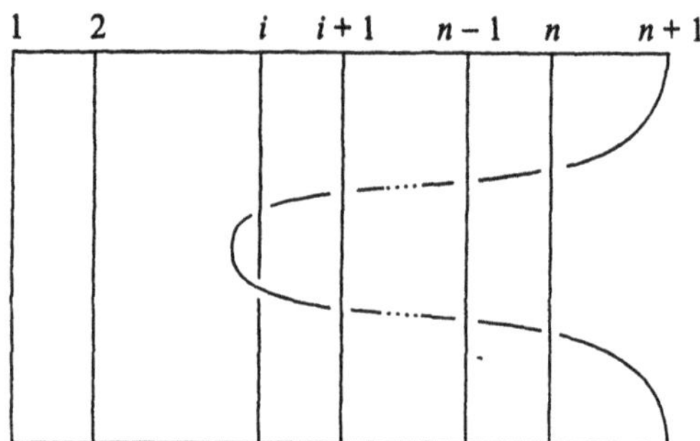

Le même raisonnement de fibration a d'autres applications. Notons $\mathcal{A}_n$ l'algèbre de cohomologie de de Rham de $\mathbf{C}_*^n$; la projection de $\mathbf{C}_*^{n+1}$ sur $\mathbf{C}_*^n$ identifie $\mathcal{A}_n$ à une sous-algèbre de $\mathcal{A}_{n+1}$; si l'on note a_i la classe de cohomologie de la forme différentielle $d\log(z_{n+1} - z_i)$ (pour $1 \leq i \leq n$), alors on a $\mathcal{A}_{n+1} = \mathcal{A}_n \oplus \mathcal{A}_n a_1 \oplus \cdots \oplus \mathcal{A}_n a_n$. Le polynôme de Poincaré de l'espace $\mathbf{C}_*^n$ est donc $\prod_{k=1}^{n-1}(1+kt)$. Un résultat dû à Fox, et généralisé par Brieskorn [E7] et Deligne [E8] affirme que le revêtement universel de $\mathbf{C}_*^n$ est contractile ; par suite, la cohomologie du groupe B_n est isomorphe à $\mathcal{A}_n$. Enfin, avec les notations du n° 4.3, l'algèbre de Lie $\mathcal{G}_{n+1}$ associée au groupe P_{n+1} est somme directe de la sous-algèbre de Lie $\mathcal{G}_n$ et d'un idéal I_n qui est une algèbre de Lie libre à n générateurs.

1.6. Relations entre tresses et nœuds. Soit D une droite dans $\mathbf{R}^3$. Un entrelac L qui ne rencontre pas D sera appelé une *tresse fermée* à n brins, d'axe D , si tout demi-plan limité par D rencontre L en n points exactement. Choisissons un demi-plan P_0 limité par D , et notons P_t (pour $0 \leq t \leq 1$) le demi-plan obtenu par la rotation ρ_t d'angle $2\pi t$ autour de D (après choix d'orientations). Posons

$$F_t = \rho_t^{-1}(P_t \cap L) . \tag{1.10}$$

Alors la famille $(F_t)_{0\leq t\leq 1}$ est un lacet dans l'espace de configuration $\Phi_n(P_0)$ et définit donc un élément du groupe de tresses B_n . On déduit de là *une bijection entre les tresses fermées à n brins* (à isotopie près) *et les classes de conjugaison dans* B_n .

À isotopie près, tout entrelac est une tresse fermée pour une droite bien choisie. On déduit de là la *classification de Markov des entrelacs.* Notons B l'ensemble somme des ensembles $B_1, B_2, \cdots$ et $\sim$ la plus petite relation d'équivalence dans B pour laquelle on ait

$$g \sim hgh^{-1} \qquad \text{et} \qquad g \sim g\sigma_n \sim g\sigma_n^{-1}$$

pour g, h dans B_n . Alors *l'ensemble $\mathcal{M}$ des classes d'équivalence classifie les entrelacs* (non sauvages, à isotopie ambiante près) *dans* $\mathbf{R}^3$ *ou dans* $\mathbf{S}^3$.

2. INVARIANTS POLYNOMIAUX DES NŒUDS

2.1. Décrivons la stratégie générale fondée sur les représentations linéaires des groupes B_n, selon Jones [C5] et Turaev [D13]. Soit V un espace vectoriel de dimension finie sur $\mathbf{C}$, et soit R un opérateur inversible agissant dans $V \otimes V = V^{\otimes 2}$; on note P l'opérateur de symétrie dans $V^{\otimes 2}$, transformant $x \otimes y$ en $y \otimes x$, et l'on pose $\hat{R} = RP$. De plus, on choisit un entier $n \geq 1$, et l'on note $\hat{R}_{i,i+1}$ l'opérateur dans $V^{\otimes n}$ défini par

$$\hat{R}_{i,i+1}(x_1 \otimes \cdots \otimes x_n) = x_1 \otimes \cdots \otimes x_{i-1} \otimes \hat{R}(x_i \otimes x_{i+1}) \otimes x_{i+2} \otimes \cdots \otimes x_n . \tag{2.1}$$

De manière générale, pour tout opérateur T dans $V^{\otimes 2}$, et des entiers i, j tels que $1 \leq i < j \leq n$, on note T_{ij} l'opérateur dans $V^{\otimes n}$ qui "agit comme T sur les facteurs de rang i et j ".

On cherche une représentation T_R du groupe B_n dans l'espace $V^{\otimes n}$ qui transforme le générateur σ_i en l'opérateur $\hat{R}_{i,i+1}$. Compte tenu de la présentation de B_n donnée par (1.3) et (1.4), la condition nécessaire et suffisante pour que ceci ait lieu est que $\hat{R}$ satisfasse à la relation

$$\hat{R}_{12}\hat{R}_{23}\hat{R}_{12} = \hat{R}_{23}\hat{R}_{12}\hat{R}_{23} \tag{2.2}$$

(égalité d'opérateurs dans $V^{\otimes 3}$). Pour l'opérateur R , cela se transcrit en l'équation de Yang-Baxter (version quantique)

$$R_{12}R_{13}R_{23} = R_{23}R_{13}R_{12} . \tag{2.3}$$

Par exemple, si l'on a $R = 1$, les opérateurs R_{ij} sont tous égaux à 1 , et l'équation de Yang-Baxter est satisfaite. Dans ce cas, l'action d'un élément g de B_n est donnée par

$$T_1(g)(x_1 \otimes \cdots \otimes x_n) = x_{\pi^{-1}(1)} \otimes \cdots \otimes x_{\pi^{-1}(n)} \tag{2.4}$$

où π est l'image de g par l'homomorphisme canonique de B_n sur $\mathbf{S}_n$.

2.2. Voici une classe de solutions de l'équation de Yang-Baxter. Soit V l'espace $\mathbf{C}^N$ avec sa base canonique $e_1, \cdots, e_N$. On note q un nombre complexe non nul. Soit R l'opérateur dans $V^{\otimes 2}$ défini par

$$\begin{cases} R(e_i \otimes e_i) = q\, e_i \otimes e_i \\ R(e_i \otimes e_j) = e_i \otimes e_j \\ R(e_j \otimes e_i) = (q - q^{-1})\, e_i \otimes e_j + e_j \otimes e_i \end{cases} \tag{2.5}$$

Alors (2.3) est satisfaite. Pour $q = 1$, l'opérateur R se réduit à l'identité. L'opérateur $\hat{R}$ a les valeurs propres q (multiplicité $N(N+1)/2$) et $-q^{-1}$ (multiplicité $N(N-1)/2$), et l'on a donc $(\hat{R} - q)(\hat{R} + q^{-1}) = 0$. On peut traduire ainsi le résultat obtenu :

a) L'algèbre de Hecke $\mathcal{H}_n(q)$ est une algèbre ayant des générateurs $T_1, \cdots, T_{n-1}$ avec les relations

$$T_i\, T_{i+1}\, T_i = T_{i+1}\, T_i\, T_{i+1} \tag{2.6}$$

$$T_i\, T_j = T_j\, T_i \tag{2.7}$$

$$(T_i - q)(T_i + q^{-1}) = 0\,. \tag{2.8}$$

Elle a une base de la forme (T_π), où π parcourt $\mathbf{S}_n$, avec $T_1 = 1$, et la règle de multiplication

$$T_\pi\, T_i = T_{\pi s_i} \tag{2.9}$$

$$T_\pi\, T_i = T_{\pi s_i} + (q - q^{-1})\, T_\pi \tag{2.10}$$

b) Il existe une représentation linéaire λ_n de l'algèbre $\mathcal{H}_n(q)$ dans l'espace $V^{\otimes n}$, telle que $\lambda_n(T_i)$ soit égal à $\hat{R}_{i,i+1}$ pour $1 \leq i \leq n-1$.

c) Il existe un homomorphisme ε_n du groupe B_n dans le groupe multiplicatif $\mathcal{H}_n(q)^\times$ qui transforme σ_i en T_i. Dans ces conditions, on a

$$T_R(g) = \lambda_n(\varepsilon_n(g)) \qquad \text{pour } g \text{ dans } B_n\,. \tag{2.11}$$

Lorsque $q = 1$, l'algèbre $\mathcal{H}_n(q)$ se réduit à l'algèbre $\mathbf{C}\mathbf{S}_n$ du groupe symétrique, de sorte qu'on ait $T_\pi = \pi$. Supposons qu'on ait $q^k \neq 1$ pour $k = 1, \cdots, n$. Alors l'algèbre $\mathcal{H}_n(q)$ est semi-simple et ses représentations simples U_μ sont paramétrées par les partitions $\mu = (\mu_1, \cdots, \mu_n)$ de n (avec la convention $\mu_1 \geq \cdots \geq \mu_n \geq 0$ et $n = \mu_1 + \cdots + \mu_n$). La représentation λ_n de l'algèbre $\mathcal{H}_n(q)$ dans l'espace $(\mathbf{C}^N)^{\otimes n}$ fait intervenir exactement les représentations U_μ correspondant aux partitions $\mu = (\mu_1, \cdots, \mu_n)$ avec $\mu_k = 0$ pour $k > N$; la multiplicité de U_μ est la même que celle du cas spécial $q = 1$ (*cf.* Hoefsmit [C17]).

2.3. On suppose maintenant donnés un espace vectoriel V (de dimension finie sur $\mathbf{C}$), un opérateur inversible R dans $V \otimes V$ satisfaisant à l'équation de Yang-Baxter, et un opérateur μ dans V satisfaisant aux conditions suivantes :

a) *les opérateurs R et $\mu \otimes \mu$ commutent* ;

b) *pour tout opérateur u dans $End(V)$, on a les relations*

$$Tr((u \otimes \mu)\,.\,\hat{R}) = \alpha\, Tr(u) \tag{2.12}$$

$$Tr((u \otimes \mu)\,.\,\hat{R}^{-1}) = \alpha^{-1}\, Tr(u)\,, \tag{2.13}$$

où α est une constante complexe convenable (avec $\alpha \neq 0$).

Notons w l'homomorphisme de B_n dans $\mathbf{Z}$ qui envoie chaque générateur σ_i sur 1, puis définissons une fonction $\phi^n_{R,\mu}$ sur le groupe B_n par la relation

$$\phi^n_{R,\mu}(g) = \alpha^{-w(g)}\, Tr(\mu^{\otimes n}\,.\,T_R(g)) \tag{2.14}$$

où la représentation T_R de B_n dans l'espace $V^{\otimes n}$ satisfait à $T_R(\sigma_i) = \hat{R}_{i,i+1}$ pour $1 \leq i \leq n-1$. Il résulte facilement des hypothèses faites qu'on a les relations

$$\phi^n_{R,\mu}(hgh^{-1}) = \phi^n_{R,\mu}(g) \tag{2.15}$$

$$\phi^{n+1}_{R,\mu}(g\sigma_n^{\pm 1}) = \phi^n_{R,\mu}(g) \tag{2.16}$$

pour tout g dans B_n. Compte tenu de la classification de Markov des entrelacs (voir n° 1.6), on peut définir un invariant $T_{R,\mu}(L)$ des classes d'isotopie d'entrelacs, telle que l'on ait

$$T_{R,\mu}(L) = \phi_{R,\mu}(g) \tag{2.17}$$

si l'entrelac L s'obtient en fermant une tresse à n brins correspondant à l'élément g de B_n.

2.4. Pour donner un exemple explicite, revenons au cas $V = \mathbf{C}^N$, l'opérateur R étant décrit par les formules (2.5). On prend pour μ la matrice diagonale d'éléments

$$(-1)^{N+1}q^{N-1}, (-1)^{N+1}q^{N-3}, \cdots, (-1)^{N+1}q^{1-N}$$

et l'on pose $\alpha = (-1)^{N+1}q^N$. L'invariant de Turaev de degré N est alors défini par

$$T_L^N(q) = (-1)^{N+1}\frac{q-q^{-1}}{q^N-q^{-N}}T_{R,\mu}(L) \tag{2.18}$$

pour tout entrelac L, avec R et μ ainsi précisés ; c'est un polynôme de Laurent en q.

Lorsque $N = 2$, *on retrouve le polynôme de Jones*

$$T_L^2(q) = V_L(q^2) \tag{2.19}$$

(avec les notations de Jones [C2]).

La suite des polynômes $T_L^N(q)$ pour $N = 2, 3, \cdots$ s'exprime au moyen d'un seul polynôme de Laurent en q et q^N. De manière précise, étant donné un entrelac orienté L, choisissons un point de croisement p de L et définissons trois entrelacs L_+, L_- et L_0 où L_+ est égal à L, et où L_- et L_0 s'obtiennent par les modifications au point p indiquées sur le schéma suivant :

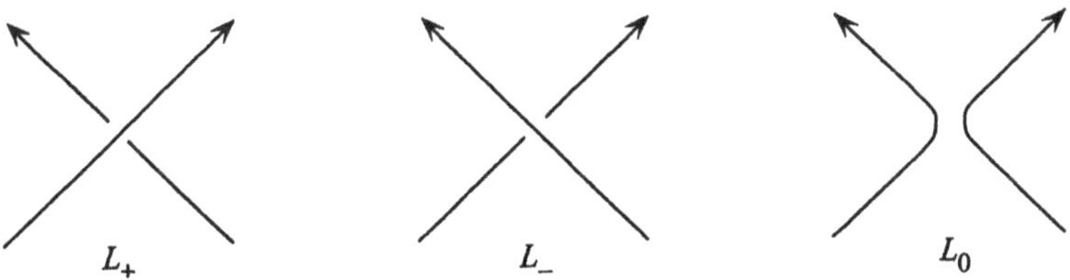

La relation $\hat{R} - \hat{R}^{-1} = q - q^{-1}$ (qui exprime le fait que $\hat{R}$ a les valeurs propres q et $-q^{-1}$) se traduit géométriquement par la relation

$$q^N T^N_{L_+}(q) - q^{-N} T^N_{L_-}(q) + (-1)^N(q-q^{-1})T^N_{L_0}(q) = 0. \tag{2.20}$$

On montre de manière combinatoire que la projection plane de tout entrelac peut être modifiée en celle d'une réunion de cercles non noués et non enlacés par des transformations de passages supérieurs en passages inférieurs et inversement. Raisonnant par récurrence sur le nombre de points de croisement de la projection de L, on arrive au résultat suivant :

THÉORÈME.— *On peut associer, de manière unique, à toute classe d'isotopie d'entrelac orienté L un polynôme $P_L(x,y,z)$ en $x, x^{-1}, y, y^{-1}, z, z^{-1}$, homogène de degré 0 et satisfaisant aux deux relations*

$$xP_{L_+} + yP_{L_-} + zP_{L_0} = 0 \tag{2.21}$$

avec les conventions ci-dessus, et

$$P_0 = 1 \tag{2.22}$$

où 0 représente un cercle. De plus, l'invariant de Turaev est donné par la formule

$$T_L^N(q) = P_L(q^N, -q^N, (-1)^N(q-q^{-1}))\,. \tag{2.23}$$

Le polynôme $P_L(x,y,z)$ n'est autre que le polynôme HOMFLY défini par les six auteurs [C12]. Il contient comme cas particuliers le polynôme d'Alexander

$$\Delta_L(q) = P(1,-1,q-q^{-1}) \tag{2.24}$$

et celui de Jones

$$V_L(q^2) = P_L(q^2, -q^{-2}, q-q^{-1})\,. \tag{2.25}$$

Remarque : Conservons les mêmes définitions de R et μ. On définit une forme linéaire τ_n sur l'algèbre de Hecke $\mathcal{H}_n(q)$ par la formule

$$\tau_n(a) = Tr(\lambda_n(a)\,.\,\mu^{\otimes n})\,; \tag{2.26}$$

la représentation λ_n de l'algèbre $\mathcal{H}_n(q)$ dans l'espace $(\mathbf{C}^N)^{\otimes n}$ est définie par $\lambda_n(T_i) = \hat{R}_{i,i+1}$. Alors τ_n est une trace, c'est-à-dire qu'on a $\tau_n(ab) = \tau_n(ba)$. D'autre part, en plongeant de manière évidente $\mathcal{H}_n(q)$ dans $\mathcal{H}_{n+1}(q)$, on a les deux relations

$$\tau_{n+1}(aT_n) = \alpha\,\tau_n(a) \tag{2.27}$$

$$\tau_{n+1}(a) = \beta\,\tau_n(a) \tag{2.28}$$

avec les constantes

$$\alpha = (-1)^{N+1}q^N \quad , \quad \beta = (-1)^{N+1}\frac{q^N - q^{-N}}{q-q^{-1}}\,.$$

Dans [C16], Vogel a déterminé *a priori* toutes les suites de traces τ_n sur les algèbres $\mathcal{H}_n(q)$ satisfaisant aux relations (2.27) et (2.28) pour des valeurs arbitraires de α et β. Le cas $\beta = 1$ est celui étudié par Jones, qui a donné le branle initial à toute cette théorie.

2.5. Un autre type d'invariant a été construit par L. Kauffman. Considérons d'abord une projection plane D d'un entrelac L. Soit V l'ensemble des points de croisement. Nous inspirant de la *Mécanique Statistique*, nous supposerons que chaque point v de V peut être dans deux états, notés α et β ; un état de D est une application ω de V dans $\{\alpha, \beta\}$. Étant donné un état ω, nous notons $m(\omega)$ le nombre de points de V dans l'état α, et $n(\omega)$ ceux dans l'état β. D'autre part, nous modifions le diagramme D simultanément en tous ses sommets selon la règle

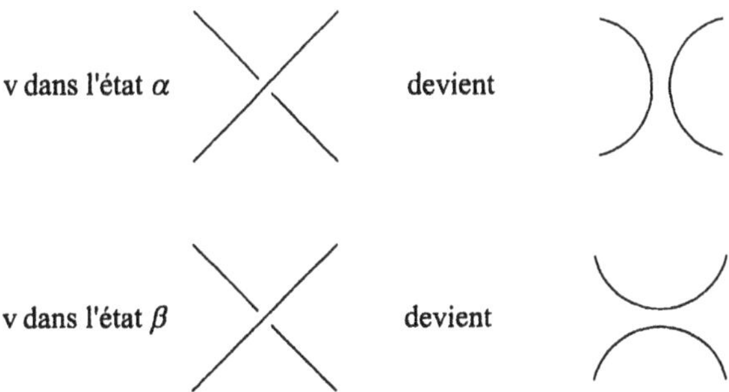

Le diagramme ainsi modifié n'a plus de points de croisement et se compose d'un nombre $p(\omega)$ de cercles disjoints. Le poids associé à un état ω est

$$\pi(\omega) = A^{m(\omega)} B^{n(\omega)} d^{p(\omega)} .$$

On associe à D le polynôme $K_D(A,B,d)$ égal à la somme des poids $\pi(\omega)$ pour les $2^{|V|}$ états possibles de D. Il s'agit là de l'analogue d'une fonction de partition en Mécanique Statistique.

Jusqu'ici, nous n'avons pas tenu compte de l'orientation de l'entrelac L ou de sa projection D. Mais on déduit de l'existence du polynôme de Kauffman $K_D(A,B,d)$ l'existence d'un nouvel invariant polynomial $Q_L(x,y)$. C'est un polynôme à coefficients entiers en x, x^{-1}, y et y^{-1} avec les propriétés caractéristiques suivantes :

a) si L est un cercle non noué, on a $Q_L = 1$;

b) soit D une projection plane de l'entrelac L et soit $w(D)$ la somme des multiplicités des points de croisement (conformément aux conventions du n° 1.2). Alors $\widetilde{Q}_D(x,y) = x^{w(D)} Q_L(x,y)$ ne dépend pas de l'orientation de L (et de D) ;

c) on a la relation

$$\widetilde{Q}_{D_+} + \widetilde{Q}_{D_-} = y\widetilde{Q}_{D_0} + y^{-1}\widetilde{Q}_{D_\infty}$$

si les quatre diagrammes D_+, D_-, D_0 et D_∞ diffèrent en un sommet selon le schéma suivant :

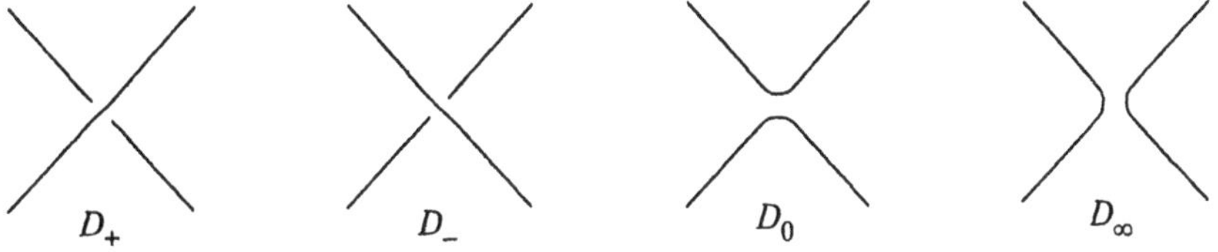

On peut retrouver le polynôme de Jones $V_L(q^2)$ comme spécialisation du polynôme de Kauffman $Q_L(x,y)$. De plus, Turaev a donné une interprétation de $Q_L(x,y)$ au moyen du schéma (V,R,μ).

3. CATÉGORIES ET ALGÈBRES DE HOPF

3.1. Considérons un ensemble E et une application φ de $E \times E$ dans E ; nous voulons formuler en général l'associativité de φ. Pour cela, on définit par récurrence une suite d'ensembles finis $\Sigma_1, \Sigma_2, \cdots$ par les règles

$$\Sigma_1 = \{0\} \quad , \quad \Sigma_n = (\Sigma_1 \times \Sigma_{n-1}) \cup (\Sigma_2 \times \Sigma_{n-2}) \cup \cdots \cup (\Sigma_{n-1} \times \Sigma_1) . \tag{3.1}$$

Par récurrence sur $n \geq 1$, on associe à chaque élément σ de Σ_n une application $\bar{\sigma}$ de E^n dans E :

a) si $n = 1$, on a $\sigma = 0$ et $\bar{\sigma}(x) = x$ pour tout $x \in E$;

b) si $n > 1$, il existe un entier p tel que $1 \leq p \leq n-1$, et deux éléments $\sigma' \in \Sigma_p$ et $\sigma'' \in \Sigma_{n-p}$ tels que $\sigma = (\sigma', \sigma'')$; on a

$$\bar{\sigma}(x_1, \cdots, x_n) = \varphi(\bar{\sigma}'(x_1, \cdots, x_p), \bar{\sigma}''(x_{p+1}, \cdots, x_n)) . \tag{3.2}$$

On peut représenter les éléments de Σ_n par des arbres binaires à n sommets libres ; donnons deux exemples d'éléments de Σ_4 et leurs représentations graphiques

$\alpha = (0, (0, (0,0)))$  $\beta = (((0, 0), 0), 0)$.

Si nous abrégeons $\varphi(a,b)$ en (ab) , les fonctions associées à α et β sont données par

$$\begin{cases} \bar{\alpha}(a,b,c,d) = (a(b(cd))) \\ \bar{\beta}(a,b,c,d) = (((ab)c)d) . \end{cases} \tag{3.3}$$

Les ensembles Σ_1 et Σ_2 ont un seul élément, et Σ_3 a deux éléments λ et μ dont les fonctions associées sont

$$\bar{\lambda}(a,b,c) = ((ab)c) \quad , \quad \bar{\mu}(a,b,c) = (a(bc)) . \tag{3.4}$$

On dit que la loi φ est *associative* si l'on a $((ab)c) = (a(bc))$ pour a,b,c dans E , c'est-à-dire $\bar{\lambda} = \bar{\mu}$. S'il en est ainsi, pour tout entier $n \geq 1$, toutes les applications de E^n dans E associées aux éléments de Σ_n coïncident. On peut par exemple vérifier de deux manières l'égalité $\bar{\alpha} = \bar{\beta}$

$$(a(b(cd))) = ((ab)(cd)) = (((ab)c)d) \tag{3.5}$$

$$(a(b(cd))) = (a((bc)d) = ((a(bc))d) = (((ab)c)d) . \tag{3.6}$$

Ces formules correspondent à deux chaînes de transformations élémentaires sur les arbres binaires.

3.2. Nous appliquons ce qui précède au cas d'une catégorie $\mathcal{C}$ et d'un foncteur Φ de $\mathcal{C} \times \mathcal{C}$ dans $\mathcal{C}$; ainsi Φ induit une loi de composition sur l'ensemble $\mathrm{Ob}(\mathcal{C})$ des objets de $\mathcal{C}$ et aussi sur l'ensemble $\mathrm{F}\ell(\mathcal{C})$ des flèches de $\mathcal{C}$. Si ces deux lois sont associatives, on dira que Φ est dit *strictement associatif*. En pratique, c'est rarement le cas ; par exemple, si $\mathcal{C}$ est la catégorie des espaces vectoriels sur $\mathbf{C}$, et qu'on prenne $\Phi(U,V) = U \otimes V$, on sait que les espaces $(U \otimes V) \otimes W$ et $U \otimes (V \otimes W)$ sont isomorphes, mais non égaux.

Supposons alors qu'on dispose d'un isomorphisme $\alpha(a,b,c)$ de $\Phi(a, \Phi(b,c))$ sur $\Phi(\Phi(a,b),c)$ fonctoriel en les objets a,b et c de $\mathcal{C}$. Par itération du foncteur Φ , on associe à chaque élément σ de Σ_n un foncteur σ_* de $\mathcal{C}^n$ dans $\mathcal{C}$. En utilisant α , on montre que les foncteurs σ_* et τ_* associés à deux éléments de Σ_n sont isomorphes. Par exemple, les deux chaînes d'égalités (3.5) et (3.6) s'interprètent en donnant deux isomorphismes ε et η du foncteur α_* sur le foncteur β_* ; on a explicitement*, en écrivant ab pour $\Phi(a,b)$:

$$\varepsilon(a,b,c,d) = \alpha(ab,c,d)\alpha(a,b,cd) \tag{3.7}$$

$$\eta(a,b,c,d) = \Phi(\alpha(a,b,c),d)\alpha(a,bc,d)\Phi(a,\alpha(b,c,d)) . \tag{3.8}$$

*Cette interprétation d'une démonstration algébrique comme une flèche d'une catégorie joue un rôle important dans certaines études récentes sur la compilation et les transformations de programmes.

Noter que les flèches dans C se composent de la droite vers la gauche. On dit que θ est une *contrainte d'associativité pour* Φ si l'on a toujours $\varepsilon(a,b,c,d) = \eta(a,b,c,d)$. S'il en est ainsi, *l'interprétation de toutes les chaînes allant de* $\sigma_*(a_1,\cdots,a_n)$ *à* $\tau_*(a_1,\cdots,a_n)$ *conduit au même isomorphisme du foncteur* σ_* *sur le foncteur* τ_* (pour σ et τ dans Σ_n). Ce théorème est dû à Mac Lane [B1].

De manière analogue, une *contrainte de commutativité* pour Φ est un isomorphisme $\rho(a,b)$ de $\Phi(a,b)$ sur $\Phi(b,a)$, fonctoriel en les objets a et b de C et tel que l'on ait

$$\Phi(\rho(a,c),b)\alpha(a,c,b)\Phi(a,\rho(b,c)) = \alpha(c,a,b)\rho(ab,c)\alpha(a,b,c)\,. \tag{3.9}$$

Cette relation traduit l'égalité des deux chaînes de transformations

$$a(bc) = a(cb) = (ac)b = (ca)b\,,$$

$$a(bc) = (ab)c = c(ab) = (ca)b\,.$$

On pourra traduire la contrainte d'associativité en un *diagramme pentagonal* et celle de commutativité en un *diagramme hexagonal.*

Si les contraintes d'associativité et de commutativité sont remplies et si $\rho(a,b)$ est inverse de $\rho(b,a)$ pour tous les objets a,b , alors pour tout σ dans Σ_n , le foncteur $\sigma_*(a_1,\cdots,a_n)$ dépend "symétriquement" des objets $a_1,\cdots,a_n$ de C .

3.3. *La dualité de Tannaka pour les groupes algébriques* repose sur les deux théorèmes suivants de Chevalley [B15], où V désigne un espace vectoriel de dimension finie (sur $\mathbf{C}$) et G un sous-groupe algébrique de $\mathrm{GL}(V)$.

THÉORÈME A.— *Toute représentation linéaire* (π,W) *de G correspondant à une application polynomiale de G dans* $GL(W)$ *peut se réaliser, à l'équivalence près, dans un sous-quotient d'un espace de la forme* $\bigoplus_{i=1}^{s} V^{\otimes n_i}\otimes(V^{\vee})^{\otimes m_i}$ *, où* $V^{\vee}$ *est le dual de* V .

THÉORÈME B.— *Il existe un nombre fini d'entiers* $r_1,\cdots,r_k$ *et des droites* D_i *dans* $V^{\otimes r_i}$ *(pour* $1\le i\le k$ *) tels que G se compose des éléments de* $GL(V)$ *qui stabilisent chacune des droites* $D_1,\cdots,D_k$.

Supposons alors que C soit une famille de représentations $(\pi_i,W_i)_{i\in I}$ du groupe G , polynomiales au sens du théorème A. On suppose de plus que C est stable par somme directe, produit tensoriel, passage au dual et aux sous-espaces et espaces quotients. Si C contient une représentation fidèle, alors toute représentation polynomiale de G est isomorphe à l'une des représentations (π_i,W_i) . Cela résulte du théorème A.

Mais, d'après le théorème B, le groupe lui-même peut être reconstruit à partir de la famille de ses représentations. Par hypothèse, la famille C est stable par produit tensoriel ; notons $i\times j$ l'élément de I tel que $\pi_{i\times j} = \pi_i\otimes\pi_j$ et $W_{i\times j} = W_i\otimes W_j$. *Supposons donnée une famille d'automorphismes* $g_i\in GL(W_i)$ *satisfaisant aux deux conditions* :

a) *on a* $fg_i = g_jf$ *pour toute application G-équivariante* f *de* W_i *dans* W_j ;

b) *on a* $g_{i\times j} = g_i\otimes g_j$.

Dans ces conditions, il existe un unique élément g de G tel que $\pi_i(g) = g_i$ *pour tout* $i\in I$.

Lorsque le groupe G est réductif, toutes ses représentations linéaires sont semi-simples (= complètement réductibles). La reconstruction précédente du groupe G peut être modifiée : on part de la classe $\mathcal{S}$ des représentations simples (= irréductibles), notées (π_λ,W_λ) pour λ dans Λ . Notons $H_{\lambda\times\mu,\nu}$ l'ensemble des applications linéaires u de $W_\lambda\otimes W_\mu$ dans W_ν telles que l'on ait

$$u\,.\,(\pi_\lambda(g)\otimes\pi_\mu(g)) = \pi_\nu(g)\,.\,u \tag{3.10}$$

pour tout g dans G . Alors toute collection d'automorphismes $g_\lambda \in \mathrm{GL}(W_\lambda)$ qui satisfait à la relation

$$u.(g_\lambda \otimes g_\mu) = g_\nu . u\,, \tag{3.11}$$

quels que soient λ, μ, ν dans Λ et u dans $H_{\lambda\times\mu,\nu}$, provient d'un unique élément g de G par la règle $g_\lambda = \pi_\lambda(g)$.

Le cas classique des groupes compacts s'obtient par les modifications suivantes : on considère seulement les représentations unitaires continues dans des espaces de Hilbert de dimension finie ; ainsi $\pi_i(g)$ est un opérateur unitaire dans W_i . Dans la reconstruction du groupe G , on impose aux g_i (ou aux g_λ) d'être des opérateurs unitaires. Les deux versions ci-dessus sont valables, et le groupe G est un groupe de Lie si et seulement si G possède une représentation fidèle, ou — ce qui revient au même — si la classe $\mathcal{C}$ est engendrée par un nombre fini de représentations au moyen des opérations permises : somme directe, produit tensoriel, dual, sous-espace et espace-quotient. Nous ajouterons deux remarques :

a) les représentations linéaires continues d'un groupe compact sont semi-simples ;

b) tout sous-groupe fermé du groupe unitaire $U(n)$ est l'intersection $G \cap U(n)$ avec $U(n)$ d'un unique sous-groupe algébrique de $\mathrm{GL}(n, \mathbf{C})$.

3.4. Il y a de nombreuses situations où le groupe n'est pas donné à l'avance, mais où on dispose de ce qui sera la classe de ses représentations. La question se pose en Géométrie algébrique où Grothendieck a inventé la "philosophie des motifs" à ce sujet, et en Physique mathématique où l'on aimerait traduire les règles de supersélection (certaines règles de conservation généralisant par exemple celle de la charge électrique) par l'existence d'un groupe de jauge. Il y a deux chaînes de recherches indépendantes : l'une commencée par Grothendieck a bénéficié des contributions de Saavedra, Milne et Deligne ; l'autre, dans le cadre des espaces de Hilbert, doit presque tout à Doplicher, Roberts et Woronowicz.

Le problème est donc de reconstruire un groupe algébrique à partir d'une catégorie de "représentations". Si G est un sous-groupe algébrique de $\mathrm{GL}(n, \mathbf{C})$, on lui associe l'algèbre $A = A(G)$ des fonctions "polynomiales" sur G , de la forme $P(g_{11}, \cdots, g_{nn})/(\det g)^s$, pour $g = (g_{ij})$, avec un polynôme P à n^2 variables et un entier $s \geq 0$. Les points de G correspondent aux homomorphismes d'algèbres de A dans $\mathbf{C}$, et la multiplication dans G se dualise en un homomorphisme Δ de A dans $A \otimes A$ tel que $\Delta(P) = \sum_i P'_i \otimes P''_i$ signifie $P(g'g'') = \sum_i P'_i(g')P''_i(g'')$. De même, l'unité de G correspond à un homomorphisme $\varepsilon : A \longrightarrow \mathbf{C}$ et l'inversion $g \mapsto g^{-1}$ dans G à un homomorphisme $S : A \longrightarrow A$. Les objets

$$A, \Delta : A \longrightarrow A \otimes A, \varepsilon : A \longrightarrow \mathbf{C}, S : A \longrightarrow A$$

satisfont à des conditions qui caractérisent axiomatiquement les *algèbres de Hopf.* Les représentations linéaires polynomiales de G correspondent aux comodules V sur A (application structurale $V \longrightarrow A \otimes V$) de dimension finie sur $\mathbf{C}$. Parmi les algèbres de Hopf, celles qui correspondent aux groupes algébriques sont celles qui satisfont à la condition (fondamentale) que l'algèbre A est commutative et la condition (secondaire) que l'algèbre A a un nombre fini de générateurs.

3.5. Le passage des groupes algébriques aux groupes "quantiques" se fait en affaiblissant l'hypothèse de commutativité de la multiplication.

Nous partons d'une catégorie abélienne $\mathcal{C}$ munie d'un foncteur exact et fidèle dans la catégorie des espaces vectoriels de dimension finie sur $\mathbf{C}$. De manière concrète, on dispose d'une famille $(V_i)_{i\in I}$ de tels espaces vectoriels, et pour chaque paire i,j d'un sous-espace $\mathcal{H}_{i,j}$ de l'ensemble $\mathrm{Hom}(V_i,V_j)$ des applications linéaires de V_i dans V_j. On suppose que la famille des V_i est stable par somme directe finie, et que le noyau et le conoyau d'un élément de $\mathcal{H}_{ij}$ sont encore dans la famille $(V_i)_{i\in I}$ (avec des conditions convenables sur la composition).

Sous ces hypothèses, la catégorie $\mathcal{C}$ se compose des comodules sur une coalgèbre A qu'on reconstruit comme suit : on note V la somme directe des V_i ; on identifie $\prod_i \mathrm{End}(V_i)$ à une sous-algèbre E de $\mathrm{End}(V)$ et la somme directe des $\mathcal{H}_{ij}$ à une autre sous-algèbre H (sans unité) de $\mathrm{End}(V)$. Soit F l'ensemble des éléments de E qui commutent à tous ceux de H ; c'est une sous-algèbre fermée de $\prod_i \mathrm{End}(V_i)$ munie de la topologie produit ; le dual topologique de F est la coalgèbre cherchée.

Le produit $m : A\otimes A \longrightarrow A$ s'obtiendra en munissant la catégorie $\mathcal{C}$ d'un foncteur Φ : $\mathcal{C}\times\mathcal{C}\longrightarrow\mathcal{C}$ avec une contrainte d'associativité ; on écrira $i\times j$ pour $\Phi(i,j)$. On suppose qu'on a $V_{i\times j}=V_i\otimes V_j$ et que la contrainte d'associativité de $i\times(j\times k)$ sur $(i\times j)\times k$ correspond à l'isomorphisme canonique de $V_i\otimes(V_j\otimes V_k)$ sur $(V_i\otimes V_j)\otimes V_k$. Grâce à cela, la multiplication dans A sera définie et associative. De manière analogue, l'unité dans A correspond à un objet particulier $\mathbf{1}$ dans $\mathcal{C}$ tel que $\mathbf{1}\times i$ et $i\times\mathbf{1}$ soient isomorphes à i. L'antipodisme $S : A\longrightarrow A$ correspond à un foncteur de dualité $i\mapsto i^\vee$ dans $\mathcal{C}$ avec un isomorphisme fonctoriel de $\mathrm{Hom}_\mathcal{C}(i\times j,k)$ sur $\mathrm{Hom}_\mathcal{C}(i,j^\vee\times k)$; on impose de plus que $V_{i^\vee}$ soit le dual de l'espace V_i.

La commutativité de l'algèbre A correspond dans ce cadre à une contrainte de commutativité $\rho : i\times j\longrightarrow j\times i$ dans la catégorie $\mathcal{C}$ compatible avec celle des espaces vectoriels, en ce sens que ρ induit l'isomorphisme $x\otimes y\mapsto y\otimes x$ de $V_{i\times j}=V_i\otimes V_j$ sur $V_{j\times i}=V_j\otimes V_i$.

Drinfeld vient d'examiner, sous le nom de *quasi-algèbres de Hopf*, ce qui se passe lorsque l'on donne dans la catégorie $\mathcal{C}$ un foncteur $\Phi : \mathcal{C}\times\mathcal{C}\longrightarrow\mathcal{C}$ tel que $V_{i\times j}=V_i\otimes V_j$, muni de contraintes d'associativité et de commutativité qui n'ont plus rien à voir avec celles du produit tensoriel. Pour chaque entier $n\geq 1$, notons $A^{\otimes n}$ la coalgèbre produit tensoriel de n facteurs égaux à A, et $\mathcal{A}_n$ l'algèbre duale de $A^{\otimes n}$. La multiplication $a\otimes b\mapsto ab$ dans A se dualise en un coproduit $\Delta : \mathcal{A}_1\longrightarrow\mathcal{A}_2$, de même Δ' est duale de la multiplication opposée $a\otimes b\mapsto ba$ dans A.

Supposons d'abord que la contrainte d'associativité soit compatible au produit tensoriel. Il existe alors un élément inversible R de $\mathcal{A}_2$ tel que l'on ait

$$\Delta'(a)=R\Delta(a)R^{-1} \qquad \text{pour tout } a\in\mathcal{A}_1 . \tag{3.12}$$

Pour i dans I, le comodule V_i sur A est aussi un module sur le dual $\mathcal{A}_1$ de A ; de même $V_{i\times j}=V_i\otimes V_j$ est un comodule sur $A^{\otimes 2}$, donc un module sur $\mathcal{A}_2$, et R définit un opérateur R_{ij} dans $V_i\otimes V_j$. La contrainte de commutativité dans la catégorie $\mathcal{C}$ est alors l'isomorphisme $P_{ij}R_{ij}$ de $V_i\otimes V_j$ sur $V_j\otimes V_i$, où $P_{ij}(x_i\otimes x_j)=x_j\otimes x_i$. La relation (3.12) exprime le défaut de commutativité de l'algèbre A, puisque $R=1$ correspond au cas où A est commutative. Le diagramme hexagonal se traduit alors par les formules

$$\Delta_{12}(R)=R_{13}R_{23} \tag{3.13}$$

$$\Delta_{23}(R)=R_{13}R_{12} \tag{3.14}$$

qui entraînent la relation de Yang-Baxter

$$R_{12}R_{13}R_{23}=R_{23}R_{13}R_{12} . \tag{3.15}$$

Noter que R_{12} par exemple est égal à R agissant sur les facteurs 1 et 2 de $A^{\otimes 3}$ par $\langle R_{12}, a_1 \otimes a_2 \otimes a_3 \rangle = \langle R, a_1 \otimes a_2 \rangle . \langle \varepsilon, a_3 \rangle$. De même, Δ_{12} est dual de l'application $a_1 \otimes a_2 \otimes a_3 \mapsto a_1 a_2 \otimes a_3$ de $A^{\otimes 3}$ dans $A^{\otimes 2}$.

La déviation éventuelle entre les contraintes d'associativité pour la catégorie $\mathcal{C}$ et pour le produit tensoriel se traduira par l'existence d'un élément inversible ϕ de $\mathcal{A}_3$. Le diagramme hexagonal et l'équation de Yang-Baxter prendront alors les formes modifiées

$$\Delta_{12}(R) = \phi_{312} R_{13} \phi_{132}^{-1} R_{23} \phi_{123} \tag{13 bis}$$

$$\Delta_{23}(R) = \phi_{231}^{-1} R_{13} \phi_{213} R_{12} \phi_{123}^{-1} \tag{14 bis}$$

$$R_{12} \phi_{312} R_{13} \phi_{132}^{-1} R_{23} \phi_{123} = \phi_{321} R_{23} \phi_{231}^{-1} R_{13} \phi_{213} R_{12} . \tag{15 bis}$$

Le diagramme pentagonal se traduit par l'identité suivante dans $\mathcal{A}_4$

$$\Delta_{34}(\phi) \Delta_{12}(\phi) = \phi_{234} . \Delta_{23}(\phi) . \phi_{123} . \tag{3.16}$$

3.6. Donnons un exemple simple pour conclure. Considérons l'espace $V = \mathbf{C}^N$ avec sa base naturelle $e_1, \cdots, e_N$; soit q un nombre complexe qui n'est pas une racine de l'unité. L'algèbre de Hecke $\mathcal{H}_n(q)$ est semi-simple et agit sur l'espace $V^{\otimes n}$ comme il a été expliqué au n° 2.2. Considérons la catégorie $\mathcal{C}$ dont les objets sont les espaces $V^{\otimes n}$, où les seuls morphismes sont les endomorphismes des $V^{\otimes n}$ fournis par les éléments de $\mathcal{H}_n(q)$. On considère le foncteur Φ qui associe à $V^{\otimes m}$ et $V^{\otimes n}$ leur produit tensoriel $V^{\otimes(m+n)}$. La contrainte d'associativité est celle des produits tensoriels; par contre, la contrainte de commutativité de $V^{\otimes m} \otimes V^{\otimes n}$ sur $V^{\otimes n} \otimes V^{\otimes m}$ est $P^{(m,n)} R^{(m,n)}$ où $P^{(m,n)}$ est la contrainte usuelle et

$$R^{(m,n)} = \prod_{i=1}^{m} \prod_{j=1}^{n} R_{i,m+j} \tag{3.17}$$

(avec les notations du n° 2.2).

L'algèbre de Hopf A_N obtenue admet les générateurs t_{ij} (pour $1 \leq i \leq N$, $1 \leq j \leq N$) satisfaisant aux relations

$$T^{(1)} T^{(2)} R = R T^{(2)} T^{(1)} \tag{3.18}$$

où T est la matrice des t_{ij} , où $T^{(1)} = T \otimes I_N$, $T^{(2)} = I_N \otimes T$. Lorsque $q = 1$, ceci exprime la commutativité des t_{ij} . Le coproduit est donné dans A_N par

$$\Delta(t_{ij}) = \sum_{k=1}^{N} t_{ik} \otimes t_{kj} . \tag{3.19}$$

Il s'agit là de la déformation de l'algèbre* des fonctions polynômes sur $\mathrm{GL}(N, \mathbf{C})$ que Faddeev et Woronowicz considèrent comme le groupe quantique associé au groupe $\mathrm{GL}(N, \mathbf{C})$. Le point de vue dual (Drinfeld, Jimbo) est celui de la déformation $U_q(\mathfrak{g}\ell(N, \mathbf{C}))$ de l'algèbre enveloppante de l'algèbre de Lie $\mathfrak{g}\ell(N, \mathbf{C})$.

*Il faudra en fait adjoindre à A_N l'inverse du déterminant quantique $\det_q T$, un polynôme qui pour $q = 1$ se réduit au déterminant de la matrice T .

4. ÉQUATIONS DIFFÉRENTIELLES ET MONODROMIE

4.1. Nous considérons le sous-espace ouvert $\mathbf{C}_*^n$ de $\mathbf{C}^n$ formé des vecteurs à composantes distinctes. On pose

$$\omega_{ij} = d\log(z_i - z_j) = \frac{dz_i - dz_j}{z_i - z_j} \tag{4.1}$$

pour $i \neq j$, en notant $z_1, \cdots, z_n$ les coordonnées sur $\mathbf{C}^n$; les ω_{ij} sont des formes différentielles holomorphes sur $\mathbf{C}_*^n$ et elles satisfont aux relations quadratiques

$$\omega_{ij} \wedge \omega_{jk} + \omega_{jk} \wedge \omega_{ki} + \omega_{ki} \wedge \omega_{ij} = 0 \tag{4.2}$$

(i, j, k distincts). De plus, on a $\omega_{ij} = \omega_{ji}$. Soit $\mathcal{A}_n$ l'algèbre de formes différentielles holomorphes engendrée (sur $\mathbf{C}$) par les ω_{ij} . Les relations (4.2) forment une présentation de cette algèbre anticommutative. De plus, les formes appartenant à $\mathcal{A}_n$ constituent des représentants de la cohomologie de de Rham de $\mathbf{C}_*^n$, d'où un isomorphisme $H^*(\mathbf{C}_*^n; \mathbf{C}) \simeq \mathcal{A}_n$ (théorème d'Arnold).

4.2. Donnons-nous un espace vectoriel W de dimension finie sur $\mathbf{C}$, et des endomorphismes A_{ij} de W. On considère la forme différentielle $\Omega = \sum_{i<j} A_{ij}\,\omega_{ij}$ sur $\mathbf{C}_*^n$, à valeurs dans $\mathrm{End}(W)$, et le système différentiel $dF = \Omega F$; les solutions sont des fonctions holomorphes $F(z) = F(z_1, \cdots, z_n)$ définies sur des ouverts de $\mathbf{C}_*^n$ à valeurs dans $\mathrm{End}(W)$. Soit γ un lacet de classe C^∞ dans $\mathbf{C}_*^n$, donné paramétriquement sous la forme $(\gamma(t))_{0\leq t\leq 1}$. Si $F_a(z)$ est une solution de $dF = \Omega F$ au voisinage de $\gamma(0)$, on peut prolonger analytiquement $F_a(z)$ le long de γ . La fonction $F_b(z)$ obtenue est une solution de l'équation $dF = \Omega F$ au voisinage de $\gamma(1) = \gamma(0)$; elle est donnée par la formule de Lappo-Danilewski

$$F_b(z) = F_a(z)\,T(\gamma) \tag{4.3}$$

avec l'expression suivante pour le transport parallèle

$$T(\gamma) = \sum_{k\geq 0} \int_{\Delta_k} A(t_1)\cdots A(t_k)\, dt_1 \cdots dt_k\,. \tag{4.4}$$

Dans cette formule intégrale, le domaine d'intégration Δ_k est défini par les inégalités $0 \leq t_1 \leq t_2 \leq \cdots \leq t_k \leq 1$ et la forme différentielle opératorielle $A(t)\,dt$ s'obtient par la substitution $z = \gamma(t)$ dans Ω .

4.3. Le système différentiel $dF = \Omega F$ est *complètement intégrable* si $T(\gamma)$ ne dépend que de la classe d'homotopie du lacet γ (d'origine fixée). Le critère classique est le suivant

$$d\Omega - \Omega \wedge \Omega = 0\,; \tag{4.5}$$

les deux membres appartiennent à $\mathcal{A}_n \otimes \mathrm{End}\,W$, et vu le théorème d'Arnold cité en 4.1, cette relation équivaut aux suivantes :

$$[A_{ij}, A_{ik} + A_{jk}] = 0 \qquad \text{pour } i, j, k \text{ distincts} \tag{4.6}$$

$$[A_{ij}, A_{k\ell}] = 0 \qquad \text{pour } i, j, k, \ell \text{ distincts}\,. \tag{4.7}$$

S'il en est ainsi, choisissons un point-base a dans $\mathbf{C}_*^n$ et réalisons le groupe de tresses pures P_n comme le groupe $\pi_1(\mathbf{C}_*^n; a)$. La formule de transport parallèle (4.4) appliquée aux lacets en

a fournit une représentation linéaire du groupe P_n dans l'espace W ; elle transforme la classe d'homotopie du lacet γ en l'automorphisme $T(\gamma)^{-1}$ de W.

Introduisons l'algèbre de Lie $\mathcal{G}_n$ quotient de l'algèbre de Lie libre de générateurs $X_{ij} = X_{ji}$ (pour $i < j$) par l'idéal engendré par les éléments :

$$[X_{ij}, X_{ik} + X_{jk}] \qquad \text{et} \qquad [X_{ij}, X_{k\ell}]$$

sous les mêmes conditions que (4.6) et (4.7). On note ξ_{ij} l'image de X_{ij} dans $\mathcal{G}_n$ et l'on introduit la forme différentielle "générique"

$$\omega = \sum_{i<j} \omega_{ij} \otimes \xi_{ij} \tag{4.8}$$

dans $\mathcal{A}_n \otimes \mathcal{G}_n$.

Le résultat précédent peut se traduire en disant que toute représentation linéaire ρ de $\mathcal{G}_n$ dans l'espace W (envoyant ξ_{ij} sur A_{ij} , donc ω sur Ω) définit une représentation linéaire T du groupe P_n dans W. Réciproquement, toute représentation du groupe P_n par des automorphismes unipotents de W provient d'une représentation ρ de $\mathcal{G}_n$ par des endomorphismes nilpotents de W *. Il est donc raisonnable de considérer *l'algèbre de Lie $\mathcal{G}_n$ comme l'algèbre de Lie du groupe P_n* (ou aussi du groupe B_n dans lequel il est d'indice fini) et les relations

$$[\xi_{ij}, \xi_{ik} + \xi_{jk}] = 0 \quad , \quad [\xi_{ij}, \xi_{k\ell}] = 0 \tag{4.9}$$

comme l'analogue infinitésimal des relations (1.3) et (1.4) de définition de B_n . Cette interprétation est renforcée par le théorème suivant (cas particulier des théorèmes sur la monodromie que j'ai exposés dans [F1]) :

Soit $(\mathbf{C}P_n)^\wedge$ le complété de l'algèbre du groupe $\mathbf{C}P_n$ pour la topologie I-adique, où I est l'idéal d'augmentation (engendré par les $g-1$ pour g dans P_n). *De manière analogue, complétons l'algèbre enveloppante $U\,\mathcal{G}_n$ en $(U\,\mathcal{G}_n)^\wedge$ pour la topologie J-adique, où J est l'idéal bilatère engendré par $\mathcal{G}_n$ dans $U\,\mathcal{G}_n$. Les algèbres $(\mathbf{C}P_n)^\wedge$ et $(U\,\mathcal{G}_n)^\wedge$ sont isomorphes, et le groupe P_n se plonge dans $(\mathbf{C}P_n)^\wedge$.*

4.4. Voici une manière de construire des solutions des équations (4.6) et (4.7). *L'équation de Yang-Baxter classique* s'écrit sous la forme

$$[r_{12}(u_{12}), r_{13}(u_{13})] + [r_{12}(u_{12}), r_{23}(u_{23})] + [r_{13}(u_{13}), r_{23}(u_{23})] = 0\,. \tag{4.10}$$

La signification est la suivante : V est un espace vectoriel de dimension finie, $r(u)$ est un opérateur dans $V \otimes V$ dépendant du paramètre complexe u de manière holomorphe ou méromorphe, on pose $u_{ij} = u_i - u_j$ et $r_{ij}(u)$ désigne $r(u)$ agissant sur les facteurs d'indices i et j (avec $i < j$) du produit tensoriel triple $V^{\otimes 3}$. D'après Belavin et Drinfeld [F2], on obtient comme suit une solution de (4.10) : choisissons une algèbre de Lie semi-simple complexe $\mathcal{G}$, de forme de Killing B , une base (I_μ) de $\mathcal{G}$ orthonormale pour B et une représentation linéaire ρ de $\mathcal{G}$ dans V . On pose alors

$$r(u) = t/u \qquad \text{avec} \qquad t = \sum_\mu \rho(I_\mu) \otimes \rho(I_\mu)\,. \tag{4.11}$$

*Dans [F5], Kohno démontre la réciproque sous l'hypothèse que la représentation de P_n est "assez voisine de l'identité". Il se réfère pour cela à un résultat de Golubeva (Math. USSR Izvest. **17** (1981), p. 227-241) qui ne s'applique malheureusement pas au cas utilisé. On peut cependant réparer la démonstration en utilisant le théorème d'approximation de M. Artin.

En utilisant le fait que l'élément $\sum_\mu I_\mu \otimes I_\mu$ de $\mathcal{G}\otimes\mathcal{G}$ est invariant par la représentation adjointe de $\mathcal{G}$, on établit les relations

$$[t_{12}, t_{13}+t_{23}] = [t_{23}, t_{12}+t_{13}] = 0 \tag{4.12}$$

et (4.10) s'ensuit immédiatement. Par ailleurs, si l'on note W l'espace $V^{\otimes n}$, les opérateurs t_{ij}, agissant dans W selon nos conventions habituelles, satisfont aux relations

$$[t_{ij}, t_{ik}+t_{jk}] = 0 \quad , \quad [t_{ij}, t_{k\ell}] = 0 \tag{4.13}$$

analogues à (4.6) et (4.7). Si l'on fait agir simultanément le groupe symétrique $\mathbf{S}_n$ sur $\mathbf{C}_*^n$ (par permutation des coordonnées) et sur $W = V^{\otimes n}$ de la manière habituelle, la forme différentielle $\Omega = \lambda \sum_{i<j} t_{ij}\,\omega_{ij}$ est invariante par $\mathbf{S}_n$ pour toute valeur complexe de λ ; on obtient alors un système complètement intégrable sur l'espace de configuration $\Phi_n(\mathbf{C}) = \mathbf{C}_*^n/\mathbf{S}_n$. Par intégration, on en déduit une représentation de monodromie T_λ de $B_n = \pi_1(\Phi_n(\mathbf{C}))$ dans W . Il résulte des raisonnements classiques de Fuchs sur les singularités des équations différentielles que l'opérateur $T_\lambda(\sigma_j)$ dans W correspondant au générateur σ_j de B_n est conjugué dans $GL(V^{\otimes n})$ à $P_j \exp \pi i \lambda t_{j,j+1}$, où P_j est l'échange des facteurs de rang j et $j+1$ dans le produit tensoriel $V^{\otimes n}$.

4.5. Drinfeld a réussi dans [D9] à identifier la représentation T_λ de B_n . Écrivons l'équation $dF = \Omega F$ sous la forme de Zamolodchikov

$$\frac{\partial F}{\partial z_j} = \lambda \sum_{\substack{k=1 \\ k\neq j}}^{n} \frac{t_{jk}}{z_j - z_k} . F \; ; \tag{4.14}$$

elle est invariante si l'on soumet chaque variable z_j à la même transformation affine $z_j \mapsto az_j + b$. Par suite, pour $n = 3$, on obtient des solutions sous la forme

$$F(z_1,z_2,z_3) = G\left(\frac{z_1 - z_2}{z_1 - z_3}\right)(z_1 - z_3)^{\lambda(t_{12}+t_{13}+t_{23})} \; ; \tag{4.15}$$

sur le revêtement universel de $\mathbf{C}_*^3$, le logarithme de $z_1 - z_3$ a un sens et l'on a posé $(z_1 - z_3)^\alpha = \exp \alpha \log(z_1 - z_3)$ (comme il se doit). La fonction G est définie sur le revêtement universel de $\mathbf{C}\backslash\{0,1\}$, à valeurs dans $\mathrm{End}(V^{\otimes 3})$, et satisfait à l'équation différentielle

$$\frac{dG}{dx} = \lambda\left(\frac{t_{12}}{x} + \frac{t_{23}}{x-1}\right) G(x) \, . \tag{4.16}$$

Conformément à la théorie de Fuchs, il y a une solution $G_1(x)$ se comportant comme $x^{\lambda t_{12}}$ pour x tendant vers 0 , et une solution $G_2(x)$ se comportant comme $(1-x)^{\lambda t_{23}}$ pour x tendant vers 1 . Ces deux solutions diffèrent par une constante, d'où un automorphisme $\phi_V(\lambda)$ de $V^{\otimes 3}$ tel que $G_1(x) = G_2(x)\,\phi_V(\lambda)$ pour tout x . On montre facilement qu'il existe une série formelle $\phi(\lambda)$ en la variable λ, à coefficients dans $(U\mathcal{G})^{\otimes 3}$, indépendante de la représentation linéaire (ρ, V) de $\mathcal{G}$, dont l'image par $\rho^{\otimes 3}$ donne le développement en série de $\phi_V(\lambda)$.

En étudiant le comportement asymptotique des solutions de l'équation de Zamolodchikov pour $n = 4$, lorsque certaines des différences $z_j - z_k$ tendent vers 0 , on établit l'analogue suivant de la relation (3.16)

$$\Delta_{34}(\phi)\,\Delta_{12}(\phi) = \phi_{234} . \Delta_{23}(\phi) . \phi_{123} \, , \tag{4.17}$$

à interpréter comme égalité de séries formelles en λ à coefficients dans $(U\mathcal{G})^{\otimes 4}$. Dans cette formule, le coproduit $\Delta : U\mathcal{G} \longrightarrow (U\mathcal{G})^{\otimes 2}$ est le classique, tel que $\Delta(x) = x \otimes 1 + 1 \otimes x$ pour x dans $\mathcal{G}$. Drinfeld démontre ensuite, par une procédure récursive, l'existence d'une série formelle $F(\lambda)$ à coefficients dans $(U\mathcal{G})^{\otimes 2}$, telle que $F(0) = 1$ et

$$\phi(\lambda) = F_{23}(\lambda)(\Delta_{23}F)(\lambda)(\Delta_{12}F)(\lambda)^{-1}F_{12}(\lambda)^{-1} . \tag{4.18}$$

On modifie le coproduit dans $(U\mathcal{G})[[\lambda]]$ en posant

$$\Delta_\lambda(a) = F(\lambda)\,\Delta(a)\,F(\lambda)^{-1} \tag{4.19}$$

pour a dans $U\mathcal{G}$. D'après les équations (4.17) et (4.18), le *nouveau coproduit* Δ_λ *est coassociatif.* Grâce à un théorème d'unicité convenable, Drinfeld montre que *l'algèbre* $(U\mathcal{G})[[\lambda]]$ *munie du coproduit* Δ_λ *est isomorphe à la quantification de Drinfeld-Jimbo de* $U\mathcal{G}$. Notons $F_{21}(\lambda)$ l'image de $F(\lambda) = F_{12}(\lambda)$ par la symétrie de $(U\mathcal{G})^{\otimes 2}$. Alors, on définit une série formelle $R(\lambda)$ à coefficients dans $(U\mathcal{G})^{\otimes 2}$ par la formule

$$R(\lambda) = F_{21}(\lambda).\left(\exp\lambda\sum_{\mu}(I_\mu \otimes I_\mu)/2\right).F_{12}(\lambda)^{-1} . \tag{4.20}$$

Elle satisfait à l'équation de Yang-Baxter

$$R_{12}(\lambda)R_{13}(\lambda)R_{23}(\lambda) = R_{23}(\lambda)R_{13}(\lambda)R_{12}(\lambda) \tag{4.21}$$

(égalité de séries formelles à coefficients dans $(U\mathcal{G})^{\otimes 3}$).

En conclusion, *la représentation* T_λ *de* B_n *par des opérateurs dans* $W = V^{\otimes n}$ *est équivalente* à la représentation de* B_n *qui applique le générateur* σ_i *sur l'opérateur*

$$\rho^{\otimes n}(\underbrace{1\otimes\cdots\otimes 1}_{i-1}\otimes R(\lambda)\otimes\underbrace{1\otimes\cdots\otimes 1}_{n-i-1}) .$$

Par exemple, si $\mathcal{G} = s\ell(N, \mathbf{C})$ avec la représentation naturelle ρ dans $\mathbf{C}^N$, on a

$$t(x\otimes y) = \frac{1}{2N}(y\otimes x) - \frac{1}{2N^2}(x\otimes y) . \tag{4.22}$$

Si l'on pose $q = e^{\pi i\lambda/N}$, la représentation obtenue est celle liée à $\mathcal{H}_n(q)$, à torsion près par une représentation de degré 1 de B_n (voir le n° 2.2).

5. CATÉGORIES ASSOCIÉES AUX TRESSES ET AUX CABLAGES

5.1. Turaev et Reshetikhin, après Freyd et Yetter, ont introduit un calcul graphique, lié aux tresses, et aux cablages qui les généralisent. Cela leur a permis de donner dans [C19] une construction rigoureuse des invariants associés par Witten aux variétés de dimension 3 ; pour ces invariants, je renvoie à la brillante description donnée dans ce même Séminaire par M. Atiyah. La méthode de Turaev et Reshetikhin fournit aussi un traitement graphique du calcul tensoriel classique, très proche de celui de Penrose dans [G8].

*Au sens des séries formelles en λ ; il faudrait établir la convergence !

5.2. Le point de départ est que *les tresses colorées forment une catégorie*. Introduisons un ensemble $\mathcal{X}$ de "couleurs". Les objets de la catégorie $Tr(\mathcal{X})$ sont les suites $(x_1,\cdots,x_n)$ d'éléments de $\mathcal{X}$; les morphismes sont les tresses "colorées", chaque brin étant associé à un élément de $\mathcal{X}$. Une telle coloration des n brins détermine évidemment une coloration des extrémités supérieures, donnant une suite $(x_1,\cdots,x_n)$ qui est l'objet de $Tr(\mathcal{X})$ dont part la tresse colorée ; définition analogue du but par les extrémités inférieures. Avec ces conventions, la composition des tresses est telle qu'un brin garde la même couleur tout au long. De plus, on définit un foncteur strictement associatif Φ dans $Tr(\mathcal{X})$: sur les objets, on a $\Phi(x,x') = (x_1,\cdots,x_n,x'_1,\cdots,x'_{n'})$ si $x = (x_1,\cdots,x_n)$ et $x' = (x'_1,\cdots,x'_{n'})$; sur les flèches, on juxtapose une tresse à n brins t à gauche et une tresse à n' brins t' à droite. Notation $x \times x'$ pour $\Phi(x,x')$ et $t \times t'$ pour $\Phi(t,t')$. Il y a une contrainte de commutativité qu'on laisse au lecteur le soin de décrire graphiquement.

Supposons alors qu'à chaque couleur $x \in \mathcal{X}$ on associe un espace vectoriel V_x et qu'on se donne des opérateurs linéaires $R_{x,y} : V_x \otimes V_y \longrightarrow V_x \otimes V_y$. On suppose que, pour tout triplet de couleurs x,y,z , on a la relation de Yang-Baxter

$$R_{12}R_{13}R_{23} = R_{23}R_{13}R_{12}$$

entre opérateurs sur $V_x \otimes V_y \otimes V_z$; par exemple, on a $R_{12} = R_{x,y} \otimes \mathrm{Id}_{V_z}$, etc...

On peut alors généraliser la construction du n° 2.1. Si t est une tresse colorée, de source $x = (x_1,\cdots,x_n)$ et de but $y = (y_1,\cdots,y_n)$, on lui associe une application linéaire $H(t)$ de $H(x) = V_{x_1} \otimes \cdots \otimes V_{x_n}$ dans $H(y) = V_{y_1} \otimes \cdots \otimes V_{y_n}$. On obtient ainsi un foncteur H de la catégorie $Tr(\mathcal{X})$ dans celle des espaces vectoriels sur $\mathbf{C}$. Il est caractérisé par les propriétés suivantes :

a) *on a* $H(x \times y) = H(x) \otimes H(y)$ *pour les objets et les flèches* ;

b) *si* $s_{x,y}$ *est la tresse colorée à* 2 *brins ci-dessous, on a* $H(S_{x,y}) = \sigma_{x,y} R_{x,y}$, *où* $\sigma_{x,y}$ *est la symétrie usuelle de* $V_x \otimes V_y$ *sur* $V_y \otimes V_x$;

c) *si* e_x *est la tresse à* 1 *brin ci-dessous,* $H(e_x)$ *est l'application identique de* V_x.

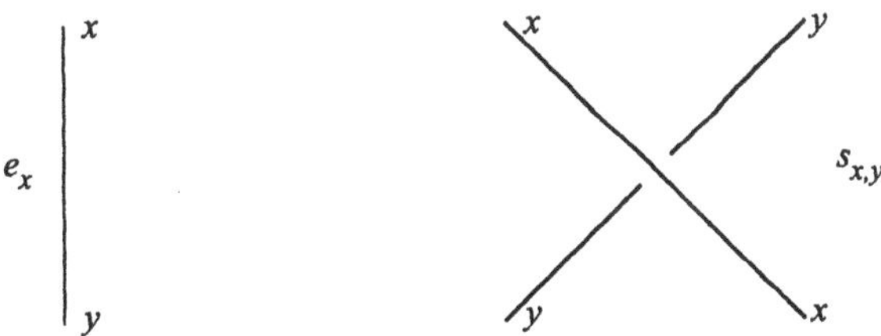

5.3. On généralise la notion de tresse en celle de *cablage*. Supposons donc donnés deux plans parallèles Π et Π' , une suite de points distincts $P_1,\cdots,P_n$ dans Π et de même $P'_1,\cdots,P'_{n'}$ dans Π'. Un cablage M ayant les extrémités $P_1,\cdots,P_n,P'_1,\cdots,P'_{n'}$ sera réunion d'un nombre fini d'arcs ou de lacets (différentiables par morceaux) ; on fait les hypothèses suivantes :

a) ces arcs et lacets sont deux à deux disjoints ;

b) les extrémités des arcs appartiennent à l'ensemble $\{P_1,\cdots,P_n,P'_1,\cdots,P'_{n'}\}$ et chacun de ces points intervient exactement une fois ;

c) si C est l'un des arcs ou lacets, il est transverse à tous les plans parallèles à Π et Π' sauf éventuellement un nombre fini ;

d) M est entièrement contenu dans la région H comprise entre les plans Π et Π'.

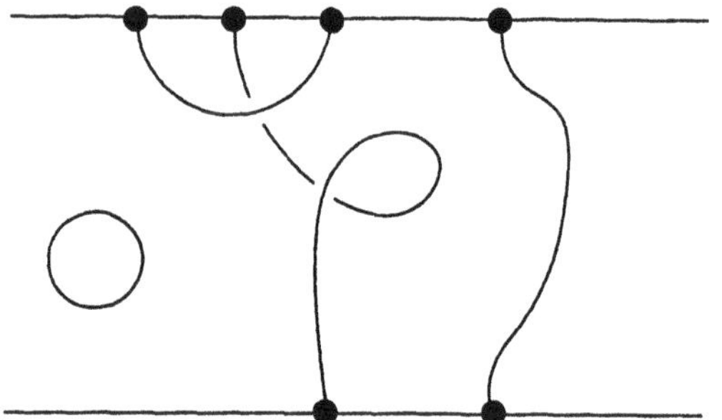

Une projection plane d'un cablage

On remarquera qu'un entrelac n'est autre qu'un cablage dont l'ensemble des extrémités est vide. De plus, on supposera chaque arc ou lacet *orienté*, et *muni d'une couleur* appartenant à un ensemble $\mathcal{X}$. Deux telles configurations sont équivalentes s'il existe un homéomorphisme de $\mathbf{R}^3$, conservant l'orientation et transformant l'une en l'autre.

Ces cablages colorés forment les flèches d'une catégorie $Cab(\mathcal{X})$ *dont les objets sont les suites* $(x_1, \varepsilon_1, \cdots, x_n, \varepsilon_n)$ *formées de couleurs* $x_1, \cdots, x_n$ *et de nombres* $\varepsilon_1, \cdots, \varepsilon_n$ *égaux à* 1 *ou* -1 . Les conventions de source et de but sont figurées ci-dessous

A côté des configurations fournies par les tresses, apparaissent deux situations élémentaires nouvelles

5.4. Décrivons maintenant une méthode très générale pour fabriquer des invariants d'isotopie des entrelacs. On reprend les hypothèses du n° 3.5. Soit $(A, \Delta, \varepsilon, S)$ une algèbre de Hopf ; on note $\mathcal{A}_n$ l'algèbre duale de la coalgèbre $A^{\otimes n}$ et l'on suppose donné un élément R de $\mathcal{A}_2$ satisfaisant aux relations (3.12) à (3.15) de Drinfeld. On considère des comodules sur la coalgèbre A , de dimension finie comme espaces vectoriels sur $\mathbf{C}$; ce sont des $\mathcal{A}_1$-modules. Si M et N sont deux tels comodules, R définit un opérateur $R_{M,N}$ dans l'espace vectoriel $M \otimes N$. De plus, l'antipodisme S de A permet d'associer à tout comodule M (à gauche sur A) un comodule $M^\vee$ (à gauche sur A) dont l'espace vectoriel sous-jacent est le dual de M . On note aussi $\mathbf{1}$ le A-module ayant $\mathbf{C}$ pour espace vectoriel sous-jacent avec le coproduit $\lambda \mapsto 1 \otimes \lambda$ de $\mathbf{C}$ dans $A \otimes \mathbf{C}$. Il existe alors deux homomorphismes de comodules

$$e_M : \mathbf{1} \longrightarrow M \otimes M^\vee \quad , \quad \varepsilon_M : M^\vee \otimes M \longrightarrow \mathbf{1}$$

définis par

$$e_M(1) = \sum_{i=1}^{N} v_i \otimes v^i \quad , \quad \varepsilon_M(x \otimes y) = \langle x, y \rangle$$

($v_1, \cdots, v_N$ forment une base de M sur $\mathbf{C}$, et $v^1, \cdots, v^N$ la base duale de $M^\vee$).

Supposons qu'on ait associé à chaque couleur $x \in \mathcal{X}$ un comodule M_x. Grâce aux opérateurs de Yang-Baxter R_{M_x,M_y} , on peut définir comme au n° 5.2 un foncteur H de la catégorie $Tr(\mathcal{X})$ dans celle des A-comodules. On peut alors étendre ce foncteur en un foncteur $\widetilde{H}$ de la catégorie $Cab(\mathcal{X})$ vers celle des comodules sur A . Il suffit d'ajouter les règles suivantes aux règles a), b) et c) du n° 5.2 :

d) *A un objet* $(x_1, \varepsilon_1, \cdots, x_n, \varepsilon_n)$ *de* $Cab(\mathcal{X})$, *on associe le comodule* $M_{x_1}^{\varepsilon_1} \otimes \cdots \otimes M_{x_n}^{\varepsilon_n}$ (par convention, on a $M^+ = M$ et $M^- = M^\vee$) ; *en particulier, l'objet vide de* $Cab(\mathcal{X})$ *donne par* $\widetilde{H}$ *le comodule* $\mathbf{1}$.

e) *Le foncteur* $\widetilde{H}$ *associe à la configuration* a_x *ci-dessous le morphisme* e_{M_x} *de* $\mathbf{1}$ *dans* $M_x \otimes M_x^\vee$ *et à* b_x *il associe le morphisme* ε_{M_x} *de* $M_x^\vee \otimes M_x$ *dans* $\mathbf{1}$.

Le point crucial est que *deux cablages isotopes fournissent le même morphisme de comodules. En particulier, un entrelac coloré et orienté L fournit un endomorphisme de* $\mathbf{1}$, *c'est-à-dire un nombre complexe* $e(L)$. *Ce nombre ne dépend que de la classe d'isotopie de* L . On comparera ce résultat aux théories quantiques des champs topologiques de Witten et Atiyah. Il restera à expliciter les invariants ainsi obtenus !!

6. BIBLIOGRAPHIE RAISONNÉE

6.1. Ouvrages généraux. Voici d'abord les textes fondateurs :

1. E. ARTIN - *Collected papers* (édités par S. Lang et J. Tate), Addison-Wesley, 1965. [Voir en particulier la partie consacrée à la Topologie (p. 416-498) qui contient les articles célèbres : *Theorie der Zöpfe* (1925), *Theory of braids* (1947), *Braids and permutations* (1947).]

2. W. BURAU - *Über Zopfinvarianten*, Hamburg Abh. **9** (1932), p. 117-124.

3. A.A. MARKOV - *Über die freie Equivalenz der geschlossenen Zöpfe*, Recueil Soc. Math. Moscou **43** (1936), p. 73-78.

4. R.H. FOX - *Free differential calculus* I : *Derivations in the free group ring*, Ann. of Math. **57** (1953), p. 547-560.

5. E. FADELL and L. NEUWIRTH - *Configuration spaces*, Math. Scand. **10** (1962), p. 111-118.

6. R.H. FOX and L. NEUWIRTH - *The braid groups*, Math. Scand. **10** (1962), p. 119-126.

Nous donnons maintenant une sélection de manuels consacrés essentiellement aux nœuds et tresses.

7. K. REIDEMEISTER - *Knotentheorie*, Erg. Math. Vol. 1, Springer (1932).

8. R.H. CROWELL and R.H. FOX - *Introduction to knot theory*, Ginn Co. (1963) (réimprimé dans GTM, vol. 57, Springer (1977)).

9. L. NEUWIRTH - *Knot groups*, Annals Math. Studies **56**, Princeton Univ. Press (1965).

10. W. MAGNUS, A. KARASS, D. SOLITAR - *Combinatorial group theory*, Interscience/ Wiley (1966).

11. J. BIRMAN - *Braids, links and mapping class groups*, Annals Math. Studies **82**, Princeton Univ. Press (1974).

12. J. ROLFSEN - *Knots and links*, Publish or Perish (1976).

13. L. KAUFFMAN - *Formal knot theory*, Math. Notes **30**, Princeton Univ. Press (1983).

14. G. BURDE and H. ZIESCHANG - *Knots*, W. de Gruyter (1985).

Mentionnons également deux rapports au Séminaire N. Bourbaki qui ont conservé une certaine actualité :

15. A. GRAMAIN - *Rapport sur la théorie classique des nœuds* (1ère partie), Sém. Bourbaki, exp. n° 485, juin 1976, Springer, Lect. Notes Math. **567** (1977), p. 222-237.

16. A. DOUADY - *Nœuds et structures de contact en dimension* 3 [*d'après D. Bennequin*], Sém. Bourbaki, exp. n° 604 (février 1983), Astérisque **105-106**, 1983, p. 129-148.

Pour terminer, mentionnons un ouvrage récent explorant la plupart des aspects liés aux groupes de tresses :

17. J. BIRMAN and A. LIBGOBER (éditeurs) - *Braids*, Proceedings of a Research Conference, Contemporary Math. vol. 78, American Math. Soc. (1988).

6.2. Catégories et dualité. Articles fondamentaux sur les conditions de cohérence dans les catégories :

1. S. MACLANE - *Natural associativity and commutativity*, Rice Univ. Studies, **49** (1963), p. 28-46.

2. S. EILENBERG and G.M. KELLY - *Closed categories*, in Proc. Conf. Categorical Algebra, Springer (1966).

3. S. MACLANE - *Categories for the working mathematician*, GTM vol. 5, Springer (1974).

4. G.M. KELLY and M.L. LAPLAZA - *Coherence for compact closed categories*, J. Pure Appl. Alg. **19** (1980), p. 193-213.

Pour la dualité de Tannaka "classique" des groupes compacts, les articles de base :

5. T. TANNAKA - *Über den Dualitätssatz der nicktkommutativen topologischen Gruppen*, Tohoku Math. J. **45** (1939), p. 1-12.

6. M.G. KREIN - *A principle of duality for a bicompact group and a square block algebra*, Dokl. Akad. Nauk SSSR **69** (1949), p. 725-728.

7. M. TAKESAKI - *A characterization of group algebras as a converse of Tannaka-Stinespring-Tatsuuma duality theorem*, Amer. J. Math. **91** (1969), p. 529-564.

8. S. DOPLICHER and J. ROBERTS - *Duals of compact Lie groups realized in the Cuntz algebras and their actions on C^*-algebras*, J. Funct. Anal. **74** (1987), p. 96-120.

9. S. DOPLICHER and J. ROBERTS - *Endomorphisms of C^*-algebras, cross products and duality for compact groups*, Ann. of Math. **130** (1989), p. 78-119.

10. S. DOPLICHER and J. ROBERTS - *A new duality theory for compact groups*, Invent. Math. **98** (1989), p. 157-218.

Pour une extension dans le cas des groupes quantiques :

11. P. GHEZ, R. LIMA and J. ROBERTS - *W^*-categories*, Pacific J. Math. **120** (1985), p. 79-109.

12. S.L. WORONOWICZ - *Duality in the C^*-algebra theory*, Proc. Int. Cong. Math., Varsovie 1983, vol. 2, p. 1347-1350.

13. S.L. WORONOWICZ - *Compact matrix pseudo-groups*, Comm. Math. Phys. **111** (1987), p. 613-665.

14. S.L. WORONOWICZ - *Tannaka-Krein duality for compact matrix pseudo-groups. Twisted $SU(N)$*, Invent. Math. **93** (1988), p. 35-76.

Enfin, l'extension de la dualité de Tannaka-Krein aux groupes algébriques :

15. C. CHEVALLEY - *Théorie des groupes de Lie, tome* II : *groupes algébriques*, Hermann (1951).

16. P. CARTIER - *Dualité de Tannaka et algèbres de Lie*, C.R. Acad. Sci. Paris, **242** (1956), p. 322-325.

17. S. SAAVEDRA - *Catégories tannakiennes*, Lect. Notes Math. vol. 265, Springer (1972).

18. P. DELIGNE et J.S. MILNE - *Tannakian categories* in *Hodge cycles, Motives and Shimura varieties*, Lect. Notes Math. vol. 900, Springer (1982).

19. P. DELIGNE - *Catégories tannakiennes*, in Grothendieck's Festschrift, Prog. in Math., Birkhäuser (1990).

6.3. Invariants polynomiaux des nœuds. Voici les articles originaux pour le polynôme de Jones :

1. V. JONES - *Index for subfactors*, Invent. Math. **72** (1983), p. 1-25.

2. V. JONES - *A polynomial invariant for knots via von Neumann algebras*, Bull. Amer. Math. Soc. **12** (1985), p. 103-112.

3. V. JONES - *Braid groups, Hecke algebras and type* II_1 *factors*, in "Geometric methods in operator algebras", Proc. U.S.-Japan Symposium, Wiley, 1986, p. 242-273.

4. V. JONES - *On a certain value of the Kauffman polynomial*, Comm. Math. Phys.,

5. V. JONES - *Hecke algebra representations of braid groups and link polynomials*, Ann. of Math. **126** (1987), p. 335-388.

6. V. JONES - *Notes on subfactors and statistical mechanics*, in "Braid group, knot theory and statistical mechanics" (C.N. Yang et M.L. Ge, éditeurs), World Scientific, 1989, p. 1-25.

7. A. OCNEANU - *Quantized groups, string algebras and Galois theory for algebras*, prépublication Penn. State Univ., 1985.

8. A. OCNEANU - *A polynomial invariant for knots : a combinatorial and an algebraic approach.*

9. H. WENZL - *Representations of Hecke algebras and subfactors*, Thèse, Univ. of Pennsylvania, 1985.

10. H. WENZL - *Hecke algebras of type A_n and subfactors*, Invent. Math. **92** (1988), p. 349-383.

11. H. WENZL - *Braid group representations and the quantum Yang-Baxter equation.*

La généralisation en un polynôme à deux variables est contenue dans l'article-annonce :

12. P. FREYD, D. YETTER, J. HOSTE, W. LICKORISH, K. MILLETT, A. OCNEANU - *A new polynomial invariant of knots and links*, Bull. Amer. Math. Soc. **12** (1985), p. 239-246.

Les articles suivants fournissent un exposé général et des extensions des résultats précédents :

13. A. CONNES - *Indice des sous-facteurs, algèbres de Hecke et théorie des nœuds*, Sém. Bourbaki, exp. n° 647, juin 1985, Astérisque **133-134** (1986), p. 289-308.

14. F. GOODMAN, P. DE LA HARPE et V. JONES - *Coxeter-Dynkin diagrams and towers of algebras*, M.S.R.I. Publ. vol. 14, Springer (1989).

15. P. DE LA HARPE, M. KERVAIRE et C. WEBER - *On the Jones polynomial*, Ens. Math. **32** (1986), p. 271-335.

16. P. VOGEL - *Représentations et traces des algèbres de Hecke, polynôme de Jones-Conway*, Ens. Math. **34** (1988), p. 333-356.

Voici la référence de base sur les représentations des algèbres de Hecke :

17. P.N. HOEFSMIT - *Representations of Hecke algebras of finite groups with BN pairs of classical type*, Thèse, Univ. of British Columbia, 1974.

Enfin, l'approche tridimensionnelle est contenue dans les travaux suivants :

18. M. ATIYAH - *Topological quantum field theories*, Publ. I.H.E.S. **68** (1988), p. 175-186.

19. N. RESHETIKHIN and V. TURAEV - *Invariants of 3-manifolds via link polynomial and quantum groups.*

20. E. WITTEN - *Quantum field theory and the Jones polynomial*, Comm. Math. Phys. **121** (1989), p. 351-399.

6.4. Groupes quantiques et invariants des nœuds. Références de base sur les groupes quantiques :

1. V. DRINFELD - *Hopf algebras and the quantum Yang-Baxter equation*, Sov. Math. Dokl. **32** (1985), p. 254-258.

2. V. DRINFELD - *Quantum groups*, Proc. Int. Cong. Math., Berkeley 1986, vol. 1, p. 798-820.

Y joindre un compte-rendu à Bourbaki :

3. J.-L. VERDIER - *Groupes quantiques [d'après V. Drinfeld]*, Sém. Bourbaki, exp. n° 685, juin 1987, Astérisque **152-153** (1987), p. 305-319.

Sur les représentations des groupes quantiques :

4. J. JIMBO - *A q-analogue of* $U(g\ell(N+1))$*, Hecke algebras and the Yang-Baxter equation*, Lett. Math. Phys. **11** (1986), p. 247-252.

5. A. KIRILLOV and N. RESHETIKHIN - *Representations of the algebra* $U_q(s\ell(2))$*, q-orthogonal polynomials and invariants of links*, LOMI preprint, Leningrad, 1988.

6. G. LUSZTIG - *Quantum deformations of certain simple modules over enveloping algebras*, Adv. Math. **70** (1988), p. 237-249.

7. M. ROSSO - *Finite dimensional representations of the quantum analog of the enveloping algebra of a complex simple Lie algebra*, Comm. Math. Phys. **117** (1988), p. 581-593. [Corrections dans la thèse de l'auteur.]

Pour conclure, voici une liste de travaux de l'École de Léningrad (applications des groupes quantiques aux invariants des nœuds) :

8. V. DRINFELD - *Sur les algèbres de Hopf presque cocommutatives*, à paraître dans "Algebra and Analysis", Leningrad, 1989 (en russe).

9. V. DRINFELD - *Quasi-algèbres de Hopf*, Algebra and Analysis, vol. 1, n° 2 (1989), p. 30-46 (en russe).

10. N. RESHETIKHIN - *Quantized universal enveloping algebras, the Yang-Baxter equation and invariants of links*, I, II, LOMI preprint, Léningrad, 1988.

11. N. RESHETIKHIN - *Algèbres de Hopf quasi-triangulaires et invariants des nœuds*, Algebra and Analysis, vol. 1, n° 2 (1989), p. 169-188 (en russe).

12. N. RESHETIKHIN, L. TAKHTADJAN et L. FADDEEV - *Groupes de Lie quantiques et algèbres de Lie*, Algebra and Analysis, vol. 1, n° 2 (1989), p. 178-205 (en russe).

13. V. TURAEV - *The Yang-Baxter equation and invariants of links*, Invent. Math. **92** (1988), p. 527-553.

14. V. TURAEV - *Algebra of loops on surfaces, algebra of knots, and quantization*, LOMI preprint, Léningrad, 1988.

Référence supplémentaire :

15. C.M. RINGEL - *Hall algebras and quantum groups*, (Bielefeld, RFA).

6.5. Topologie des espaces de configuration. Le point de départ est fourni par les travaux d'Arnold et Brieskorn :

1. V.I. ARNOLD - *The cohomology ring of the colored braid group*, Mat. Zametki **5** (1969), p. 227-231.

2. V.I. ARNOLD - *Topological invariants of algebraic functions* II, Funct. Anal. Appl. **4** (1970), p. 91-98.

3. E. BRIESKORN - *Die Fundamentalgruppe des Raumes der regulären Orbits einer endlichen komplexen Spiegelungsgruppe*, Invent. Math. **12** (1971), p. 57-61.

Il y a eu déjà deux rapports au Séminaire Bourbaki sur ce sujet :

4. E. BRIESKORN - *Sur les groupes de tresses [d'après V.I. Arnold]*, Sém. Bourbaki, exp. n° 401, novembre 1971, Springer, Lect. Notes Math. **317** (1973), p. 21-44.

5. P. CARTIER - *Arrangements d'hyperplans : un chapitre de Géométrie combinatoire*, Sém. Bourbaki, exp. n° 582, novembre 1980, Springer, Lect. Notes Math. **901** (1982), p. 1-22.

Quant à la structure des groupes de tresses, elle est élucidée dans ce qui suit :

6. F.A. GARSIDE - *The braid groups and other groups*, Quart. J. Math. Oxford **20**, (1969), p. 235-254.

7. E. BRIESKORN and K. SAITO - *Artin-Gruppen und Coxeter-gruppen*, Invent. Math. **17**, (1972), p. 245-271.

8. P. DELIGNE - *Les immeubles des groupes de tresses généralisés*, Invent. Math. **17**, (1972), p. 273-302.

9. M. FALK and R. RANDELL - *Pure braid groups and products of free groups*, in "Braids", Contemporary Math., Amer. Math. Soc. (1988), vol. 78, p. 217-228.

6.6. Équations différentielles et monodromie. Pour un rapport général sur la monodromie, voir mon exposé récent :

1. P. CARTIER - *Jacobiennes généralisées, monodromie unipotente et intégrales itérées*, Sém. Bourbaki, exp. n° 687, novembre 1987, Astérisque **161 – 162** (1988), p. 31-52.

Voici quelques références complétant la bibliographie de cet exposé :

2. A.A. BELAVIN and V. DRINFELD - *Solution of the classical Yang-Baxter equation for simple Lie algebras*, Funct. Anal. Appl. **16** (1982), p. 1-29.

3. T. KOHNO - *Homology of a local system on the complement of hyperplanes*, Proc. Japan Acad. Sci. **62**, série A (1986), p. 144-147.

4. T. KOHNO - *One-parameter family of linear representations of Artin's braid groups*, Adv. Stud. in Pure Math. **12** (1987), p. 189-200.

5. T. KOHNO - *Linear representations of braid groups and classical Yang-Baxter equations*, in "Braids", Contemporary Math., Amer. Math. Soc. (1988), vol. 78, p. 339-363.

6. A.B. GIVENTAL and V.V. SHEKHTMAN - *Monodromy groups and Hecke algebras*, Usp. Math. Nauk **12** n° 4 (1987), p. 138- .

7. K. AOMOTO - *A construction of integrable differential system associated with braid groups*, in "Braids", Contemporary Math., Amer. Math. Soc. (1988), vol. 78, p. 1-11.

8. A.B. GIVENTAL - *Twisted Picard-Lefschetz formulas*, Funct. Anal. Appl. **22** (1988), p. 10-18.

Donnons enfin quelques articles fondamentaux provenant de la théorie conforme des champs quantiques :

9. A.A. BELAVIN, A.N. POLYAKOV and A.B. ZAMOLODCHIKOV - *Infinite dimensional symmetries in two-dimensional quantum field theory*, Nucl. Phys. **B241** (1984), p. 333-380.

10. V.G. KNIZHNIK and A.B. ZAMOLODCHIKOV - *Current algebras and Wess-Zumino models in two dimensions*, Nucl. Phys. **B247** (1984), p. 83-103.

11. A. TSUCHIYA and Y. KANIE - *Vertex operators in conformal field theory on* $\mathbf{P}^1$ *and monodromy representations of braid group*, Adv. Studies in Pure Math. **12** (1988), p. 297-372.

6.7. Mécanique statistique et théorie des nœuds. Deux collections d'articles sur le sujet viennent de paraître :

1. C.N. YANG and M.L. GE (éditeurs) - *Braid group, knot theory and statistical mechanics*, Adv. Series Math. Phys. vol. 9, World Scientific (1989).

2. M.L. GE and N.C. SONG (éditeurs) - *Conformal field theory and braid group* (Nankai lectures in mathematical physics, 1988), World Scientific (1989).

On doit à Kauffman trois exposés de synthèse sur l'interprétation des invariants des nœuds :

3. L. KAUFFMAN - *Statistical mechanics and the Jones polynomial*, in "Braids", Contemporary Math., Amer. Math. Soc. (1988), vol. 78, p. 263-297.

4. L. KAUFFMAN - *New invariants in the theory of knots*, Astérisque **163-164** (1988), p. 137-219.

5. L. KAUFFMAN - *Knots, abstract tensors and the Yang-Baxter equation*, prépublication IHES/M/89/24 (avril 1989).

Deux références classiques, de physiciens, sur les modèles de Mécanique Statistique :

6. R.J. BAXTER - *Exactly solved models in statistical mechanics*, Acad. Press (1982).

7. F.Y. WU - *The Potts model*, Rev. Mod. Phys. vol. 54, n° 1, Janvier 1982.

Sur l'approche par les "modèles à vertex", voir :

8. R. PENROSE - *Applications of negative dimensional tensors*, in Comb. Math. and its Appl. (Welsch éditeur), Acad. Press (1971).

9. L. KAUFFMAN - *State models and the Jones polynomial*, Topology **26** (1987), p. 395-407.

10. V. JONES - *On knot invariants related to statistical mechanics models*, Pac. J. Math. **138** (1989), p. .

11. M. ROSSO - *Groupes quantiques et modèles à vertex de V. Jones en théorie des nœuds*, C.R. Acad. Sci. Paris, Série I, vol. 307 (1988), p. 207-210.

6.8. Applications des tresses à la physique. Voici un échantillon, portant surtout sur les travaux de l'école de Zürich :

1. V. DOTSENKO and V. FATEEV - *Conformal algebra and multipoint correlation functions in 2D statistical mechanics*, Nucl. Phys. **B240** (1984), p. 312-.

2. J. FRÖHLICH - *Statistics of fields, the Yang-Baxter equation and the theory of braids and links*, Cargèse lectures 1987, G.'t Hooft (éditeur), Plenum Press (1988).

3. J. FRÖHLICH - *Statistics and monodromy in two- and three-dimensional quantum field theory*, in "Differential geometrical methods in mathematical physics", K. Bleuler et M.Werner (éditeurs), Dordrecht (1988).

4. K.H. REHREN and B. SHROER - *Einstein causality and Artin braids*, prépublication, Freie Universität Berlin, 1988.

5. J. FRÖHLICH and C. KING - *Two-dimensional conformal field theory and three-dimensional topology*, prépublication ETH/TH/89-9, Zürich, 1989.

6. G. FELDER, J. FRÖHLICH and G. KELLER - *Braid matrices and structure constants for minimal conformal models*, Comm. Math. Phys. **124** (1989), p. 647-664.

7. J. FRÖHLICH, G. GABBIANI and P.A. MARCHETTI - *Braid statistics in three-dimensional local quantum theory*, in the Proceedings of the Banff Summer School in Theoretical physics (août 1989) on "Physics, Geometry and Topology".

6.9. Applications arithmétiques des groupes de tresses.

1. A. GROTHENDIECK - *Esquisse d'un programme*, Notes miméographiées, Univ. Montpellier, 1982.

2. R. COLEMAN - *Dilogarithms, regulators and p-adic L-functions*, Invent. Math. **69** (1982), p. 171-208.

3. Y. IHARA - *Profinite braid groups, Galois representations and complex multiplication*, Ann. of Math. **123** (1986), p. 3-106.

4. Y. IHARA - *On Galois representations arising from towers of coverings of* $\mathbf{P}^1\backslash\{0,1,\infty\}$, Invent. Math. **86** (1986), p. 427-459.

5. B.H. MATZAT - *Konstruktive Galois Theorie*, Lect. Notes Math. **1284**, Springer (1987).

6. T. KOHNO and T. ODA - *The lower central series of the pure braid group of an algebraic curve*, Adv. Studies in Pure Math. **12** (1988).

7. P. DELIGNE - *Le groupe fondamental de la droite projective moins trois points*, in "Galois groups over **Q**", M.S.R.I. Publ. vol. 16, Springer, 1989.

9

AN INTRODUCTION TO KNOT THEORY AND FUNCTIONAL INTEGRALS

L. Kauffman

Department of Mathematics
Statistics and Computer Science
University of Illinois at Chicago
851 South Morgan Street
Chicago, IL

ABSTRACT

This paper is an expanded version of a course of lectures on knots and functional integration delivered in Corsica during September 1996.

1. INTRODUCTION

Knot theory has its origins in the arts of weaving and practical knot tying. When topology emerged as a mathematical discipline, it was inevitable that knots would become part of the subject matter of topology. It is surprising, however that knot theory got its first big push from an intimation that knots were the patterns of elementary particles in three dimensional space. At that time the elementary particles were atoms and the theory was Lord Kelvin's (William Thompson) theory of atoms as vortices in the ether. Kelvin encouraged the mathematicians Tait, Kirkwood and Little to make tables of knots in the hope that these tables would provide the key to the structure of the vortex atoms. It is a fascinating image: swirling knotted vortices interacting in the fluid ether of the vacuum. The theory had a certain popularity but eventually disappeared along with those notions of the ether. The knot tables survived and became the first empirical evidence for many properties of knots and links, some of which have been proved only recently.

The actual mathematics of knots began with the application of Poincare's fundamental group to the complement of the knot in three dimensional space. This began a long investigation, continuing to the present day, of the relationship of algebraic properties of the fundamental group and topological properties of knots and links. In the 1920's J. W. Alexander discovered, with the help of the fundamental group, a polynomial invariant of knots and links. This Alexander polynomial could be understood combinatorially, and it led to a whole era of

Functional Integration: Basics and Applications
Edited by Cécile DeWitt-Morette, Plenum Press, New York, 1997

research into its properties and into other sources for its definition. The idea that there might be other "polynomial" (The polynomial is in quotes because even in the Alexander polynomial one allows negative powers of the variable(s).) invariants of knots and links did not occur to anyone until 1984.

In 1984 Vaughan Jones discovered a new polynomial invariant of knots and links by using properties of Von Neumann algebras, a subject emanating from statistical mechanics and quantum theory. There ensued a quick discovery of many new invariants. The whole notion of knot theory changed overnight. The change was heralded by a realization that knot invariants could be produced by a kind of "higher curve fitting" technique. Define a function on link diagrams that is well defined at the diagrammatic level. Adjust the parameters of this function in such a way that it becomes invariant under the appropriate topological moves. It turns out that functions of this sort are just analogous to the partition functions in two dimensional discrete statistical mechanics. The analogy is so strong that it is possible to take the results of statistical mechanics (solutions to the Yang–Baxter Equation) and use them to make many new knot invariants including the Jones polynomial and its generalizations.

It had been possible to express "polynomial" invariants of links as state summations analogous to partition functions in statistical mechanics. After the landmark paper of Witten in 1988 it became possible to express these invariants as functional integrals in a quantum field theoretic context. These integrals, void of traditional mathematical existence, have cast an enchanted spell on the subject of low dimensional topology from which it may hope to wake through either the means of combinatorics or perhaps in the ultimate justification of these integrals.

The present paper is a much expanded version of [1], and it follows the structure of the author's Corsica lectures in September of 1996. We outline some of the first steps in dealing with the formalism of the Witten integral, and how these steps shed light on the structure of the Vassiliev invariants of knots and links. There are many other aspects of the functional integral that could have been explored (such as the stationary phase approximation). The scope of this paper is restricted, but within that scope there are insights to be gained about knot theory and quantum field theory.

The paper is organized as follows: Section 2 is a review of basic knot theory. Section 3 is a review of state sum invariants of knots and links with an emphasis on the bracket state sum for the Jones polynomial. Section 4 is an exposition of quantum link invariants. Section 5 discusses the Gauss linking integral. Section 6 introduces a Level 0 heuristic that shows how integration could smooth out a switching relation to produce a global skein relation. This heuristic contains some of the ingredients of Section 7 where we discuss the formalism of the Witten integral and produce the general form of the switching relation for that integral in terms of Casimir insertion. Section 8 discusses the combinatorics of Vassiliev invariants and how to produce weight systems for Vassiliev top rows from Lie algebras. The end of Section 8 and the contents of Section 9 point out how these top row evaluations are foreshadowed in our work with the Witten integral in Section 7. Section 10 is a very quick sketch of the Kontsevich work on Vassiliev invariants. Section 11 is an appendix on square dancing.

ACKNOWLEDGMENTS

It gives the author pleasure to thank Maria Elisa Sarraf Borelli for her intelligent comments on the course of lectures in Corsica upon which this paper is based, and Cécile DeWitt-

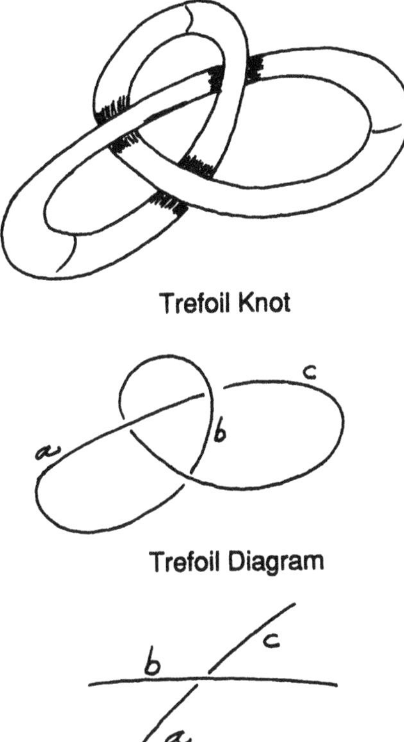

Figure 1. Trefoil knot.

Morette and Pierre Cartier for the invitation to deliver those lectures. The author thanks the National Science Foundation for support of this research under NSF Grant DMS -2528707.

2. COMBINATORIAL KNOT THEORY — REIDEMEISTER MOVES, QUANDLE, SKEIN POLYNOMIALS

In this section we review some basic facts of combinatorial knot theory. A knot is an embedding of a circle into three dimensional space. It is convenient to represent a knot by a planar drawing with conventions that allow the reconstruction of the knot in three space. These drawings can be regarded as depictions of the corresponding embeddings, or as planar graphs with special structure. The depth of the combinatorial approach to knots and links turns on the interaction of ideas from these two domains.

A diagram for the trefoil knot is illustrated in Figure 1.

Each crossing in the diagram consists locally in three arcs a, b, c with endpoints of two of these arcs (say a and c) correlated across the third arc (b) passing between them. This is interpreted as the crossing of the arc b in space over the arc obtained from a and c by identifying the endpoints near b by a segment that passes underneath b. It is a common pictorial convention to indicate an overcrossing of one line by another via the erasure of part of the undercrossing line.

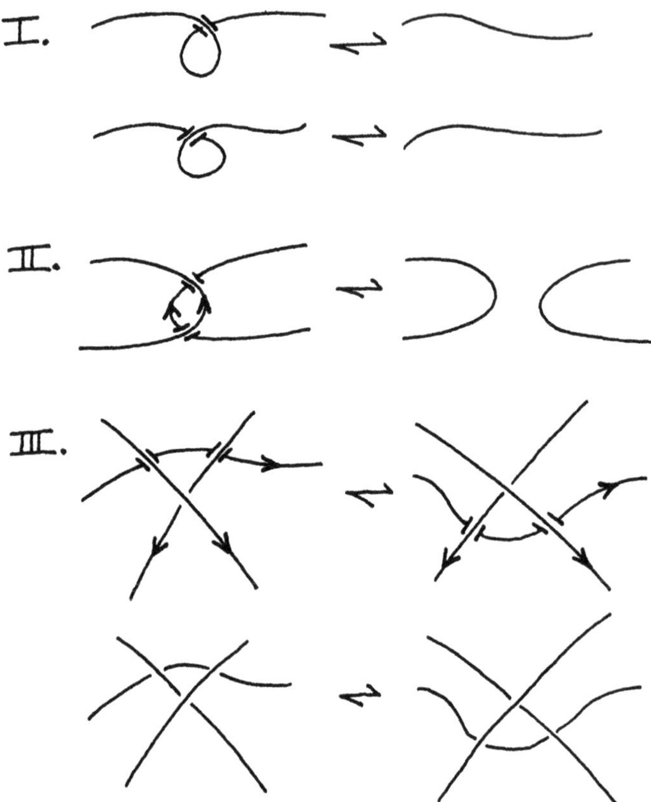

Figure 2. Reidemeister moves.

A knot is said to be *tame* if for every point on it there is a homeomorphism of three dimensional space that takes a neighborhood of that point and an intersection of that neighborhood with the knot to a standard pair of ball and straight arc. Knots defined in a finite piecewise linear fashion and knots defined by differentiable functions are tame. With these conventions, we can represent all tame knots and links by diagrams with a finite number of crossings.

Two knots or links are said to be *ambient isotopic* if there is a continuous family of embeddings starting from one knot and ending at the other. The basic problem of knot theory is the question of the classification of knots up to ambient isotopy. Thus we should prove that the trefoil knot is inequivalent to the unknotted circle.

The *mirror image* of a knot is obtained by reversing all the crossings. We should also prove that the trefoil knot is inequivalent to its mirror image.

Neither of these problems is trivial, and the second one is actually quite subtle.

Reidemeister [2] discovered a simple set of moves on the diagrams that captures the concept of ambient isotopy of knots in three dimensional space. There are three basic Reidemeister moves. Two diagrams represent ambient isotopic knots (or links) if and only if there is a sequence of Reidemeister moves taking one diagram to the other. The Reidemeister moves are illustrated in Figure 2.

Reidemeister's three moves are interpreted as performed on larger diagram in which the small diagram shown is a literal part. Each move is performed without disturbing the rest of the diagram. Note that this means that each move occurs, up to topological deformation, just

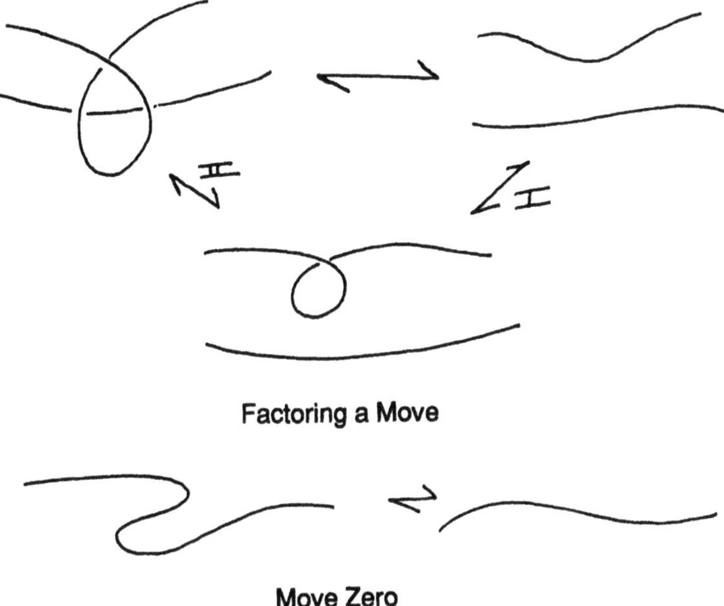

Figure 3. Factorable move, move zero.

as it is shown in the diagrams in Figure 2. There are no extra lines in the local diagrams. For example, the equivalence (A) in Figure 3 is **not** an instance of a single first Reidemeister move. Taken literally, it factors into a move II followed by a move I.

Diagrams are always subject to topological deformations in the plane that preserve the structure of the crossings. These deformations could be designated as "Move Zero." See Figure 3.

A few exercises with the Reidemeister moves are in order. First of all, view the diagram in Figure 4. It is unknotted and you can have a good time finding a sequence of Reidemeister moves that will do the trick. Diagrams of this type are produced by tracing a curve and always producing an undercrossing at each return crossing. This type of knot is called a *standard unknot*. Of course we see clearly that a standard unknot is unknotted by just *pulling* on it, since it has the same structure as a coil of rope that is wound down onto a flat surface.

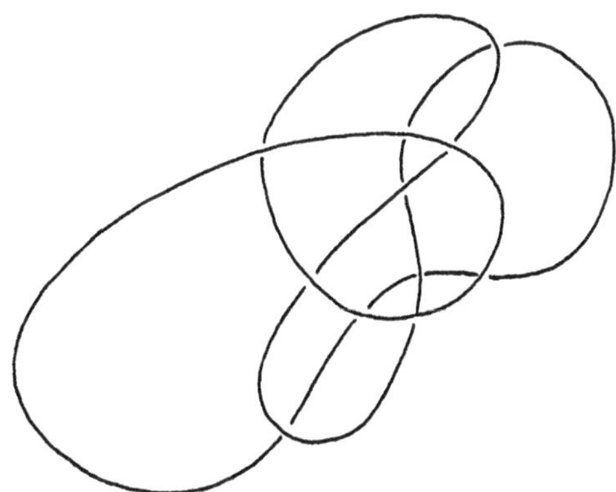

Figure 4. Standard unknot.

Figure 5. A demon.

You might wonder whether one can recognize unknots by simply looking for sequences of Reidemeister moves that undo them. This would be easy if it were not for the case that there are examples of unknots that require some moves that increase the number of crossings before they can be subsequently decreased. Such an demonic example is illustrated in Figure 5.

It is generally not so easy to recognize unknots. However, here is a tip: Look for *macro moves* of the type shown in Figure 6. In a macro move, we identify an arc that passes entirely under some piece of the diagram (or entirely over) and shift this part of the arc, keeping it under (or over) during the shift. Macro moves often allow a reduction in the number of crossings even though the number of crossings will increase during a sequence of Reidemeister moves that generates the macro move.

As shown in Figure 6, the macro-move includes as a special case both the second and the third Reidemeister moves, and it is not hard to verify that a macro move can be generated by a sequence of type *II* and type *III* Reidemeister moves. You should also prove to yourself that the type *I* moves can be left to the end of any deformation. If you now return to the demon of Figure 5, you will see that it is easily demolished by macro moves, and that from the point of view of macro moves the diagram never gets more complicated.

Lets say that an knot can be *reduced* by a set of moves if it can be transformed by these moves to the unknotted circle diagram through diagrams that never have more crossings than the original diagram. Then we have shown that there are diagrams representing the unknot that cannot be reduced by the Reidemeister moves. On the other hand, *I* do not know whether unknotted diagrams can always be reduced by the macro moves in conjunction with the first Reidemeister move. If this were true it would give a combinatorial way to recognize the unknot.

It may seem a bit mysterious that all the complexity of knotting and linking of curves in three dimensional space can be reduced to this combinatorial game of the Reidemeister moves. Lets reduce the mystery by indicating how Reidemeister proved this Theorem.

An embedding of a knot or link in three dimensional space is said to be *piecewise linear* if it consists in a collection of straight line segments joined end to end. Reidemeister started with a *single* move in three dimensional space for piecewise linear knots and links. Consider a point in the complement of the link, and an edge in the link such that the surface of the triangle formed by the end points of that edge and the new point is not pierced by any other edge in the link. Then one can replace the given edge on the link by the other two edges of the triangle, obtaining a new link that is ambient isotopic to the original link. Conversely, one can remove two consecutive edges in the link and replace them by a new edge that goes directly from initial to final points, whenever the triangle spanned by the two consecutive edges is not pierced by any other edge of the link. This triangle replacement constitutes Reidemeister's three dimensional move. See Figure 7. It can be shown that two piecewise linear knots or

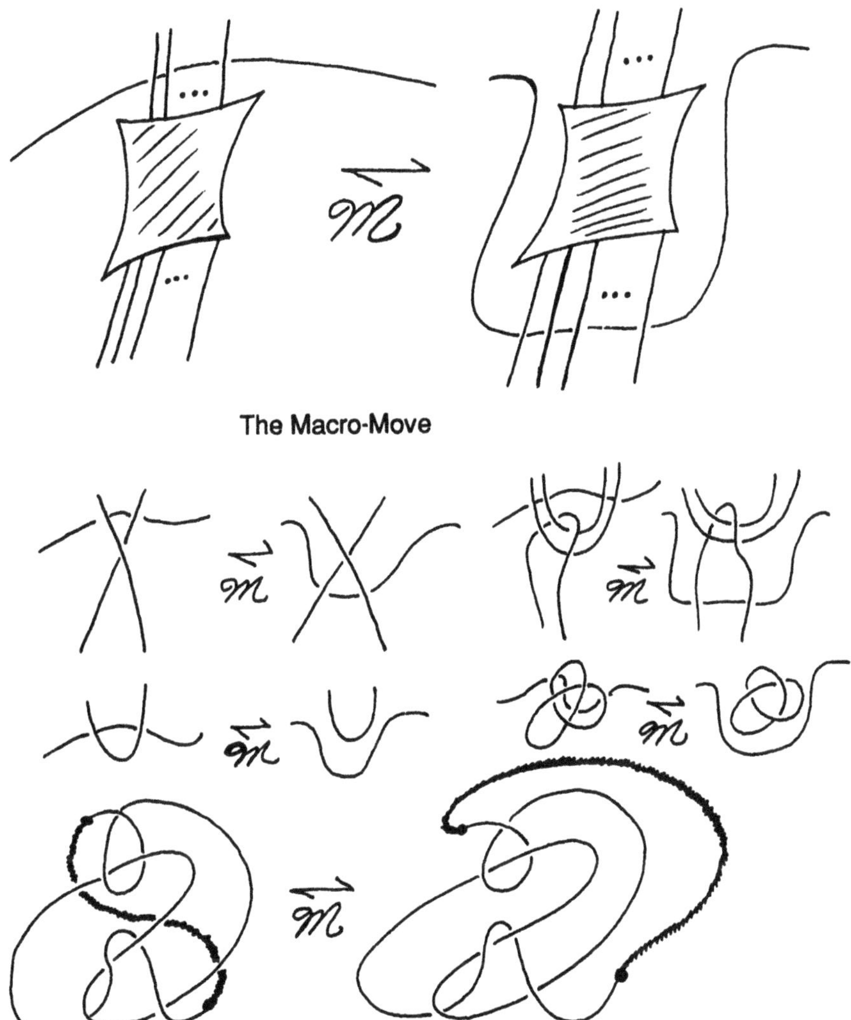

Figure 6. Macro move.

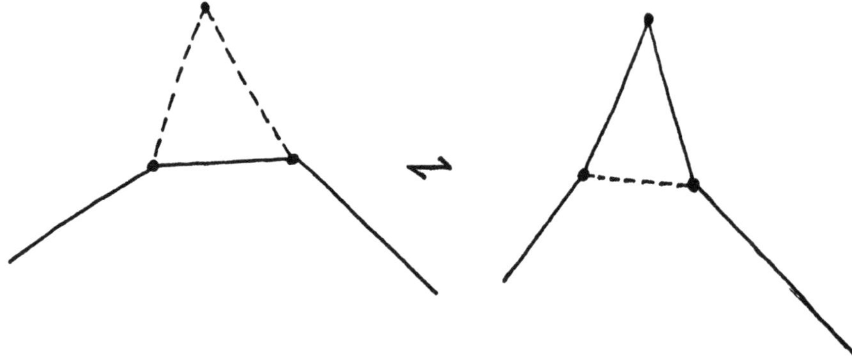

Figure 7. Triangle move.

links are ambient isotopic in three dimensional space if and only if there is a sequence of Reidemeister triangle moves from one to the other.

It can also be shown that tame knots and links have piecewise linear representatives in their ambient isotopy class. It is sufficient for our purposes to work with piecewise linear knots and links. Reidemeister's planar moves then follow from an analysis of the shadows projected into the plane by Reidemeister triangle moves in space. Figure 8 gives a hint of this analysis. The result is a reformulation of the three dimensional problems of knot theory to a combinatorial game in the plane.

It should be mentioned that one often works with *oriented equivalence* of knots and links. Here each component of the link is endowed with a direction, and two links are considered to be equivalent if one can be transformed into the other in such a way that these orientations are preserved. In this domain there are some particularly subtle problems. For example, the knot

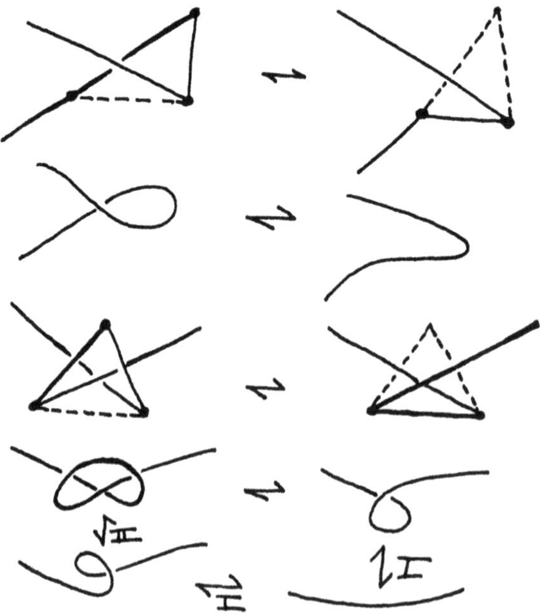

Figure 8. Shadows.

Figure 9. The knot 8_{17}.

8_{17} (shown in Figure 9) from the Reidemeister tables is known to be *inequivalent* to its own reverse [3], but there is no known proof of that fact using the sorts of invariants of links that we will discuss in this paper. In fact it is an open problem whether there exist Vassiliev invariants (see Section 8) of finite type that can detect the difference between 8_{17} and its reverse.

In studying links, the notion of orientation is quite natural, since we want to measure the number of times one component winds around another. This is the *linking number* of a two component link. If the link has oriented components A and B, then we let $lk(A,B)$ denote the linking number of A and B. Linking number is an invariant of ambient isotopy and has many definitions such as "the sum of the oriented intersection numbers of A with an oriented surface spanning B that is placed in general position with respect to A." In other words, one can count the number of times that one curve pierces a surface that spans the other curve. A description of this kind for the linking number is very useful for spatial understanding, but involves more topology than the Reidemeister moves for its full explication. An equivalent definition using diagrams is the following: Let p be a crossing in an oriented link diagram. Let $\varepsilon(p)$ be $+1$ or -1 according to the type of the crossing as shown in Figure 10.

If a link L consists in two components A and B, define the linking number of L to be

$$lk(A,B) = \sum_{p \in C(A,B)} \varepsilon(p)/2.$$

Here $C(A,B)$ is the set of crossings between A and B. Self-crossings of A and B are not included in this calculation. It follows easily that $lk(A,B)$ is invariant under the Reidemeister moves. It is an interesting exercise to see that this combinatorial definition of the linking number coincides with the first definition that we gave in terms of surfaces and piercing numbers.

It is also useful to define the *writhe*, $wr(K)$, of an oriented link as the sum of the signs of *all* the crossings of the link. Thus we let

$$wr(K) = \sum_{p \in C(K)} \varepsilon(p)$$

where $C(K)$ denotes all the crossings in the diagram K. The writhe is invariant under Moves *II* and *III* but changes by $+1$ or -1 under move number *I*. This leads to the definition of *regular isotopy* as the equivalence relation on link diagrams that is generated by moves *II* and *III* in conjunction with topological equivalence of diagrams. Regular isotopy is quite important because it nearly captures the notion of ambient isotopy of framed links. A framed link is a link such that every component is equipped with a normal vector field. This can be visualized by replacing each component with a band whose core is the original component and such that the twisting of the band matches the twisting of the normal vector field. Any framed link can be represented by a diagram by resolving 360 degree twists into curls as shown in Figure 11.

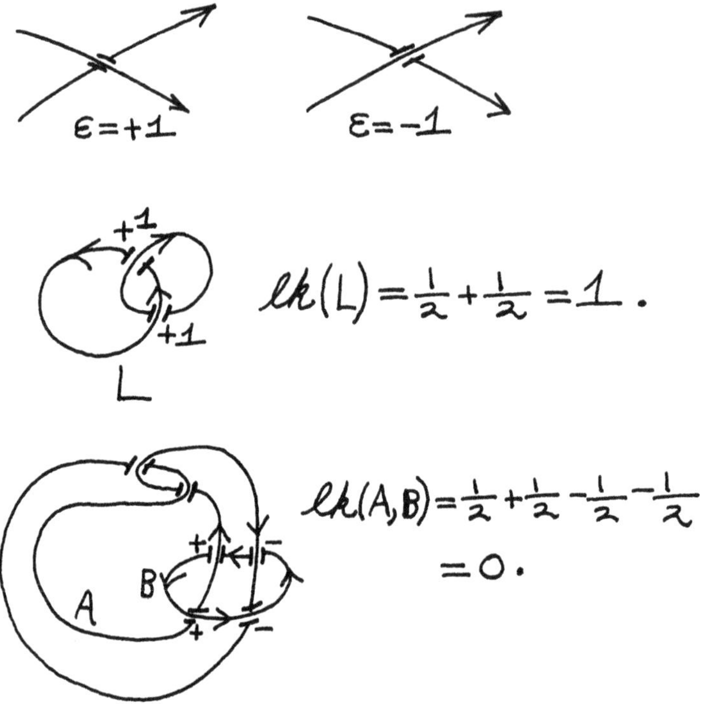

Figure 10. Crossing type and linking number.

Writhe and twist interchange under a topological deformation. This fact is also the source of a famous formula due to Jim White (in the differential geometry context) [4] that is used in studying DNA. White's formula relates the linking number of two sides of a band to the twisting of the band and the writhing number of the core of the band:

$$lk = tw + wr.$$

The White formula is illustrated in Figure 11 for the simple case of a nearly flat band.

Reidemeister's theorem is fertile ground for the construction of many invariants of knots and links. For example, suppose that we decide to label all the arcs of an oriented link with labels from a set S with local rules governing the labellings at crossings denoted by $c = a * b$ or $c = a \bar{*} b$ where a undercrosses b to emerge as c and the choice of $*$ or $\bar{*}$ is determined by whether b proceeds to the right or to the left as one approaches b along a. See Figure 12.

Then it is easy to see (Figure 12) that the following equations correspond directly to invariance of this labelling scheme under moves I, II and III:

$$\begin{aligned} &I. && a * a = a, a \bar{*} a = a \\ &II. && (a * b) \bar{*} b = a, (a \bar{*} b) * b = a \\ &III. && (a * b) * c = (a * c) * (b * c), (a \bar{*} b) \bar{*} c = (a \bar{*} c) \bar{*} (b \bar{*} c). \end{aligned}$$

An abstract algebra with these axioms is called a *quandle*. The quandle generated by labels for the arcs and the indicated relations at each crossing of a diagram K is called the *quandle*, $Q(K)$, of the link K [5]. This algebra is a powerful invariant of the link K, generalizing the fundamental group of the complement of K. For more information on the quandle, see [6], [7], [8].

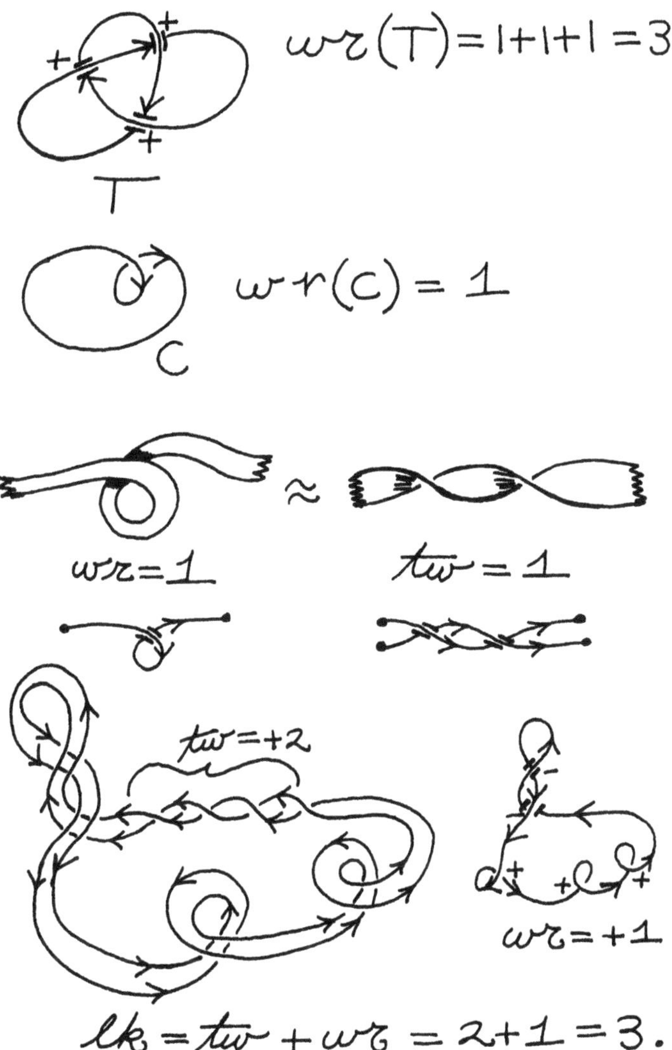

Figure 11. Framing and writhe.

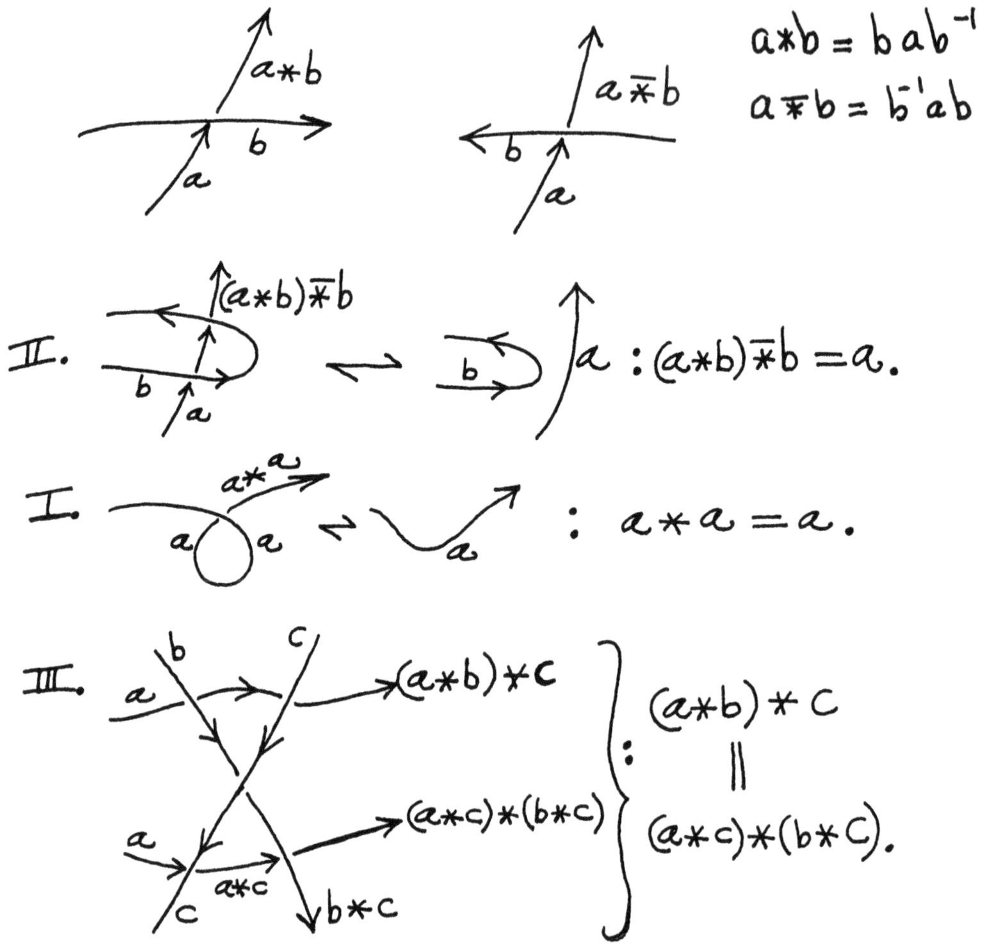

Figure 12. Quandle conventions.

For a description of the fundamental group itself, recall that the fundamental group of a topological space X is generated by the loops on that space based at a given point p in the space. A loop α based at p is a mapping of the unit interval $I = [0,1]$ into X such that $\alpha(0) = \alpha(1) = p$. Two loops are said to be *homotopic relative to p* if there is a continuous family of loops at p joining them. Loops are multiplied by taking the composite loop consisting of doing the first loop and then the second (suitably reparametrizing each loop so that the first occurs on the interval $[0,1/2]$ and the second on the interval $[1/2,1]$). The fundamental group, $\pi_1(X,p)$, is the set of homotopy classes of loops relative to the base point p.

In the case of the complement of an oriented knot, the knot diagram provides a convenient set of generators of the fundamental group: Choose a basepoint p in the complement, say above the plane of the diagram. Choose one loop for each arc of the diagram such that this loop encircles the arc with linking number one as in Figure. It is then easy to see (Figure 13) that if $[a]$, $[b]$ and $[c]$ denote the loops in the fundamental group of the knot complement corresponding to arcs a,b,c meeting at a positive crossing of the diagram so that a undercrosses b to meet c, then $[b][a] = [c][b]$, hence $[c] = [b][a][b]^{-1}$. The fundamental group of the knot complement is completely described by the arc generators and these crossing relations. Letting $[a] * [b] = [b][a][b]^{-1}$ and $[a] \bar{*} [b] = [b]^{-1}[a][b]$, we obtain the special case of

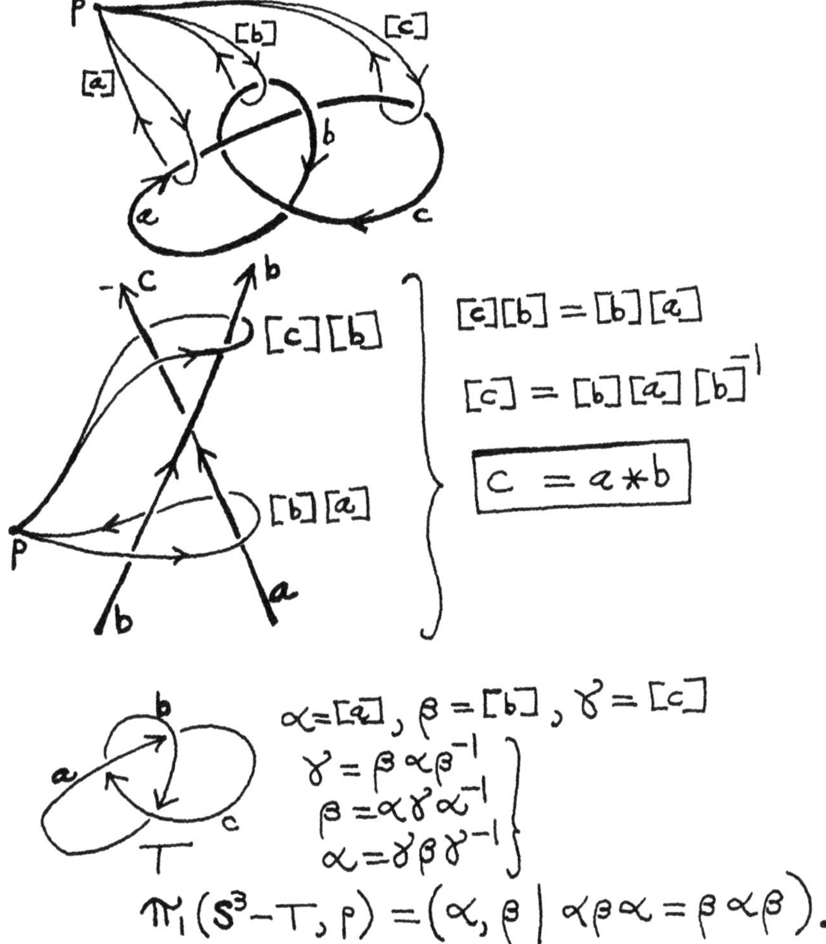

Figure 13. Fundamental group of a knot.

the quandle structure that produces the fundamental group.

The *involutory quandle*, $IQ(K)$ is the quotient of the quandle $Q(K)$ by the extra axiom $(a*b)*b=a$ for any a and b in $Q(K)$. This makes the second operation equal to the first. The involutory quandle is often easy to compute, often a finite algebraic system.

We are now in a position to verify that the trefoil knot is knotted. View Figure 14. The quandle of the trefoil T can be described by generators a, b, c and the relations $a*b=c, b*c=a$, $c*a=b$. In the involutory quandle $IQ(T)$, we have $a=(a*b)*b=c*b, b=(b*c)*c=a*c$ and $c=(c*a)*a=b*a$. Thus $IQ(T)$ is a finite set consisting in the three elements a, b, c. The $*$ product of any two distinct elements yields the third element, while the $*$ product of any element with itself yields the same element again. The non-triviality of this little involutory quandle proves the knottedness of the trefoil knot.

The products in the three element system for $IQ(T)$ can be regarded as rules for coloring the arcs of the trefoil with three colors so that an any given crossing there are either three colors ($a*b=c$) or only one color ($a*a=a$). You can use just this idea of coloring to produce a completely elementary proof that the trefoil is knotted: Verify that the application of a Reidemeister move to an already three-colored diagram yields a new diagram that can

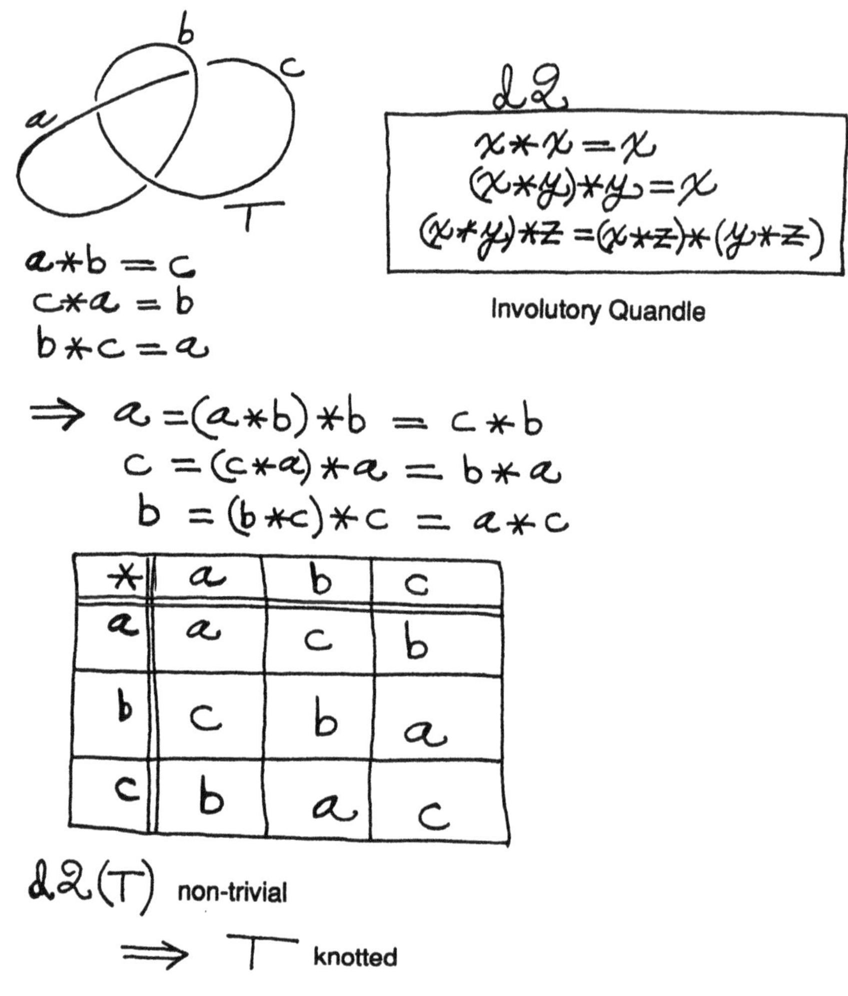

*	a	b	c
a	a	c	b
b	c	b	a
c	b	a	c

Figure 14. *IQ* of the trefoil.

also be three colored — making only local changes or additions to the coloring corresponding to the given Reidemeister move. (This is an exercise for the reader!) Now note that the fact that the standard trefoil diagram is three colored implies that all diagrams obtained from the standard diagram by Reidemeister moves are necessarily also three colored. However, there is no way to color the standard diagram of the unknot with more than one color. The only conclusion is that there is no sequence of Reidemeister moves starting from the standard trefoil diagram and ending with the unknot. The trefoil is knotted.

A particularly useful quandle is obtained from any module M over the ring $Z[t,t^{-1}]$. Define

$$a*b = ta+(1-t)b$$

and

$$a\bar{*}b = t^{-1}a+(1-t^{-1})b.$$

It is not hard to see that this defines the structure of a quandle on M. If we take M to be freely generated by labels for the arcs of a link diagram with one arc labelled 0, and let $M(K)$ denote

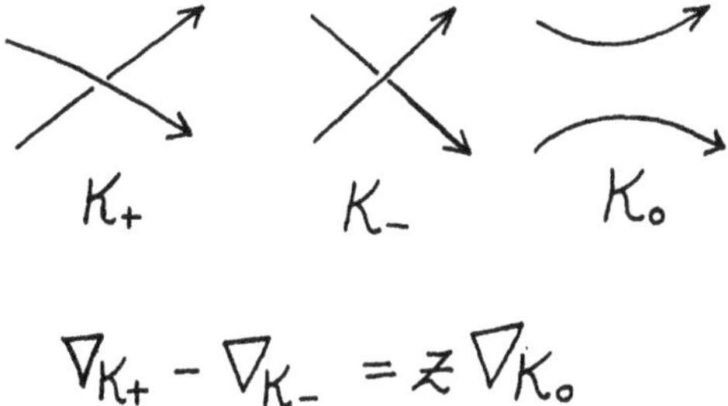

Figure 15. Skein relation for Conway polynomial.

the quotient of M by the submodule of relations generated from the crossings according to the above formulas, then we get a special quandle of the form

$$M(K) = M/\langle \Delta_K(t) \rangle$$

where $\Delta_K(t)$ generates the principal ideal in $Z[t,t^{-1}]$ of Laurent polynomials that annihilate elements of $M(K)$. $\Delta_K(t)$ is the classical *Alexander polynomial* of K written in this quandle context. Note that $\Delta_K(t)$ is defined only up to sign and powers of t.

In [9] John H. Conway found a remarkable reformulation of the Alexander polynomial that makes it independent of the algebraic context via an identity that allows recursive computation from oriented link diagrams. Conway's version of the Alexander polynomial is denoted ∇_K. The Conway polynomial is normalized to the value 1 on the unknot, it is an invariant of ambient isotopy and it satisfies the *skein relation*

$$\nabla_{K_+} - \nabla_{K_-} = z\nabla K_0.$$

Here K_+, K_- and K_0 denote three diagrams such that K_+ has a positive crossing that is replaced by a negative crossing in K_- and a smoothed crossing in K_0. The local forms of K_+, K_- and K_0 are indicated in Figure 15.

The relation between the Conway polynomial and the Alexander polynomial is given by the formula

$$\nabla_K(t^{1/2} - t^{-1/2}) \doteq \Delta_K(t)$$

where $\doteq$ denotes equality up to sign and powers of t. The Conway polynomial can detect the difference between many links and their mirror images. However, neither the Conway nor the Alexander polynomial can detect any difference between a knot and its mirror image.

Vaughan Jones [10] discovered in 1984 a new Laurent polynomial invariant of knots and links satisfying an identity quite similar to that for the Conway polynomial. The Jones polynomial, denoted by $V_K(t)$ for a given oriented link K, can detect the difference between many knots and their mirror images. The Jones polynomial is an invariant of ambient isotopy, normalized to 1 on the unknot and satisfying the skein relation

$$t^{-1}V_{K_+} - tV_{K_-} = (t^{1/2} - t^{-1/2})V_{K_0}.$$

Although the Conway and Jones polynomials satisfy very similar skein relations their properties are quite different. There does not seem to be a simple method of modelling the Jones polynomial in terms of an augmented version of the fundamental group of the knot complement or in any of the other well-known ways of relating the Alexander polynomial to algebraic topology. Jones defined his polynomial via a representation of Artin braid group to a certain von Neumann algebra — identical in structure to the Temperley Lieb algebra in statistical mechanics [11]. This representation is in fact directly related to a state summation model for the polynomial that we shall discuss in the next section. It turns out that the Jones polynomial, and its generalizations are susceptible to a new kind of algebraic topology consisting in a "sum over states" of the link diagram that is analogous to the sum over states of a physical system. Such sums are made in calculating a partition function in statistical mechanics.

A striking generalization of both the Jones and Conway polynomials occurs in the *Homfly* polynomial, $P_K(a,z)$, [12], a two variable Laurent polynomial in the variables a and z, invariant under ambient isotopy, normalized to 1 on the unknot and satisfying the skein identity

$$a^{-1}P_{K_+} - aP_{K_-} = zP_{K_0}.$$

There is yet another two variable invariant, denoted $Y_K(a,z)$, called the Dubrovnik version of the Kauffman polynomial (having been discovered in Dubrovnik, Yugoslavia) [13]. We shall refer to this invariant as the Dubrovnik polynomial. The Dubrovnik polynomial is an invariant of oriented links and it is normalized to 1 on the unknot. However, this polynomial is best explained as a writhe normalization of an invariant of regular isotopy that we denote by $D_K(a,z)$. That is,

$$Y_K(a,z) = a^{-wr(K)}D_K(a,z)$$

where $wr(K)$ is the writhe of the link K as defined earlier in this section. D_K is a regular isotopy invariant of *unoriented* links satisfying the skein relation

$$D_K - D_{EK} = z(D_{SK} - D_{S'K})$$

and the relations

$$D_{R_+K} = aDK,$$
$$D_{R_-K} = a^{-1}DK.$$

In these equations EK denotes the result of switching a single crossing in K while SK denotes one of the smoothings of that same crossing and $S'K$ denotes the opposite smoothing of that crossing. Thus, K, EK, SK and $S'K$ differ only at the site of one crossing in K. Since K is an unoriented diagram, there are two possible smoothings of a given crossing. By convention, *we take SK to be that smoothing that connects the two regions swept out at the crossing by turning the overcrossing line counterclockwise.* Note that $EEK = K$ and that $SEK = S'K$. See the illustration below.

R_+K is the result of adding a positive curl anywhere in the diagram K. Similarly, R_-K is the result of adding a negative curl anywhere in the diagram K. Note that this performs a Reidemeister move of type I on the diagram. The corresponding equations show that D_K is not only an invariant of regular isotopy, it is in fact an invariant of ribbon equivalence (that is it gives the same result for curls of the same writhe that rotate left or right in the plane) and hence an invariant of framed links in 3-space (We may view diagrams up to regular isotopy as shorthand for framed links as long as we identify left and right rotating curls of the same writhe.).

The Dubrovnik polynomial is also good at detecting knots from their mirror images and it is a generalization of the original Jones polynomial that is distinct from the Homfly polynomial. A good way to see how the Dubrovnik polynomial is a generalization of the original Jones polynomial is to use the bracket state sum model alluded to in the discussion of the Jones polynomial. We shall do this in the next section of the paper.

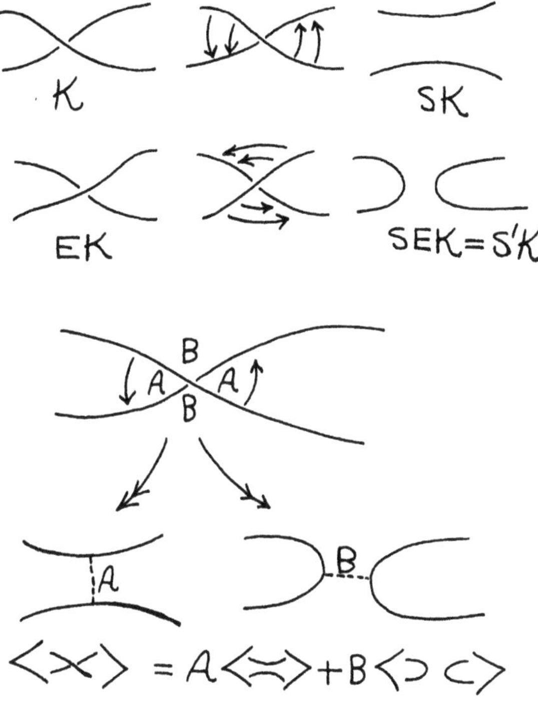

Figure 16. Smoothing conventions.

3. STATE SUMMATION MODELS FOR LINK POLYNOMIALS

The *bracket polynomial* [14], [15], $\langle K\rangle = \langle K\rangle(A)$, assigns to each unoriented link diagram K a Laurent polynomial in the variable A such that

1. If K and K' are regularly isotopic links, then $\langle K\rangle = \langle K'\rangle$.

2. If $K \quad O$ denotes the disjoint union of K with an extra unknotted and unlinked component O, then
$$\langle K \quad O\rangle = d\langle K\rangle$$
where
$$d = -A^2 - A^{-2}.$$

3. $\langle K\rangle$ satisfies the following formula where the small diagrams represent parts of larger diagrams that are identical except at the site indicated in the bracket.
$$\langle \times \rangle = A\langle \asymp \rangle + A^{-1}\langle\,\rangle\,\langle\,\rangle.$$

It is easy to verify that the bracket is invariant under the second and third Reidemeister moves. That is, *property 1. is a direct consequence of properties 2. and 3.* The second two properties define the bracket on arbitrary link diagrams. See Figure 17.

In fact we could have begun with the following more general definition: Let K be any unoriented link diagram. Define a *state* of K to be a choice of smoothings for all the crossings of K. There are 2^N states of a diagram with N crossings. A *smoothing* of a crossing is a

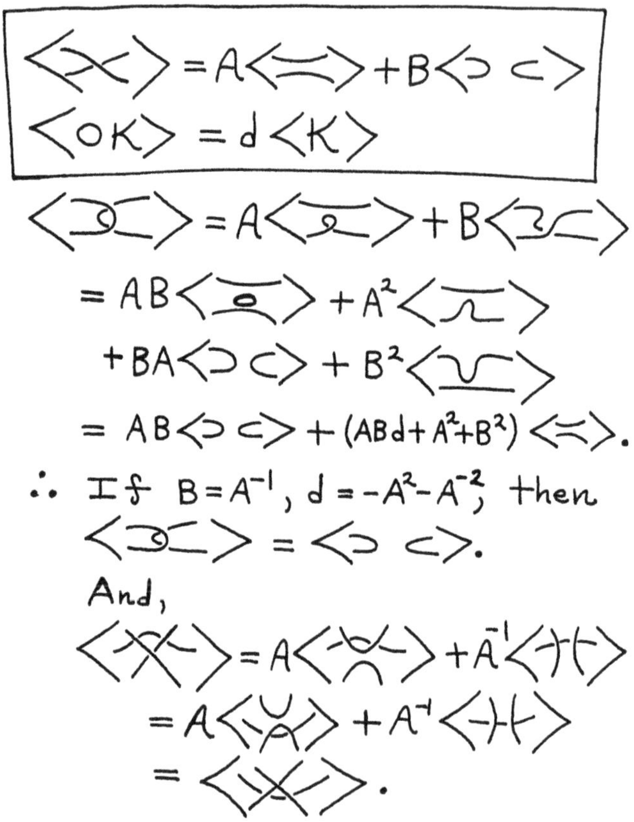

$\therefore$ If $B = A^{-1}$, $d = -A^2 - A^{-2}$, then

And,

Figure 17. Invariance of the bracket.

local replacement of that crossing with two arcs that do not cross one another, as shown below. There are two choices for smoothing a given crossing. In illustrating a state it is convenient to label the smoothing with A or B to indicate the crossing from which it was smoothed. The A or B is called a *vertex weight* of the state. See Figure 18.

Label each state with *vertex weights* A or B as illustrated in Figure 18. Here A and B are commuting polynomial variables. Define two evaluations related to the state: The first evaluation is the product of the vertex weights, denoted

$$[K|S].$$

The second evaluation is the number of loops (Jordan curves) in the state S, denoted

$$||S||.$$

Define the *state summation*, $[K]$, by the formula

$$[K] = \sum_S [K|S] d^{||S||-1}.$$

It follows from this definition, that $[K]$ satisfies the formulas

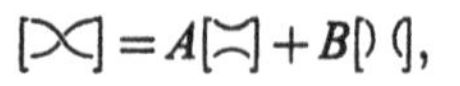

$$[\asymp\!\!\!\!\times] = A[\asymp] + B[)\ (],$$
$$[O\quad K] = d[K],$$

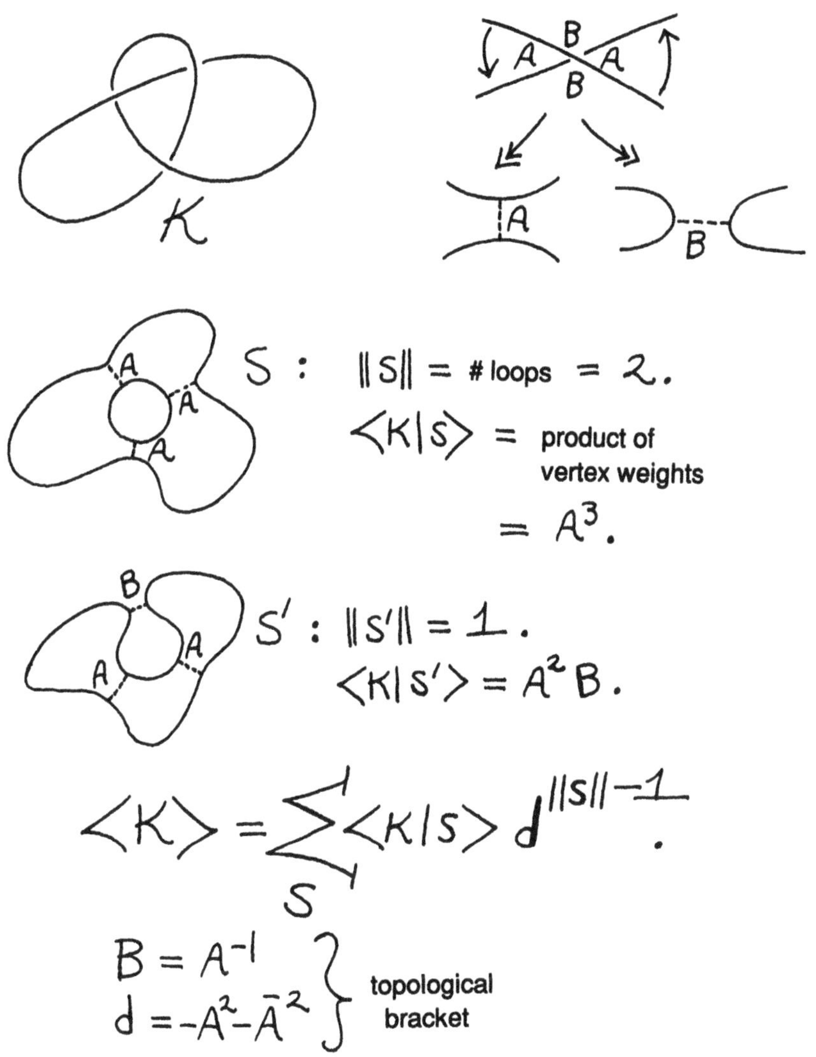

Figure 18. Bracket states.

and

$$[O] = 1.$$

The demand that $[K]$ be invariant under the *second* Reidemeister move leads to the conditions $B = A^{-1}$ and $d = -A^2 - A^{-2}$. This specialization is easily seen to be invariant under the third Reidemeister move. Calling this specialization the *topological* bracket, and denoting it (as above) by $\langle K \rangle$ one finds the following behavior under the first Reidemeister move

$$\langle \text{curl} \rangle = -A^3$$

and

$$\langle \text{curl} \rangle = -A^{-3}.$$

The topological bracket is invariant under regular isotopy and can be normalized to an invariant of ambient isotopy by the definition

$$f_K(A) = (-A^3)^{-wr(K)} \langle K \rangle (A)$$

where $wr(K)$ is the sum of the crossing signs of the oriented link K. $wr(K)$ is called the writhe of K as in the previous section.

If K^* denotes the mirror image of K, obtained by switching all the crossings of K, then it is easy to see that $\langle K^* \rangle (A) = \langle K \rangle (A^{-1})$ and $f_{K^*}(A) = f_K(A^{-1})$. Thus if K is ambient isotopic to its mirror image K^*, then $f_K(A) = f_{K^*}(A^{-1}) = f_K(A^{-1})$. If $f_K(A)$ and $f_K(A^{-1})$ are not equal, then K cannot be ambient isotopic to its mirror image.

A knot is said to be *chiral* if it is not ambient isotopic to its mirror image.

We are now in a position to calculate the bracket polynomial of the trefoil knot and to show that this knot is not equivalent to its mirror image. The diagrammatic calculation is shown in Figure 19. Note that our end result shows that for the trefoil T, $f_T(A)$ is not equal to $f_T(A^{-1})$.

Is there a simpler proof of the chirality of the trefoil? The proof we just saw depends upon nothing but the Reidemeister moves and the elementary construction of the bracket polynomial. Perhaps there is an even simpler way, as yet unknown to us.

By a change of variables one obtains the original Jones polynomial, $V_K(t)$ [10] from the normalized bracket:

$$V_K(t) = f_K(t^{-1/4}).$$

It is useful to reformulate the bracket state sum so that it is a state sum on oriented diagrams, giving the Jones polynomial directly. This is now easily done by translating the unoriented state summation through the normalization to the oriented case. The result is the following expansion formulas. We will use these formulas when we discuss Vassiliev invariants in Section 8.

$$V_{K_+} = -t^{1/2} V_{K_0} - t V_{K_\infty}$$
$$V_{K_-} = -t^{-1/2} V_{K_0} - t^{-1} V_{K_\infty}.$$

The value of the loop is

$$\delta = -(t^{1/2} + t^{-1/2}).$$

In these formulas we have used the previously decided conventions for positive and negative crossings and oriented smoothings. The notation K_∞ stands for the reverse smoothing at an oriented crossing. In the reverse smoothing two oriented arcs aim at one point and two other

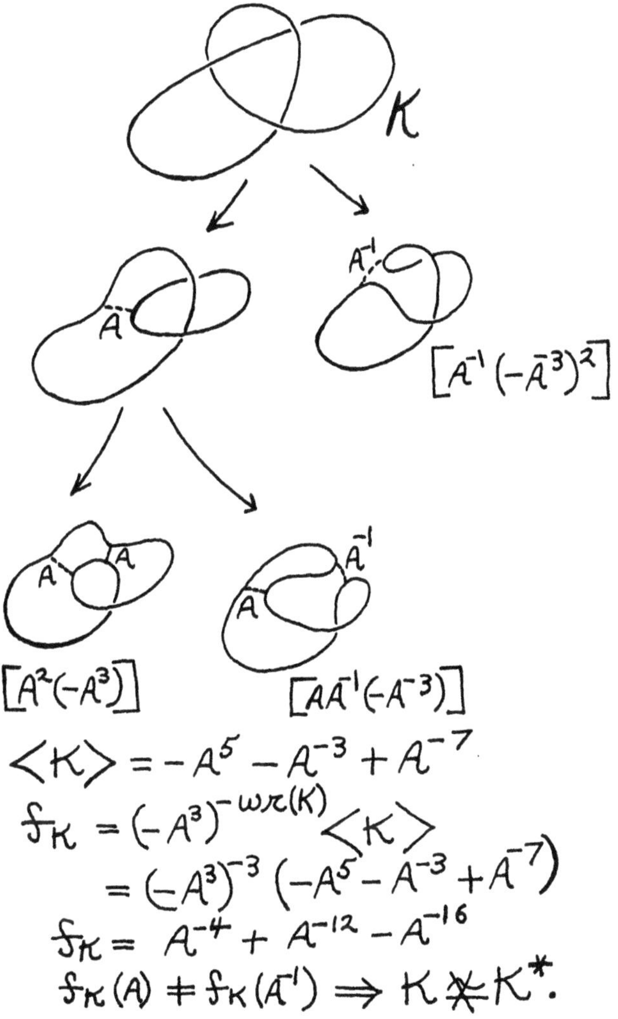

Figure 19. Bracket of the trefoil knot.

oriented arcs aim away from this point. The resulting state summation respects these local orientations but evaluates all loops in the final sum by the same loop value. To see the pattern inherent in these formulas, note that the basic structural formula for the Jones polynomial

$$t^{-1}V_{K_+} - tV_{K_-} = (t^{1/2} - t^{-1/2})V_{K_0}$$

follows at once from these two state expansion formulas by multiplying one by the inverse of t, the other by t and subtracting.

Using the notation at the end of the last section, we can express the basic equation for the bracket by the formula

$$\langle K\rangle = A\langle SK\rangle + A^{-1}\langle S'K\rangle$$

from which it follows that

$$\langle EK\rangle = A\langle S'K\rangle + A^{-1}\langle SK\rangle,$$

whence

$$\langle K\rangle - \langle EK\rangle = (A - A^{-1})(\langle SK\rangle - \langle S'K\rangle).$$

Thus $\langle K\rangle$ is a special case of the polynomial D_K and it then follows at once (through the normalization described above) that the original Jones polynomial is a special case of the Dubrovnik polynomial Y_K.

The bracket polynomial is a good example of a state summation model for a link polynomial. By using solutions to the Yang–Baxter equation and quantum groups [16], [17], [6], [18], [19] just about all of the new link invariants are captured in the form of generalized state summations. The bracket model is singular in its simplicity and provides our discussion with a concrete example of this kind of model. All models of this type can be written in the form of a summation

$$\langle K\rangle = \sum_{\sigma}\langle K|\sigma\rangle\lambda^{\|\sigma\|}$$

where $\langle K|\sigma\rangle$ denotes a product of vertex weights and σ runs over a finite collection of states or configurations of the system associated with the link diagram with $\|\sigma\|$ an appropriate auxiliary functional on the states. In this formulation the summation $\sum_{\sigma}\lambda^{\|\sigma\|}$ is analogous to the bare partition function and $\langle K\rangle$ is the analogue of the expectation value of an observable for the system where the knot becomes the means for creating this observable.

3.1. The Conway Polynomial

It turns out that the Conway–Alexander polynomial (see the end of the previous section) has an interesting state summation model all its own [44]. One way to describe this model is as a sum over all paths, of a certain type, on the link diagram. In this sense the Conway–Alexander polynomial is a discrete path integral. This fits right into our theme of knots and functional integration. This model for the Conway polynomial is illustrated in Figure 20.

On viewing Figure 20 the reader will see that in this model the states of a connected link diagram consist in local choices at the crossings. Each crossing has four local regions incident to it. A state consists in one such choice, designated by one marker, per crossing. With no further restrictions there would be 4^n states where n is the number of crossings in the diagram. However, we only count those states where there is no more than one marker in any full region of the diagram, and the two regions that are unmarked are adjacent to one another. The unmarked regions are designated by stars in the diagram. We then

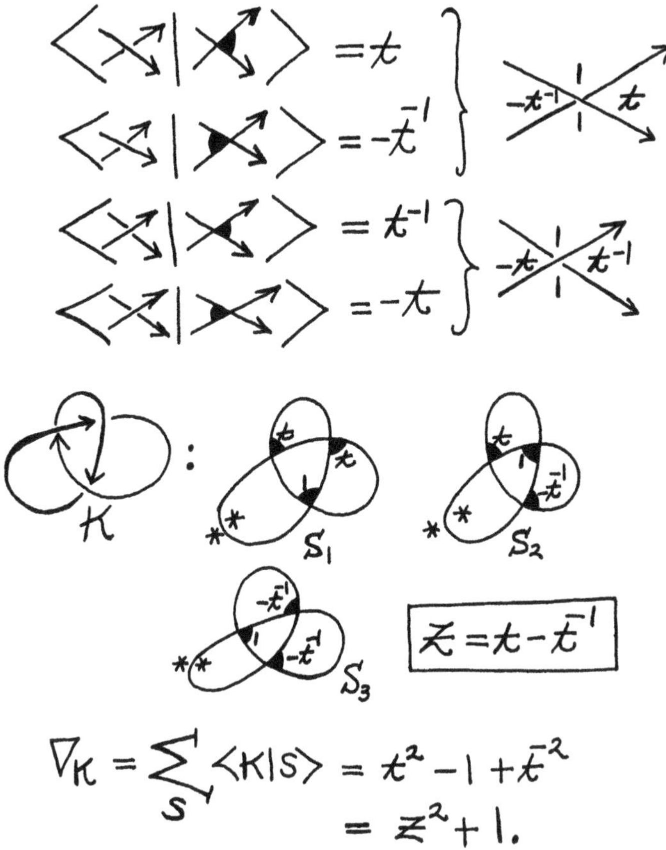

Figure 20. FKT model for Conway polynomial.

define vertex weights in relation to a state by comparing the location of the marker with the corresponding crossing. In the case of a crossing of positive type, a forward marker gets t while a backward marker gets $-t^{-1}$, and up or down markers get 1. The designations forward, back, up and down refer to the marker being located between outgoing, ingoing or mixed direction lines of the diagram. With this notion of vertex weight, the partition function is defined as $\nabla_K = \Sigma_S\langle K|S\rangle$ where $\langle K|S\rangle$ denotes the product of the vertex weights for the state S.

Another way to view this model of the Conway polynomial is to think of the knot as a string that extends from one wall to another (a *tangle*). Then a diagram of the knot has two ends, and each state in the state sum described above can be regarded (see Figure 21) as a self-avoiding walk from one end of the diagram to the other. The model is constructed so that ∇_K is invariant under all three Reidemeister moves and hence is an invariant of ambient isotopy. Thus ∇_K can be regarded as a discrete path integral with properties of topological invariance.

There is another formula for the Conway–Alexander polynomial in terms of the Seifert Pairing, a matrix of linking numbers of curves on a surface whose boundary is the given knot or link. The interested reader can find an account of this formula in [20] or [21]. In Figure 22 we have illustrated this formula

$$\nabla_K = \mathrm{Det}(t\theta - t^{-1}\theta^T).$$

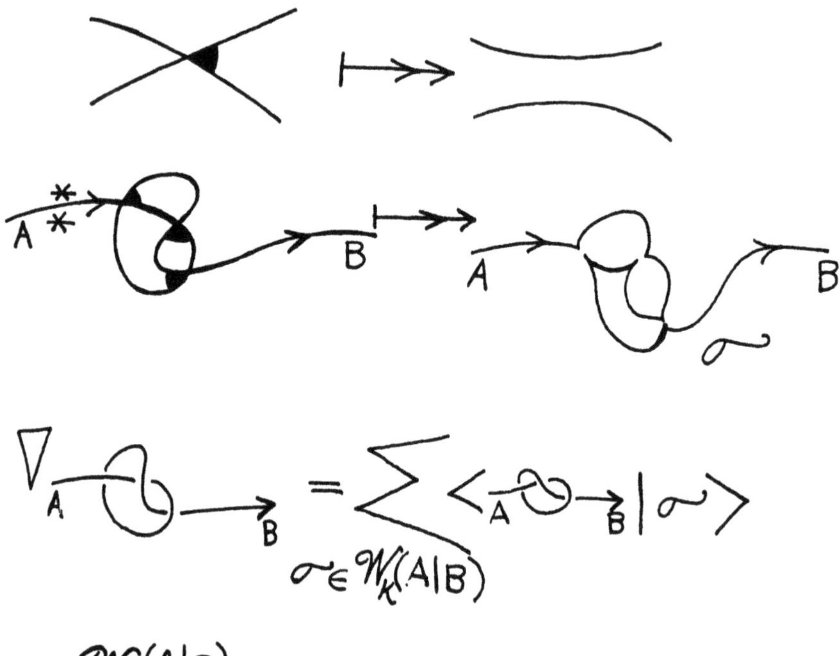

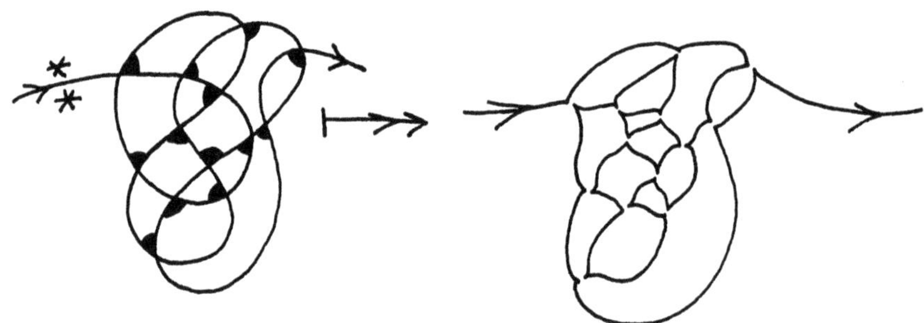

Figure 21. States as self-avoiding walks.

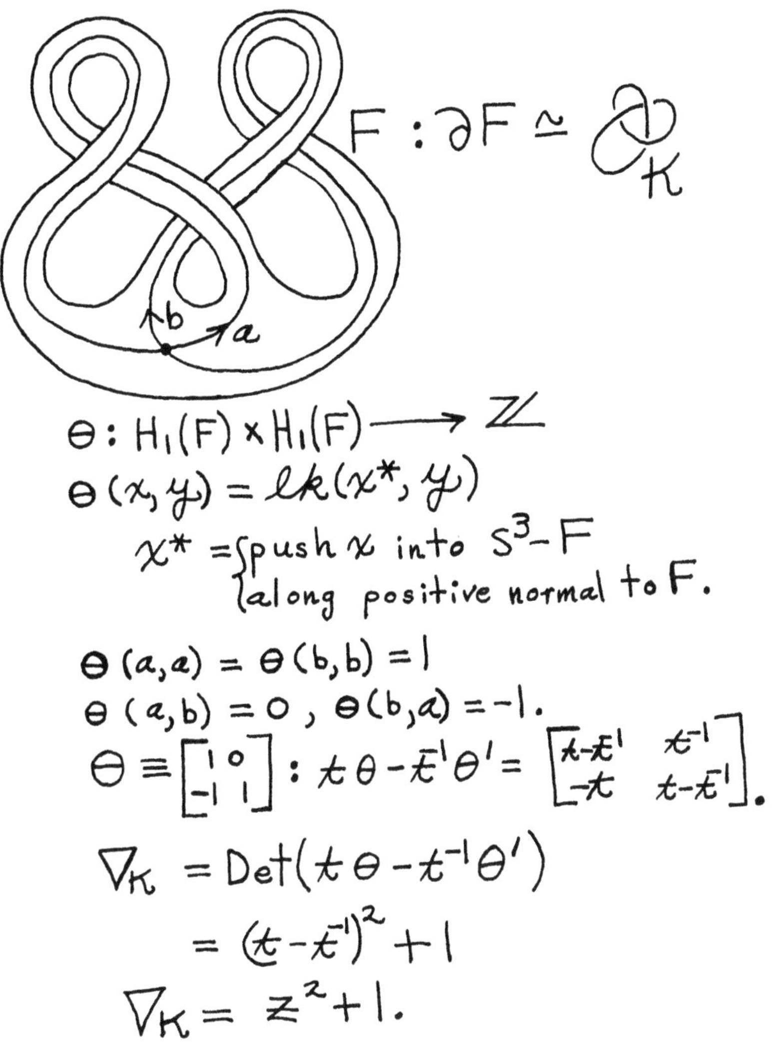

Figure 22. Seifert pairing and Conway polynomial.

The symbol θ denotes the Seifert pairing, defined on the first homology group of a surface F whose boundary is the knot or link K.

$$\theta(x,y) = lk(x^*,y)$$

where lk denotes linking number and a superscript $*$ on a homology class denotes its translation into the complement of the surface along the positive normal vector.

The formula for the Conway polynomial has many good consequences. For example, it can be used to study knots that bound smooth disks in the four dimensional ball. To date there is no known counterpart of the Seifert Pairing formula for the Jones polynomial and its generalizations.

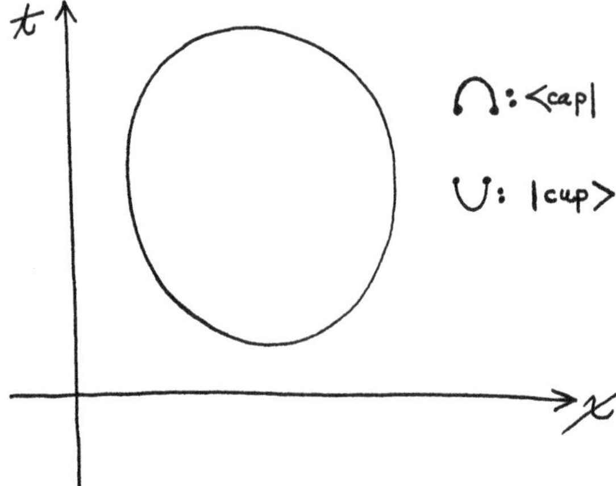

Figure 23. Space–time circle.

4. QUANTUM LINK INVARIANTS

In this section we describe the construction of quantum link invariants from knot and link diagrams that are arranged with respect to a given direction in the plane. This special direction will be called "time." Arrangement with respect to the special direction means that perpendiculars to this direction meet the diagram transversely (at edges or at crossings) or tangentially (at maxima and minima). The designation of the special direction as time allows the interpretation of the consequent evaluation of the diagram as a generalized scattering amplitude.

4.1. Knot Amplitudes

Consider first a circle in a space–time plane with time represented vertically and space horizontally as in Figure 23.

The circle represents a vacuum to vacuum process that includes the creation of two "particles" and their subsequent annihilation. We could divide the circle into these two parts (creation "cup" and annihilation "cap") and consider the amplitude $\langle \mathrm{cap}|\mathrm{cup}\rangle$. Since the diagram for the creation of the two particles ends in two separate points, it is natural to take a vector space of the form $V \otimes V$ as the target for the bra and as the domain of the ket. We imagine at least one particle property being catalogued by each factor of the tensor product. For example, a basis of V could enumerate the spins of the created particles.

Any non-self-intersecting differentiable curve can be rigidly rotated until it is in general position with respect to the vertical. It will then be seen to be decomposed into an interconnection of minima and maxima. We can evaluate an amplitude for any curve in general position with respect to a vertical direction. Any simple closed curve in the plane is isotopic to a circle, by the Jordan Curve Theorem. If these are topological amplitudes, then the value for any simple closed curve should be equal to the original amplitude for the circle. What condition on creation (cup) and annihilation (cap) will insure topological amplitudes? The answer derives from the fact that isotopies of the simple closed curves are generated by the cancellation of adjacent maxima and minima as illustrated in Figure 24.

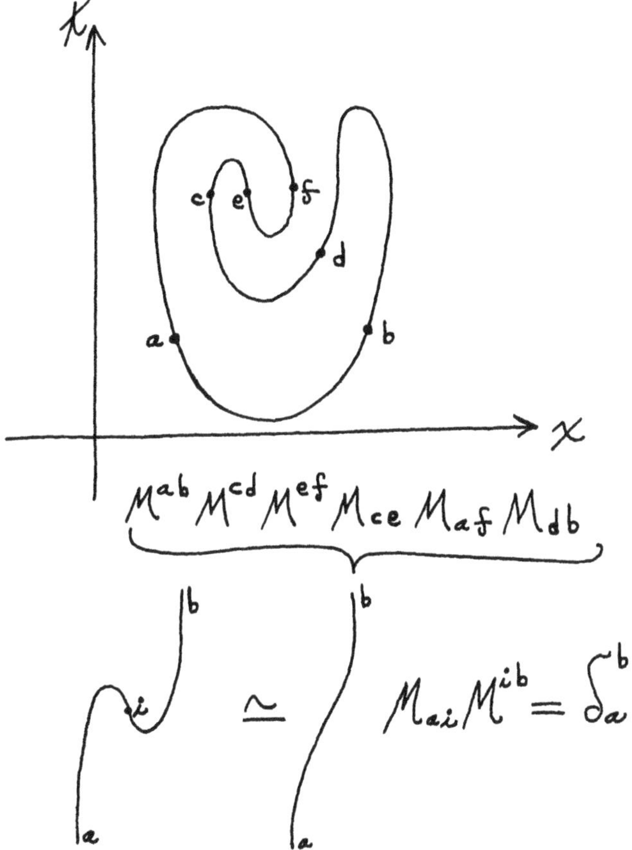

Figure 24. Space–time Jordan curve.

This condition is articulated by taking a matrix representation for the corresponding operators. Specifically, let $\{e_1, e_2, \ldots, e_n\}$ be a basis for V. Let $e_{ab} = e_a \otimes e_b$ denote the elements of the tensor basis for $V \otimes V$. Then there are matrices M_{ab} and M^{ab} such that

$$|\text{cup}\rangle(1) = \sum M^{ab} e_{ab}$$

with the summation taken over all values of a and b from 1 to n. Similarly, $\langle\text{cap}|$ is described by

$$\langle\text{cap}|(e_{ab}) = M_{ab}.$$

Thus the amplitude for the circle is

$$\begin{aligned}\langle\text{cap}|\text{cup}\rangle(1) &= \langle\text{cap}| \sum M^{ab} e_{ab} \\ &= \sum M^{ab} \langle\text{cap}|(e_{ab}) = \sum M^{ab} M_{ab}.\end{aligned}$$

In general, the value of the amplitude on a simple closed curve is obtained by translating it into an "abstract tensor expression" in the M^{ab} and M_{ab}, and then summing over these products for all cases of repeated indices.

Returning to the topological conditions we see that they are just that the matrices M^{ab} and M_{cd} are inverses in the sense that

$$\sum_i M^{ai} M_{ib} = \delta^a_b$$

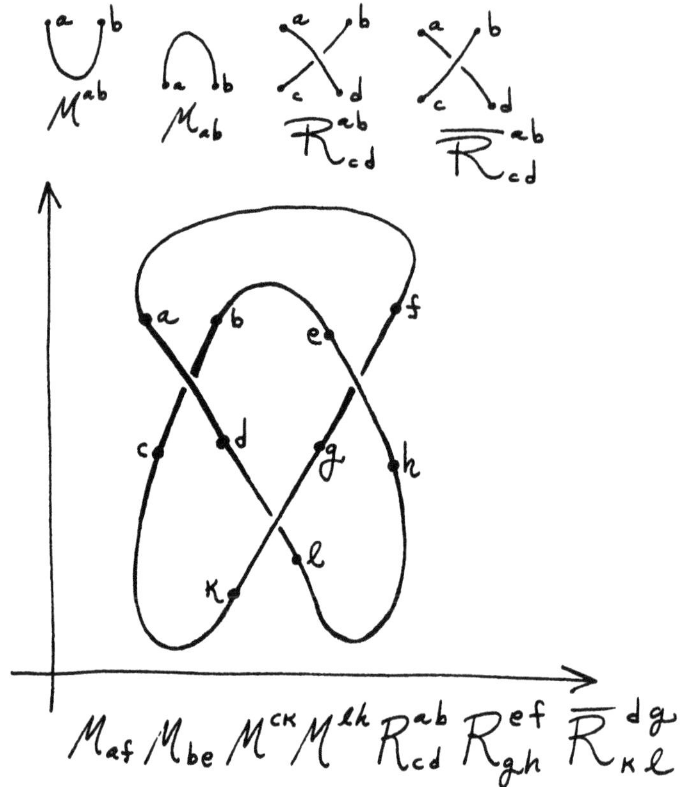

Figure 25. Cups, caps and crossings.

where δ^a_b denotes the identity matrix. See Figure 24.

One of the simplest choices is to take a 2×2 matrix M such that $M^2 = I$ where I is the identity matrix. Then the entries of M can be used for both the cup and the cap. The value for a loop is then equal to the sum of the squares of the entries of M:

$$\langle \text{cap}|\text{cup}\rangle(1) = \sum M^{ab}M_{ab} = \sum M_{ab}M_{ab} = \sum M^2_{ab}.$$

In particular, consider the following choice for M. It has square equal to the identity matrix and yields a loop value of $d = -A^2 - A^{-2}$, just the right loop value for the bracket polynomial model for the Jones polynomial [15], [14].

$$M = \begin{bmatrix} 0 & iA \\ -iA^{-1} & 0 \end{bmatrix}$$

Any knot or link can be represented by a picture that is configured with respect to a vertical direction in the plane. The picture will decompose into minima (creations) maxima (annihilations) and crossings of the two types shown in Figure 25. Here the knots and links are unoriented. These models generalize easily to include orientation.

Next to each of the crossings we have indicated mappings of $V \otimes V$ to itself, called R and R^{-1} respectively. These mappings represent the transitions corresponding to elementary braiding. We now have the vocabulary of cup, cap, R and R^{-1}. Any knot or link can be written as a composition of these fragments, and consequently a choice of such mappings determines an amplitude for knots and links. In order for such an amplitude to be topological (i.e. an

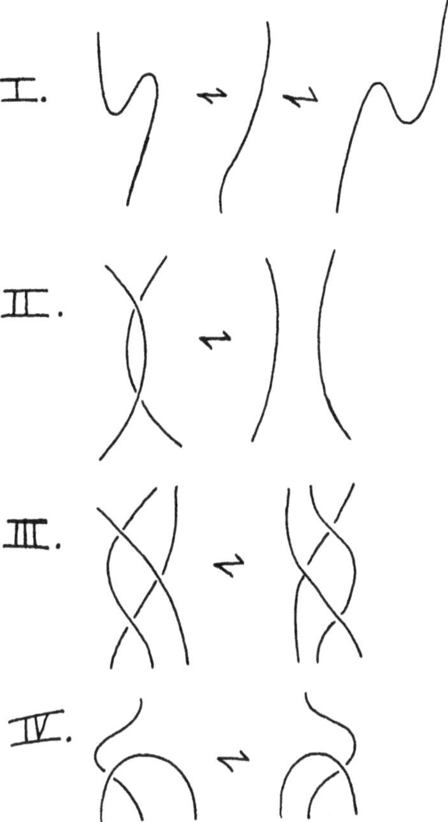

Figure 26. Regular isotopy with respect to a vertical direction.

invariant of regular isotopy the equivalence relation generated by the second and third of the classical Reidemeister moves) we want it to be invariant under a list of local moves on the diagrams as shown in Figure 26. These moves are an augmented list of Reidemeister moves, adjusted to take care of the fact that the diagrams are arranged with respect to a given direction in the plane. In this context Move *III* is the Yang–Baxter equation that occurred for the first time in problems of exactly solved models in statistical mechanics [11].

All the moves taken together are directly related to the axioms for a quasi-triangular Hopf algebra (aka quantum group). We shall not take up this theme here. Many seeds of the structure of Hopf algebras are prefigured in the patterns of link diagrams and the structure of the category of tangles. The interested reader can consult [29], [24], [18] and [6], [22], [23] for more information on this point.

Here is the list of the algebraic versions of the topological moves. Move 0 is the cancellation of maxima and minima. Move *II* corresponds to the second Reidemeister move. Move *III* is the Yang–Baxter equation. Move *IV* expresses the relationship of switching a line across a maximum. (There is a corresponding version of *IV* where the line is switched across a minimum.)

$$I. \qquad M^{ai}M_{ib} = \delta^a_b$$

$$II. \qquad R^{ab}_{ij}\overline{R^{ij}_{cd}} = \delta^a_c\delta^b_d$$

$$III. \qquad R^{ab}_{ij}R^{jc}_{kf}R^{ik}_{de} = R^{bc}_{ij}R^{ai}_{dk}R^{kj}_{ef}$$

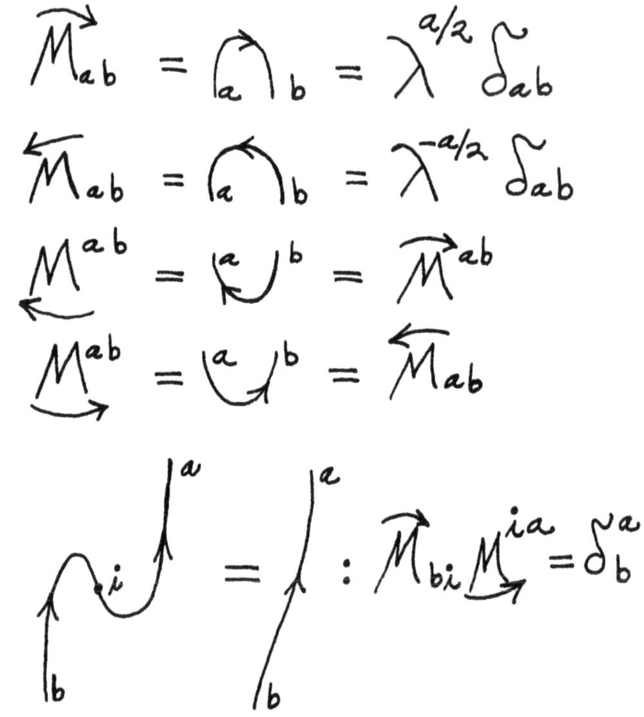

$$\overrightarrow{M}_{ab} = \lambda^{a/2}\tilde{\delta}_{ab}$$

$$\overleftarrow{M}_{ab} = \lambda^{-a/2}\tilde{\delta}_{ab}$$

$$\underleftarrow{M}^{ab} = \overrightarrow{M}^{ab}$$

$$\underrightarrow{M}^{ab} = \overleftarrow{M}_{ab}$$

$$\overrightarrow{M}_{bi}\underrightarrow{M}^{ia} = \tilde{\delta}^{a}_{b}$$

Figure 27. Right and left cups and caps.

$$IV. \qquad R^{ai}_{bc}M_{id} = M_{bi}R^{ia}_{cd}$$

In the case of the Jones polynomial we have all the algebra present to make the model. It is easiest to indicate the model for the bracket polynomial: Let cup and cap be given by the 2×2 matrix M, described above so that $M_{ij} = M^{ij}$. Let R and R^{-1} be given by the equations

$$R^{ab}_{cd} = AM^{ab}M_{cd} + A^{-1}\delta^a_c\delta^b_d,$$
$$R^{-1}{}^{ab}_{cd} = A^{-1}M^{ab}M_{cd} + A\delta^a_c\delta^b_d.$$

This definition of the R-matrices exactly parallels the diagrammatic expansion of the bracket, and it is not hard to see, either by algebra or diagrams, that all the conditions of the model are met.

4.2. Oriented Amplitudes

Slight but significant modifications are needed to write the oriented version of the models we have discussed in the previous section. See [6], [17], [19], [25]. In this section we sketch the construction of oriented topological amplitudes.

The generalization to oriented link diagrams naturally involves the introduction of right and left oriented caps and cups. These are drawn as shown in Figure 27 below.

A right cup cancels with a right cap to produce an upward pointing identity line. A left cup cancels with a left cap to produce a downward pointing identity line.

Just as we considered the simplifications that occur in the unoriented model by taking the cup and cap matrices to be identical, lets assume here that right caps are identical with left

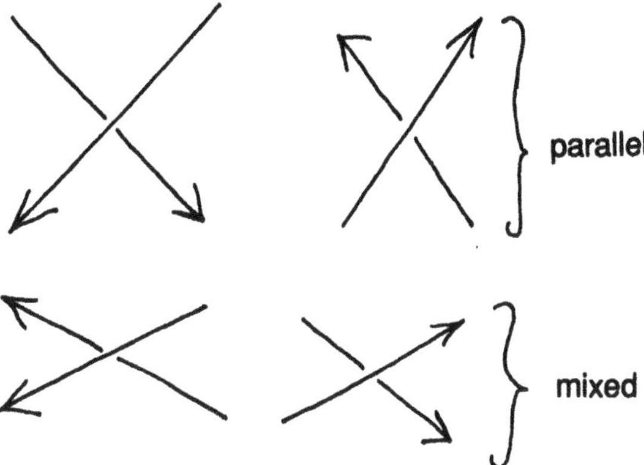

Figure 28. Oriented crossings.

cups and that consequently left caps are identical with right cups. In fact, let us assume that the right cap and left cup are given by the matrix

$$M_{ab} = \lambda^{a/2}\delta_{ab}$$

where λ is a constant to be determined by the situation, and δ_{ab} denotes the Kronecker delta. Then the left cap and right cup are given by the inverse of M:

$$M^{-1}_{ab} = \lambda^{-a/2}\delta_{ab}.$$

We assume that along with M we are given a solution R to the Yang–Baxter equation, and that in an oriented diagram the specific choice of R^{ab}_{cd} is governed by the local orientation of the crossing in the diagram. Thus a and b are the labels on the lines going into the crossing and c and d are the labels on the lines emanating from the crossing.

Note that with respect to the vertical direction for the amplitude, the crossings can assume the aspects: both lines pointing upward, both lines pointing downward, one line up and one line down (two cases). See Figure 28.

Call the cases of one line up and one line down the *mixed* cases and the upward and downward cases the *parallel* cases. A given mixed crossing can be converted, in two ways, into a combination of a parallel crossing of the same sign plus a cup and a cap. See Figure 29.

This leads to an equation that must be satisfied by the R matrix in relation to powers of λ (again we use the Einstein summation convention):

$$\lambda^{a/2}\delta^{ai}R^{ci}_{jb}\lambda^{-d/2}\delta_{jd} = \lambda^{-c/2}\delta_{ic}R^{ia}_{dj}\lambda^{b/2}\delta^{jb}.$$

This simplifies to the equation

$$\lambda^{a/2}R^{ca}_{db}\lambda^{-d/2} = \lambda^{-c/2}R^{ca}_{db}\lambda^{b/2},$$

from which we see that R^{ca}_{db} is necessarily equal to zero unless $b+d=a+c$. We say that the R matrix is *spin preserving* when it satisfies this condition. Assuming that the R matrix is spin preserving, the model will be invariant under all orientations of the second and third Reidemeister moves just so long as it is invariant under the anti-parallel version of the second Reidemeister move as shown in Figure 30.

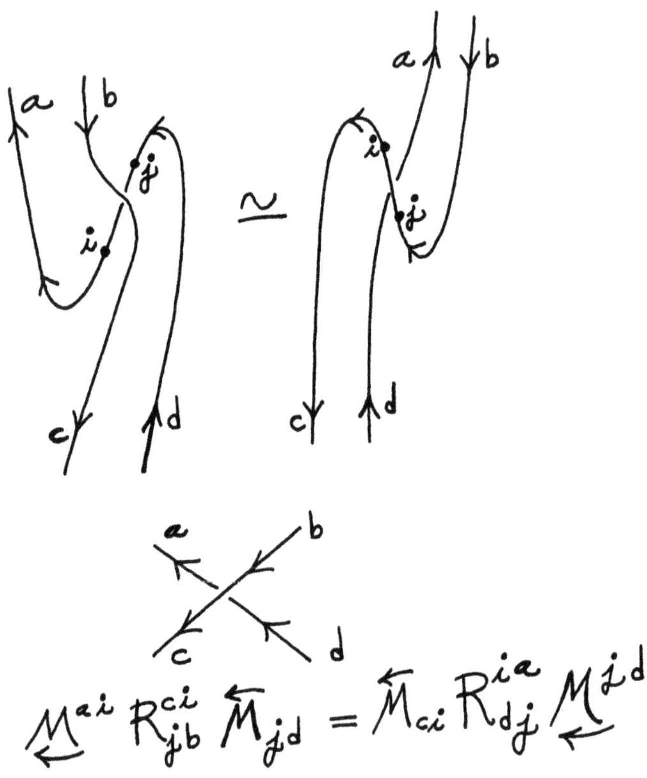

Figure 29. Conversion.

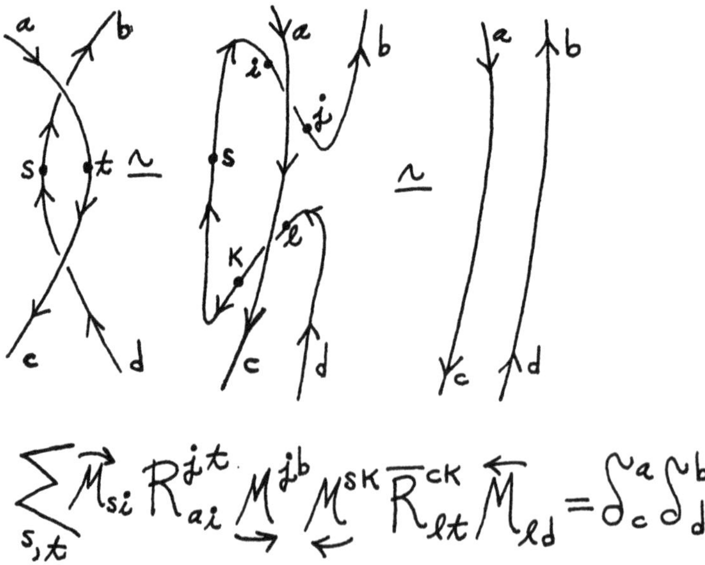

Figure 30. Antiparallel second move.

This antiparallel version of the second Reidemeister move places the following demand on the relation between λ and R:

$$\sum_{st} \lambda^{(s-b)/2} \lambda^{(t-c)/2} R^{bt}_{as} \overline{R^{cs}_{dt}} = \delta^a_c \delta^b_d.$$

Call this the *$R-\lambda$ equation.* The reader familiar with [10] or with the piecewise linear version as described in [6] will recognize this equation as the requirement for regular homotopy invariance in these models.

5. THE GAUSS LINKING INTEGRAL

The first integral to appear in knot theory is due to Carl Friedrich Gauss. Gauss devised an integral that measures the linking number of two space curves and gives a "self-linking number" or writhe to a single space curve. The writhe is not a topological invariant of the single curve, but it is a useful tool for studying the differential geometry of space curves. Suppose A and B are two curves in Euclidean three space R^3. Define the (Gauss) map

$$\phi : A \times B \longrightarrow S^2$$

by the equation

$$\phi(a,b) = (a-b)/\|a-b\|.$$

Here S^2 denotes the unit two-sphere of vectors in three-space with length one, and $\|v\|$ denotes the length of a vector v.

Gauss showed that

$$lk(A,B) = (1/4\pi)\int_{A\times B} \phi^*(\omega)$$

where ω denotes the area two-form on S^2. It is not hard to see that this integral has the specific form

$$lk(A,B) = (1/4\pi)\int_{A\times B} e_{AB} \bullet (T_A \times T_B)/(\|e_{AB}\|^3) ds_A ds_B.$$

Here T_A denotes a unit tangent vector to the curve A, $\bullet$ denotes the dot product of vectors in three-space and ds_A denotes differential arclength along the curve A. The expression

$$G = e_{AB} \bullet (T_A \times T_B)/(\|e_{AB}\|^3)$$

is called the "Gauss Kernel."

This expression for the linking number due to Gauss is beautiful and tantalizing. (Exercise: Show by thinking about links that lie nearly flat in the plane that Gauss' integral is proportional to the linking number of the two curves. Show also that Gauss' integral can be defined for a single curve and that the result for nearly flat knots is proportional to the writhe of the diagram.) Surely other invariants of knots and links can be expressed in terms of integrals!

The answer to the speculation in the last paragraph is: Yes indeed! For the rest of the paper we will look at how integrals came back into knot theory via the route of functional integration. In the end, Gauss's integral stands generalized and a vast array of invariants are built from direct generalizations of the Gauss integral. These generalizations correspond to Feynman diagrams based on the knot. We will get just to the beginning of *that* story by the end of these notes.

6. LEVEL ZERO INTEGRAL HEURISTICS

This section gives a self-contained and highly simplified heuristic derivation of the skein relation for the Homfly polynomial in an integral formalism. It is useful to have a formal pattern to compare with the full approach using gauge theory.

Our approach is what it is — a heuristic. However it is closely related in structure to the the Witten functional integral. We discovered this formalism in the course of conversations with Lee Smolin. It was his idea [26] to compare the variation of the Chern–Simons Lagrangian with the change in holonomy induced by switching a crossing. What is remarkable about this heuristic is that it does not assume any details about the Lagrangian, nor does it involve the concept of holonomy. These are the structures that will be added in the next section.

Recall that the Homfly polynomial may be expressed in regular isotopy form via a skein identity and a specification of behavior under the twist of a type I Reidemeister move.

1. Skein Identity.

$$H_{K_+} - H_{K_-} = zH_{K_0}.$$

2. Twist Behavior.

$$H_{R_+K} = aH_K,$$
$$H_{R_-K} = a^{-1}H_K.$$

The functional H is normalized to be equal to one on an unknotted and twist-free circle in the plane.

Now suppose that W is a functional on knots and links that is represented in the form of an integral:

$$W_K = \int dAe^L T_K(A)$$

where we assume that T is not an invariant of the loop K, but that it satisfies a *local skein identity* of the form

$$T_{K_+} - T_{K_-} = -zF\tilde{T}_{K_0}.$$

Here F is a function that depends upon the point where the strands are switched. You can think of F as an analog of the curvature of the gauge field at this point if you have already read the next section. (The point being that the difference between over and undercrossing is a little loop around the switch point, and one can send a test particle around this point to measure the local curvature.)

On the right-hand side of this identity we have written T with a tilde over it to indicate a modification that takes care of the dependence on the location of the switch point. We shall *assume* that $\tilde{T}$ satisfies the differential equation

$$d\tilde{T}_K/dA = T_K.$$

This equation says that the local change in T due to moving the loop is compensated by a corresponding change in the "field" A.

We further assume that the "curvature" F and the "Lagrangian" L are related by the equation

$$dL/dA = F.$$

Finally, we assume that in an integration by parts involving the integral

$$\int dAe^L T_K(A)$$

the boundary terms vanish.

Then the skein relation for W is a formal consequence:

Proposition.

$$W_{K_+} - W_{K_-} = zW_{K_0}.$$

Proof.

$$\begin{aligned}
W_{K_+} - W_{K_-} &= \int dAe^L T_{K_+} - \int dAe^L T_{K_-} \\
&= \int dAe^L [T_{K_+} - T_{K_-}] \\
&= \int dAe^L [-zF\tilde{T}_{K_0}] \\
&= -z \int dA[e^L F]\tilde{T}_{K_0} \\
&= -z \int dA[de^L/dA]\tilde{T}_{K_0} \\
&= z \int dAe^L [d\tilde{T}_{K_0}/dA] \\
&= z \int dAe^L T_{K_0} \\
&= zW_{K_0}.
\end{aligned}$$

This completes the proof.

This heuristic simply shows how it is possible for an integration process to convert locally changing switching relations to one global relation. It suggests further investigation of this theme, and indeed the Witten functional integral discussed in the next section embodies these themes in a more sophisticated and effective way.

7. THE WITTEN FUNCTIONAL INTEGRAL

In [27] Edward Witten proposed a formulation of a class of 3-manifold invariants as generalized Feynman integrals taking the form $Z(M)$ where

$$Z(M) = \int dA \exp[(ik/4\pi)S(M,A)].$$

Here M denotes a 3-manifold without boundary and A is a gauge field (also called a gauge potential or gauge connection) defined on M. The gauge field is a one-form on a trivial G-bundle over M with values in a representation of the Lie algebra of G. The group G corresponding to this Lie algebra is said to be the gauge group. In this integral the "action" $S(M,A)$ is taken to be the integral over M of the trace of the Chern–Simons three-form $CS = AdA + (2/3)AAA$. (The product is the wedge product of differential forms.)

The integral $Z(M)$ integrates over all gauge fields modulo gauge equivalence (See [28] for a discussion of the definition and meaning of gauge equivalence.)

The Witten integral $Z(M)$ is, in its form, a typical integral in quantum field theory. In its content $Z(M)$ is highly unusual. The formalism and internal logic of Witten's integral supports the existence of a large class of topological invariants of 3-manifolds and associated invariants of knots and links in these manifolds.

The invariants associated with this integral have been given rigorous combinatorial descriptions [29], [30], [31], [32], [33], [34], but questions and conjectures arising from the integral formulation are still outstanding. (See for example [35], [36], [37], [38], [39].) Specific conjectures about this integral take the form of just how it implicates invariants of links and 3-manifolds, and how these invariants behave in certain limits of the coupling constant k in the integral. Many conjectures of this sort can be verified through the combinatorial models. On the other hand, the really outstanding conjecture about the integral is that it exists! At the present time there is no measure theory or generalization of measure theory that supports it. It is a fascinating exercise to take the speculation seriously, suppose that it does really work like an integral and explore the formal consequences. This is what we are going to do in the remainder of this section. Here is a formal structure of great beauty. It is also a structure whose consequences can be verified by a remarkable variety of alternative means. Perhaps in the course of the exploration there will appear a hint of the true nature of this form of integration.

We now look at the formalism of the Witten integral in more detail and see how it implicates invariants of knots and links corresponding to each classical Lie algebra. In order to accomplish this task, we need to introduce the Wilson loop. The Wilson loop is an exponentiated version of integrating the gauge field along a loop K in three space that we take to be an embedding (knot) or a curve with transversal self-intersections. For this discussion, the Wilson loop will be denoted by the notation $W_K(A) = \langle K|A\rangle$ to denote the dependence on the loop K and the field A. It is usually indicated by the symbolism $\mathrm{tr}\,(P\exp\,(\int_K A))$. Thus

$$W_K(A) = \langle K|A\rangle = \mathrm{tr}\left(P\exp\left(\int_K A\right)\right).$$

Here the P denotes path ordered integration — we are integrating and exponentiating matrix valued functions, and so must keep track of the order of the operations. The symbol tr denotes the trace of the resulting matrix.

With the help of the Wilson loop functional on knots and links, Witten writes down a functional integral for link invariants in a 3-manifold M:

$$\begin{aligned} Z(M,K) &= \int dA\exp[(ik/4\pi)S(M,A)]\,\mathrm{tr}\left(P\exp\left(\int_K A\right)\right) \\ &= \int dA\exp[(ik/4\pi)S]\langle K|A\rangle. \end{aligned}$$

Here $S(M,A)$ is the Chern–Simons Lagrangian, as in the previous discussion.

We abbreviate $S(M,A)$ as S and write $\langle K|A\rangle$ for the Wilson loop. Unless otherwise mentioned, the manifold M will be the three-dimensional sphere S^3

An analysis of the formalism of this functional integral reveals quite a bit about its role in knot theory. This analysis depends upon key facts relating the curvature of the gauge field to both the Wilson loop and the Chern–Simons Lagrangian. The idea for using the curvature in this way is due to Lee Smolin [26] (See also [40]). To this end, let us recall the local coordinate structure of the gauge field $A(x)$, where x is a point in three-space. We can write $A(x) = A_a^k(x)T^a dx_k$ where the index a ranges from 1 to m with the Lie algebra basis $\{T^1, T^2, T^3, \ldots, T^m\}$. The index k goes from 1 to 3. For each choice of a and k, $A_a^k(x)$ is a smooth function defined on three-space. In $A(x)$ we sum over the values of repeated indices. The Lie algebra generators T^a are matrices corresponding to a given representation of the Lie algebra of the gauge group G. We assume some properties of these matrices as follows:

1. $[T^a, T^b] = if_{abc}T^c$ where $[x,y] = xy - yx$, and f_{abc} (the matrix of structure constants) is totally antisymmetric. There is summation over repeated indices.
2. $\mathrm{tr}(T^a T^b) = \delta^{ab}/2$ where δ^{ab} is the Kronecker delta ($\delta^{ab} = 1$ if $a = b$ and zero otherwise).

We also assume some facts about curvature. (The reader may enjoy comparing with the exposition in [6]. But note the difference of conventions on the use of i in the Wilson loops and curvature definitions.) The first fact is the relation of Wilson loops and curvature for small loops:

Fact 1. The result of evaluating a Wilson loop about a very small planar circle around a point x is proportional to the area enclosed by this circle times the corresponding value of the curvature tensor of the gauge field evaluated at x. The curvature tensor is written

$$F_a^{rs}(x)T^a dx_r dy_s.$$

It is the local coordinate expression of $AdA + AA$.

Application of Fact 1. Consider a given Wilson line $\langle K|S\rangle$. Ask how its value will change if it is deformed infinitesimally in the neighborhood of a point x on the line. Approximate the change according to Fact 1, and regard the point x as the place of curvature evaluation. Let $\delta\langle K|A\rangle$ denote the change in the value of the line. $\delta\langle K|A\rangle$ is given by the formula

$$\delta\langle K|A\rangle = dx_r dx_s F_a^{rs}(x)T^a\langle K|A\rangle.$$

This is the first order approximation to the change in the Wilson line.

In this formula it is understood that the Lie algebra matrices T^a are to be inserted into the Wilson line at the point x, and that we are summing over repeated indices. This means that each $T^a\langle K|A\rangle$ is a new Wilson line obtained from the original line $\langle K|A\rangle$ by leaving the form of the loop unchanged, but inserting the matrix T^a into that loop at the point x. A Lie algebra generators is diagrammed by a little box with a single index line and two input/output lines which correspond to its role as a matrix (hence as mappings of a vector space to itself). See Figure 31.

Remark. In thinking about the Wilson line $\langle K|A\rangle = \mathrm{tr}(P\exp(\int_K A))$, it is helpful to recall Euler's formula for the exponential:

$$e^x = \lim_{n\to\infty}(1 + x/n)^n.$$

The Wilson line is the limit, over partitions of the loop K, of products of the matrices $(1 + A(x))$ where x runs over the partition. Thus we can write symbolically,

$$\langle K|A\rangle = \prod_{x\in K}(1 + A(x)) = \prod_{x\in K}(1 + A_a^k(x)T^a dx_k).$$

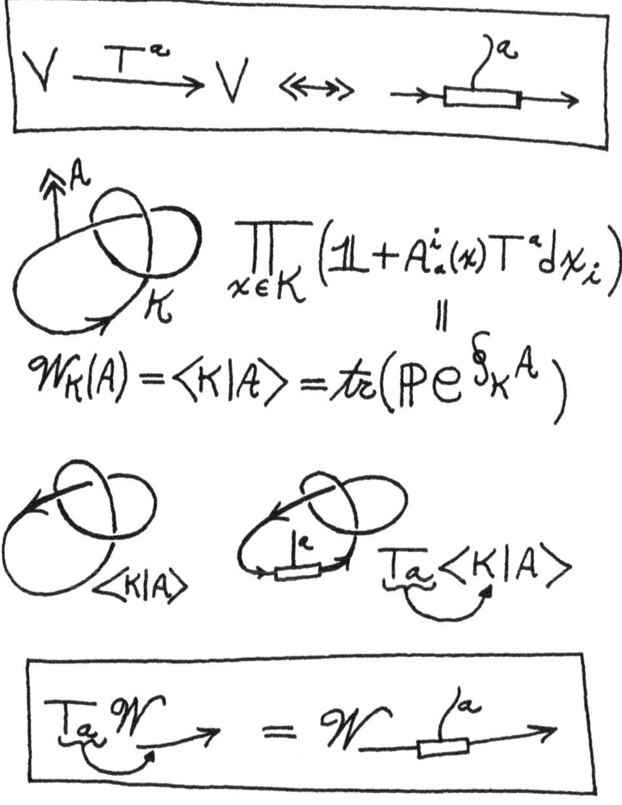

Figure 31. Wilson loop insertion.

It is understood that a product of matrices around a closed loop connotes the trace of the product. The ordering is forced by the one dimensional nature of the loop. Insertion of a given matrix into this product at a point on the loop is then a well-defined concept. If T is a given matrix then it is understood that $T\langle K|A\rangle$ denotes the insertion of T into some point of the loop. In the case above, it is understood from context in the formula

$$dx_r dx_s F_a^{rs}(x) T^a \langle K|A\rangle$$

that the insertion is to be performed at the point x indicated in the argument of the curvature.

Remark. The previous remark implies the following formula for the variation of the Wilson loop with respect to the gauge field:

$$\delta\langle K|A\rangle/\delta(A_a^k(x)) = dx_k T^a \langle K|A\rangle.$$

Varying the Wilson loop with respect to the gauge field results in the insertion of an infinitesimal Lie algebra element into the loop.

Proof.

$$\delta\langle K|A\rangle/\delta(A_a^k(x)) = \delta \prod_{y\in K}(1 + A_a^k(y)T^a dy_k)/\delta(A_a^k(x))$$

$$= \prod_{y<x\in K} (1+A_a^k(y)T^a dy_k)[T^a dx_k] \prod_{y>x\in K} (1+A_a^k(y)T^a dy_k)$$
$$= dx_k T^a \langle K|A\rangle .$$

Fact 2. The variation of the Chern–Simons Lagrangian S with respect to the gauge potential at a given point in three-space is related to the values of the curvature tensor at that point by the following formula:

$$F_a^{rs}(x) = \varepsilon_{rst}\delta S/\delta(A_a^t(x)).$$

Here ε_{abc} is the epsilon symbol for three indices, i.e. it is $+1$ for positive permutations of 123 and -1 for negative permutations of 123 and zero if any two indices are repeated.

With these facts at hand we are prepared to determine how the Witten integral behaves under a small deformation of the loop K.

Proposition 1. (Compare [6]) All statements of equality in this proposition are up to order $(1/k)^2$.

1. Let $Z(K) = Z(S^3, K)$ and let $\delta Z(K)$ denote the change of $Z(K)$ under an infinitesimal change in the loop K. Then

$$\delta Z(K) = (4\pi i/k)\int dA \exp[(ik/4\pi)S][\mathrm{Vol}]T^aT^a\langle K|A\rangle$$

where $\mathrm{Vol} = \varepsilon_{rst}dx_r dx_s dx_t$.

The sum is taken over repeated indices, and the insertion is taken of the matrix products T^aT^a at the chosen point x on the loop K that is regarded as the "center' of the deformation. The volume element $\mathrm{Vol} = \varepsilon_{rst}dx_r dx_s dx_t$ is taken with regard to the infinitesimal directions of the loop deformation from this point on the original loop.

2. The same formula applies, with a different interpretation, to the case where x is a double point of transversal self intersection of a loop K, and the deformation consists in shifting one of the crossing segments perpendicularly to the plane of intersection so that the self-intersection point disappears. In this case, one T^a is inserted into each of the transversal crossing segments so that $T^aT^a\langle K|A\rangle$ denotes a Wilson loop with a self intersection at x and insertions of T^a at $x+\varepsilon_1$ and $x+\varepsilon_2$ where ε_1 and ε_2 denote small displacements along the two arcs of K that intersect at x. In this case, the volume form is nonzero, with two directions coming from the plane of movement of one arc, and the perpendicular direction is the direction of the other arc.

Proof.

$$\begin{aligned}
\delta Z(K) &= \int dA \exp[(ik/4\pi)S][\delta\langle K|A\rangle] \\
&= \int dA \exp[(ik/4\pi)S][dx_r dy_s F_a^{rs}(x)T^a\langle K|A\rangle] \\
&= \int dA \exp[(ik/4\pi)S]dx_r dy_s[\varepsilon_{rst}\delta S/\delta(A_a^t(x))]T^a\langle K|A\rangle \\
&= \int dA[\exp[(ik/4\pi)S]\delta S/\delta(A_a^t(x))]\varepsilon_{rst}dx_r dy_s T^a\langle K|A\rangle
\end{aligned}$$

$$= (-4\pi i/k) \int dA[\delta \exp[(ik/4\pi)S]/\delta(A^t_a(x))]\varepsilon_{rst} dx_r dy_s T^a \langle K|A\rangle$$

$$= (4\pi i/k) \int dA \exp[(ik/4\pi)S]\varepsilon_{rst} dx_r dy_s [\delta[T^a \langle K|A\rangle]/\delta(A^t_a(x))]$$

(integration by parts assuming the boundary terms vanish)

$$= (4\pi i/k) \int dA \exp[(ik/4\pi)S][\mathrm{Vol}]T^a T^a \langle K|A\rangle.$$

This completes the formalism of the proof. In the case of part 2., a change of interpretation occurs at the point in the argument when the Wilson line is differentiated. Differentiating a self intersecting Wilson line at a point of self intersection is equivalent to differentiating the corresponding product of matrices with respect to a variable that occurs at two points in the product (corresponding to the two places where the loop passes through the point). One of these derivatives gives rise to a term with volume form equal to zero, the other term is the one that is described in part 2. This completes the proof of the proposition.

7.1. Diagrams

In the lecture corresponding to this paper the proof of the Proposition took on a different aspect, as we had set up a diagrammatic method to do the calculation. Since diagrams of this sort do help in keeping track of the logic of a computation involving many indices, I have decided to include that diagrammatic method in this subsection of the paper. The translation to diagrams is accomplished with the aid of Figure 32 and Figure 33. In Figure 32 we give diagrammatic equivalents for the component parts of our machinery. Tensors become labelled boxes. Indices become lines emanating from the boxes. Repeated indices that we intend to sum over become lines from one box to another. (The eye can immediately apprehend the repeated indices and the tensors where they are repeated.) Note that we use a capital D with lines extending from the top and the bottom for the partial derivative with respect to the gauge field, a capital W with a link diagrammatic subscript for the the Wilson loop, a cubic vertex for the three index epsilon, little triangles with emanating arcs for the differentials of the space variables.

The Lie algebra generators are little boxes with single index lines and two input/output lines which correspond to their roles as matrices (hence as mappings of a vector space to itself). The Lie algebra generators are, in all cases of our calculation, inserted into the Wilson line either through the curvature tensor or through insertions related to differentiating the Wilson line.

We shall have more to say about diagrammatic notation when we discuss the Vassiliev invariants.

7.2. Applying Proposition 1.

The formula of Proposition 1 shows that the integral $Z(K)$ is unchanged if the movement of the loop does not involve three independent space directions (since [Vol] computes a volume.). This means that $Z(K) = Z(S^3, K)$ *is invariant under moves that slide the knot along a plane.* In particular, this means that if the knot K is given in the nearly planar representation of a knot diagram, then $Z(K)$ is invariant under regular isotopy of this diagram. That is, it is invariant under the Reidemeister moves II and III. We expect more complicated behavior under move I since this deformation does involve three spatial directions. This will be discussed momentarily.

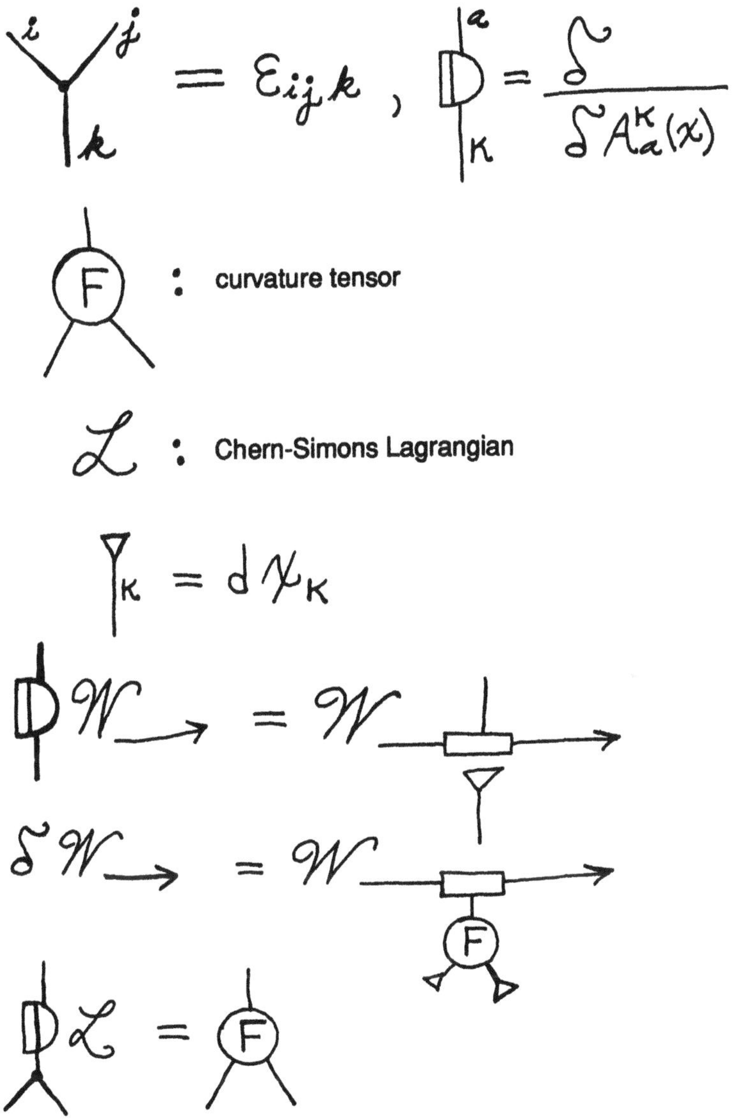

$= \varepsilon_{ijk}$, $= \dfrac{\delta}{\delta A^{K}_{a}(x)}$

: curvature tensor

$\mathcal{L}$: Chern-Simons Lagrangian

$= d\psi_K$

$\delta \mathcal{W} = \mathcal{W}$

Figure 32. Notation.

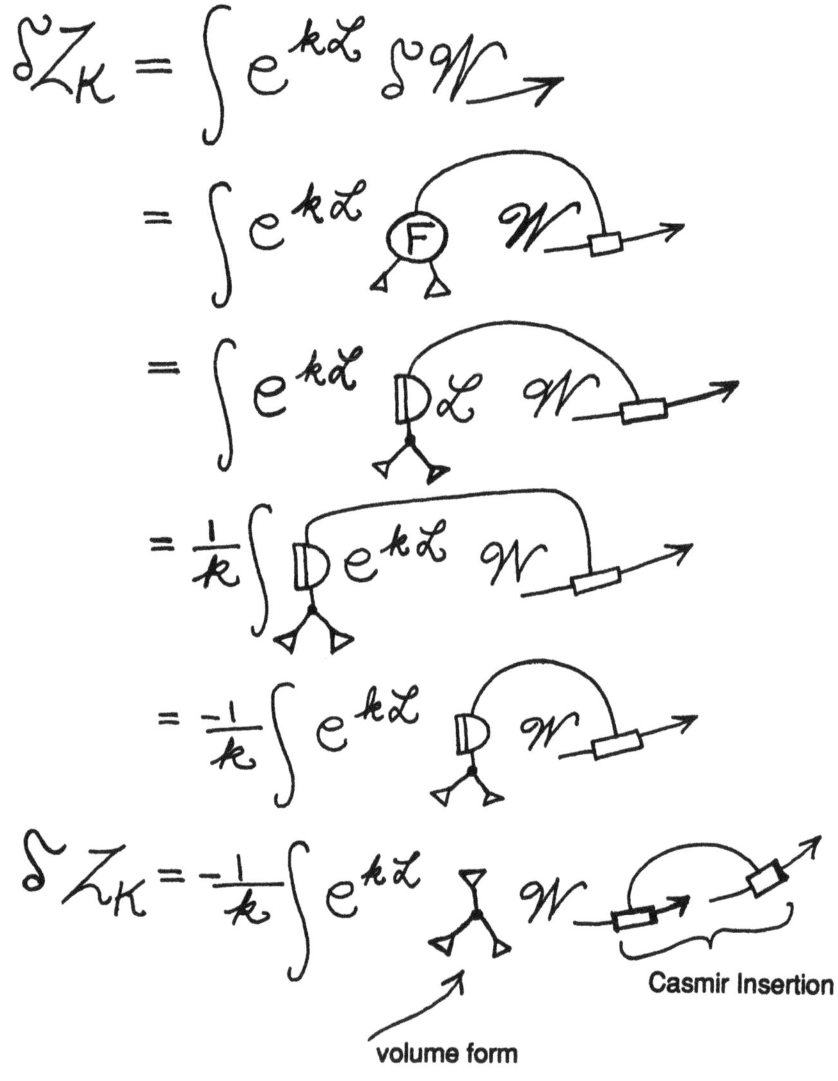

Figure 33. Derivation.

Remark. The proposition and the comments above point out that the integral $Z(K)$ is *not* an ambient isotopy invariant of links. This is usually discussed by saying that the integral requires a framing compensation. This can require the replacement of the Wilson loops by a functional that keeps track of normal framing on the embedding of the link in question. Here we are looking directly at the formalism of an integral that is not an invariant of ambient isotopy of links. This means that *the imaginary measure that we are contemplating does not have to produce complete topological invariance.* It is possible that this fact loosens the restrictions on the measure and allows it to come into existence! We only require from this "measure" that it allow us to do the operations that lead to the conclusion of the Proposition. It is important to note that while $Z(K)$ as described by the Proposition is not an ambient isotopy invariant of links, it *is* sufficient to obtain regular isotopy information about knots and links and this can be used to obtain true topological information. The important knot theoretic point is that *invariants of regular isotopy are sufficient to do knot theory.* If we wish to utilize an invariant, Z, of regular isotopy to determine that two knots are not ambient isotopic, we simply set up representatives of these knots as knot diagrams with with same writhe and Whitney degree (i.e. the same total turn of the tangent vector relative to the diagram plane). If the regular isotopy invariant Z declares different values on these representatives, then they are not ambient isotopic. Thus by shifting the labor to the preparation of the sample knots, we can use regular isotopy invariants in the place of ambient isotopy invariants. In this regard it is quite fascinating that the bare integral implicates regular isotopy invariants by restricting the motions of loops to planes, while it is globally not an invariant of ambient isotopy at all.

Switching Formula. We first determine the difference between $Z(K_+)$ and $Z(K_-)$ where K_+ and K_- denote knots that differ only by switching a single crossing. We take the given crossing in K_+ to be of positive type and the crossing in K_- to be of negative type.

The strategy for computing this difference is to use K_{**} as an intermediate, where K_{**} is the link with a transversal self-crossing replacing the given crossing in K_+ or K_-. Thus we must consider $\Delta_+ = Z(K_+) - Z(K_{**})$ and $\Delta_- = Z(K_-) - Z(K_{**})$. The second part of the Proposition applies to each of these differences and gives

$$\Delta_+ = (4\pi i/k)\int dA \exp[(ik/4\pi)S][\mathrm{Vol}]T^a T^a \langle K_{**}|A\rangle$$

where this evaluation is taken along the loop with the singularity, K_{**}, and the T^aT^a insertion occurs along the two transversal arcs at the singular point. The sign of the volume element will be opposite for Δ_- and consequently we have that

$$\Delta_+ + \Delta_- = 0.$$

(The volume element [Vol] must be given a conventional value in our calculations. There is no reason to assign it different absolute values for the cases of Δ_+ and Δ_- since they are symmetrical except for the sign.)

Therefore $Z(K_+) - Z(K_{**}) + (Z(K_-) - Z(K_{**})) = 0$. Hence

$$Z(K_{**}) = (1/2)(Z(K_+) + Z(K_-)).$$

This result is central to further calculations. It tells us that the evaluation of a singular Wilson line can be replaced with the average of the results of resolving the singularity in the two possible ways. Note that it depends on our hypotheses about symmetry in the calculation.

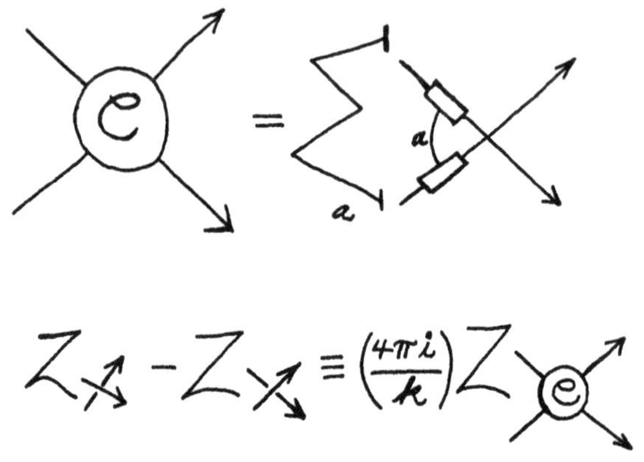

Figure 34. Difference formula.

Now we are interested in the difference $Z(K_+) - Z(K_-)$:

$$Z(K_+) - Z(K_-) = \Delta_+ - \Delta_- = 2\Delta_+$$
$$= (8\pi i/k)\int dA \exp[(ik/4\pi)S][\mathrm{Vol}]T^a T^a \langle K_{**}|A\rangle.$$

Volume Convention. It is useful to make a specific convention about the volume form. We take $[\mathrm{Vol}] = 1/2$ for Δ_+ and $[\mathrm{Vol}] = -1/2$ for Δ_-.

Thus

$$Z(K_+) - Z(K_-) = (4\pi i/k)\int dA \exp[(ik/4\pi)S]T^a T^a \langle K_{**}|A\rangle.$$

Integral Notation. Let $Z(T^a T^a K_{**})$ denote the integral

$$Z(T^a T^a K_{**}) = \int dA \exp[(ik/4\pi)S]T^a T^a \langle K_{**}|A\rangle.$$

Difference Formula. Write the difference formula in abbreviated form

$$Z(K_+) - Z(K_-) = (4\pi i/k)Z(T^a T^a K_{**}).$$

This formula is the key to unwrapping many properties of knot invariants. The rest of this section will be devoted to discussion of these matters. For diagrammatic work it is convenient to rewrite the difference equation in the form shown below. The crossings denote small parts of otherwise identical larger diagrams, and the Casimir insertion $T^a T^a K_{**}$ is indicated with crossed lines entering a disk labelled C. See Figure 34.

The Lie algebra element $\sum_a T^a T^a$ is called the Casimir. Its key property is that it is in the center of the algebra and so can have common eigenvalues with other elements. Note that by our conventions $\mathrm{tr}(\sum_a T^a T^a) = \sum_a \delta^{aa}/2 = d/2$ where d is the dimension of the Lie algebra. This implies that an unknotted loop with one singularity and a Casimir insertion will have Z-value $d/2$. See Figure 34.

In fact, for the classical semi-simple Lie algebras one can choose a basis so that the Casimir is a diagonal matrix with identical values $(d/2D)$ on its diagonal. D is the dimension of the representation. We then have the general formula:

$$Z(T^a T^a K_{**\mathrm{loc}}) = (d/2D)Z(K)$$

Figure 35. Local loops.

for any knot K. Here $K_{**\text{loc}}$ denotes the singular knot obtained by placing a local self-crossing loop on K as shown in figure 35.

Note that $Z(K_{**\text{loc}}) = Z(K)$. (Let the flat loop shrink to nothing. The Wilson line is still defined on a loop with an isolated cusp and it is equal to the Wilson loop obtained by smoothing that cusp.) Let $K_{+\text{loc}}$ denote the result of adding a positive local curl to the knot K, and $K_{-\text{loc}}$ the result of adding a negative local curl to K.

By the above discussion and the difference formula, we have

$$\begin{aligned} Z(K_{+\text{loc}}) &= Z(K_{**\text{loc}}) + (2\pi i/k)Z(T^aT^aK_{**\text{loc}}) \\ &= Z(K) + (2\pi i/k)(d/2D)Z(K). \end{aligned}$$

Thus,

$$Z(K_{+\text{loc}}) = (1 + (\pi i/k)(d/D))Z(K).$$

Similarly,

$$Z(K_{-\text{loc}}) = (1 - (\pi i/k)(d/D))Z(K).$$

These calculations show how the difference equation, the Casimir, and properties of Wilson lines determine the framing factors for the knot invariants. In some cases we can use special properties of the Casimir to obtain skein relations for the knot invariant.

For example, in the fundamental representation of the Lie algebra for $SU(N)$ the Casimir obeys the following equation [41]:

$$\sum_a (T^a)_{ij}(T^a)_{kl} = (1/2)\delta_{il}\delta_{jk} - (1/2N)\delta_{ij}\delta_{kl}$$

Hence

$$Z(T^aT^aK_{**}) = (1/2)Z(K_0) - (1/2N)Z(K_{**})$$

where K_0 denotes the result of smoothing a crossing as discussed in section 2.

Using $Z(K_{**}) = (1/2)(Z(K_+) + Z(K_-))$ and the difference identity, we obtain

$$Z(K_+) - Z(K_-) = (4\pi i/k)(1/2)Z(K_0) - (1/2N)[(Z(K_+) + Z(K_-))/2].$$

Hence

$$(1+\pi i/Nk)Z(K_+)-(1-\pi i/Nk)Z(K_-)=(2\pi i/k)Z(K_0)$$

or

$$q^{1/N}Z(K_+)-q^{-1/N}Z(K_-)=[q-q^{-1}]Z(K_0)$$

where $q=\exp(\pi i/k)$ taken up to $O(1/k^2)$.

Here $d=N^2-1$ and $D=N$, so the framing factor is

$$a=(1+(\pi i/k)((N^2-1)/N))=q^{N-(1/N)}.$$

Therefore, if $P(K)=a^{-wr(K)}Z(K)$ denotes the normalized invariant of ambient isotopy associated with $Z(K)$ (with $wr(K)$ the sum of the crossing signs of K), then

$$aq^{1/N}P(K+)-a^{-1}q^{-1/N}P(K_-)=[q-q^{-1}]P(K_0).$$

Hence

$$q^N P(K_+)-q^{-N}P(K_-)=[q-q^{-1}]P(K_0).$$

This last equation shows that at this level of approximation $P(K)$ is a specialization of the Homfly polynomial for arbitrary N, and that for $N=2$ ($SU(2)$) it is a specialization of the Jones polynomial.

In the next section we look at the relationship between this level of integral heuristics and the existence and construction of Vassiliev invariants.

8. GRAPH INVARIANTS AND VASSILIEV INVARIANTS

If $V(K)$ is a (Laurent polynomial valued, or more generally—commutative ring valued) invariant of knots, then it can be naturally extended to an invariant of rigid vertex graphs by defining the invariant of graphs in terms of the knot invariant via an 'unfolding" of the vertex. That is, we can regard the vertex as a "black box" and replace it by any tangle of our choice. Rigid vertex motions of the graph preserve the contents of the black box, and hence implicate ambient isotopies of the link obtained by replacing the black box by its contents. Invariants of knots and links that are evaluated on these replacements are then automatically rigid vertex invariants of the corresponding graphs. If we set up a collection of multiple replacements at the vertices with standard conventions for the insertions of the tangles, then a summation over all possible replacements can lead to a graph invariant with new coefficients corresponding to the different replacements. In this way each invariant of knots and links implicates a large collection of graph invariants. See [42], [43].

The simplest tangle replacements for a 4-valent vertex are the two crossings, postive and negative, and the oriented smoothing. Let $V(K)$ be any invariant of knots and links. Extend V to the category of rigid vertex embeddings of 4-valent graphs by the formula (See Figure 36)

$$V(K_*)=aV(K_+)+bV(K_-)+cV(K_0)$$

Here K_* indicates an embedding with a transversal 4-valent vertex. We use the symbol $*$ to distinguish this choice of vertex designation from the previous one we used involving a self-crossing Wilson line.

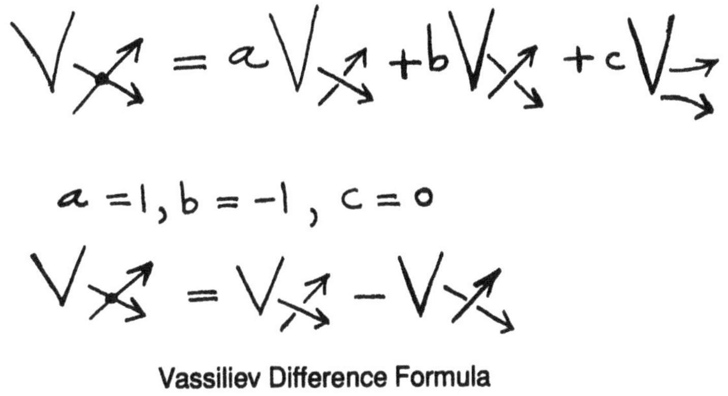

$a = 1, b = -1, c = 0$

Vassiliev Difference Formula

Figure 36. Graphical vertex formulas.

This formula means that we define $V(G)$ for an embedded 4-valent graph G by taking the sum

$$V(G) = \sum_{S} a^{i_+(S)} b^{i_-(S)} c^{i_0(S)} V(S)$$

with the summation over all knots and links S obtained from G by replacing a node of G with either a crossing of positive or negative type, or with a smoothing (denoted 0). It is not hard to see that if $V(K)$ is an ambient isotopy invariant of knots, then, this extension is an rigid vertex isotopy invariant of graphs. In rigid vertex isotopy the cyclic order at the vertex is preserved, so that the vertex behaves like a rigid disk with flexible strings attached to it at specific points.

There is a rich class of graph invariants that can be studied in this manner. The Vassiliev Invariants [45], [46], [47] constitute the important special case of these graph invariants where $a = +1, b = -1$ and $c = 0$. Thus $V(G)$ is a Vassiliev invariant if

$$V(K_*) = V(K_+) - V(K_-).$$

Call this formula the *exchange identity* for the Vassiliev invariant V. V is said to be of finite type k if $V(G) = 0$ whenever $|G| < k$ where $|G|$ denotes the number of 4-valent nodes in the graph G. The notion of finite type is of extraordinary significance in studying these invariants. The reason for this is the following basic Lemma.

Lemma. If a graph G has exactly k nodes, then the value of a Vassiliev invariant v_k of type k on G, $v_k(G)$, is independent of the embedding of G.

Proof. The different embeddings of G can be represented by link diagrams with some of the 4-valent vertices in the diagram corresponding to the nodes of G. It suffices to show that the value of $v_k(G)$ is unchanged under switching of a crossing. However, the exchange identity for v_k shows that this difference is equal to the evaluation of v_k on a graph with $k+1$ nodes and hence is equal to zero. This completes the proof.

The upshot of this Lemma is that Vassiliev invariants of type k are intimately involved with certain abstract evaluations of graphs with k nodes. In fact, there are restrictions (the four-term relations) on these evaluations demanded by the topology (we shall articulate these restrictions shortly) and it follows from results of Kontsevich [47] that such abstract evaluations actually determine the invariants. The invariants derived from classical Lie algebras are all built from Vassiliev invariants of finite type. All this is directly related to Witten's functional integral.

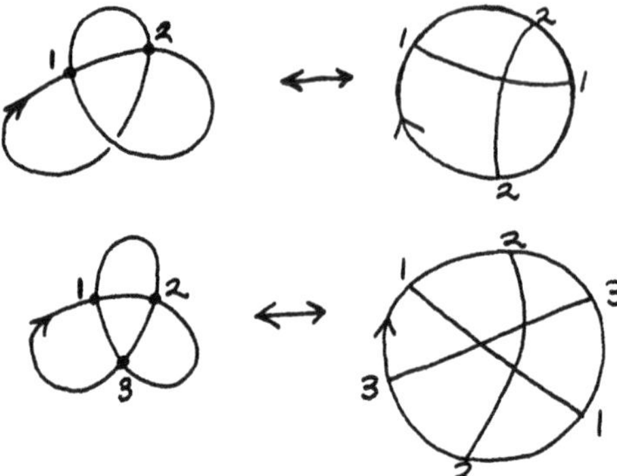

Figure 37. Chord diagrams.

Definition. Let v_k be a Vassiliev invariant of type k. The *top row* of v_k is the set of values that v_k assigns to the set of (abstract) 4-valent graphs with k nodes. If we concentrate on Vassiliev invariants of knots, then these graphs are all obtained by marking $2k$ points on a circle, and choosing a pairing of the $2k$ points. The pairing can be indicated by drawing a circle and connecting the paired points with arcs. Such a diagram is called a *chord diagram.* Some examples are indicated in Figure 37.

Note that a top row diagram cannot contain any isolated pairings since this would correspond to a difference of local curls on the corresponding knot diagram (and these curls, being isotopic, yield the same Vassiliev invariants.

The Four-Term Relation. (Compare [48].) Consider a single embedded graphical node in relation to another embedded arc, as illustrated in Figure 38. The arc underlies the lines incident to the node at four points and can be slid out and isotoped over the top so that it overlies the four nodes. One can also switch the crossings one-by-one to exchange the arc until it overlies the node. Each of these four switchings gives rise to an equation, and the left-hand sides of these equations will add up to zero, producing a relation corresponding to the right-hand sides. Each term in the right-hand side refers to the value of the Vassiliev invariant on a graph with two nodes that are neighbors to each other. See Figure 38.

There is a corresponding 4-term relation for chord diagrams. This is the 4-term relation for the top row. In chord diagrams the relation takes the form shown at the bottom of Figure 38. Here we have illustrated only those parts of the chord diagram that are relevant to the two nodes in question (indicated by two pairs of points on the circle of the chord diagram). The form of the relation shows the points on the chord diagram that are immediate neighbors. These are actually neighbors on any chord diagram that realizes this form. Otherwise there can be many other pairings present in the situation.

As an example, consider the possible chord diagrams for a Vassiliev invariant of type 3. There are two possible diagrams as shown in Figure 39. One of these has the projected pattern of the trefoil knot and we shall call it the *trefoil graph.* These diagrams satisfy the 4-term relation. This shows that one diagram must have twice the evaluation of the other. Hence it suffices to know the evaluation of one of these two diagrams to know the top row of a Vassiliev invariant of type 3. We can take this generator to be the trefoil graph

Now one more exercise: Consider any Vassiliev invariant v and let's determine its value on the the trefoil graph as in Figure 40.

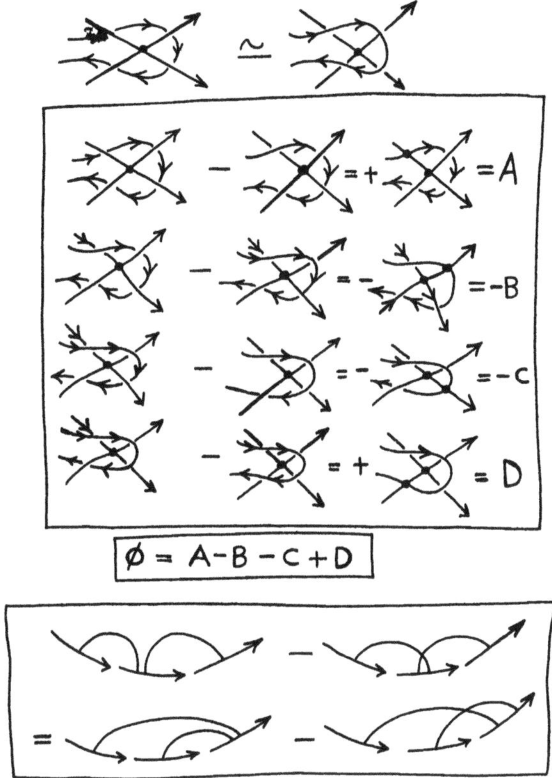

Figure 38. The four term relation.

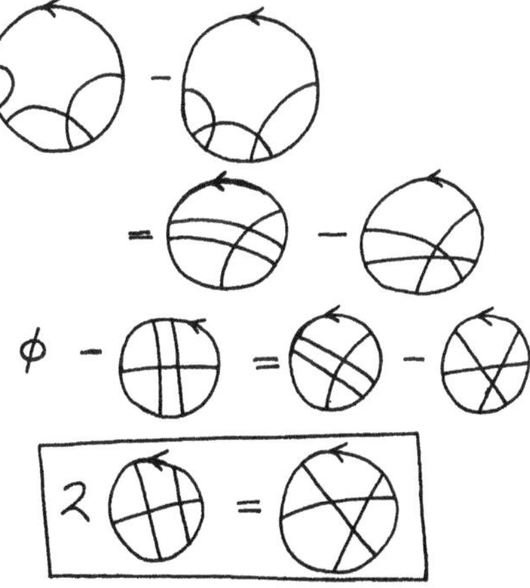

Figure 39. Four term relation for type three.

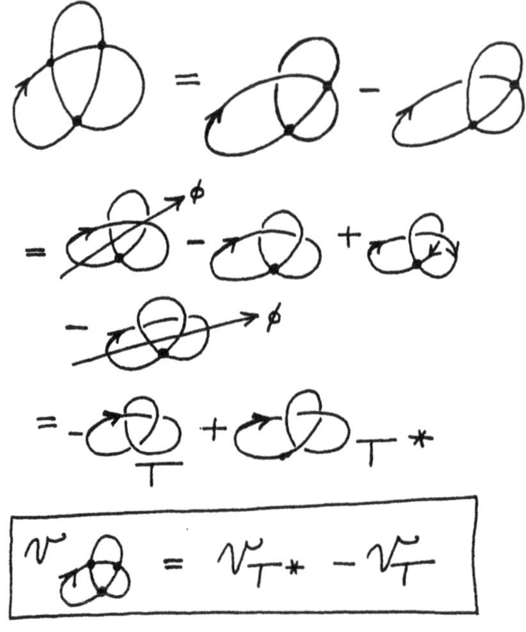

Figure 40. Trefoil graph.

The value of this invariant on the trefoil graph is equal to the difference between its values on the trefoil knot and its mirror image. Therefore any Vassiliev invariant that assigns a non-zero value to the trefoil graph can tell the difference between the trefoil knot and its mirror image.

Example. This example shows how the original Jones polynomial is composed of Vassiliev invariants of finite type. Let $V_K(t)$ denote the original Jones polynomial, as described in Section 3. Recall that at the end of Section 3 we described an oriented state expansion for the Jones polynomial with the basic formulas (δ is the loop value.)

$$\begin{aligned} V_{K_+} &= -t^{1/2}V_{K_0} - tV_{K_\infty} \\ V_{K_-} &= -t^{-1/2}V_{K_0} - t^{-1}V_{K_\infty}. \\ \delta &= -(t^{1/2}+t^{-1/2}). \end{aligned}$$

Let $t = e^x$. Then

$$\begin{aligned} V_{K_+} &= -e^{x/2}V_{K_0} - e^{x}V_{K_\infty} \\ V_{K_-} &= -e^{-x/2}V_{K_0} - e^{-x}V_{K_\infty}. \\ \delta &= -(e^{x/2}+e^{-x/2}). \end{aligned}$$

Thus

$$V_{K_*} = V_{K_+} - V_{K_-} = -2\sinh(x/2)V_{K_0} - 2\sinh(x)V_{K_\infty}.$$

Thus x divides V_{K_*}, and therefore x^k divides V_G whenever G is a graph with at least k nodes. Letting

$$V_G(e^x) = \sum_{k=0}^{\infty} v_k(G)x^k,$$

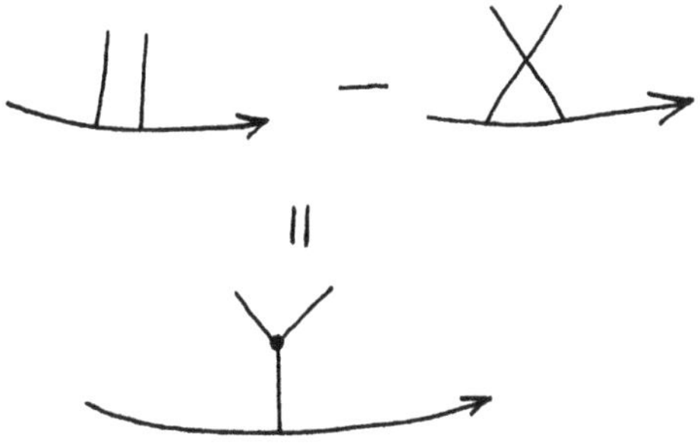

Figure 41. The STU relation.

we see that this condition implies that $v_k(G)$ vanishes whenever G has more than k nodes. Hence *the coefficients of the powers of x in the expansion of $V_K(e^x)$ are Vassiliev invariants of finite type!* This result was first observed by Birman and Lin [46] by a different argument.

Let's look a little deeper and see the structure of the top row for the Vassiliev invariants related to the Jones polynomial. By our previous remarks the top row evaluations correspond to the leading terms in the power series expansion. Since

$$\begin{aligned} \delta &= -(e^{x/2}+e^{-x/2}) = -2+[\text{higher}], \\ -e^{x/2}+e^{-x/2} &= -x+[\text{higher}], \\ -e^{x}+e^{-x} &= -2x+[\text{higher}], \end{aligned}$$

it follows that the top rows for the Jones polynomial are computed by the recursion formulas

$$\begin{aligned} v(K_*) &= -v(K_0) - 2V(K_\infty), \\ v([\text{loop}]) &= -2, \\ v\times &= -v\asymp - 2v\succ\!\prec \\ v\circledast &= 24. \end{aligned}$$

The reader can easily check that this recursion formula for the top rows of the Jones polynomial implies that v_3 takes the value 24 on the trefoil graph and hence it is the Vassiliev invariant of type 3 in the Jones polynomial that first detects the difference between the trefoil knot and its mirror image.

This example gives a good picture of the general phenomenon of how the Vassiliev invariants become building blocks for other invariants. In the case of the Jones polynomial, we already know how to construct the invariant and so it is possible to get a lot of information about these particular Vassiliev invariants by looking directly at the Jones polynomial. This, in turn, gives insight into the structure of the Jones polynomial itself.

Mirror–Mirror. Now we go straight through the looking glass. Consider the diagrammatic relation shown in Figure 41. Call it (after Bar-Natan [47]) the STU relation.

Lemma. STU implies the 4-term relation.

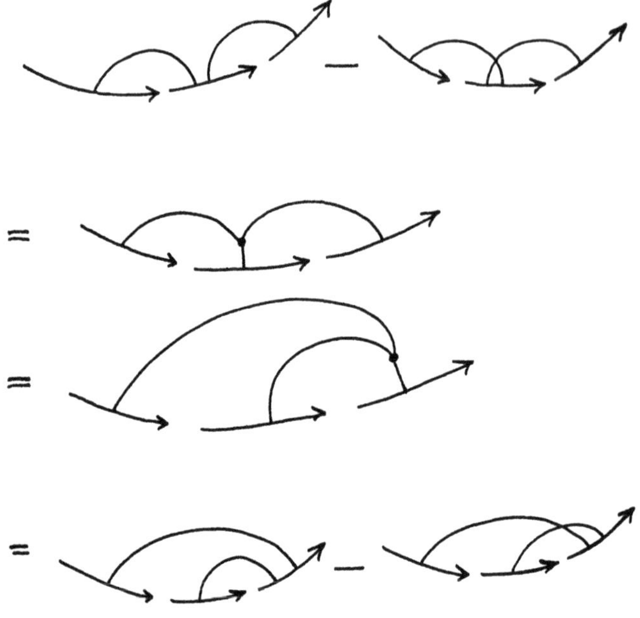

Figure 42. A diagrammatic proof.

Proof. View Figure 42.

STU is the smile of the Cheshire cat. The cat is a generalization of the idea of a Lie algebra. Take a Lie algebra with generators T^a. Then

$$T^aT^b - T^bT^a = if_{abc}T^c$$

expresses the closure of the Lie algebra under commutators. Translate this equation into diagrams as shown in Figure 42, and see that this translation is *STU* with Lie algebraic clothing!

Here the structure tensor of the Lie algebra has been assumed (for simplicity) to be invariant under cyclic permutation of the indices. This invariance means that our last Lemma applies to this Lie algebraic interpretation of *STU*. The upshot is that we can manufacture weight systems for graphs that satisfy the 4-term relation by replacing paired points on the

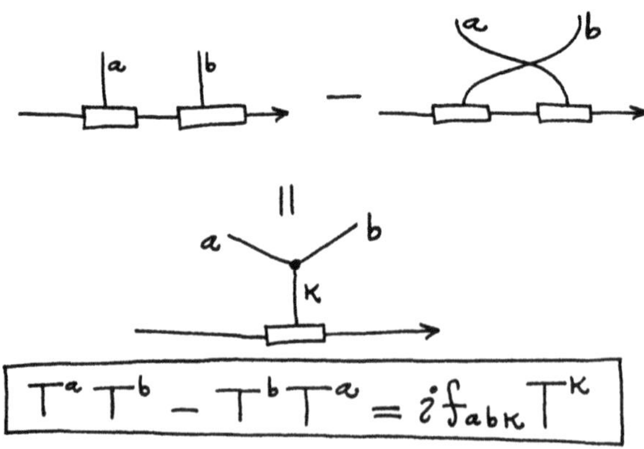

Figure 43. Algebraic clothing.

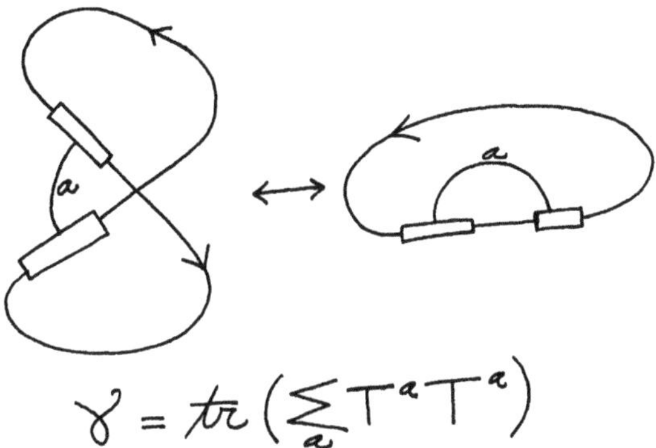

Figure 44. Weight system and Casimir insertion.

chord diagram by an insertion of T^a in one point of the pair and a corresponding insertion of T^a at the other point in the pair and summing over all a. The result of all such insertions on a given chord diagram is a big sum of specific matrix products along the circle of the diagram, each of which (being a circular product) is interpreted as a trace.

Lets say this last matter more precisely: Regard a graph with k nodes as obtained by identifying k *pairs* of points on a circle. Thus a code such as 1212 taken in cyclic order specifies such a graph by regarding the points 1, 2, 1, 2 as arrayed along a circle with the first and second 1's and 2's identified to form the graph. Define, for a code $a_1a_2\ldots a_m$

$$wt(a_1a_2\ldots a_m) = \text{trace}(T^{a_1}T^{a_2}T^{a_1}\ldots T^{a_m})$$

where the Einstein summation convention is in place for the double appearances of indices on the right-hand side. This gives the weight system.

The weight system described by the above procedure satisfies the 4-term relation, but does not necessarily satisfy the vanishing condition for isolated pairings. This is because the framing compensation for converting an invariant of regular isotopy to ambient isotopy has not yet been introduced. We will show how to do this in the course of the discussion in the next paragraph. The main point to make here is that by starting with the idea of extending an invariant of knots to a Vassiliev invariant of embedded graphs and searching out the conditions on graph evaluation demanded by the topology, we have inevitably re-entered the domain of relations between Lie algebras and link invariants. Since the STU does not demand Lie algebras for its satisfaction we see that the landscape has widened beyond that original context. Nevertheless, this situation demands to be related to the approach that was described in the last section with Witten's functional integral. Such a weight system, generating Vassiliev invariants, can be regarded as an abstract Wilson loop evaluation.

In fact, we can line up this weight system with the formalism related to the functional integral from the last section by writing the Lie algebra insertions back on the 4-valent graph.

We then get a Casimir insertion at the node just as we described it in the last section in relation to the difference formula for the functional integral. See Figure 44.

To get the framing compensation, note that an isolated pairing corresponds to the trace of the Casimir. Let γ denote this trace. See Figure 44.

$$\gamma = \text{tr}(\sum_a T^a T^a)$$

Let D be the trace of the identity. Then it is easy to see that we must compensate the given weight system by subtracting (γ/D) multiplied by the result of dropping the identification of the two given points. We can diagram this by drawing two crossed arcs without a node drawn to bind them. Then the modified recursion formula becomes

$$V[\text{diagram}] = V[\text{diagram}] - \frac{\gamma}{D} V\times$$

For example, in the case of $SU(N)$ we have $D = N$, $\gamma = (N^2 - 1)/2$ so that

$$V[\text{diagram}] = V[\text{diagram}] - \frac{N^2-1}{2N} V\times$$

and since the Casimir insertion satisfies the Fierze identity

$$[\text{diagram}] = \frac{1}{2}\asymp - \frac{1}{2N}\times$$

we obtain

$$V[\text{diagram}] = \frac{1}{2}\asymp - \frac{N}{2} V\times$$

For $N = 2$ this is, up to a multiple, exactly the top row formula that we deduced for the Jones polynomial from its combinatorial structure.

9. VASSILIEV INVARIANTS AND WITTEN'S INTEGRAL

We return to the invariants derived from the functional integral $Z(K)$. We have shown that $Z(K_{+\text{loc}}) = aZ(K)$ with

$$a = q^{d/D}.$$

Hence

$$P(K) = a^{-wr(K)} Z(K)$$

is an ambient isotopy invariant. The equation

$$Z(K_*) = Z(K_+) - Z(K_-) = (4\pi i/k) Z(T^a T^a K_{**})$$

implies that if $wr(K_+) = w + 1$, then we have the ambient isotopy difference formula:

$$P(K_*) = P(K_+) - P(K_-) = a^{-w}(4\pi i/k)[Z(T^a T^a K_{**}) - (d/2D)Z(K_{**})].$$

We leave the proof of this formula as an exercise for the reader.

This formula tells us that for the Vassiliev invariant associated with P we have

$$P(K_*) = a^{-w}(4\pi i/k)[Z(T^a T^a K_{**}) - (d/2D)Z(K_{**})].$$

Furthermore, if $V_j(K)$ denotes the coefficient of $(4\pi i/k)^j$ in the expansion of $P(K)$ in powers of $1/k$, then the ambient difference formula implies that $(1/k)^j$ divides $P(G)$ when G has j or more nodes. Hence $V_j(G) = 0$ if G has more than j nodes. Therefore $V_j(K)$ is a Vassiliev invariant of finite type.

The fascinating thing is that the ambient difference formula, appropriately interpreted, actually tells us how to compute $V_k(G)$ when G has k nodes. Under these circumstances each node undergoes a Casimir insertion, and because the Wilson line is being evaluated abstractly, independent of the embedding, this suggests that we insert nothing else into the line. Thus we take the pairing structure associated with the graph (the chord diagram) and use it as a prescription for obtaining a trace of a sum of products of Lie algebra elements with T^a and T^a inserted for each pair, or a simple crossover for the pair multiplied by (d/D). This yields the graphical evaluation implied by the recursion

$$V(G_*) = V(T^a T^a G_{**}) - (d/2D)V(G_{**}).$$

This is exactly the weight system that we ended up with at the end of the last section in discussing Vassiliev invariants in general. Thus we know that not only is this weight system implicated by Witten's integral, but we have independently verified that it satisfies the 4-term relations and is correctly compensated for framing. This result is equivalent to Bar-Natan's result. It is very interesting to see how it follows from a minimal approach to the Witten integral.

It must be mentioned that this brings us to the core question about Vassiliev invariants: Are there non-trivial Vassiliev invariants of knots and links that cannot be constructed through combinations of Lie algebraic invariants? There are many other open questions in this arena, all circling this basic problem.

9.1. Quantum Link Invariants and Vassiliev Invariants

Vassiliev invariants can be used as building blocks for all the presently known quantum link invariants.

It is this result that we can now make clear in the context of the models given in our section on quantum link invariants. Suppose that λ is written as a power series in a variable h, say $\lambda = \exp(h)$ to be specific. Suppose also, that the R-matrices can be written as power series in h with matrix coefficients so that $R = P + r_+ h + O(h^2)$ and $R^{-1} = P + r_- h + O(h^2)$ where P denotes the map of $V \otimes V$ that interchanges the tensor factors. Let $Z(K)$ denote the value of the oriented amplitude described by this choice of λ and R. Then we can write

$$Z(K) = Z_0(K) + Z_1(K)h + Z_2(K)h^2 + \ldots$$

where each $Z_n(K)$ is an invariant of regular isotopy of the link K. Furthermore, we see at once that *h divides the series for $Z(K_+) - Z(K_-)$*. By the definition of the Vassiliev invariants this implies that h^k divides $Z(G)$ if G is a graph with k nodes. Therefore $Z_n(G)$ vanishes if n is less than the number of nodes of G. Therefore Z_n is a Vassiliev invariant of finite type n. Hence the quantum link invariant is built from an infinite sequence of interlocked Vassiliev invariants.

It is an open problem whether the class of finite type Vassiliev invariants is greater than those generated from quantum link invariants. It is also possible that there are quantum link invariants that can not be generated by Vassiliev invariants.

10. VASSILIEV INVARIANTS AND THE KONTSEVICH INTEGRAL

Kontsevich [49], [47] proved that a weight assignment for a Vassiliev top row that satisfies the 4-term relation and the framing condition (that the weights vanish for graphs with isolated

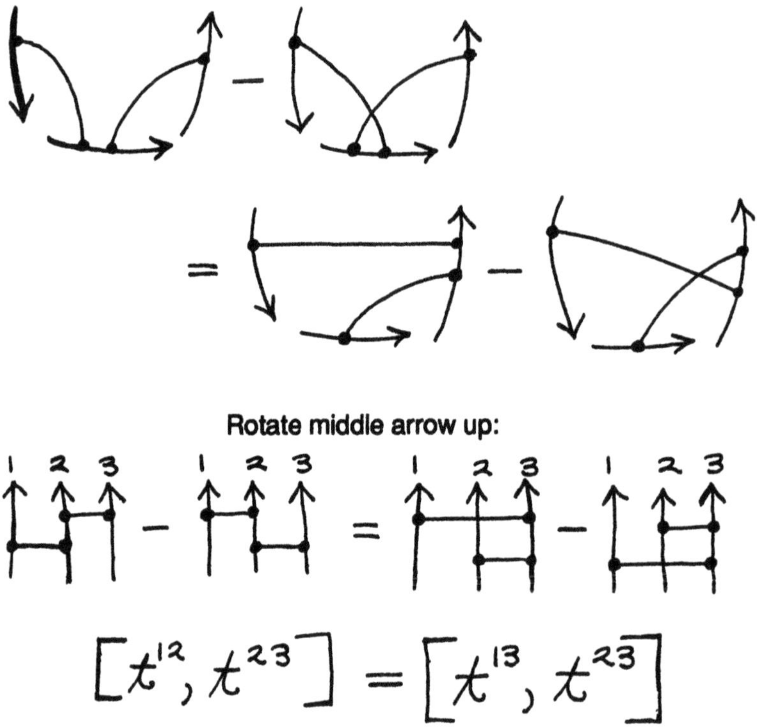

Rotate middle arrow up:

$$[t^{12}, t^{23}] = [t^{13}, t^{23}]$$

Figure 45. Infinitesimal braiding.

double points) actually extends to a Vassiliev invariant defined on all knots. His method is motivated by the perturbative expansion of the Witten integral and by Witten's interpretation of the integral in terms of conformal field theory. This section will give a brief description of the Kontsevich approach and the questions that it raises about the functional integral itself.

The key to this approach is to see that the 4-term relations are a kind of "infinitesimal braid relations." That is, we can re-write the 4-term relations in the form of tangle operators as shown in Figure 45.

In fact the 4-term relations translate exactly into the infinitesimal braid relations studied by Kohno [50]. Kohno showed that his version of infinitesimal braid relations corresponded to a flatness condition for a certain connection (the Knizhnik–Zamolodchikov connection) and that this meant that these relations constituted an integrability condition for making representations of the braid group via monodromy. Others have verified that the braid group representations related to the Chern–Simons form and the Witten integral arise in this same way from the Knizhnik–Zamolodchikov equations. In the case of Chern–Simons theory the weights in the K–Z equations come from the Casimir of a classical Lie algebra, just as we have discussed. Kontsevich observed that since the arbitrary 4-term relations could also be regarded as an integrability condition it was possible to use them in a generalization of Kohno's ideas to produce braid group representations via iterated integration. He then generalized the process of producing these braid group representations to the production of knot invariants and these become realizations of Vassiliev invariants that have given admissible weight systems for their top rows.

The upshot of the Kontsevich work is a very specific integral formula for the Vassiliev invariants. See [47] for the specifics. It is clear from the nature of the construction that the

Kontsevich formula captures the various orders of perturbative terms in the Witten integral. At this writing there is no complete published description of this correspondence.

10.1. Perturbation Heuristics

Reconsider the Witten functional integral:

$$Z(M,K) = \int dA \exp[(ik/4\pi)S(M,A)] \operatorname{tr}\left(P\exp\left(\int_K A\right)\right)$$

Here $S(M,A)$ is taken to be the integral over M of the trace of the Chern–Simons three-form $CS = AdA + (2/3)A^3$.

$$S(M,A) = \int_M \operatorname{tr}(AdA + (2/3)A^3)$$

Replace A by $A/\sqrt{k}$. Then $Z(M,K)$ is replaced by

$$\int dA \exp[(i/4\pi)\int_M \operatorname{tr}(AdA) \exp[(i/6\pi\sqrt{k})\int_M \operatorname{tr}(A^3)] \operatorname{tr}(P\exp((1/\sqrt{k})\int_K A)$$

The $\exp[(i/4\pi)\int_M \operatorname{tr}(AdA)$ can take the formal role of the quadratic Gaussian and one can view the whole integral as an expansion in powers of $1/k$ of terms related in the series expansions of the remaining exponentials in the integral. This leads to a special Feynman diagram expansion for the functional integral [51], [52], [53], [54] where the diagrams are the embedding of the knot or link decorated with extra bits of Feynman graphs. We will describe these graphs in a moment. To each such graph there will be an associated actual integral, and the coefficient of $(1/k)^n$ in this expansion is a sum of such integrals. This coefficient represents a Vassiliev invariant of type n. The full story for these perturbation terms is quite intricate —see the references cited above.

The Feynman diagrams consist in the knot diagram with a finite selection of points on the diagram itself, plus a set of points in the complement of the knot diagram. Each of these points is a trivalent vertex. For the points on the knot diagram this means that along with the two edges on either side of the point on the diagram, there is a third edge leading off into the complementary space. Each point in space has three edges emanating from it. The graph has no other edges or vertices. Each spatial vertex corresponds to an A^3 term in the perturbation expansion and hence contributes a factor of $(1/k)^{1/2}$. Each vertex on the knot diagram corresponds to an A from the Wilson loop and also contributes a factor of $(1/k)^{1/2}$. The simplest case is one extra edge attached to the knot diagram. The contribution is at order $1/k$ and the associated integral is the Gauss self-linking integral. The next case has two graphs, one consisting in two extra edges attached to the knot diagram, the other consisting of an external tri-valent vertex attached by three legs to the knot diagram. An appropriate sum of these two contributions gives the second Vassiliev invariant. See Figure 46.

11. APPENDIX ON SQUARE DANCING

In Corsica we had square dancing at the beginning of the third lecture. Perhaps you think it strange that Cécile DeWitt-Morette and Pierre Cartier would indulge in square dancing during an academic lecture! The dance was instigated in the cause of fractions and topology,

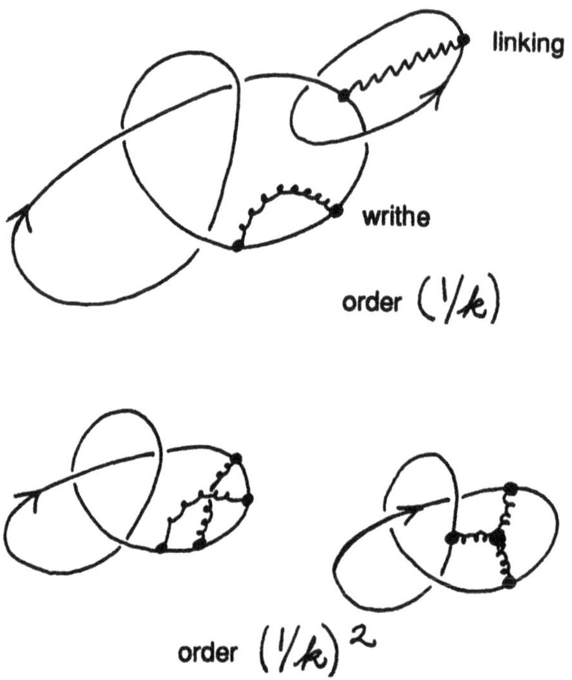

Figure 46. Feynman knot diagrams.

for it turns out that the arithmetic of rational fractions has an interpretation in terms of four square dancers holding two lengths of tangled rope.

Lets call the person who calls out the moves of the dance the *barker*. The barker in this square dance makes two kinds of calls. Sometimes he/she says "negative reciprocate!" and sometimes she says "add 1!." Sometimes, but rarely, she/he calls out the present value of the dance. She might cry out "22/7!" or "Very Fine: 1/137!" (In square dancing all the factorials are literary. We will have to devise a special square dance notation for quantum factorials.)

Initially the four dancers form a square with a forward pair (nearest the audience) holding a rope between them, and the backward pair holding a rope between them. The ropes are unknotted and initially untangled with each other. At the beginning of the dance the barker calls out the value "zero!"

The dancers have been informed that "negative reciprocate" means that everybody turns counterclockwise by ninety degrees *while holding onto the ends of the rope together*. If this little abstract tensor

$$S^{AB}_{CD}$$

represents the the square of dancers, then the dancers are represented (without their ropes) by the imaginatively chosen indices A, B, C, D. When the barker says "negative reciprocate!" the dancers move to the new configuration

$$S^{DA}_{CB}.$$

When the barker says "add 1" then the dancers representing *forward right* and *backward right* positions with respect to the audience exchange places. They do so in such a way that the forward dancer brings his/her rope *over* the rope of the backward dancer. This operation adds a twist to the tangle. Without indicating the ropes, "add one!" performs the operation shown

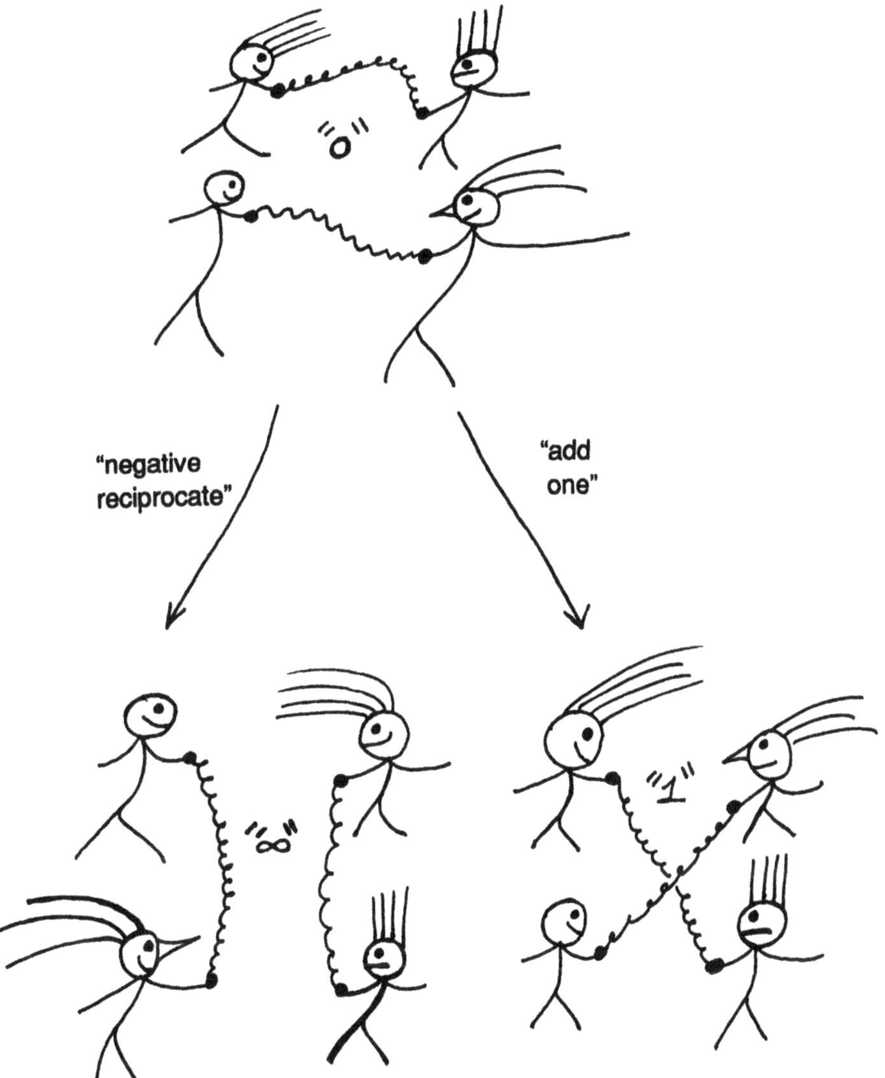

Figure 47. Beginning the dance.

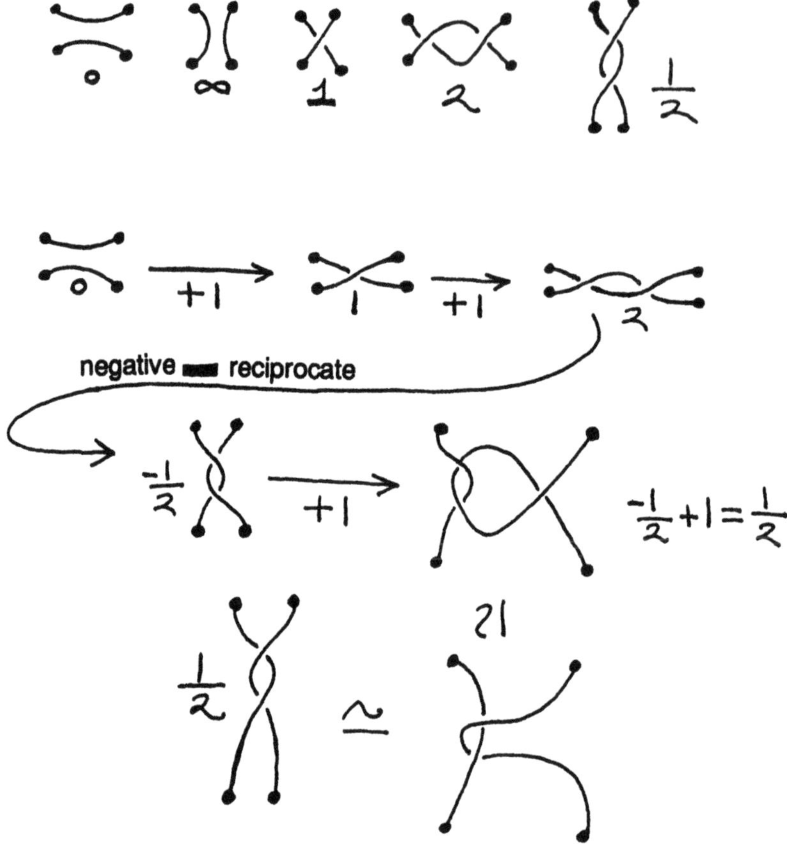

Figure 48. Calling the dance.

below on the dancer/indices:

$$S^{AB}_{CD}$$

becomes

$$S^{AD}_{CB}.$$

In Figure 48 we have illustrated a dance or two, and if you read this figure carefully you will begin to see that it does make sense to identify the barkers call of "negative reciprocate!" with that operation on rational fractions.

The square dance provides an abstract illustration of facts about $SL(2, Z)$ *and* it illustrates the association, due to John Horton Conway, of a rational fraction with each so-called *rational tangle*. For our purposes the rational tangles are exactly those tangles that the dancers can produce in the course of their dance. Conway devised the game of square dancing to illustrate his ideas about tangles.

And what is a *tangle*? Well a tangle is a pattern of rope with four free ends (a *four tangle* to be exact) such that the four free ends emanate from a ball in three dimensional space. Two tangles are *equivalent* if there is an ambient isotopy within their respective balls, leaving the free ends fixed, that makes them identical. Thus this version of square dancing is actually a form of ball room dancing. But we digress.

John Conway stated the beautiful Theorem [9] that *two rational tangles are equivalent if and only if they have the same fraction.* Recently, the author and Jay Goldman [55] found a very neat proof of Conway's Theorem. The main part of our proof involves showing how two rational tangles with the same fraction must be equivalent. For the converse, we show the fraction is an invariant of tangle equivalence by using the bracket polynomial. This is an appendix on square dancing — an inappropriate place for a mathematical proof. The barker suggests that you consult our paper for the details.

REFERENCES

[1] L. H. Kauffman, Functional Integration and the theory of knots, J. Math. Physics, Vol. 36 (5), May 1995, pp. 2402–2429.

[2] K. Reidemeister, *Knotentheorie*, Chelsea Pub. Co., New York, 1948, Copyright 1932, Julius Springer, Berlin.

[3] H. Trotter, Non-invertible knots exist, *Topology*, Vol. 2, 1964, pp. 275–280.

[4] J. H. White, Self-linking and the Gauss integral in higher dimensions. *Amer. J. Math.*, **91** (1969), pp. 693–728.

[5] D. Joyce, A classifying invariant for knots:the knot quandle, *J. Pure and Appl. Alg.*, Vol. 23, 1982, pp. 37–65.

[6] L. H. Kauffman, *Knots and Physics*, World Scientific Pub., 1991 and 1993.

[7] R. Fenn and C. Rourke, Racks and links in codimension two, *J. of Knot Theory and Its Ramif.*, Vol. 1, No. 4, 1992, pp. 343–406.

[8] L. H. Kauffman and S. K. Winker, *Quandles, Crystals and Racks — A New Approach to Knot Theory*, World Scientific Pub., 1997 (to appear).

[9] J. H. Conway, An enumeration of knots and links and some of their algebraic properties, in *Computational Problems in Abstract Algebra*, Pergammon Press, N. Y., 1970, pp. 329–358.

[10] V. F. R. Jones, A polynomial invariant for links via von Neumann algebras, *Bull. Amer. Math. Soc.*, Vol. 129, 1985, pp.103–112.

[11] R. J. Baxter, *Exactly Solved Models in Statistical Mechanics*, Acad. Press, 1982.

[12] V. F. R. Jones, Hecke algebra representations of braid groups and link polynomials, *Ann. of Math.*, Vol. 126, 1987, pp. 335–388.

[13] L. H. Kauffman, An invariant of regular isotopy, *Trans. Amer. Math. Soc.*, Vol. 318. No. 2, 1990, pp. 417–471.

[14] L. H. Kauffman, State Models and the Jones Polynomial, *Topology*, Vol. 26, 1987, pp. 395–407.

[15] L. H. Kauffman, Statistical mechanics and the Jones polynomial, *AMS Contemp. Math. Series*, Vol. 78, 1989, pp. 263–297.

[16] V. F. R. Jones, On knot invariants related to some statistical mechanics models, *Pacific J. Math.*, Vol. 137, No. 2, 1989, pp. 311–334.

[17] V. G. Turaev, The Yang–Baxter equations and invariants of links, LOMI preprint E-3–87, Steklov Institute, Leningrad, USSR., *Inventiones Math.*, Vol. 92, Fasc.3, pp. 527–553.

[18] L. H. Kauffman, Gauss codes, quantum groups and ribbon Hopf algebras, *Reviews in Mathematical Physics*, Vol. 5, No. 4., 1993, pp. 735–773.

[19] N. Y. Reshetikhin, Quantized universal enveloping algebras, the Yang–Baxter equation and invariants of links, I and II, *LOMI reprints E-4-87 and E-17-87*, Steklov Institute, Leningrad, USSR.

[20] L. H. Kauffman, The Conway polynomial, *Topology*, **20** (1980), pp. 101–108.

[21] L. H. Kauffman, *On Knots*, Annals Study No. 115, *Princeton University Press* (1987)

[22] L. H. Kauffman and D. Radford, Invariants of 3-Manifolds derived from finite dimensional Hopf algebras, *Journal of Knot Theory and Its Ramifications*. Vol. 4, No. 1 (1995), 131–162.

[23] L. H. Kauffman, Hopf algebras and 3-Manifold invariants, *Journal of Pure and Applied Algebra*. Vol. 100 (1995), 73–92.

[24] D. Yetter, Quantum groups and representations on monoidal categories. *Math. Proc. Camb. Phil. Soc.*, Vol. 108 (1990), pp. 197–229.

[25] M. A. Hennings, Hopf algebras and regular isotopy invariants for link diagrams, *Math. Proc. Camb. Phil. Soc.*, Vol. 109 (1991), pp. 59–77

[26] Lee Smolin, Link polynomials and critical points of the Chern–Simons path integrals, *Mod. Phys. Lett. A*, Vol. 4, No. 12, 1989, pp. 1091–1112.

[27] E. Witten, Quantum field theory and the Jones polynomial, *Commun. Math. Phys.*, Vol. 121, 1989, pp. 351–399.

[28] M. F. Atiyah, *Geometry of Yang–Mills Fields*, Accademia Nazionale dei Lincei Scuola Superiore Lezioni Fermiare, Pisa, 1979.

[29] N. Y. Reshetikhin and V. Turaev, Invariants of Three Manifolds via link polynomials and quantum groups, *Invent. Math.*, Vol. 103, 1991, pp. 547–597.

[30] V. G. Turaev and H. Wenzl, Quantum invariants of 3-manifolds associated with classical simple Lie algebras, *International J. of Math.*, Vol. 4, No. 2, 1993, pp. 323–358.
[31] R. Kirby and P. Melvin, On the 3-manifold invariants of Reshetikhin–Turaev for sl(2, C), *Invent. Math.* 105, 473–545, 1991.
[32] W. B. R. Lickorish, The Temperley Lieb Algebra and 3-manifold invariants, *Journal of Knot Theory and Its Ramifications*, Vol. 2, 1993, pp. 171–194.
[33] K. Walker, On Witten's 3-Manifold Invariants, (preprint 1991).
[34] L. H. Kauffman and S. Lins, *Temperley Lieb Recoupling Theory and Invariants of 3-Manifolds*, Annals of Mathematics Studies Number 134, Princeton University Press, 1994.
[35] M. F. Atiyah, *The Geometry and Physics of Knots*, Cambridge University Press, 1990.
[36] S. Garoufalidis, Applications of TQFT to invariants in low dimensional topology, (preprint 1993).
[37] D. Freed and R. Gompf, Computer calculations of Witten's 3-manifold invariants, *Comm. Math. Phys.*, Vol. 41, pp. 79–117, 1991.
[38] L. C. Jeffrey, On Some Aspects of Chern–Simons Gauge Theory, Thesis — Oxford, 1991.
[39] L. Rozansky, A large k asymptotics of Witten's invariant of Seifert manifolds, (preprint 1993).
[40] P. Cotta-Ramusino, E. Guadagnini, M. Martellini, M. Mintchev, Quantum field theory and link invariants, (preprint 1990)
[41] T. Cheng and L Li, *Gauge Theory of Elementary Particle Physics*, Clarendon Press, Oxford, 1988.
[42] L. H. Kauffman, New invariants in the theory of knots, *Amer. Math. Monthly*, Vol. 95, No. 3, March 1988. pp. 195–242.
[43] L. H. Kauffman and P. Vogel, Link polynomials and a graphical calculus, *Journal of Knot Theory and Its Ramifications*, Vol. 1, No. 1, March 1992.
[44] L. H. Kauffman, *Formal Knot Theory*, Princeton University Press, Lecture Notes Series #30 (1983).
[45] V. Vassiliev, Cohomology of knot spaces, In *Theory of Singularities and Its Applications*, V. I. Arnold, ed., Amer. Math. Soc., 1990, pp. 23–69.
[46] J. Birman and X. S. Lin, Knot polynomials and Vassiliev's invariants, (to appear in Invent. Math.).
[47] D. Bar-Natan, On the Vassiliev knot invariants, (to appear in Topology).
[48] T. Stanford, Finite-type invariants of knots, links and graphs, (preprint 1992).
[49] M. Kontsevich, Graphs, homotopical algebra and low dimensional topology, (preprint 1992).
[50] T. Kohno, Linear representations of braid groups and classical Yang–Baxter equations, *Contemporary Mathematics*, Vol. 78, Amer. Math. Soc., 1988, pp. 339–364.
[51] E. Guadagnini, M. Martellini and M. Mintchev, Chern–Simons model and new relations between the Homfly coefficients, *Physics Letters B*, Vol. 238, No. 4, Sept. 28 (1989), pp. 489–494.
[52] Dror Bar-Natan, *Perturbative Aspects of the Chern–Simons Topological Quantum field Theory*, Ph. D. Thesis, Princeton University, June 1991.
[53] Raoul Bott and Clifford Taubes, On the self-linking of knots, Jour. Math. Phys. **35** (1994), pp. 5247–5287.
[54] Daniel Altshuler and Laurent Friedel, Vassiliev knot invariants and Chern–Simons perturbation theory to all orders, preprint (1996).
[55] J. Goldman and L. H. Kauffman, Rational Tangles, (to appear in Advances in Appl. Math.)

10

LOCALLY SELF-AVOIDING WALKS

D. Iagolnitzer[1] and J. Magnen[2]

[1]Service de Physique Théorique
CEA, CE-Saclay F-91191 Gif sur Yvette Cedex
France
[2]de Physique Théorique, CNRS, URA 14
Ecole Polytechnique — 91128 Palaiseau Cedex
France

1. INTRODUCTION

The Edwards model [1] of (possibly weakly) self-avoiding walks plays an important role in polymer theory. In that model, a polymer is represented by a continuous walk subject to a self-avoiding constraint imposed via a δ-potential between any pair of points (the polymer cannot occupy twice the same position in space). Relevant quantities, such as the partition and 2-point correlation functions for a given length T of the polymers, are integrals over all walks of length T starting from a given point or going from points x to y. They are *à priori* ill-defined in view of short-distance problems, but can be well defined (after renormalization) in space dimensions 2 and 3 [2,3]. Nevertheless, there is so far no result on the main problem of physical interest, namely the behavior of such quantities in the $T \to \infty$ limit. Results on that problem have been obtained only in space dimensions $d \geq 4$, in which case some ultraviolet cut-off, regularizing interactions at short distances, has to be introduced. Besides perturbative results [4] and results in the hierarchical approximation [5], some rigorous results have been obtained in that direction in the eighties through probabilistic methods or correlation inequalities [6]. More complete results have then been achieved in space dimensions ≥ 5 through lace expansions [7] or, at weak coupling, in the marginal dimension 4 through renormalization group analysis [8]. More precisely, results of [8] apply to a related model, also due to Edwards, in which quantities treated correspond to integrals over all possible lengths, and then give information for large separation of endpoints. Results can probably be established for the model of self-avoiding walks itself in the $T \to \infty$ limit either [9] through an extension of the analysis of [8], or in a direct way [10].

Our purpose here is different. We consider, in any space dimension, a simplified version of the Edwards model. First, an ultraviolet cut-off is again introduced: a polymer of length T will be represented by a discrete walk (=chain) of $T+1$ points at distances $0, 1, \cdots, T$ from the origin along the walk, with a related regularization of the δ-potential. The crucial simplification is then the following: the self-avoiding constraint is imposed only between

Functional Integration: Basics and Applications
Edited by Cécile DeWitt-Morette, Plenum Press, New York, 1997

points whose distance along the walk is at most equal to some fixed N. In view of the local character of this constraint, basic modifications of the behavior of the theory at large T are not expected with respect to the standard free case. The model, which is not gaussian, will become very quickly gaussian asymptotically (much quicker than in usual super-renormalizable or asymptotically free renormalizable models), with however renormalized effective parameters. Explicit expressions for the latter will be given at $N = 1$ and 2.

This model has been extensively studied, mainly as a toy model for self avoiding walks [11].

In this lecture, we will illustrate in a simple way some of the general methods used in statistical mechanics and constructive field theory. As matter of fact, in the case $N = 1$ the expansion is exactly the lace expansion. The model and results are presented in Sect. 2 and proofs are given in Sect. 3. The ultraviolet cutoff can be removed in dimensions 2 and 3 (for example using a kind of lace expansion; in preparation).

This model is one of the simplest example of a non Gaussian functional integral, that is of an integral (non Gaussian) over an infinite number of degrees of freedom.

On this example one sees some of the general features of these integrals. First they generates infinities:

The partition function, for a coupling constant λ, is $Z(\lambda,T) \sim e^{-a_N(\lambda)T}$, which is only the renormalized quantity $Z(\lambda,T)\, e^{a_N(\lambda)T}$ which has a finite limit.

The mean value of all the (locally self-avoiding) path starting at the origin and being at point x at time T is given asymptotically as $T \longrightarrow \infty$ by (theorem 1)

$$(2\pi)^{-d/2}\chi_N(\lambda)e^{-a_N(\lambda)T}(C_N(\lambda)T)^{-d/2}\, e^{-C_N(\lambda)^{-1}x^2/(2T)}$$

where $\chi_N(\lambda)$, $(C_N(\lambda) \longrightarrow 1$ as $\lambda \longrightarrow 0$. More generally the asymptotic behavior of the correlation functions of the locally self-avoiding walks is similar to the ones of the free theory ($\lambda = 0$), but with non trivial effective parameters

$$a_N(\lambda), \quad C_N(\lambda), \quad \chi_N(\lambda).$$

This is described by stating that the parameters of the free theory $a_N(0) = 0$, $C_N(0) = 1$, $\chi_N(0) = 1$ have been renormalized i.e. their value is λ dependent. For more singular interactions these renormalizations become infinite; e.g. for the self-avoiding walks (on a lattice) in 4 dimensions $C_N(\lambda) \sim (lnT)^{1/4}$ as $T \longrightarrow \infty$. For even more singular cases such as the self avoiding walks in three dimensions, the asymptotic theory has not the scaling laws of the free theory.

2. THE MODEL AND RESULTS

We consider walks σ, $t \longrightarrow \sigma(t)$, where t is the length parameter (measured from the origin along the walk) and $\sigma(t)$ is the position of the walk in space for length t; $\sigma(t) \in \mathbb{R}^d$ where d is the space dimension. It will also be denoted σ_t in the discretized version introduced below. An example of a walk is shown in Figure 1: In the discretized version treated in Sect. 3, the length parameter t will take the integer values $0, 1, 2, \cdots, T$ where T is the total length of the walk. The Edwards model of self-avoiding walks introduces a factor $\exp[-\lambda\ \delta(\sigma(t) - \sigma(t'))]$ for each pair $t, t', t \neq t'$. In the discretized version, δ will be replaced by a regularized function $\hat{\delta}$ such as a normalized gaussian with a small fixed width (otherwise, the theory would be trivial: the above factors would take value zero, instead of 1 in the free case, but only on subsets of

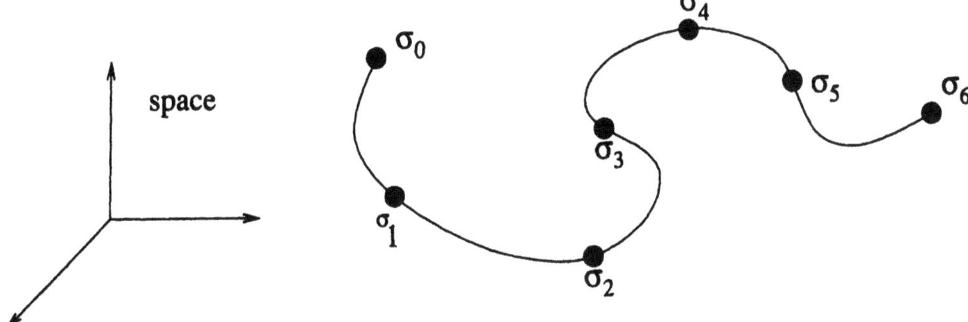

Figure 1. A walk σ.

zero measure in the space of variables $\sigma_0, \sigma_1, \ldots, \sigma_T$, so that the values of integrals defining partition or correlation functions would not be modified). The factors $\exp[-\lambda\ \hat{\delta}(\sigma_t - \sigma_{t'})]$ will nevertheless entail that σ_t cannot be close to $\sigma_{t'}$ at different lengths t, t', at least in a weak sense at small λ. In the local version of the model to be treated, the above constraint applies only when $|t-t'| \leq N$, N fixed finite.

For definiteness, we shall consider the partition function Z_N and the 2-point correlation F_N which are integrals over all walks of length T, starting at a given point x at $t=0$ in the case of Z_N, or going from points x to y in the case of F_N (Z_N does not depend on x and F_N depends only on $y-x$):

$$Z_N(\lambda, T) = \int \delta(\sigma_0 - x)\{\exp -\lambda[\sum_{\substack{t,t'=0,\cdots,T \\ t<t';t'-t\leq N}} \hat{\delta}(\sigma_{t'} - \sigma_t)]\}d\mu(\sigma) \tag{1}$$

$$F_N(x,y;\lambda,T) = \int \delta(\sigma_0 - x)\delta(\sigma_T - y)\{\exp -\lambda[\sum_{\substack{t,t'=0,\cdots,T \\ t<t';t'-t\leq N}} \hat{\delta}(\sigma_{t'} - \sigma_t)]\}d\mu(\sigma) \tag{2}$$

where $d\mu(\sigma)$ is the free measure

$$d\mu(\sigma) = \Big(\prod_{i=1}^{T} c_0 \exp -\frac{(\sigma_i - \sigma_{i-1})^2}{2}\Big)\prod_{i=0}^{T} d\sigma_0 d\sigma_1 \cdots d\sigma_T \tag{3}$$

with $c_0 = (2\pi)^{-d/2}$ (normalization constant).

At $\lambda = 0$ (free random walks), F_N and Z_N (which then do not depend on N) are equal respectively to:

$$F(x,y\,0,T) = (2\pi\ T)^{-d/2} e^{-(y-x)^2/(2T)} \tag{4}$$

$$Z(0,T) = 1 \tag{5}$$

At $\lambda > 0$, we shall mainly restrict our attention to $N=1$ and $N=2$ but it is clear that the methods apply similarly for other values of N. It will be useful to use the variables

$$b_i = \sigma_i - \sigma_{i-1} \quad i = 1, \cdots, T \tag{6}$$

The functions Z_1 and Z_2 then take the form:

$$\begin{aligned} Z_1(\lambda, T) &= \int \prod_{i=1}^{T} e^{-\lambda\hat{\delta}(b_i)} \times c_0 e^{-b_i^2/2} db_i \\ &= \Big[\int e^{-\lambda\hat{\delta}(b)} c_0 e^{-b^2/2} db\Big]^T \end{aligned} \tag{7}$$

$$Z_2(\lambda,T)=\int\{\exp-\lambda[\sum_{i=1}^{T}\hat{\delta}(b_i)+\sum_{i=1}^{T-1}\hat{\delta}(b_i+b_{i+1})]\}\prod_{i=1}^{T}(c_0e^{-b_i^2/2})db_i$$
$$=\int\exp-[\lambda\hat{\delta}(b_1)+\lambda\hat{\delta}(b_1+b_2)+\lambda\hat{\delta}(b_2)+\cdots+\lambda\hat{\delta}(b_{T-1}+b_T)+\lambda\hat{\delta}(b_T)]$$
$$\times\prod_{i=1}^{T}(c_0e^{-b_i^2/2})db_i \tag{8}$$

The functions F_1 and F_2 are defined similarly with the inclusion of the further factor $\delta(b_1+b_2+\cdots+b_T-x)$ in the integrand:

$$F_1(0,x\,\lambda,T)=\int\delta(b_1+b_2+\cdots+b_T-x)\prod_{i=1}^{T}e^{-\lambda\hat{\delta}(b_i)}c_0e^{-b_i^2}db_i \tag{9}$$

$$F_2(0,x\,\lambda,T)=\int\delta(b_1+b_2+\cdots+b_T-x)\exp-[\lambda\hat{\delta}(b_1)+\lambda\hat{\delta}(b_1+b_2)+\cdots]$$
$$\times\prod_{i=1}^{T}c_0e^{-b_i^2/2}db_i \tag{10}$$

The following result (which reduces to (4) (5) at $\lambda=0$) is established in Sect. 3 for $N=1$ and 2 and can be established similarly for any given N:

Theorem 1: There exist $\lambda_N>0$, and functions $\chi_N(\lambda)$, $a_N(\lambda)$, $C_N(\lambda)$ (where a_N and C_N are given explicitly below at $N=1$ and 2), $a_N(\lambda)$ of first order in $\lambda, a_N(\lambda)>0$ at $\lambda>0$, $\chi_N(\lambda)$ and $C_N(\lambda)$ close to 1 at small λ (and equal to 1 at $\lambda=0$), such that, for $\lambda<\lambda_N$:

(i)

$$Z_N(\lambda,T)=\chi_N(\lambda)e^{-a_N(\lambda)T}(1+R_N(\lambda,T)) \tag{11}$$
$$|R_N(\lambda,T)|<(cst_N\,\lambda)^{T/(N-1)} \tag{12}$$

(ii)

$$F_N(0,x\,\lambda,T)=(2\pi)^{-d/2}\chi_N(\lambda)e^{-a_N(\lambda)T}(C_N(\lambda)T)^{-d/2}\left[e^{-C_N(\lambda)^{-1}x^2/(2T)}+R_N'(x,\lambda,T)\right] \tag{13}$$

$$|R_N'(x,\lambda,T)|<\frac{cst_N\times\lambda}{T} \tag{14}$$

Notes

(i) The coefficient $a_N(\lambda)$ corresponds, from the viewpoint of field theory, to mass renormalization; $C_N(\lambda)$ is related to wave function renormalization.

(ii) The bound (14) is of interest when x^2 is not too large with respect to T. A more detailed analysis of the remainder would give similarly an expansion as a sum of well defined terms in powers of $1/T$, up to any order n, with a remainder satisfying a bound in $1/T^{n+1}$.

(iii) Theorem 1 applies in any space dimension d. The more refined methods of [7] can be used to get results valid for all N (finite or infinite) and all λ in dimensions ≥ 5. Those of [8–10] will give results valid for all N, in dimension 4, at λ sufficiently small independently of N. In the latter case, the error term R in (11) will then satisfy a bound

of the form $cst/\ell nT$ in the Edwards model (with ultraviolet cut-off), or more generally $[\lambda^{-1}+cst\ \ell nN(T)]^{-T/N(T)}$ for any N arbitrarily large, with $N(T)=\mathrm{Inf}(N,T)$).
In dimensions 2 or 3, methods of [8–10] should give results valid with λ_N of the form $cst/N^{(2-d/2)}$ in agreement with heuristic arguments. Values of λ_N following from methods of Sect. 3 are smaller.

(iv) In space dimensions 2 and 3, Th. 1 can also be proved, as already mentioned in the Introduction, for a model in which t takes continuous values and with a true delta function.

We give below three proofs. The first one is via a standard cluster expansion which doesn't take advantage of the one time dimension of the time axis; it is given here as an illustration of this technique in a very simple example. The second one is an adaptation of the Mayer procedure to this one dimensional problem (the time dimension). The third one is also one dimensional and is also standard [7,12].

3. FIRST PROOF OF THEOREM 1

3.1. Explicit Forms of the Coefficients

At $N=1$, one has trivially (see (7)):

$$a_1(\lambda)=|\log I_1(\lambda)| \tag{15}$$

where

$$I_1(\lambda)=\int e^{-\lambda\hat{\delta}(b)}c_0e^{-b^2/2}db \qquad (<1 \text{ at } \lambda>0) \tag{16}$$

At $N=2$, one has

$$\begin{aligned} a_2(\lambda)&=|\log z_0(\lambda)| \\ &=|\log I_1(\lambda)|-K(\lambda) \end{aligned} \tag{17}$$

K is given by the following explicit expansion, convergent at small λ (and is alternatively [12], as $z_0(\lambda)$, the solution of an implicit equation indicated later):

$$K(\lambda)=\sum_{p\geq 1}\ \sum_{\substack{\alpha_1\cdots\alpha_p \\ 2\leq\alpha_1\leq\alpha_2\cdots\leq\alpha_p}} c_{\alpha_1,\cdots,\alpha_p}J_{\alpha_1}(\lambda)\cdots J_{\alpha_p}(\lambda) \tag{18}$$

where

$$J_\alpha(\lambda)=I_\alpha(\lambda)/I_1(\lambda)^\alpha \tag{19}$$

$$\begin{aligned} I_\alpha(\lambda)=(-\lambda)^{\alpha-1}\int db_1\ldots db_\alpha\int_0^1\cdots\int_0^1 dh_{1,2}dh_{2,3}\cdots dh_{\alpha-1,\alpha}\prod_{i=1}^{\alpha-1}\hat{\delta}(b_i+b_{i+1}) \\ \times\exp-\lambda[\hat{\delta}(b_1)+h_{1,2}\ \hat{\delta}(b_1+b_2)+\hat{\delta}(b_2)+\cdots+h_{\alpha-1,\alpha}\ \hat{\delta}(b_{\alpha-1}+b_\alpha)+\hat{\delta}(b_\alpha)] \\ \times\prod_{i=1}^{\alpha}c_0\ e^{-b_i^2/2} \end{aligned} \tag{20}$$

$$c_{\alpha_1,\cdots,\alpha_p} = \frac{1}{p!} \sum_{\substack{\text{connected graphs } G \\ \text{linking the indices } (1,\cdots,p)}} (-1)^{|G|} \times$$

$$\times\{\text{Number of sets of intervals } L_1,\cdots L_p \text{ of} > 0 \text{ integers such that}$$

$$|L_i| \geq 2,\ i=1,\cdots,p,\ \{|L_i|\} \equiv \{\alpha_i\}, \bigcup_{i=1}^{p} L_i \text{ contains the integer 1 and}$$

$$L_i \cap L_j \text{non empty if } (i,j) \in G\} \tag{21}$$

In (21), an interval L is a set of successive integers, $|L|$ denotes the number of integers (=sites) in L, for instance $|L| = 3$ if $L = (4,5,6)$, graphs G are sets of lines linking the indices $1,\cdots,p$ with at most one line between any pair of indices and $|G|$ is the number of lines in G. An example of a graph G and of a set of intervals $L_1,\cdots,L_p$ satisfying the conditions stated in (21) is shown in Fig. 2 at $p=4$. It follows from the analysis of Sect. 3 that the series (18) is convergent and defines $K(\lambda)$ at small λ. (In view of the easy bounds $|I_\alpha(\lambda)| < (cst\ \lambda)^{\alpha-1}$ and of the fact that $I_1(\lambda)$ is close to 1, the term of order p in (18) will be smaller than $(cst\ \lambda)^p$).

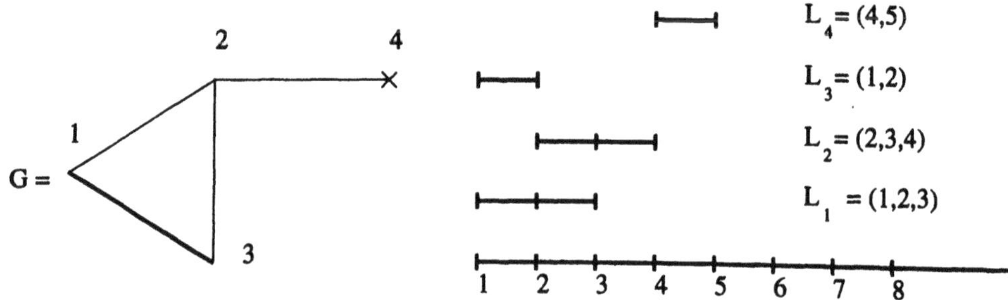

Figure 2. A graph G and a set $(L_1,\cdots,L_p)$ contributing to (21) and (38) at $p=4$, $\alpha_1=\alpha_2=2$, $\alpha_3=\alpha_4=3$.

We finally give the explicit form of $C_1(\lambda)$ and $C_2(\lambda)$

$$C_1(\lambda) = J_{2,1}(\lambda) = I_{2,1}(\lambda)/I_1(\lambda) \tag{22}$$

$$C_2(\lambda) = \frac{\sum_{\alpha\geq 1} J_{2,\alpha}(\lambda) K'_{(\alpha)}(\lambda)}{\sum_{\alpha\geq 1} \alpha J_\alpha(\lambda) K'_{(\alpha)}(\lambda)} \tag{23}$$

where $J_{2,\alpha}(\lambda) = I_1(\lambda)^{-\alpha} I_{2,\alpha}(\lambda)$ and $I_{2,\alpha}(\lambda)$ is defined in a way similar to $I_\alpha(\lambda)$ in (16) at $\alpha=1$ or in (20) at $\alpha \geq 2$, except that the further factor $\sum_{i=1}^{\alpha} b_i^2$ is included in the integrand of the right-hand side; $K'_{(\alpha)}(\lambda)$ is defined by an expansion similar to (18) with coefficients $c'(\alpha;\alpha_1,\cdots,\alpha_p)$, where $c' = \lim_{n\to\infty} c'_n$ and c'_n has the same definition as $c_{\alpha,\alpha_1,\cdots,\alpha_p}$ except that the condition $1 \in \cup L_i$ is replaced by $L_1 = [n, n+\alpha]$.

As explained at the end of Sect. 3, the series in the numerator and denominator of (24) are convergent (and both are close to 1) at small λ.

3.2. Proof of Theorem 1 at $N=1$

Part (i) is trivial as already stated.

To prove Part (ii), one may replace

$$\delta(b_1+\cdots+b_T-x)$$

in (9) by

$$(2\pi)^{-d}\int e^{i\tau(b_1+\cdots+b_T-x)}d\tau,$$

so that

$$F(0,x;\lambda,T)=(2\pi)^{-d}\int e^{-i\tau x}(I_1(\lambda,\tau))^T d\tau \tag{24}$$

where

$$I_1(\lambda,\tau)=\int e^{-\lambda\hat{\delta}(b)}c_0 e^{-b^2/2}e^{i\tau b}db \tag{25}$$

We first show:

Lemma 1

$I_1(\lambda,\tau)$ can be written in the form

$$I_1(\lambda,\tau)=I_1(\lambda)[e^{-C_1(\lambda)\tau^2}+h(\lambda,\tau)] \tag{26}$$

where $C_1(\lambda)$ is defined in Eq. (22) and

$$|h(\lambda,\tau)|<cst\ \lambda\,\mathrm{Inf}(\tau^4,(1+|\tau|)^{-1}) \tag{27}$$

Proof of Lemma 1

Eq. (27) defines $h(\lambda,\tau)$ as $I_1(\lambda,\tau)/I_1(\lambda)-e^{-C_1(\lambda)\tau^2}$. As easily seen, it vanishes at $\lambda=0$ and terms of order $0,1,2,3$ in its expansion in τ also vanish in view of the definition of $C_1(\lambda)$ in (22). The precise bound (27) with the fall-off factor in $(|\tau|)^{-1}$ at large τ, can be obtained by considering $h(\lambda,\tau)$ as the Fourier transform with respect to b (up to a power of 2π) of the function

$$g(\lambda,b)=I_1(\lambda)^{-1}e^{-\lambda\hat{\delta}(b)-b^2/2}-\frac{1}{C_1(\lambda)^{d/2}}e^{-b^2/2C_1(\lambda))} \tag{28}$$

Since $\hat{\delta}$ is regular, e.g. C^∞, g is also regular and the result follows easily.

We now come back to the proof of part (ii) of Th. 1, starting from Eqs. (24),(26). Eq. (26) gives:

$$(I_1(\lambda,\tau))^T=I_1(\lambda)^T[e^{-C_1(\lambda)\tau^2T}+\sum_{p=1}^{T}\binom{p}{T}(e^{-C_1(\lambda)\tau^2})^{T-p}(h(\lambda,\tau))^p] \tag{29}$$

The Fourier transform in τ corresponding to the first term in the bracket gives the dominant term in (13). The Fourier transform of remaining terms will be bounded in modulus by replacing $e^{-i\tau x}$ by 1 in the integrand. The sum from 1 to a fraction of T, say $T/10$, is easily bounded at small λ by using the bound $cst\ \lambda\tau^4$ on $|h|$, while the sum from $T/10$ to T is easily bounded (at small λ) by using the bound $cst\ \lambda(1+|\tau|)^{-1}$. In the first sum, one may bound $\binom{p}{T}$ by $T^p/p!$, the integral $\int e^{-cst\ T\tau^2}(\tau^4)^p d\tau$ by $(cst\ T)^{-2p-d/2}p!^2$, and finally $p!$ by $(cst\ T)^{p-1}$. The bound on h in τ^4 (rather than τ^2) is at the origin of the factor $1/T$ (besides $T^{-d/2}$) in the bound. For the second sum, one may bound $\binom{p}{T}$ by 2^T. This sum decreases exponentially in T at small λ. The bound (14) follows. Q.E.D.

3.3. Proof of Part (i) of Theorem 1 at $N = 2$

We start from the definition of Z_2 given in Eq. (8) (and leave the index 2 implicit below). We introduce auxiliary variables $h_{1,2}, h_{2,3}, \cdots, h_{T-1,T}$ and define $Z(\lambda, T; \{h\})$ as follows: each factor $\lambda\hat{\delta}(b_i + b_{i+1})$ in the exponent of the right-hand side of (8) is replaced by $\lambda h_{i,i+1}\hat{\delta}(b_i + b_{i+1})$, so that $Z(\lambda, T)$ is equal to $Z(\lambda, T; \{1\})$ while putting some of the variables $h_{i,i+1}$ at zero amounts to remove corresponding interactions. We then use successive first order Taylor expansions around the origin in each variable $h_{i,i+1}$. Equivalently, we directly apply formula (A-1) of the Appendix to $Z(\lambda, T)$ with respect to the functions $f_\alpha = \hat{\delta}(b_\alpha + b_{\alpha+1}), \alpha = 1, \cdots, T-1$. This gives easily the following cluster expansion:

$$Z(\lambda, T) = \sum_{p=1}^{T} \sum_{\substack{L_1, \cdots, L_p \\ L_1 < L_2 < \ldots < L_p; L_1 \cup L_2 \cdots \cup L_p = (1, \cdots, T)}} \prod_{j=1}^{p} I_{|L_j|}(\lambda) \tag{30}$$

where $I_\alpha(\lambda)$ is defined in (20) and the sum runs over all sets of successive (non overlapping) intervals whose union covers $(1, \cdots, T)$. As already mentioned below (21), an interval is here a set of successive integers; in agreement with explanations just given, the notation $L_i < L_j$ in (31) means that all integers in L_i are strictly smaller than those of L_j. Each term in the right-hand side of (30) corresponds to a choice of those variables $h_{i,i+1}$ which are fixed at zero, others being integrated from 0 to 1: fixing $h_{i,i+1}$ at zero entails that i and $i+1$ belong to different intervals. An example of a set $L_1, \cdots, L_p$ is shown in Figure 3 when $h_{3,4} = 0$, $h_{5,6} = h_{6,7} = h_{7,8} = 0$ while all others are integrated from 0 to 1:

Figure 3. A set $(L_1, \cdots, L_p)$ contributing to (31) at $p = 5$: $L_1 = (1,2,3), L_2 = (4,5), L_3 = (6), L_4 = (7), L_5 = (8, \cdots, T)$.

The expansion (30) is first obtained with functions $I_{L_j}(\lambda)$ defined for each interval L_j: if L is the interval $(\ell+1, \cdots, \ell+\alpha)$, this function is defined as in (20) with all indices i replaced by $i+\ell$: there is in particular one integration variable h and one regularized delta-function for each pair of successive integers i, $i+1$ that belong to L. However, I_L clearly depends only on the number of sites in L and coincides with $I_{|L|}$.

We prove below Part (ii) of Th. 1 with the explicit form of $a_2(\lambda)$ given in Sect. 2.

It is convenient to first write (30) in the form:

$$Z(\lambda, T) = I_1(\lambda)^T \sum_{p=1}^{T} \sum_{\substack{L_1, \cdots, L_p \\ L_1 < L_2 \cdots < L_p \\ |L_i| \geq 2, i = 1, \cdots, p \\ \cup_i L_i \subset (1, \cdots, T)}} \prod_{i=1}^{p} J_{|L_i|}(\lambda) \tag{31}$$

where $J_\alpha(\lambda)$ is defined in (19) and the sum Σ now runs over all possible sets of successive (non overlapping) intervals $L_1, \cdots, L_p$ such that $|L_i| \geq 2, \quad i = 1, \cdots, p$; but there may now be one or more integers between two successive intervals and the union of these intervals does not necessarily cover $(1, \cdots, T)$: intervals of length 1 have been extracted.

As already mentioned, $I_1(\lambda)$ is close to 1 at small λ, with $I_1(\lambda) < 1$ at $\lambda > 0$. We note on the other hand that I_α and J_α satisfy bounds of the form:

$$|I_\alpha(\lambda)|, |J_\alpha(\lambda)| < (cst\ \lambda)^{\alpha-1} \tag{32}$$

Eq. (31) is then rewritten in the form

$$Z(\lambda,T) = I_1(\lambda)^T \sum_{p=1}^{\infty} \frac{1}{p!} \sum_{\substack{L_1 \cdots L_p \\ |L_i|\geq 2, \cup L_i \subset (1,\cdots,T)}} \prod_{i=1}^{p} J_{|L_i|}(\lambda) \times$$
$$\times \prod_{\substack{(i,j)\, i;j=1,\cdots,p \\ i\neq j}} (1+v(L_i,L_j)) \tag{33}$$

where $v(L_i,L_j) = -1$ if L_i, L_j overlap (i.e. have one or more common integers) so that $1+v=0$ and $v(L_i,L_j) = 0$ otherwise. The sum Σ now runs over all ways of choosing the (possibly overlapping) intervals $L_1,\cdots,L_p$, $|L_i| \geq 2, i = 1,\cdots,p$ independently. The sum over p is written from 1 to infinity, although all terms with $p > T$ clearly vanish at this stage.

An expansion of the last factors and easy regroupings of terms (putting together intervals linked in a connected way by non zero factors $v(L_i,L_j)$) gives in turn, so far formally since this procedure now introduces infinite series:

$$Z(\lambda,T) = I_1(\lambda)^T \sum_{q=1}^{\infty} \frac{1}{q!} \sum_{\substack{L'_1 \cdots L'_q \\ L'_i \subset [1,T], |L'_i|\geq 2}} \prod_{j=1}^{q} K_{|L'_j|}(\lambda) \tag{34}$$

$$= I_1(\lambda)^T e^{\sum_{L\subset[1,T],|L|\geq 2} K_{|L|}(\lambda)} \tag{35}$$

$$= I_1(\lambda)^T e^{\sum_{\alpha;2\leq\alpha\leq T} K_\alpha(\lambda)(T-\alpha+1)} \tag{36}$$

where

$$K_\alpha(\lambda) = \sum_{p\geq 1} \frac{1}{p!} \sum_{\substack{L_1,\cdots,L_p \\ L_i\subset[1,\alpha],|L_i|\geq 2 \\ \cup L_i = (1,\cdots,\alpha)}} \prod_{i=1}^{p} J_{|L_i|}(\lambda)$$
$$\times \sum_{\substack{\text{connected graphs } G \\ \text{linking the indices } 1,\cdots,p}} \prod_{\ell\in G} v_\ell \tag{37}$$

In (38), the sum Σ runs again over independent intervals $L_1,\cdots,Lp$ satisfying the conditions indicated and $v_\ell \equiv v(L_i,L_j)$ if line ℓ joins indices i,j. As in (21), graphs G have at most one line between any pair of indices. An example of a set $L_1,\ldots,L_p$is shown in Fig. 2 at $p=4$.

Formula (38) is not yet satisfactory for showing (by taking absolute values) bounds and convergence properties of the series involved: the number of graphs G is too large. Some further regroupings of terms are first needed. Various methods allowing one to transform the sum over graphs into a sum over trees can be used to that purpose. The most elegant one to our knowledge is the use of formula (A.3) [12] of the Appendix (a somewhat more direct alternative proof is given in the Appendix). Applied in (34) to the product of the factors $1+v(L_i,L_j)$, it directly allows one to replace in (38) the sum over connected graphs by a sum over connected trees, up to factors bounded by 1 in modulus in our situation. As a matter of fact, whatever the functions $v_{i,j}$ are, as a byproduct of the same analysis:

$$\sum_{\substack{\text{conn. graphs } G \\ \text{linking } 1,\cdots,p}} \prod_{\ell\in G} v_\ell = \sum_{\substack{\text{conn. trees } \mathcal{T} \\ \text{linking } 1,\cdots,p}} \prod_{\ell\in\mathcal{T}} v_\ell \times$$
$$\times \int_0^1 \cdots \int_0^1 dh_1 \cdots dh_{p-1} \prod_{(i,j)\notin\mathcal{T}} (1+h_{(i,j)}(\mathcal{T})v_{i,j}) \tag{38}$$

where there is one variable h_l for each line of the tree, the last product runs over pairs (i,j) of indices that are not joined by a line of $\mathcal{T}$ and $h_{(i,j)}(\mathcal{T})$ is the minimal value of all variables $h_{\ell'}$ along the path in the tree $\mathcal{T}$ allowing are to join indices i,j.

The bounds (33) then yield:

$$|K_\alpha(\lambda)| < \sum_p \frac{1}{p!} \sum_{\mathcal{T}} \sum_{\substack{L_1,\cdots,L_p |L_i| \geq 2 \\ \cup L_i = [1,\alpha] \\ L_i \cap L_j \neq \emptyset \text{ if } (i,j) \in \mathcal{T}}} (cst\ \lambda)^{\Sigma_i(|L_i|-1)} \tag{39}$$

To show convergence of the series in the right-hand side of (40), one may e.g. use Cayley theorem: the number of trees joining p indices with coordination numbers $d_1,\cdots,d_p$ is equal to $(p-2)!/\Pi(d_i-1)!$. (The coordination number is the number of lines involved at a given site). All trees can be written in the form shown in Figure 4: Given a site i with $d_i - 1$ lines

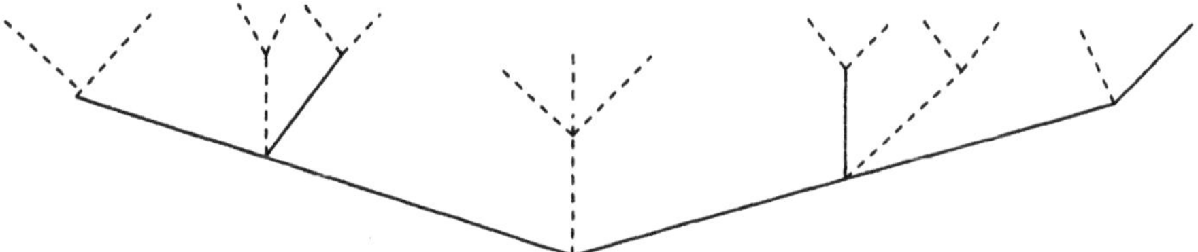

Figure 4. A general tree (only the first strata are shown).

issued from it and going down, each one of intervals L_j associated to a next site j must overlap L_i. This gives at most $|L_i|^{|L_j|}$ possibilities for placing L_j. Taking all sites into account amounts to attribute a factor $|L_i|^{d_i}$ to each site i. Summation over all possible d_i's will give a factor $\Sigma_{d_i} \frac{|L_i|^{d_i}}{(d_i-1)!}$ where the factor $(d_i-1)!$ in the denominator arises from the number of trees with given coordination numbers. This factor is bounded by $cst^{|L_i|}$ and is thus controlled at small λ by the factors $(cst\ \lambda)^{\Sigma_i(|L_i|-1)}$ in (40) ($|L_i| \geq 2$). The convergence of the series (40) follows. A slightly more careful analysis, taking into account the fact that $\Sigma_i(|L_i|-1) \geq \alpha + p - 1$ (in view of the overlap conditions) gives a bound on $|K_\alpha(\lambda)|$ analogous to (33):

$$|K_\alpha(\lambda)| < (cst\ \lambda)^{\alpha-1} \tag{40}$$

Results stated in Sect. 1 at $N=2$, i.e. Eqs. (11), (12), (17), and (18) are obtained easily in turn from (36) with

$$\chi(\lambda) = exp - [\Sigma_{\alpha\geq 2}(\alpha-1)K_\alpha(\lambda)],\ 1+R(\lambda,T) = exp - [\Sigma_{\alpha \geq T+1}(T-\alpha+1)K_\alpha(\lambda)] \tag{41}$$

and

$$K(\lambda) = \Sigma_{\alpha\geq 2} K_\alpha(\lambda) \tag{42}$$

3.4. Proof of Part (ii) of Theorem 1 at $N=2$ (sketch)

As in Sect. 3, $F_2(0,x;\lambda T)$ is expressed as the Fourier transform in τ of a function $F_2(\tau;\lambda,T)$. The latter admits a cluster expansion similar to (30):

$$F_2(\tau;\lambda,T) = \sum_p \sum_{L_1,\cdots,L_p} \prod_{i=1}^{p} I_{|L_i|}(\lambda,\tau) \tag{43}$$

with the same conditions on the second sum as in (31).
The analogue of (26) is now, for each L_i:

$$I_{Li}(\lambda,\tau) = e^{-\tau^2 C_2(\lambda)|L_i|}\left(I_{L_i}(\lambda,0) - \tau^2\left\{I_{2,L_i}(\lambda,0) - C_2(\lambda)|L_i|I_{L_i}(\lambda,0)\right\}\right) + \text{remainder in } \tau^4. \tag{44}$$

In turn:

$$\begin{aligned} F_2(\tau;\lambda,T) = I_1(\lambda,0)^T e^{-\tau^2 T C_2(\lambda)} & \Bigg[\sum_{\substack{L_1,\cdots,L_p \\ |L_i|\geq 2 \\ \text{non overlapping}}} \prod_{i=1}^{p} J_{L_i}(\lambda,0) - \tau^2 \sum_{L,|L|\geq 1} \sum_{\substack{L_1,\cdots,L_p \\ |L_i|\geq 2 \\ L,L_1,\ldots,L_p \text{non overlapping}}} \\ & \times \left[J_{2,|L|}(\lambda,0) - C_2(\lambda)|L|J_{|L|}(\lambda,0)\right] \prod_{i=1}^{p} J_{|L_i|}(\lambda,0)\Bigg] \\ & + \text{remainder in } \tau^4 \end{aligned} \tag{45}$$

Using the same procedure as that used for Z_2 (introduction of factors $1+v_{ij}$ and expansions), one obtains in turn the following expression for the coefficient of τ^2:

$$\sum_{\alpha} \sum_{L,|L|=\alpha} \left(J_{2,\alpha}(\lambda)K'_{(\alpha)}(\lambda) - C_2(\lambda)\alpha J_\alpha(\lambda)K'_{(\alpha)}(\lambda)\right) \tag{46}$$

where the sum over α runs from 2 to T. The choice of $C_{\in}(\lambda)$ given in Sect. 1 ensures that the leading term, proportional to T, arising from (50) vanishes. The error terms in (49) are bounded as in the Z_2 case, so that they contribute to corrections in $1/T^{(d/2)+1}$.

4. SECOND PROOF OF THEOREM 1 (ONLY PART I))

It is convenient to introduce

$$I'_1(\lambda) = I_1(\lambda) - I_1(0) = I_1(\lambda) - 1 = 0(\lambda); I'_\alpha(\lambda) = I_\alpha(\lambda) \text{ for } \alpha > 1 \tag{47}$$

We start from the expression (30) of $Z(\lambda,T)$ which is with a slight change in the notations due to the fact that we replace I_1 by $1+I'_1$

$$Z(\lambda,T) = 1 + \sum_{p=1}^{T} \sum_{\substack{t_1,t'_1\cdots,t_p,t'_p \\ t_i<t'_i\leq t_{i+1} \quad i=1,\ldots,p-1 \\ 0\leq t_1 \ t_{p+1}<t'_p\leq T}} \prod_{j=1}^{p} I'_{|t'_j-t_j|}(\lambda) \tag{48}$$

For each sum on the t'_j's, starting with $j=p-1$ we write

$$\sum_{t'_j;\ t_j<t'_j;t'_j\leq t_{j+1}} I'_{|t'_j-t_j|} = \sum_{t'_j;\ t_j<t'_j\leq T} I'_{|t'_j-t_j|} - \sum_{t'_j;\ t_{j+1}<t'_j\leq T} I'_{|t'_j-t_j|} \tag{49}$$

We are thus led to introduce the notations

$$\hat{I}(t,T) = \sum_{t',t<t'\leq T} I'_{t'-t}$$

$$R(t_j,t_{j+1};T) = -\sum_{t'_j;\ t_{j+1}<t'_j\leq T} I'_{t'_j-t_j} \tag{50}$$

Then (47) is reexpressed as

$$Z(\lambda,T) = 1+\sum_{t_1<T}\hat{I}(t_1,T)+\sum_{p\geq 2}\sum_{t_1<t'_1\leq\ldots t_{p-1}<t'_{p-1}<t_p} I'_{t'_1-t_1}\cdots I'_{t'_{p-1}-t_{p-1}}\hat{I}(t_p,T)$$

$$= 1+\sum_{t_1<T}\hat{I}(t_1,T)+\sum_{p\geq 2}\sum_{t_1<t'_1\leq\ldots t_{p-1}<t_p} I'_{t'_1-t_1}\cdots I'_{t'_{p-2}-t_{p-2}}$$

$$\left\{\hat{I}(t_{p-1},T)+R(t_{p-1},t_p;T)\right\}\hat{I}(t_p,T)$$

and one obtains finally

$$Z(\lambda,T) = 1+\sum_{k=1}^{T}\sum_{\substack{t_1,\ldots,t_k\\ t_1<\ldots<t_k<T}}\prod_{j=1}^{k}\hat{R}(t_k;T) \tag{51}$$

with

$$\hat{R}(t;T) = \hat{I}(t,T)+\sum_{n\geq 1}\sum_{t=t_1<\ldots<t_n<t_{n+1}<T}\left[\prod_{j=1}^{n}R(t_j,t_{j+1};T)\right]\hat{I}(t_{n+1},T) \tag{52}$$

Thus

$$Z(\lambda,T) = \prod_{t=1}^{T}\left[1+\hat{R}(t;T)\right] \tag{53}$$

and

$$\lim_{T\longrightarrow\infty}\hat{R}(t;T) = \hat{R}_\infty \tag{54}$$

So that

$$Z(\lambda,T) = \left[1+\hat{R}_\infty\right]^T\prod_{t=1}^{T}\left[1+\frac{\hat{R}(t;T)-\hat{R}_\infty}{1+\hat{R}_\infty}\right] \tag{55}$$

where with

$$\delta R(T-t) = \hat{R}(t;T)-\hat{R}_\infty \tag{56}$$

Then

$$\prod_{\alpha=1}^{\infty}\left[1+\frac{\delta R(\alpha)}{1+\hat{R}_\infty}\right] = \left[1+\hat{R}_\infty\right]^T\left\{\prod_{\alpha=1}^{\infty}\left[1+\frac{\delta R(\alpha)}{1+\hat{R}_\infty}\right]\right\}B(T) \tag{57}$$

$$B(T) = \prod_{\alpha>T}^{\infty}\left[1+\frac{\delta R(\alpha)}{1+\hat{R}_\infty}\right] \tag{58}$$

From the bounds of section 3 we have the bounds

$$|I'_\alpha|\leq cst\ \lambda^{sup(1,\alpha-1)};\ |\hat{I}(t,T)\leq cst\ \lambda\ \ |R(t,t';T)|\leq cst\ \lambda^{sup(1,t'-t-1)}$$

$$|\hat{R}(t;T)|\leq cst\lambda \tag{59}$$

In the same way

$$|\hat{R}(t;T) - \hat{R}(t;T')| \leq cst\lambda^{|T-T'|} \tag{60}$$

So that $\hat{R}_\infty \sim \lambda$ exists and

$$|\delta R(T-t)| \leq cst\ \lambda^{T-t} \tag{61}$$

Thus

$$\prod_{\alpha=1}^{\infty} [1 + \frac{\delta R(\alpha)}{1+\hat{R}_\infty}] \text{ exists and is of order } 1 + O(\lambda) \tag{62}$$

and

$$B(T) = 1 + O(\lambda^T) \tag{63}$$

We have then obtained the behavior of $Z(\lambda, T)$

$$Z(\lambda, T) = [1 + \hat{R}_\infty]^T \left\{ \prod_{\alpha=1}^{\infty} \frac{[1 + \delta R(\alpha)}{1+\hat{R}_\infty}] \right\} \left(1 + O(\lambda^T)\right) \tag{64}$$

that is theorem 1 part i).

5. THIRD PROOF OF THEOREM 1 PART (I)

The terms $z_0(\lambda)$ and $K(\lambda)$ satisfy on the other hand (and are in fact the unique solutions at small λ of) the respective equations

$$1 - \sum_{\alpha \geq 1} I_\alpha(\lambda) z^\alpha = 0 \tag{22a}$$

$$e^{-K(\lambda)} + \sum_{\alpha \geq 2} J_\alpha(\lambda) e^{-\alpha K(\lambda)} = 1 \tag{65}$$

We start from the expansion (31) of Z, and define on the other hand the function

$$f(z, \lambda) = [1 - \sum_{\alpha \geq 1} I_\alpha(\lambda) z^\alpha]^{-1} \tag{66}$$

We first check (formally) that

$$f(z, \lambda) = \sum_T Z(\lambda, T) z^T \tag{67}$$

so that $Z(\lambda, T)$ will be equal, for each T, to the Tth derivative of f in z at $z = 0$ (times $T!$).

Proof of (67)

In view of (30), the right-hand side of (45) can be written in the form

$$\sum_p \sum_{\substack{\alpha_1, \cdots, \alpha_p \\ \alpha_i \geq 1}} \sum_{\substack{L_1, \cdots, L_p; \{|L_i|\} = \{\alpha_i\} \\ L_1 < L_2 < \ldots < L_p; L_1 \cup \ldots \cup L_p \text{ is connected and contains } 1}} \prod_{i=1}^{p} (I_{\alpha_i} z^{\alpha_i}) \tag{68}$$

On the other hand, $f(z, \lambda)$ is equal (formally), in view of (66), to

$$\sum_p \left(\sum_{\alpha \geq 1} I_\alpha z^\alpha \right)^p \tag{69}$$

Identification of coefficients of each product $\prod_i I_{\alpha_i} z^{\alpha_i}$ follows easily. Q.E.D.

>From the bounds (32) ($|I_\alpha(\lambda)| \leq (cst\ \lambda)^{\alpha-1}, \alpha \geq 2$), the fact that $I_1(\lambda)$ is close to 1 at small λ and the implicit function theorem, it is seen that f is analytic in z, at small λ, in a large region around the origin with radius of the order of cst/λ, apart from a pole at $z_0(\lambda)$ close to 1, which is the (unique) solution of the equation (65):

$$1 - \sum_{\alpha \geq 1} I_\alpha(\lambda) z^\alpha \equiv 1 - I_1(\lambda) z - \sum_{\alpha \geq 2} I_\alpha(\lambda) z^\alpha = 0 \tag{70}$$

Results analogous to Eq. (11) follow, with $a_2(\lambda)$ of the form (17) and $z_0(\lambda)$, $K(\lambda)$ unique solutions of (65), through a Cauchy formula expressing each Tth derivative of f in z at the origin as an integral, over the contour $C_1 \cup C_2$ shown in Figure 5, of the function $\frac{f(z)}{z^{T+1}}$. Here, C_2 is a circle of radius cst/λ, centered at the origin; C_1 is a circle centered at $z_0(\lambda)$ with radius tending to zero.

The integral over C_2 is easily bounded by $(cst\ \lambda)^T$. It corresponds to the remainder $\chi_N(\lambda) e^{-a_N(\lambda)T} R_N(\lambda, T)$ in (11). The integral over C_1 is equal to $cst/z_0(\lambda)^{T+1}$ times the residue of the pole at $z_0(\lambda)$, which is equal, as easily seen, to $[\sum_\alpha \alpha I_\alpha(\lambda) z_0(\lambda)^{\alpha-1}]^{-1}$. The result is equal to $cst\ z_0(\lambda)^{-T} \times [\sum_\alpha I_\alpha(\lambda) z_0(\lambda)^\alpha]^{-1}$, so that $a_2(\lambda) = |\ln z_0(\lambda)|$.

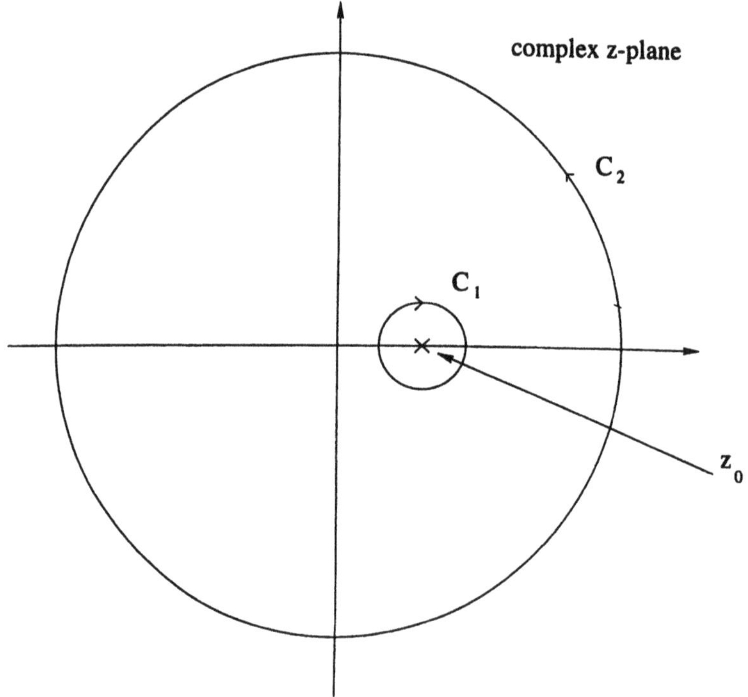

Figure 5. The integration contour $C_1 \cup C_2$.

ACKNOWLEDGMENTS

We thank D. Brydges and B. Duplantier for interesting discussions. We also thank B. Duplantier for a careful reading of the manuscript and suggestions, and D. Brydges for indicating to us the alternative method described at the end of Sect. 3.2. (After this work was achieved, we have learnt from him that he had also considered the model treated here in collaboration with T. Spencer.)

APPENDIX

The following formulae (A.1), (A.3) are useful in Sect. 3. Formula (A.3) is due to Ref. 12. An alternative proof [13] is given below.

They apply respectively to functions $f_\alpha, \alpha = 1, \cdots, n$, and functions $v_{i,j}, i, j = 1, \cdots, n$, $i \neq j$. The functionals F may also depend on other functions which are left implicit.

Lemma 1

$$F(\{f_\alpha\}) = \sum_{S \subset (1,\cdots,n)} \int_0^1 \cdots \int_0^1 \left(\prod_{\alpha \in S} \frac{d}{ds_\alpha}\right) F(\{s_\alpha f_\alpha\})_{|\substack{s_\alpha=0 \\ \text{if } \alpha \notin S}} \times \prod_{\alpha \in S} ds_\alpha \tag{A.1}$$

Proof: Eq. (A.1) can e.g. be proved by induction. At $n = 1$, it is the usual formula:

$$F(f) = F(sf)_{|s=1} = F(0) + \int_0^1 ds \frac{d}{ds} F(sf) \tag{A.2}$$

We next assume that (A.1) applies for a given n and prove it for $n+1$. Let F be a functional of $f_1, f_2, \cdots, f_n, f_{n+1}$. (A.1) is first applied to F with respect to $f_1, \cdots, f_n$ with f_{n+1} "spectator". Then formula (A.2) is applied with respect to f_{n+1}, for each one of the terms obtained. The result follows easily. Q.E.D.

Lemma 2

$$F(\{v_{i,j}\}) = \sum_{\substack{\text{trees } \mathcal{T} \text{ (connected} \\ \text{or not connected}}} \int_0^1 \cdots \int_0^1 dh_1 \cdots dh_{|\mathcal{T}|} \times \left(\prod_{\ell \in \mathcal{T}} \frac{d}{dh_\ell}\right) F(\{h_{i,j} v_{i,j}\})_{|h_{i,j} = h_{i,j}(\mathcal{T})} \tag{A.3}$$

where $|\mathcal{T}|$ is the number of lines of $\mathcal{T}$, there is one variable h_ℓ for each line of $\mathcal{T}$ and

$$\begin{aligned} h_{i,j}(\mathcal{T}) &= 0 \quad \text{if } i, j \text{ cannot be linked by a path of } \mathcal{T} \\ h_{i,j}(\mathcal{T}) &= \text{Inf}(\{h_\ell, \ell \in \mathcal{P}\}) \quad \text{if they can be linked by a path } \mathcal{P} \end{aligned} \tag{A.0}$$

Note: The sum over trees in (A.3) includes the empty tree, with the contribution $F(\{0\})$. Non connected trees are more generally composed of connected trees linking subsets of one or more indices.

As indicated at the end of this Appendix, Lemma 2 will be an easy consequence of Lemma 3 proved below. Lines in trees considered in Lemma 3 are ordered from 1 to $|\mathcal{T}|$, different orders leading to different ordered trees.

Lemma 3

$\forall r \leq p-1$:

$$F(\{v_{ij}\}) = \sum_{\substack{\text{ordered trees of length} \leq r \\ \text{(conn. or non conn.)}}} \int_0^1 dh_1 \int_0^{h_1} dh_2 \cdots \int_0^{h_{|\mathcal{T}|-1}} dh_{|\mathcal{T}|}$$
$$\times \left(\prod_{\ell \in \mathcal{T}} \frac{d}{dh_\ell} \right) F(\{h_{ij} v_{ij}\})_{|h_{i,j}=h'_{i,j}(\mathcal{T})} \tag{A.1}$$

where

$$h'_{i,j}(\mathcal{T}) = h_{i,j}(\mathcal{T}) \quad = \text{if } i,j \text{ can be linked by a path of } \mathcal{T} \text{ or if } |\mathcal{T}| < r \tag{A.2}$$

$$h'_{i,j}(\mathcal{T}) = h_r \quad \text{otherwise } (|\mathcal{T}| = r,\ i,j \text{ not linked by a path of } \mathcal{T}) \tag{A.3}$$

Proof of Lemma 3

To show Lemma 3 at $r = 1$, one may write:

$$\begin{aligned} F(\{v_{i,j}\}) &\equiv F(\{hv_{ij}\})_{|h=1} \\ &= F(\{0\}) + \int_0^1 \frac{dF}{dh}(\{hv_{ij}\}) dh \\ &\equiv F(\{0\}) + \sum_{\mathcal{T}, |\mathcal{T}|=1} \int_0^1 \frac{d}{dh_\ell} F(\{h_{ij} v_{ij}\})_{\big| \substack{h_{ij}=h \\ \forall i,j}} \end{aligned} \tag{A.4}$$

where ℓ in (A.4) denotes the pair of indices corresponding to the unique line of $\mathcal{T}$. Q.E.D.

To show Lemma 2 at $r+1$, assuming it holds at a given value of r, we consider in (A.1) the ordered trees of length r. For each one, the integrand $\left(\prod \frac{d}{dh_\ell}\right) F$ can be written, in view of the definition of $h'_{i,j}(\mathcal{T})$, in the form

$$\left(\prod_{\ell \in \mathcal{T}} \frac{d}{dh_\ell} \right) F \left(\{h_{ij} v_{ij}\}_{\substack{i,j \text{ linked by} \\ \text{a path of } \mathcal{T}}}, \{hv_{ij}\}_{\substack{i,j \text{ not linked} \\ \text{by a path of } \mathcal{T}}} \right)$$

where (after derivation) the $h_{i,j}$'s in the first class are fixed at $h_{i,j}(\mathcal{T})$ and the variable h occurring in the second one is fixed at h_r. (Note that derivatives apply only with respect to parameters $h_{i,j}$ of the first class). A first order Taylor expansion around the origin in this variable h gives the announced result at $r+1$. Q.E.D.

Proof of Lemma 2

Lemma 3 is applied at $r = p-1$ in which case $h'_{ij}(\mathcal{T}) \equiv h_{i,j}(\mathcal{T})$ (since, for $|\mathcal{T}| = p-1$, there is always a path linking any pair i,j: $\mathcal{T}$ is then a connected tree). Lemma 2 then follows by regrouping terms associated to a common (non ordered) tree. Q.E.D.

REFERENCES

[1] Edwards, S. F., Proc. Phys. Soc. **85**, 613 (1965), 88, 265 (1966)
[2] Varadhan, S. R. S., Appendix to Symanzik, K., Local Quantum Theory, Jost R. ed., Acad. Press (1970), 285
[3] Westwater, M. J., Commun. Math. Phys. **72**, 131 (1980)
[4] De Gennes, P. G., Phys. Lett.**38A**, 339 (1972); *Scaling Concepts in Polymer Physics*, Cornell Univ. Press., Ithaca, N. Y. (1979)
Des Cloiseaux, J., J. Phys. **42**, 635 (1981)
Duplantier, B., C. R. Acad. Sci. Paris **290B**, 199 (1980), Nucl. Phys. **275B**, 319 (1986), Commun. Math. Phys.**117**, 279 (1988), J. Stat. Phys. **54**, 581 (1989)
Des Cloizeaux, J., Jannink G., *Polymers in Solution, their Modelling and Structure*, Oxford Univ. Press, Oxford (1989)
For related results in field theory, Zinn-Justin J., *Quantum Field Theory and Critical Phenomena*, Oxford Univ. Press 1989, Ch. 24, and references therein.
[5] Brydges, D., Evans, S. N., Imbrie, J. Z., Ann. Prob. **20**, 82 (1992)
[6] Lawler, G. F., Commun. Math. Phys. **86**, 539 (1982), Aizenman, M., Commun. Math. Phys. **97**, 91 (1985)
[7] Brydges, D., Spencer, T. Commun. Math. Phys. **97**, 125 (1985), Hara, T., Slade, G., Commun. Math. Phys. **147**, 101 (1992)
[8] Iagolnitzer, D., Magnen, J. Commun. Math. Phys. **162**, 85 (1994)
[9] Golowich S. E., *Constructive Physics*, Rivasseau V.,ed., Lecture Notes in Physics 446, Springer-Verlag, Heidelberg (1995), 301
[10] Iagolnitzer, D., Magnen, J., in preparation
[11] Madras, N., Slade, G., *Self Avoiding Walks* , Birkhäuser, Boston (1993)
[12] Brydges, D., private communication (the argument is attributed by him to M. Fisher)
[13] Brydges, D., Kenned, T., Journ. Stat. Phys. **48**, 19 (1987)
[14] Ecole Polytechnique group (unpublished)

11

GAUGE THEORY WITHOUT GHOSTS

B. DeWitt and C. Molina-París

Center for Relativity and Department of Physics
University of Texas
Austin, TX

ABSTRACT

A quantum effective action for gauge field theories is constructed that is gauge invariant and independent of the choice of gauge breaking terms in the functional integral that defines it. The loop expansion of this effective action leads to new Feynman rules, involving new vertex functions but without diagrams containing ghost lines. The new rules are given in full for the Yang–Mills field, both with and without coupling to fermions, and renormalization procedures are described. No BRST arguments are needed. Implications are briefly discussed.

1. SUMMARY OF NOTATION

Φ	Space of field histories (a principal fiber bundle)
$\mathcal{G}$	Gauge group (the typical fiber)
$\Phi/\mathcal{G}$	Base space (the space in which real physics takes place)
$\boldsymbol{\gamma}$	Ultralocal, gauge-invariant metric on Φ
$\mathbf{Q}_\alpha$	Fiber generators (pointwise linearly independent vertical vector fields)
$c^\gamma{}_{\alpha\beta}$	Structure constants of $\mathcal{G}$
S	Classical action functional
$F\Phi$	Frame bundle over Φ
$\Gamma_\gamma{}^i{}_{jk}$	Pullback to Φ of Riemannian connection on $F\Phi$
$\omega^\alpha{}_i$	Connection 1-form on Φ (defined uniquely by $\boldsymbol{\gamma}$ and the $\mathbf{Q}_\alpha$)
$\Pi^i{}_j$	Horizontal projection operator
I^A, K^α	Fiber-adapted coordinates
$\hat{\mathfrak{F}}, \hat{\mathfrak{G}}$	Ghost operator and ghost propagator
$\boldsymbol{g}$	Projection of $\boldsymbol{\gamma}$ onto $\Phi/\mathcal{G}$
$F(\Phi/\mathcal{G})$	Frame bundle over $\Phi/\mathcal{G}$
$\Gamma^A{}_{BC}$	Pullback to $\Phi/\mathcal{G}$ of Riemannian connection on $F(\Phi/\mathcal{G})$
$\Gamma^i{}_{jk}$	Vilkovisky's connection (extension of $\Gamma^A{}_{BC}$ to Φ)
φ^i	Arbitrary local field variables

Functional Integration: Basics and Applications
Edited by Cécile DeWitt-Morette, Plenum Press, New York, 1997

ϕ^a	Gaussian normal fields
Γ	Effective action
$\mu[\varphi_*, \phi]$	Measure for the functional integral defining Γ
Σ	$\Gamma - S$ (loop contribution to Γ)
$A^\alpha{}_\mu$	Gauge potentials for Yang–Mills theory
$\gamma_{\alpha\beta}$	Cartan–Killing metric
$F^\alpha{}_{\mu\nu}$	Yang–Mills curvature 2-form
$f^\alpha{}_{\beta\gamma}$	Structure constants of the Yang–Mills Lie group
f_α	Generators of the adjoint representation: $(f^\beta{}_{\alpha\gamma})$
$Q^\alpha{}_{\mu\beta'}$	Fiber generators for the Yang–Mills field
$\gamma_\alpha{}^\mu{}_{\beta'}{}^{\nu'}$	Ultralocal invariant metric for the Yang–Mills field
$\Pi^\alpha{}_{\mu\beta'}{}^{\nu'}$	Horizontal projection operator for the Yang–Mills field
Z	Yang–Mills field renormalization constant
Y	Renormalization constant for the three-pronged vertex
X	Renormalization constant for the four-pronged vertex
Ξ	Coefficient of the nonlocal counter term needed for mopping up trace residues coming from the divergent parts of graphs having four external lines
μ	Auxiliary mass for dimensional regularization
$\tilde{Z}$	Spinor field renormalization constant
$\tilde{\Xi}$	Coefficient of the nonlocal counter term needed to complete the renormalization of the fermion vertex
$\cdot$	Denotes covariant functional differentiation based on the Riemannian connection $\Gamma_\gamma{}^i{}_{jk}$
;	Denotes covariant functional differentiation based on Vilkovisky's connection $\Gamma^i{}_{jk}$

2. INTRODUCTION

Ghosts were invented by Feynman [1] in 1962 while trying to construct a quantum theory of gravity. Having convinced himself that there was no way in which the gravitational field could consistently escape quantization in a universe where everything else is subject to the laws of quantum mechanics, he was trying to see how these laws would work when applied to space–time curvature. The first obstacle he faced was the non-Abelian character of the diffeomorphism group (the gauge group of gravity) which forces the gravitational field to act partly as its own source. In the language of Feynman graphs this means that gravitational charge (stress-energy) is carried by graviton lines as well as by all other lines and hence leaks all over every graph. The easy proof of gauge invariance in quantum electrodynamics, which makes use of the fact that those lines that carry charge do not bifurcate, does not work here.

Feynman's key idea for solving the problem was to replace every Feynman propagator by its equivalent, an advanced Green's function minus a positive-frequency Wightman function, and to throw away all noncausal loops of advanced Green's functions [2], obtaining thereby a mode sum over tree functions. It is easy to show that tree functions are gauge invariant provided the external lines bear only physical mode functions. Feynman therefore proposed to restrict the mode sums to physical modes, a procedure that not only secures gauge invariance but unitarity as well. But there is a difficulty: Because the physical mode functions are defined in a special frame, the procedure is not manifestly Lorentz invariant [3]. Feynman was able to show that deletion of the nonphysical modes is equivalent to subtracting, from the contribution of every closed loop, that of another (Lorentz invariant) loop propagating a particle having spin 1 (or one less than that of the gauge field). This is the ghost. Because its contribution is subtracted, it is a fermion.

In the classical theory ghost propagators are used only when a specific choice of gauge is made and then only to check consistency of the gauge choice. The theory itself is gauge invariant. Feynman's discovery, and the work that it stimulated, made it seem as if the quantum theory of gauge fields cannot even be formulated without ghosts. It is the purpose of this paper to show that ghosts are in fact not needed, even in a manifestly covariant quantum theory, and that a gauge invariant quantum effective action can be introduced which, like the classical action from which one starts, is independent of ghosts and gauge-breaking terms.

3. GEOMETRY OF THE SPACE OF HISTORIES

It is ironic that the developments in quantum gravity stimulated by Feynman's work had their greatest impact on Yang–Mills theories. (Even Feynman chose to use the Yang–Mills field to display his arguments.) Because of their immediate relevance to experimental physics these developments were carried out rapidly, without much concern for the geometry of the space in which the gauge fields themselves take their values, the so-called *space of histories*. This paper will make use of the geometry in a way that leads to new methods of calculation and renormalization, in which ghosts and BRST arguments play no role.

The space of histories, denoted here by Φ, is a principal fiber bundle having for its typical fiber the gauge group $\mathcal{G}$. Real physics takes place in the base space of this bundle, denoted by $\Phi/\mathcal{G}$. Since $\mathcal{G}$ is a group manifold (infinite dimensional) it admits a group invariant Riemannian or pseudo-Riemannian metric. This metric can be extended in an infinity of ways to an invariant metric on Φ. But it turns out that if one requires the extended metric to be *ultralocal* (a requirement that is probably essential for the success of the renormalization program described in this paper) then, up to a scale factor, it is unique in the case of the Yang–Mills field and belongs to a one-parameter family in the case of gravity.

To be precise denote the metric tensor by $\boldsymbol{\gamma}$ and its components in a chart by γ_{ij}. Denote the chart coordinates (i.e., the field variables, which are usually the gauge potentials $A^{\alpha}{}_{\mu}(x)$ in Yang–Mills theory or the space–time metric components $g_{\mu\nu}(x)$ in gravity theory) generically by φ^i. Let Latin indices i,j,k, etc., be understood to range over a continuum of values that specify not only discrete components but space–time points x,x',x'', etc. as well. This allows a condensed notation to be introduced [4] in which summations over repeated indices include integrations over space–time. A symbol like γ_{ij} then specifies two space–time points, x and x' say, and the condition of ultralocality says simply that γ_{ij} is equal to the *undifferentiated* δ-function $\delta(x,x')$ times a coefficient that involves no space–time derivatives of fields.

The fibers of Φ are generated by a set of vector fields $\mathbf{Q}_\alpha$ on Φ that are pointwise linearly independent over Φ and satisfy

$$\mathbf{Q}_\alpha S = 0, \tag{1}$$

$$[\mathbf{Q}_\alpha, \mathbf{Q}_\beta] = -c^{\gamma}{}_{\alpha\beta}\, \mathbf{Q}_\gamma. \tag{2}$$

Here S is the scalar field on Φ known as the *action functional* (often denoted by $S[\varphi]$) and the c's in the closure condition (2) are the structure constants of $\mathcal{G}$. The Greek indices α,β,γ, like the Latin indices i,j,k, specify both discrete components and space–time points. These points are always understood to lie between the "in" and "out" regions of space–time where the boundary conditions of a given problem (which may contribute boundary terms to $S[\varphi]$) are specified.

An immediate corollary of (1) is

$$\delta S = S_{,i}\,\delta\varphi^i = 0, \tag{3}$$

$$\delta\varphi^i = Q^i{}_\alpha\,\delta\xi^\alpha, \tag{4}$$

where the comma denotes functional differentiation, $Q^i{}_\alpha$ are the components of $\mathbf{Q}_\alpha$ in the chart with coordinates φ^i and the $\delta\xi^\alpha$ are infinitesimal *gauge parameters*. Equation (1) (or (3)) is the statement of gauge invariance, and Eq. (4) is the infinitesimal gauge transformation law representing the action of $\mathcal{G}$ on Φ.

Group invariance of the metric $\boldsymbol{\gamma}$ is the statement

$$\pounds_{\mathbf{Q}_\alpha}\boldsymbol{\gamma} = 0, \tag{5}$$

which, in component language, is equivalent to

$$Q_{i\alpha\cdot j} + Q_{j\alpha\cdot i} = 0, \qquad Q_{i\alpha} = \gamma_{ij}\,Q^j{}_\alpha, \tag{6}$$

the dot denoting *covariant functional differentiation* based on the pullback to Φ of the Riemannian connection on the frame bundle $F\Phi$:

$$\Gamma_\gamma{}^i{}_{jk} = \frac{1}{2}\gamma^{il}(\gamma_{jl,k} + \gamma_{kl,j} - \gamma_{jk,l}), \qquad \gamma_{ik}\,\gamma^{kj} = \delta_i{}^j. \tag{7}$$

The $\mathbf{Q}_\alpha$ are *Killing vector fields* for the metric $\boldsymbol{\gamma}$ and *vertical vector fields* for the principal bundle Φ. Moreover, a choice of invariant metric on Φ immediately singles out a unique connection 1-form $\boldsymbol{\omega}^\alpha$ on Φ:

$$\omega^\alpha{}_i = \mathfrak{G}^{\alpha\beta}\,Q_{i\beta}. \tag{8}$$

Here $\mathfrak{G}^{\alpha\beta}$ is the symmetric Green's function appropriate to the boundary conditions at hand (typically Feynman boundary conditions) of the operator

$$\mathfrak{F}_{\alpha\beta} = -Q_{i\alpha}\,Q^i{}_\beta \tag{9}$$

[5]. In view of the relation $\mathfrak{F}_{\alpha\gamma}\mathfrak{G}^{\gamma\beta} = -\delta^\beta_\alpha$ one easily sees that

$$\omega^\alpha{}_i\,Q^i{}_\beta = \delta^\alpha{}_\beta \tag{10}$$

and that horizontal vectors on Φ are those that are perpendicular (under the metric $\boldsymbol{\gamma}$) to the fibers. A horizontal vector may be obtained from any vector by application of the *horizontal projection operator*:

$$\Pi^i{}_j = \delta^i{}_j - Q^i{}_\alpha\,\omega^\alpha{}_j. \tag{11}$$

4. FIBER-ADAPTED COORDINATE PATCHES

When dealing with Φ it is convenient to make a conceptual transformation

$$\varphi^i \longrightarrow I^A, K^\alpha \tag{12}$$

to a set of *fiber-adapted coordinates* I^A, K^α. Here the I's label the fibers (i.e., the points in $\Phi/\mathcal{G}$) and are gauge invariant:

$$\mathbf{Q}_\alpha I^A = 0. \tag{13}$$

The K's label the points within each fiber. Because there is no canonical way of associating points on one fiber with those on another, transformations between fiber-adapted coordinate patches have the general structure

$$I'^A = I'^A[I], \qquad K'^\alpha = K'^\alpha[I,K], \tag{14}$$

which is still special enough so that the Jacobian of the transformation splits into factors:

$$\frac{\delta(I',K')}{\delta(I,K)} = \frac{\delta(I')}{\delta(I)}\frac{\delta(K')}{\delta(K)}. \tag{15}$$

Although the I's usually remain purely conceptual it is in practice (e.g., when choosing a gauge) necessary to make specific choices for the K's. One must single out a *base point* φ_* in Φ and choose the K's so that the matrix

$$\hat{\mathfrak{F}}^\alpha{}_\beta = \mathbf{Q}_\beta K^\alpha = K^\alpha{}_{,i}\, Q^i{}_\beta \tag{16}$$

is a nonsingular differential or (pseudo-differential) operator at and in a neighborhood of φ_*. Typical convenient choices for φ_* are $A^\alpha{}_\mu(x)_* = 0$ in Yang–Mills theory and $g_{\mu\nu}(x)_* =$ some well studied background metric (Minkowski, Friedman–Robertson–Walker, black hole, etc.) in gravity theory.

$\hat{\mathfrak{F}}$ is a possible *ghost operator* for the theory. If it is nonsingular one may easily show that

$$\frac{\delta}{\delta K^\alpha} = -\hat{\mathfrak{G}}^\beta{}_\alpha \mathbf{Q}_\beta, \tag{17}$$

where $\hat{\mathfrak{G}}$ is a Green's function of $\hat{\mathfrak{F}}$ (i.e., a *ghost propagator*), typically having boundary conditions similar to the "in" and "out" boundary conditions of the problem. Equation (17) has two consequences that need to be stressed:

1. Despite the fact that the K^α are explicitly chosen, and hence well defined, the meaning of $\delta/\delta K^\alpha$ depends on context (i.e., on boundary conditions).

2. When $\mathcal{G}$ is non-Abelian it is impossible for the K^α to be valid coordinates globally, for if they were then the $\delta/\delta K^\alpha$, which are vertical vector fields that commute with each other, would generate Abelian orbits (fibers). This is an aspect of the well-known Gribov phenomenon and means that $\hat{\mathfrak{F}}$, unlike the operator $\mathfrak{F}$ of Eq. (9), cannot be everywhere nonsingular.

These consequences imply that when ghosts are being used one is barred from carrying out global studies (e.g., lattice functional integration) and must be content with perturbation theory.

5. THE VILKOVISKY CONNECTION

Since the metric γ on Φ is group invariant its horizontal component may be projected down to yield a unique metric g on the base space $\Phi/\mathcal{G}$:

$$g_{AB} = \bar{\gamma}_{ij}\, \varphi^i{}_{,A}\, \varphi^i{}_{,B}, \tag{18}$$

$$\bar{\gamma}_{ij} = \gamma_{kl}\, \Pi^k{}_i\, \Pi^l{}_j = \gamma_{ik}\, \Pi^k{}_j = \gamma_{ij} - Q_{i\alpha}\, \mathfrak{G}^{\alpha\beta} Q_{j\beta}. \tag{19}$$

In Eq. (18) we have extended the comma notation thus:

$$B_{,A} = \delta B/\delta I^A, \qquad B_{,\alpha} = \delta B/\delta K^\alpha, \tag{20}$$

where B is any functional on Φ. It is not difficult to verify the following:

$$\bar{\gamma}_{ij} = g_{AB} I^A{}_{,i} I^B{}_{,j}, \tag{21}$$

$$g^{AB} = I^A{}_{,i}\, I^B{}_{,j}\, \gamma^{ij}, \tag{22}$$

$$g_{AB},_{\alpha} = 0, \qquad g^{AB}{}_{,\alpha} = 0, \tag{23}$$

where g^{AB} and γ^{ij} are the inverses (contravariant versions) of g_{AB} and γ_{ij} respectively. Corresponding to the metric g there is a Riemannian connection on the frame bundle $F(\Phi/\mathcal{G})$, with pullback to $\Phi/\mathcal{G}$ given by:

$$\Gamma^A{}_{BC} = \frac{1}{2} g^{AD}(g_{DB,C} + g_{DC,B} - g_{BC,D}), \tag{24}$$

$$\Gamma^A{}_{BC},_{\alpha} = 0. \tag{25}$$

Up to this point we have introduced three different principal bundles: $\Phi, F(\Phi/\mathcal{G})$ and $F\Phi$. To G. A. Vilkovisky [6] goes the credit for emphasizing the importance of the third and for endowing it with a remarkable fiber-adapted connection. We have already called attention to the pullback (7) of the Riemannian connection on $F\Phi$. The pullback of Vilkovisky's connection is obtained by starting in a fiber-adapted coordinate patch, choosing $\Gamma^A{}_{BC}$ as above, and setting

$$\Gamma^A{}_{B\alpha} = 0, \qquad \Gamma^A{}_{\alpha\beta} = 0. \tag{26}$$

Conditions (25) and (26) remain invariant under transformations from one fiber-adapted coordinate patch to another and have the following consequences:

1. Geodesics in Φ project onto geodesics in $\Phi/\mathcal{G}$.
2. If a functional B is gauge invariant then, in a fiber-adapted coordinate system $\{I^A, K^\alpha\}$, the components of the nth covariant functional derivative of B vanish unless all indices are capital Latin. In a general coordinate system φ^i it therefore follows that

$$B_{;i_1 \ldots i_n} = B_{;A_1 \cdots A_n} I^{A_1}{}_{,i_1} \ldots I^{A_n}{}_{,i_n}, \tag{27}$$

$$B_{;i_1 \ldots i_k \ldots i_n}\, Q^{i_k}{}_\alpha = 0, \tag{28}$$

where the semicolon denotes covariant functional differentiation based on Vilkovisky's connection.

What about the connection components $\Gamma^{\alpha}{}_{AB}, \Gamma^{\alpha}{}_{A\beta}, \Gamma^{\alpha}{}_{\beta\gamma}$? These components play no role in the computation of covariant functional derivatives of gauge invariant quantities and can be chosen arbitrarily. Vilkovisky makes a particularly convenient choice by first noting that the components $\Gamma^{i}{}_{jk}$ in an arbitrary coordinate system are changed by linear combinations of the $Q^{i}{}_{\alpha}$ (or of the $\varphi^{i}{}_{,\alpha}$) under changes in $\Gamma^{\alpha}{}_{AB}, \Gamma^{\alpha}{}_{A\beta}, \Gamma^{\alpha}{}_{\beta\gamma}$. He then carries out an intricate series of steps (given in Appendix A) to show that, modulo such linear combinations, the choices (24) and (26) imply

$$\begin{aligned}\Gamma^{i}{}_{jk} &= \Gamma_{\gamma}{}^{i}{}_{jk} - Q^{i}{}_{\alpha\cdot j}\,\omega^{\alpha}{}_{k} - Q^{i}{}_{\alpha\cdot k}\,\omega^{\alpha}{}_{j} \\ &+\frac{1}{2}\omega^{\alpha}{}_{j}\,Q^{i}{}_{\alpha\cdot l}\,Q^{l}{}_{\beta}\,\omega^{\beta}{}_{k} + \frac{1}{2}\omega^{\alpha}{}_{k}\,Q^{i}{}_{\alpha\cdot l}\,Q^{l}{}_{\beta}\,\omega^{\beta}{}_{j}.\end{aligned} \tag{29}$$

This is Vilkovisky's connection. It is not Riemannian. However, it, like $\Gamma_{\gamma}{}^{i}{}_{jk}$, is invariant under rescalings of $\boldsymbol{\gamma}$ ($\boldsymbol{\gamma} \longrightarrow \zeta\boldsymbol{\gamma}$).

When Vilkovisky's connection is used the following remarkable results hold:

1. Let λ be a geodesic based on it. If the tangent vector to λ is horizontal at one point along λ then it is horizontal everywhere along λ, so that λ is a horizontal lift of a geodesic in $\Phi/\mathcal{G}$ based on the Riemannian connection $\Gamma^{A}{}_{BC}$. A horizontal geodesic is also a geodesic based on the Riemannian connection $\Gamma_{\gamma}{}^{i}{}_{jk}$.
2. If a geodesic is tangent to a fiber at one point then it lies in that fiber.
3. For all α
$$Q^{i}{}_{\alpha;j} = \frac{1}{2}Q^{i}{}_{\gamma}c^{\gamma}{}_{\alpha\beta}\omega^{\beta}{}_{j}. \tag{30}$$
4. If B is a gauge invariant functional (e.g., $B = S$) then for all n
$$B_{;(i_1\ldots i_n)} = B_{\cdot(j_1\ldots j_n)}\Pi^{j_1}{}_{i_1}\ldots\Pi^{j_n}{}_{i_n}, \tag{31}$$
where the parentheses indicate that a symmetrization of the indices they embrace is to be performed

These results are proved in Appendices B, C, D and E.

6. GAUSSIAN NORMAL FIELDS

In discussing the conditions necessary for choosing fiber-adapted coordinates K^{α} we introduced a base point φ_*, typically some convenient background field. Introduction of a base point is also useful for a quite different and independent purpose. Let φ be an arbitrary point of Φ and λ a Vilkovisky geodesic connecting it to φ_*. Let s and s_* be the values, at φ and φ_* respectively, of an affine parameter along λ. Define

$$\boldsymbol{\phi} = (s - s_*)\left(\frac{\partial}{\partial s}\right)_{\lambda(s_*)}. \tag{32}$$

$\boldsymbol{\phi}$ is a vector at φ_* (i.e. an element of $T_{\varphi_*}\Phi$) invariant under rescaling of the affine parameter. Its components ϕ^{a} in an arbitrary frame at φ_* are known as *Gaussian normal coordinates* or, as we shall call them, *Gaussian normal fields*. They are functionals of the φ^{i}_{*} and φ^{i}

$$\phi^{a} = \phi^{a}[\varphi_*, \varphi], \tag{33}$$

multivalued if there is more than one geodesic between φ_* and φ. With their aid one can carry out covariant functional Taylor expansions about the point φ_*, of scalar fields $A[\varphi]$ on Φ:

$$A[\varphi] = \sum_{n=0}^{\infty} \frac{1}{n!} A_{;(a_1 \ldots a_n)}[\varphi_*]\, \phi^{a_1} \ldots \phi^{a_n}. \tag{34}$$

(See Appendix F for a derivation of this result.) In what follows we shall abuse notation slightly by using the same symbol A to represent the scalar field, whether it is regarded as a functional of the φ^i or of the φ^i_* and ϕ^a:

$$A[\varphi] = A[\varphi_*, \phi]. \tag{35}$$

We call attention to the convention of using indices from the first part of the alphabet to label the components of tensors at φ_*. More generally, the components of any geometrical quantity at *any* point of Φ will be labeled by indices from the first part of the alphabet when viewed in the coordinate system defined by the ϕ^a. For example, the components of the $\mathbf{Q}_\alpha$ in this coordinate system will be denoted by $Q^a{}_\alpha$.

The ϕ^a comprise a *preferred coordinate system* in Φ. In the case of Yang–Mills theory one might suppose that the potentials $A^\alpha{}_\mu$, being components of a Lie-algebra-valued 1-form in space–time, are already sufficiently special to constitute a preferred set of variables. (In gravity theory the choice of the $g_{\mu\nu}$ as preferred variables is problematical.) But we shall see that the ϕ^a, *despite being nonlocally and nonlinearly related to the* φ^i, are more elegant and useful for certain purposes. One property of the ϕ^a, in particular, is of prime importance in the theory of the effective action (see below). Because of the fact that geodesics in Φ project onto geodesics in $\Phi/\mathcal{G}$, in fiber-adapted coordinates ϕ^A depends only on the $I^A{}_*$ and I^A:

$$\phi^A = \phi^A[I_*, I], \tag{36}$$

$$\phi^\alpha = \phi^\alpha[I_*, K_*, I, K]. \tag{37}$$

7. THE QUANTUM EFFECTIVE ACTION

It is well known that all physical processes in quantum field theory are in principle determined if the so-called *quantum effective action*, or *effective action* for short, is given. We shall assume the reader to be already familiar with the functional integral representation of the effective action and its formal construction via external sources and the Legendre transform. When no gauge group is present and the fields are sections of vector bundles, the natural metric on the space of histories is flat and the Riemannian connection components vanish. The field variables are already Gaussian normal, and the effective action for the theory, denoted here by $\Gamma[\bar{\varphi}]$, may be constructed by an iterative process starting from the functional integral equation

$$e^{i\Gamma[\bar{\varphi}]} = C \int e^{iS[\varphi] + i\Gamma_1[\bar{\varphi}](\bar{\varphi} - \varphi)}\, [d\varphi], \tag{38}$$

where $\Gamma_1[\bar{\varphi}]$ is a supercondensed notation representing the first functional derivative of $\Gamma[\bar{\varphi}]$, and C is an essentially arbitrary normalization constant. (Changes in C add irrelevant constants to $\Gamma[\bar{\varphi}]$). The iterative process allows one to bypass the introduction of external sources and go directly to their conjugates under the Legendre transform, namely the variables $\bar{\varphi}^i$. It also insures that all loop graphs generated by Gaussian expansion of the functional integral will be 1-particle irreducible.

The effective action, which is also known as the *generator of full vertex functions*, is of great theoretical importance, playing major roles in renormalization theory and the theory of spontaneous symmetry breaking as well as in other basic areas. It would be very helpful if it could be introduced also when the fields are *not* sections of vector bundles (e.g., in the case of nonlinear sigma models). Unfortunately, $\bar{\varphi}-\varphi$ is then not a vector and Eq. (38) does not make good geometrical sense. However, this difficulty is easily overcome by switching to Gaussian normal fields (based on some natural connection on the space of histories) and writing

$$e^{i\Gamma[\varphi_*,\bar{\phi}]} = N\int e^{iS[\varphi_*,\phi]+i\Gamma_1[\varphi_*,\bar{\phi}](\bar{\phi}-\phi)}\mu[\varphi_*,\phi][d\phi], \tag{39}$$

where μ is an appropriate measure functional and Γ_1 denotes the functional derivative with respect to $\bar{\phi}$.

When a gauge group is present there is a natural analog of Eq. (39), which recognizes that real physics takes place in the base space $\Phi/\mathcal{G}$, namely

$$e^{i\Gamma[I_*,\bar{I}]} = C\int e^{iS[I_*,I]+i\Gamma_1[I_*,\bar{I}](\bar{I}-I)}\mu[I_*,I][dI]. \tag{40}$$

Here the I^A are assumed to be already Gaussian normal; i.e., they are the ϕ^A of Eq. (36).

Although Eq. (40) is a reasonable starting point for the quantum theory of gauge fields it is of formal validity only because the I^A are purely conceptual. For explicit calculations one must pass from $\Phi/\mathcal{G}$ to Φ by introducing variables K^α. By virtue of the fact that a geodesic tangent to a fiber lies in the fiber, the K^α too may be chosen to be Gaussian normal, complementary to the I^A. (They may be identified with the ϕ^α of Eq. (37).) Being Gaussian normal, both the I^A and the K^α are necessarily *linearly related* to the Gaussian normal variables ϕ^a in an arbitrary frame:

$$I^A = P^A{}_a[\varphi_*]\phi^a, \qquad K^\alpha = P^\alpha{}_a[\varphi_*]\phi^a. \tag{41}$$

Here we have made the P-coefficients depend on the base point φ_*, which can be any point in the fiber over I_*. Since $\phi^a=0$ when $\varphi=\varphi_*$ it is clear that the $I^A{}_*$ themselves actually vanish.

With introduction of the K^α one may rewrite Eq. (40) in the form

$$\begin{aligned} e^{i\Gamma[I_*,\bar{I}]} = N\int[dI]\int[dK]e^{i(S[I_*,I]+\frac{1}{2}\kappa_{\alpha\beta}[\varphi_*]K^\alpha K^\beta)+i\Gamma_1[I_*,\bar{I}](\bar{I}-I)} \\ \times(\det\kappa[\varphi_*])^{\frac{1}{2}}\mu[I_*,I], \end{aligned} \tag{42}$$

where $\kappa_{\alpha\beta}$ is typically an ultralocal symmetric matrix. The term in $\kappa_{\alpha\beta}$ in the exponent is often called a gauge "fixing" term, but it should strictly be called a gauge *breaking* term because it breaks the gauge invariance of the exponent. We hasten to add that Γ, defined by (42), is still gauge invariant as well as independent of the choice of the $\kappa_{\alpha\beta}$ and the $P^\alpha{}_a$.

Since $\phi^A = I^A$ and $\bar{\phi}^A = \bar{I}^A$ one can immediately transform the variables of integration in (42) to Gaussian normal fields ϕ^a in an arbitrary frame, noting, by virtue of (41), that the Jacobian $J[\varphi_*]$ for the transformation is a *constant.* We shall again abuse notation slightly by writing, for all gauge invariant functionals B,

$$B[I_*,I] = B[\varphi_*,\phi], \qquad B[I_*,\bar{I}] = B[\varphi_*,\bar{\phi}], \tag{43}$$

where the $\bar{\phi}^a$ are such that $\bar{I}^A = P^A{}_a[\varphi_*]\bar{\phi}^a$. $\Gamma[\varphi_*,\bar{\phi}]$, as defined by the transformed integral, continues to be gauge invariant and independent of the choice of κ's and P's *provided* the $\bar{\phi}^a$

are held at values such that $\bar{K}^\alpha = P^\alpha{}_a[\varphi_*]\bar{\phi}^a$ vanishes. If the $\bar{\phi}^a$ are *not* held at such values but are allowed to run freely then Γ suffers the (fortunately simple) modification $\Gamma \longrightarrow \hat{\Gamma}$ where

$$\hat{\Gamma}[\varphi_*, \bar{\phi}] = \Gamma[\varphi_*, \bar{\phi}] + \frac{1}{2}\kappa_{\alpha\beta}[\varphi_*]P^\alpha{}_a[\varphi_*]P^\beta{}_b[\varphi_*]\bar{\phi}^a\bar{\phi}^b. \tag{44}$$

Hence finally

$$e^{i\hat{\Gamma}[\phi_*,\bar{\phi}]} = C' \int e^{i(S[\varphi_*,\phi]+\frac{1}{2}\kappa_{\alpha\beta}[\varphi_*]P^\alpha{}_a[\varphi_*]P^\beta{}_b[\varphi_*]\phi^a\phi^b)+i\hat{\Gamma}_{,i}[\varphi_*,\bar{\phi}](\bar{\phi}-\phi)}$$
$$\times(\det\kappa[\varphi_*])^{\frac{1}{2}}J[\varphi_*]\mu[\varphi_*,\phi][d\phi], \tag{45}$$

8. THE GHOST

When the K's are chosen as above the ghost operator (16) becomes

$$\hat{\mathfrak{F}}^\alpha{}_\beta[\varphi_*, \phi] = P^\alpha{}_a[\varphi_*]\, Q^a{}_\beta[\varphi_*, \phi]. \tag{46}$$

Suppose the $P^\alpha{}_a$ suffer an infinitesimal variation $\delta P^\alpha{}_a$. Then $\ln J$ suffers the change

$$\delta(\ln J) = \phi^a{}_{,\alpha}\,\delta K^\alpha{}_{,a} = -Q^a{}_\beta\,\hat{\mathfrak{G}}^\beta{}_\alpha\,\delta P^\alpha{}_a$$
$$= -\hat{\mathfrak{G}}^\beta{}_\alpha\,\delta\hat{\mathfrak{F}}^\alpha{}_\beta = -\delta(\ln\det\hat{\mathfrak{G}}). \tag{47}$$

Evidently $J\det\hat{\mathfrak{G}}$ is independent of the $P^\alpha{}_a$. Hence, as far as assuring the P-independence of Γ is concerned, J may be replaced by $(\det\hat{\mathfrak{G}})^{-1}$ in the functional integral (45). This is the well known ghost determinant.

It is perfectly permissible to use $(\det\hat{\mathfrak{G}})^{-1}$ in place of J and to develop the theory along traditional lines. One must first compute vertices for the interaction of the ghost with the fields ϕ^a, by carrying out a covariant functional Taylor expansion of $Q^a{}_\alpha$ (see Appendix I). One can then compute counter terms that renormalize both the ghost fields and the ghost vertices, and make use of standard BRST arguments. But if one does this, using dimensional regularization, one finds that, because of the special forms that the ghost vertices take, *every Feynman graph containing a ghost line vanishes*. Explicit examples of this fact are displayed in Appendix J.

It is not hard to adduce formal arguments for the vanishing of those graphs that contain internal ghost lines. Internal ghost lines always occur in closed loops arising from $\ln\det\hat{\mathfrak{G}}$. To insert non-ghost lines into these loops one simply differentiates $\ln\det\hat{\mathfrak{G}}$ with respect to the ϕ^a and attaches non-ghost propagators. Now since J is independent of the ϕ^a it follows from Eq. (47) that $\delta(\ln\det\hat{\mathfrak{G}})$ is independent of the ϕ^a. This implies that $\ln\det\hat{\mathfrak{G}}$ itself must have the structure

$$\ln\det\hat{\mathfrak{G}} = -\ln J[\varphi_*] + B[\varphi_*, \phi], \tag{48}$$

where the term $B[\varphi_*, \phi]$, which is the only part of $\ln\det\hat{\mathfrak{G}}$ that contributes to the internal loops, is independent of the $P^\alpha{}_a$.

What can be said about the functional derivatives of $B[\varphi_*, \phi]$? From Eq. (48) it follows that $B_{,a} = (\ln\det\hat{\mathfrak{G}})_{,a} = \hat{\mathfrak{G}}^\alpha{}_\beta\,P^\beta{}_b\,Q^b{}_{\beta,a}$, so one is led to ask what can be said about $Q^b{}_{\beta,a}$, and this in turn leads one to refer to Eq. (G.3) of Appendix G. It is easy to see that this equation is true if S is replaced by any gauge invariant quantity $I[\varphi_*, \phi]$. Differentiating the equation $I_{,b}Q^b{}_\alpha = 0$, one gets $0 = I_{,ba}Q^b{}_\alpha + I_{,b}Q^b{}_{\alpha,a} = I_{,b}Q^b{}_{\alpha,a}$. Since this is true for *all* invariants I it follows that $Q^b{}_{\alpha,a}$ (and also $Q^b{}_{\alpha,ac}, Q^b{}_{\alpha,acd}, \ldots$) must be a linear combination of the $Q^b{}_\gamma$'s. That is, $Q^b{}_{\alpha,a} = Q^b{}_\gamma A^\gamma{}_{\alpha a}$ where the A's are certain coefficients. Inserting this into the expression for $B_{,a}$ one gets $B_{,a} = \hat{\mathfrak{G}}^\alpha{}_\beta \hat{\mathfrak{F}}^\beta{}_\gamma A^\gamma{}_{\alpha a} = -A^\alpha{}_{\alpha a}$, an expression from which the P's have disappeared, as they must.

Now what can be said about $A^\alpha{}_{\beta a}$ or, using the less condensed, more explicit notation, $A^\alpha{}_{\beta' a''}$? Typically $A^\alpha{}_{\beta' a''}$ will be proportional to $\delta(x,x')$ or its derivative, times a function of x and x' (cf. Appendix I). Therefore $A^\alpha{}_{\alpha a''}\, d^N x$ will be proportional to the δ-function or its derivative, *with coincident arguments.* Such a quantity, when expressed as a momentum integral, vanishes under dimensional regularization. From this one may infer that if B is expressed as a Taylor expansion in powers of the ϕ's, the Fourier transforms of the expansion coefficients will be functions of the momenta associated with the ϕ's, multiplied by integrals like $\int (k^2)^m\, d^N k$, which vanish under dimensional regularization. This is indeed what one finds. (See Appendix J.)

As for the vanishing of those graphs that contain external ghost lines, and its implication that *neither the ghost fields nor the ghost vertices get renormalized*, note that every such graph is a subgraph of a graph with ghost lines that are purely internal, which already vanishes. In the last analysis one may simply appeal to the ϕ-independence of J. Because J is a constant it is clear from Eq. (45) that the ghost can play no real graphical role. It has become, as it were, a ghost of itself.

9. THE LOOP EXPANSION

In what follows we shall absorb $(\det \kappa)^{1/2}$ and J into the normalization constant C' of Eq. (45). We shall also drop the measure μ since its chief role can be shown to be simply to justify throwing away nonvanishing contributions from arcs at infinity in the Wick rotation procedure. (See reference [2].) The loop perturbation series is then obtained by expanding the integrand of (45) about $\bar{\phi}$, writing

$$\phi = \bar{\phi} + \chi, \tag{49}$$

and using

$$S[\varphi_*, \bar{\phi} + \chi] = \sum_{n=0}^{\infty} \frac{1}{n!} S_{,a_1 \ldots a_n}[\varphi_*, \bar{\phi}]\, \chi^{a_1} \ldots \chi^{a_n}, \tag{50}$$

where the ordinary derivatives of $S[\varphi_*, \bar{\phi}]$ with respect to the $\bar{\phi}^a$ are given (see Eqs. (34) and (35)) by

$$S_{,a_1 \ldots a_n}[\varphi_*, \bar{\phi}] \sum_{m=0}^{\infty} \frac{1}{m!} S_{;(a_1 \ldots a_n b_1 \ldots b_m)}[\varphi_*]\, \bar{\phi}^{b_1} \ldots \bar{\phi}^{b_m}. \tag{51}$$

The loop graphs themselves are embodied in the functional

$$\Sigma[\varphi_*, \bar{\phi}] = \Gamma[\varphi_*, \bar{\phi}] - S[\varphi_*, \bar{\phi}], \tag{52}$$

which is generated by the functional integral representation (equivalent to (45))

$$e^{i\Sigma[\varphi_*, \bar{\phi}]} = N \int e^{i(-\Sigma_1[\varphi_*, \bar{\phi}]\chi + \frac{1}{2} F[\varphi_*, \bar{\phi}]\chi\chi + \frac{1}{6} S_3[\varphi_*, \bar{\phi}]\chi\chi\chi + \ldots)} [d\chi]. \tag{53}$$

Here $S_n[\varphi_*, \bar{\phi}]$ is a supercondensed notation for the nth ordinary derivative of $S[\varphi_*, \bar{\phi}]$ with respect to the $\bar{\phi}$'s, $\Sigma_1[\varphi_*, \bar{\phi}]$ denotes the first derivative of $\Sigma[\varphi_*, \bar{\phi}]$, and $F[\varphi_*, \bar{\phi}]$ is the operator

$$F_{ab}[\varphi_*, \bar{\phi}] = S_{,ab}[\varphi_*, \bar{\phi}] + \kappa_{\alpha\beta}[\varphi_*] P^\alpha{}_a[\varphi_*] P^\beta{}_b[\varphi_*] \tag{54}$$

with indices suppressed.

If the functional $\Sigma[\varphi_*, \bar{\phi}]$ is computed from (53), by expanding the integrand, using $e^{\frac{i}{2}F[\varphi_*,\bar{\phi}]\chi\chi}$ as a Gaussian integrator, and carrying out an obvious iterative process, an infinite series of terms is obtained. Each term can be represented graphically, with the functions $S_n[\varphi_*, \bar{\phi}]$ being represented by n-pronged vertices, and Green's functions $G^{ab}[\varphi_*, \bar{\phi}]$ of $F_{ab}[\varphi_*, \bar{\phi}]$ (which arise from the Gaussian integrations and have the boundary conditions appropriate to the functional integral) being represented by lines joining the vertices to one another through index linkages. The Green's-function lines and the vertices form a pattern of closed loops because all the indices are dummies. This is the loop expansion.

The functions S_n are the *bare* vertex functions of the theory. The *full* vertex functions are the functional derivatives Γ_n of Γ, which are obtained by attaching external lines to the loop graphs. In calculating physical amplitudes that involve Γ_n one often sets $\bar{\phi} = 0$. The vertex functions (51) then reduce to

$$\begin{aligned} S_{,a_1\dots a_n}[\varphi_*, 0] &= S_{;a(_1\dots a_n)}[\varphi_*] \\ &= S_{\cdot(b_1\dots b_n)}[\varphi_*]\Pi^{b_1}{}_{a_1}[\varphi_*]\dots\Pi^{b_n}{}_{a_n}[\varphi_*], \end{aligned} \tag{55}$$

and these are easily calculated (see below). The rules for obtaining the Γ_n when $\bar{\phi} = 0$ are seen to be very simple: *Replace all traditional vertex functions by* (55) *and throw away all loop graphs containing ghosts.*

It is shown in Appendix G that each of the remaining graphs is *individually* gauge invariant, which means, of course, that $\Sigma[\varphi_*, \bar{\phi}]$ is gauge invariant. It is also shown that each of the remaining graphs, with the exception of the one-loop graph, is P- and κ-invariant. The reason the one-loop graph is not invariant is that $(\det \kappa)^{1/2}$ and J have been absorbed into the renormalization constant C'. If these factors are brought back and allowed to contribute to the one-loop graph, then full invariance is restored. Of course, since neither of these factors depends on the $\bar{\phi}^a$ they disappear anyway as soon as external lines are attached. Hence Σ is *effectively* fully P- κ- and gauge invariant.

10. PURE YANG–MILLS THEORY

The classical action for the Yang–Mills field in a Minkowski space–time of N dimensions and signature $-+++\dots$ is

$$S = -\frac{1}{4}\int \gamma_{\alpha\beta}F^{\alpha}{}_{\mu\nu}F^{\beta\mu\nu}d^N x \tag{56}$$

where $\gamma_{\alpha\beta}$ is the Cartan–Killing metric of the associated Lie algebra (assumed compact, simple and non-Abelian) and

$$F^{\alpha}{}_{\mu\nu} = A^{\alpha}{}_{\nu,\mu} - A^{\alpha}{}_{\mu,\nu} + g_0 f^{\alpha}{}_{\beta\gamma}A^{\beta}{}_{\mu}A^{\gamma}{}_{\nu}, \tag{57}$$

$f^{\alpha}{}_{\beta\gamma}$ being the structure constants of the algebra and g_0 the *bare* coupling constant, and a comma followed by a Greek index from the middle of the alphabet denoting differentiation with respect to a space–time coordinate. The $f^{\alpha}{}_{\beta\gamma}$ and $\gamma_{\alpha\beta}$ are related by

$$\mathrm{tr}(f_\alpha f_\beta) = -\lambda\,\gamma_{\alpha\beta}, \qquad \mathrm{tr}(f_\alpha f_\beta f_\gamma) = -\frac{1}{2}\lambda\, f_{\alpha\beta\gamma} \tag{58}$$

where λ is a positive constant that depends on the choice of scale for the algebra basis, the f_α are the generators of the adjoint representation, i.e., the matrices built out of the structure constants,

$$f_\alpha = (f^\gamma{}_{\alpha\delta}), \tag{59}$$

and indices on the structure constants are lowered by the metric $\gamma_{\alpha\beta}$.

The gauge transformation law (4) takes the form

$$\delta A^\alpha{}_\mu = \int Q^\alpha{}_{\mu\beta'}\, \delta\xi^{\beta'} d^N x', \tag{60}$$

where

$$Q^\alpha{}_{\mu\beta'} = \left(-\delta^\alpha{}_\beta \frac{\partial}{\partial x^\mu} + g_0 f^\alpha{}_{\beta\gamma} A^\gamma{}_\mu\right) \delta(x,x'), \tag{61}$$

and the structure constants of the gauge group itself are

$$c^\alpha{}_{\beta'\gamma''} = g_0 f^\alpha{}_{\beta\gamma}\, \delta(x,x')\, \delta(x,x''). \tag{62}$$

A^α_μ and g_0 have mass dimensions $\frac{1}{2}N-1$ and $2-\frac{1}{2}N$ respectively.

The unique (up to a scale factor) ultralocal invariant metric that has previously been denoted by γ_{ij} takes here the form

$$\gamma_\alpha{}^\mu{}_{\beta'}{}^{\nu'} = \gamma_{\alpha\beta}\, \eta^{\mu\nu} \delta(x,x') \tag{63}$$

where $\eta^{\mu\nu}$ is the Minkowski metric. Expression (63) is independent of the $A^\alpha{}_\mu$ and hence constant on Φ. This is a *flat* metric and, in the coordinates $A^\alpha{}_\mu$, the associated Riemannian connection components $\Gamma_\gamma{}^i{}_{jk}$ vanish. This means that covariant functional differentiation based on (63) reduces to ordinary functional differentiation, i.e., that the dots in Eqs. (6), (29), (31) and (55) can be replaced by commas [7]. The fact that the dot in Eq. (55) can be replaced by a comma is of great practical importance. It means that there are only two distinct vertex functions, S_3 and S_4, just as in the traditional formalism. It also means that when used in the explicit expression for S the $A^\alpha{}_\mu$ and their associated Gaussian normal fields may be treated *as if* they were linearly related, although they may by no means be so treated in general.

The $P^\alpha{}_a$ and $\kappa_{\alpha\beta}$ for Yang–Mills fields are conveniently chosen to be

$$P^\alpha{}_{\beta'}{}^{\mu'} = -\delta^\alpha{}_\beta\, \eta^{\mu\nu} \frac{\partial}{\partial x^\nu} \delta(x,x'), \tag{64}$$

$$\kappa_{\alpha\beta'} = -\gamma_{\alpha\beta}\, \delta(x,x'). \tag{65}$$

The operator (54) then takes the simple form

$$F_\alpha{}^\mu{}_{\beta'}{}^{\nu'} \longrightarrow -\gamma_{\alpha\beta}\, \eta^{\mu\nu} p^2, \tag{66}$$

where "$\longrightarrow$" means "set $A^\alpha{}_\mu = 0$, pass to the Fourier transform, and drop momentum-space δ-function factors as well as factors of 2π." The corresponding Green's function is

$$G^\alpha{}_\mu{}^{\beta'}{}_{\nu'} \longrightarrow \gamma^{\alpha\beta}\, \eta_{\mu\nu}/p^2 \qquad (\gamma_{\alpha\delta}\, \gamma^{\delta\beta} = \delta_\alpha{}^\beta). \tag{67}$$

Expression (66), as well as expressions for the vertex functions and horizontal projection operator in momentum space, are derived in Appendix H.

Traditionally one inserts a factor $1/\zeta$ into the definition of $\kappa_{\alpha\beta'}$ and hence into the gauge breaking term for the functional integral, complicating expression (67) for the Green's function a bit. But there is no need to do this here. The horizontal projection operator (11) takes the form

$$\Pi^{\alpha}{}_{\mu\beta'}{}^{\nu'} \longrightarrow \delta^{\alpha}{}_{\beta}(\delta_{\mu}{}^{\nu} - p_{\mu}p^{\nu}/p^2). \tag{68}$$

(see Appendix H), which allows one to restate the calculational rules for Yang–Mills theory as follows:

1. Retain only those traditional graphs that contain no ghost lines and, in these, use the traditional vertices.
2. For the internal lines use the Green's function in the Landau gauge ($\zeta = 0$):

$$G_L{}^{\alpha}{}_{\mu}{}^{\beta'}{}_{\nu'} \longrightarrow \gamma^{\alpha\beta}(\eta_{\mu\nu} - p_{\mu}p_{\nu}/p^2)/p^2 \tag{69}$$

3. Apply the operator (68) to all external prongs.

Thus one arrives at the Landau gauge no matter what one initially chooses for ζ.

11. RENORMALIZATION

Renormalization in the ghost-free formalism, although technically requiring as much work as in the traditional formalism, is conceptually simpler. There are only two renormalization constants instead of three, a field renormalization constant Z and a constant Y that renormalizes the three-pronged vertex. The constant, call it X, that renormalizes the four-pronged vertex is fixed by gauge invariance to be

$$X = Z^{-1}Y^2. \tag{70}$$

To see this, introduce renormalized fields $A_R{}^{\alpha}{}_{\mu}$, renormalized gauge parameters $\delta\xi_R{}^{\alpha}$, and a renormalized coupling constant g by

$$A^{\alpha}{}_{\mu} = Z^{1/2}A_R{}^{\alpha}{}_{\mu}, \qquad \delta\xi^{\alpha} = Z^{1/2}\delta\xi_R{}^{\alpha}, \qquad g_0 = \mu^{2-N/2}Z^{-3/2}Yg, \tag{71}$$

where μ is the usual auxiliary mass used in dimensional regularization. The classical action then becomes

$$\begin{aligned} S = & -\frac{1}{4}Z\int \gamma_{\alpha\beta}(A_R{}^{\alpha}{}_{\nu,\mu} - A_R{}^{\alpha}{}_{\mu,\nu})(A_R{}^{\beta\nu,\mu} - A_R{}^{\beta\mu,\nu})d^Nx \\ & -\frac{1}{2}\mu^{2-N/2}gY\int \gamma_{\alpha\beta}f^{\alpha}{}_{\gamma\delta}A_R{}^{\gamma}{}_{\mu}A_R{}^{\delta}{}_{\nu}(A_R{}^{\beta\nu,\mu} - A_R{}^{\beta\mu,\nu})d^Nx \\ & -\frac{1}{4}\mu^{4-N}g^2X\int \gamma_{\alpha\beta}f^{\alpha}{}_{\gamma\delta}f^{\beta}{}_{\varepsilon\zeta}A_R{}^{\gamma}{}_{\mu}A_R{}^{\delta}{}_{\nu}A_R{}^{\varepsilon\mu}A_R{}^{\zeta\nu}d^Nx \end{aligned} \tag{72}$$

and is invariant under

$$\delta A_R{}^{\alpha}{}_{\mu} = \int Q^{\alpha}{}_{\mu\beta'}\,\delta\xi_R{}^{\beta'}d^Nx' \tag{73}$$

only if Eq. (70) is indeed satisfied. Here $Q^{\alpha}{}_{\mu\beta'}$ is the same as before (Eq. (61)) but may be rewritten

$$Q^{\alpha}{}_{\mu\beta'} = \left(-\delta^{\alpha}{}_{\beta}\frac{\partial}{\partial x_{\mu}} + \mu^{2-N/2}Z^{-1}Ygf^{\alpha}{}_{\beta\gamma}A_R{}^{\gamma}{}_{\mu}\right)\delta(x,x'). \tag{74}$$

To examine the details of the renormalization program it is helpful first to note some special features of the traditional three- and four-point vertex functions, given (see Appendix H) respectively by

$$i\,\mu^{2-N/2}\,g\,f_{\alpha\beta\gamma}[\eta^{\mu\nu}(p^{\sigma}-p'^{\sigma})+\eta^{\nu\sigma}(p'^{\mu}-p''^{\mu})+\eta^{\sigma\mu}(p''^{\nu}-p^{\nu})] \tag{75}$$

and

$$\begin{aligned}&-\mu^{4-N}g^{2}[f_{\alpha\beta\varepsilon}\,f^{\varepsilon}{}_{\gamma\delta}(\eta^{\mu\sigma}\eta^{\nu\tau}-\eta^{\mu\tau}\eta^{\nu\sigma})\\&+f_{\alpha\gamma\varepsilon}\,f^{\varepsilon}{}_{\delta\beta}(\eta^{\mu\tau}\eta^{\sigma\nu}-\eta^{\mu\nu}\eta^{\sigma\tau})\\&+f_{\alpha\delta\varepsilon}\,f^{\varepsilon}{}_{\beta\gamma}(\eta^{\mu\nu}\eta^{\tau\sigma}-\eta^{\mu\sigma}\eta^{\tau\nu})].\end{aligned} \tag{76}$$

It is not difficult to show that, up to a constant overall factor, (75) is the only expression that can be built out of the structure constants, the momentum components and the Minkowski metric, which is linear in the momenta, has three indices from the first part of the Greek alphabet and three from the middle, and is invariant under permutations of the momentum-index triplets $(\alpha,\mu,p),(\beta,\nu,p')$ and (γ,σ,p'') subject to the constraint $p+p'+p''=0$. Expression (76), by contrast, is not the only one that can be built out of the structure constants and the Minkowski metric, which is independent of the momenta, has four indices from the first part of the Greek alphabet and four from the middle, and is invariant under permutations of the index pairs $(\alpha,\mu),(\beta,\nu),(\gamma,\sigma)$ and (δ,τ). Two other expressions also have these properties, namely

$$\mathcal{P}_6\ \mathrm{tr}(f_\alpha\,f_\beta\,f_\gamma\,f_\delta)\,(\eta^{\mu\nu}\eta^{\sigma\tau}+\eta^{\mu\sigma}\eta^{\nu\tau}+\eta^{\mu\tau}\eta^{\nu\sigma}) \tag{77}$$

and

$$(\gamma_{\alpha\beta}\,\gamma_{\gamma\delta}+\gamma_{\alpha\gamma}\,\gamma_{\beta\delta}+\gamma_{\alpha\delta}\,\gamma_{\beta\gamma})\,(\eta^{\mu\nu}\eta^{\sigma\tau}+\eta^{\mu\sigma}\eta^{\nu\tau}+\eta^{\mu\tau}\eta^{\nu\sigma}), \tag{78}$$

where "$\mathcal{P}_6$" means "sum over the six essentially different permutations of indices in the trace." However, these expressions cannot appear as basic (i.e., traditional) vertex functions because they violate gauge invariance.

The situation is different with the new vertex functions S_3 and S_4, which are obtained from expressions (75) and (76) by multiplying them by $\Pi_\mu{}^{\bar\mu}(p),\Pi_\nu{}^{\bar\nu}(p'),\Pi_\sigma{}^{\bar\sigma}(p''),\Pi_\tau{}^{\bar\tau}(p''')$, where

$$\Pi_\mu{}^{\nu}(p)=\delta_\mu{}^{\nu}-p_\mu p^\nu/p^2. \tag{79}$$

Expression (75), with its Π-factors attached, is still unique. This means that the divergent part (pole part) of any graph having three external prongs, after all the subdivergences have been cancelled by counter terms, must be proportional to (75) with Π-factors attached and can itself be cancelled by a counter term proportional to S_3.

On the other hand, the divergent part of the sum of all graphs (of a given loop order) having four external prongs, after all subdivergences have been cancelled by counter terms, will *not* necessarily be simply proportional to (76) with Π-factors attached, but may have an additional part proportional to (77) with Π-factors attached. (A part proportional to (78) with Π-factors attached could arise only if the graph were disconnected, so it need not be considered.) To cancel both parts one must not only introduce a counter term proportional to S_4 but also add to the classical action a *nonlocal* counter term of the form

$$\begin{aligned}\Delta S=&\frac{\Xi}{24}\,\mathrm{tr}(f_\alpha f_\beta f_\gamma f_\delta)(\eta^{\mu\nu}\eta^{\sigma\tau}+\eta^{\mu\sigma}\eta^{\nu\tau}+\eta^{\mu\tau}\eta^{\nu\sigma})\\&\times\int\bar{A}_{R\,\mu}{}^{\alpha}\,\bar{A}_{R\,\nu}{}^{\beta}\bar{A}_{R\,\sigma}{}^{\gamma}\,\bar{A}_{R\,\tau}{}^{\delta}\,d^{N}x,\end{aligned} \tag{80}$$

where

$$\bar{A}_{R\mu}{}^{\alpha} = \int (\Pi^{\alpha}{}_{\mu\beta'}{}^{\nu'})_{A=0} A_R{}^{\beta'}{}_{\nu'} \, d^N x'. \tag{81}$$

Divergences requiring cancellation by (80) can and do arise, for two reasons:

1. Because the $A_R{}^{\alpha}{}_{\mu}$ are nonlocally related to the Gaussian normal fields.
2. Because expression (80) is gauge invariant by itself. To see this, carry out an infinitesimal gauge transformation (73) on it. The terms involving gradients of the gauge parameters are annihilated by the projection operator in Eq. (81). The terms involving the structure constants add up to zero because of the group invariance of $\mathrm{tr}(f_\alpha f_\beta f_\gamma f_\delta)$.

It is to be stressed that the nonlocal counter term makes no contribution to the particle content of the theory and plays no role in determining the β-function (see below).

In addition to ΔS and the counter terms for S_3 and S_4, counter terms contributing to the field renormalization constant Z are needed. These arise from graphs having two external prongs. The divergent parts of these graphs, after subdivergences have been subtracted, are quadratic in the external momenta and, because of the Π-factors attached to all vertices, are necessarily proportional to expression (H.1) of Appendix H, namely $-\gamma_{\alpha\beta}\,(\eta^{\mu\nu} p^2 - p^\mu p^\nu)$. Hence no problem arises here.

Note, finally, that each graph in the ghost-free formalism has the same degree of divergence as the corresponding graph in the traditional formalism. Therefore no counter terms are needed beyond those already mentioned [8].

12. ONE-LOOP GRAPHS

Computation of the values of graphs in the ghost-free formalism is just as tedious as in the traditional formalism. The only difference is that because the rules are different the divergent and finite parts turn out to be different. The value of the one-loop graph with two external prongs is found to be

$$-\frac{25}{6}\frac{g^2\lambda}{(4\pi)^2}\frac{1}{N-4}\gamma_{\alpha\beta}\,(\eta^{\mu\nu} p^2 - p^\mu p^\nu) + \text{finite part}, \tag{82}$$

whence

$$Z = 1 - \frac{25}{6}\frac{\lambda}{(4\pi)^2}\frac{1}{N-4}g^2 + O(g^4). \tag{83}$$

The value of the one-loop graph with three distinct external prongs is

$$-\frac{g^2\lambda}{(4\pi)^2}\frac{1}{N-4}S_3 + \text{finite part}, \tag{84}$$

whereas the value when two of the prongs emanate from the same point, and a permutation is carried out over all three prongs, is found to be

$$\frac{15}{4}\frac{g^2\lambda}{(4\pi)^2}\frac{1}{N-4}S_3 + \text{finite part}. \tag{85}$$

The sum of (84) and (85) is

$$\frac{11}{4}\frac{g^2\lambda}{(4\pi)^2}\frac{1}{N-4}S_3+\text{finite part} \tag{86}$$

yielding

$$Y=1-\frac{11}{4}\frac{\lambda}{(4\pi)^2}\frac{1}{N-4}g^2+O(g^4). \tag{87}$$

The total value of the one-loop graphs with four distinct external prongs is

$$\begin{aligned}&-\mathcal{P}_3\frac{g^4}{(4\pi)^2}\frac{1}{N-4}[4\,\mathrm{tr}(f_\alpha f_\beta f_\gamma f_\delta)+\lambda f_{\alpha\beta\varepsilon}f^{\varepsilon}{}_{\gamma\delta}](\eta^{\bar\mu\bar\nu}\eta^{\bar\sigma\bar\tau}+\eta^{\bar\mu\bar\sigma}\eta^{\bar\nu\bar\tau}+\eta^{\bar\mu\bar\tau}\eta^{\bar\nu\bar\sigma})\\&\times\,\Pi_{\bar\mu}{}^{\mu}(p)\,\Pi_{\bar\nu}{}^{\nu}(p')\,\Pi_{\bar\sigma}{}^{\sigma}(p'')\,\Pi_{\bar\tau}{}^{\tau}(p''')+\text{finite part,}\end{aligned} \tag{88}$$

where "$\mathcal{P}_3$" means "sum over cyclic permutations of any three of the indices $\alpha,\beta,\gamma,\delta$." Because of the Jacobi identity satisfied by the structure constants the term in λ may be dropped, but it is included here to make the expression in the square brackets symmetric in α and β as well as in γ and δ. When two of the external prongs are made to emanate from a single point the total value of the graphs becomes

$$\begin{aligned}&-\mathcal{P}_3\,\frac{g^4}{(4\pi)^2}\,\frac{1}{N-4}[4\,\mathrm{tr}(f_\alpha f_\beta f_\gamma f_\delta)+\lambda\,f_{\alpha\beta\varepsilon}f^{\varepsilon}{}_{\gamma\delta}\\&\times(\frac{13}{3}\,\eta^{\bar\mu\bar\nu}\eta^{\bar\sigma\bar\tau}+\frac{1}{3}\,\eta^{\bar\mu\bar\sigma}\eta^{\bar\nu\bar\tau}+\frac{1}{3}\,\eta^{\bar\mu\bar\tau}\eta^{\bar\nu\bar\sigma})\\&\times\,\Pi_{\bar\mu}{}^{\mu}(p)\,\Pi_{\bar\nu}{}^{\nu}(p')\,\Pi_{\bar\sigma}{}^{\sigma}(p'')\,\Pi_{\bar\tau}{}^{\tau}(p''')+\text{finite part,}\end{aligned} \tag{89}$$

where "$\mathcal{P}_3$" now means "sum over cyclic permutations of the index pairs $(\beta,\nu),(\gamma,\sigma)$ and (δ,τ)." Finally, when the other two external prongs are also made to emanate from a single point the total value of the graphs is found to be

$$\begin{aligned}&-\mathcal{P}_3\,\frac{g^4}{(4\pi)^2}\,\frac{1}{N-4}\,\{\frac{1}{24}[4\,\mathrm{tr}(f_\alpha f_\beta f_\gamma f_\delta)+\lambda\,f_{\alpha\beta\varepsilon}f^{\varepsilon}{}_{\gamma\delta}]\\&\times(37\,\eta^{\bar\mu\bar\nu}\eta^{\bar\sigma\bar\tau}+7\,\eta^{\bar\mu\bar\sigma}\eta^{\bar\nu\bar\tau}+7\,\eta^{\bar\mu\bar\tau}\eta^{\bar\nu\bar\sigma})\\&+\frac{9}{4}\,\lambda f_{\alpha\beta\varepsilon}f^{\varepsilon}{}_{\gamma\delta}(\eta^{\bar\mu\bar\sigma}\,\eta^{\bar\nu\bar\tau}-\eta^{\bar\mu\bar\tau}\,\eta^{\bar\nu\bar\sigma})\}\\&\times\,\Pi_{\bar\mu}{}^{\mu}(p)\,\Pi_{\bar\nu}{}^{\nu}(p')\,\Pi_{\bar\sigma}{}^{\sigma}(p'')\,\Pi_{\bar\tau}{}^{\tau}(p''')+\text{finite part.}\end{aligned} \tag{90}$$

The sum of (88), (89) and (90) is

$$\begin{aligned}&\mathcal{P}_3\,\frac{g^4}{(4\pi)^2}\,\frac{1}{N-4}\,\{\frac{1}{24}[4\,\mathrm{tr}(f_\alpha f_\beta f_\gamma f_\delta)+\lambda\,f_{\alpha\beta\varepsilon}f^{\varepsilon}{}_{\gamma\delta}]\\&\times(43\,\eta^{\bar\mu\bar\nu}\eta^{\bar\sigma\bar\tau}-23\,\eta^{\bar\mu\bar\sigma}\eta^{\bar\nu\bar\tau}-23\,\eta^{\bar\mu\bar\tau}\eta^{\bar\nu\bar\sigma})\\&-\frac{9}{4}\,\lambda f_{\alpha\beta\varepsilon}f^{\varepsilon}{}_{\gamma\delta}(\eta^{\bar\mu\bar\sigma}\,\eta^{\bar\nu\bar\tau}-\eta^{\bar\mu\bar\tau}\,\eta^{\bar\nu\bar\sigma})\}\\&\times\,\Pi_{\bar\mu}{}^{\mu}(p)\,\Pi_{\bar\nu}{}^{\nu}(p')\,\Pi_{\bar\sigma}{}^{\sigma}(p'')\,\Pi_{\bar\tau}{}^{\tau}(p''')+\text{finite.}\end{aligned} \tag{91}$$

This can be simplified with the aid of the easily verified identity

$$\begin{aligned}&\mathcal{P}_3\,[4\,\mathrm{tr}(f_\alpha f_\beta f_\gamma f_\delta)+\lambda f_{\alpha\beta\varepsilon}f^{\varepsilon}{}_{\gamma\delta}\,]\,(2\,\eta^{\mu\nu}\eta^{\sigma\tau}-\eta^{\mu\sigma}\eta^{\nu\tau}-\eta^{\mu\tau}\eta^{\nu\sigma})\\&=\mathcal{P}_3\,\lambda f_{\alpha\beta\varepsilon}f^{\varepsilon}{}_{\gamma\delta}(\eta^{\mu\sigma}\eta^{\nu\tau}-\eta^{\mu\tau}\eta^{\nu\sigma}),\end{aligned} \tag{92}$$

the proof of which makes use of

$$\mathrm{tr}(f_\alpha f_\beta f_\gamma f_\delta) - \mathrm{tr}(f_\alpha f_\gamma f_\delta f_\beta) = \frac{1}{2}\,\lambda\,(f_{\alpha\delta\varepsilon}f^{\varepsilon}{}_{\beta\gamma} - f_{\alpha\gamma\varepsilon}f^{\varepsilon}{}_{\beta\delta}). \tag{93}$$

Expression (91) becomes

$$\begin{aligned}&\frac{4}{3}\frac{g^2\lambda}{(4\pi)^2}\frac{1}{N-4}S_4 - \mathcal{P}_3\,\frac{1}{6}\frac{g^4}{(4\pi)^2}\frac{1}{N-4}\mathrm{tr}(f_\alpha f_\beta f_\gamma f_\delta)\\ &\times(\eta^{\bar\mu\bar\nu}\eta^{\bar\sigma\bar\tau}+\eta^{\bar\mu\bar\sigma}\eta^{\bar\nu\bar\tau}+\eta^{\bar\mu\bar\tau}\eta^{\bar\nu\bar\sigma})\\ &\times\Pi_{\bar\mu}{}^{\mu}(p)\,\Pi_{\bar\nu}{}^{\nu}(p')\,\Pi_{\bar\sigma}{}^{\sigma}(p'')\,\Pi_{\bar\tau}{}^{\tau}(p''')+\text{finite},\end{aligned} \tag{94}$$

from which it may be inferred that

$$X = 1 - \frac{4}{3}\frac{\lambda}{(4\pi)^2}\frac{1}{N-4}g^2 + O(g^4) \tag{95}$$

and

$$\Xi = \frac{1}{2}\frac{\lambda}{(4\pi)^2}\frac{1}{N-4}g^2 + O(g^4). \tag{96}$$

Comparison of expressions (83), (87) and (95) shows that Eq. (70) is satisfied up to one-loop order. Given the purely formal character of the earlier proofs of gauge invariance of the effective action, such consistency is remarkable and gives one confidence that Eq. (70) will hold to all orders.

13. DISCUSSION

It would be very nice if the calculational rules thus far obtained sufficed for calculating physical amplitudes, in particular scattering amplitudes. Thus, if one could just attach physical mode functions to the external prongs of loop graphs one would have a formalism that is easier to work with than the traditional one. Unfortunately, things are not so simple. In the Yang–Mills case this follows from the fact that the above rules amount to working in the Landau gauge and ignoring ghosts, a procedure that does *not* lead to a unitary S-matrix. The problem is that the Gaussian normal fields are not acceptable interpolating fields. They are nonlocally related to the local fields $A_R{}^{\alpha}{}_{\mu}$ and fail to satisfy the asymptotic requirements for use in the *LSZ* reduction formula. One must first express the Gaussian normal fields in terms of the $A_R{}^{\alpha}{}_{\mu}$ and then work with the latter. The steps involved in this procedure will be described elsewhere. Here we merely note that the present rules give the wrong β-function. Thus, Eqs. (71), (83) and (87) yield

$$g_0 = \mu^{2-N/2}\left[1+\frac{7}{2}\frac{\lambda}{(4\pi)^2}\frac{1}{N-4}g^2+O(g^4)\right]g, \tag{97}$$

whence, by standard arguments,

$$\beta = \mu\frac{dg}{d\mu} = \frac{7}{2}\frac{\lambda}{(4\pi)^2}g^3 + O(g^5). \tag{98}$$

When the Gaussian normal fields are replaced by the $A_R{}^\alpha{}_\mu$ additional pole terms appear, which contribute to the expression for the bare coupling constant. The constant g must therefore be replaced by a new constant, $\hat{g}$, and one finds

$$g_0 = \mu^{2-N/2}\left[1 + \frac{11}{3}\frac{\lambda}{(4\pi)^2}\frac{1}{N-4}\hat{g}^2 + O(\hat{g}^4)\right]\hat{g} \tag{99}$$

as one should. The relation between the new and old constants is evidently

$$\hat{g} = g - \frac{1}{6}\frac{\lambda}{(4\pi)^2}\frac{1}{N-4}g^3 + O(g^5) \tag{100}$$

14. COUPLING TO FERMION FIELDS

In order to isolate as much as possible the purely fermionic aspects of the additional contributions to the effective action that arise from the presence of fermion fields, we shall work in a Majorana representation (in which the elements of the Dirac matrices are real valued) with just one Majorana spinor field ψ. The presence of this field adds to the classical action (56) a real commuting Grassmann-numbered-valued (or *supernumber*-valued) integral of the form

$$S_\psi = \frac{1}{2} i \int \psi^\sim \eta[\gamma^\mu(\psi_{,\mu} + g_0 G_\alpha A^\alpha{}_\mu \psi) + m_0 \psi]\, d^N x. \tag{101}$$

Here "$\sim$" denotes the transpose and the components ψ^i of the spinor field are real *anti*commuting supernumber-valued variables having mass dimension $\frac{1}{2}(N-1)$. Strictly speaking, a Majorana representation is possible only when N is a positive integer satisfying $\sin(N-1)\pi/4 > 0$. But it is to be understood that the analytic continuation in dimension implied by letting N be variable is to be carried out only when graphs are being evaluated, and then according to standard rules (see, for example, Collins [9]).

The γ^μ, G_α and η are real matrices having the structure

$$\gamma^\mu = \hat{\gamma}^\mu \times \mathbb{1}_n, \qquad G_\alpha = \mathbb{1}_{2^{N/2}} \times \hat{G}^\alpha, \qquad \eta = \hat{\eta} \times g, \tag{102}$$

where "$\mathbb{1}_m$" denotes the m-dimensional unit matrix, the $\hat{\gamma}^\mu$ and $\hat{G}_\alpha$ are square matrices, of dimension $2^{N/2}$ and n respectively, satisfying

$$\{\hat{\gamma}^\mu, \hat{\gamma}^\nu\} = 2\,\eta^{\mu\nu}\,\mathbb{1}_{2^{N/2}}, \qquad [\hat{G}_\alpha, \hat{G}_\beta] = \hat{G}_\gamma f^\gamma{}_{\alpha\beta}, \tag{103}$$

and $\hat{\eta}$ and g are square matrices, of corresponding dimension, satisfying

$$\hat{\eta}\hat{\gamma}^\mu\hat{\eta}^{-1} = -\hat{\gamma}^{\mu\sim}, \qquad \hat{\eta}^\sim = -\hat{\eta}, \tag{104}$$

$$g\hat{G}_\alpha g^{-1} = -\hat{G}_\alpha{}^\sim, \qquad g^\sim = g. \tag{105}$$

The $\hat{G}_\alpha$ are evidently generators of an n-dimensional matrix representation of the Lie algebra of the Yang–Mills field and will be assumed to be irreducible. They necessarily satisfy

$$\gamma_{\alpha\beta}\hat{G}_\alpha\hat{G}_\beta = -\Lambda\mathbb{1}_n, \qquad \mathrm{tr}(\hat{G}_\alpha\hat{G}_\beta) = -\frac{n}{r}\Lambda\gamma_{\alpha\beta},$$
$$\mathrm{tr}(\hat{G}_\alpha\hat{G}_\beta\hat{G}_\gamma) = -\frac{1}{2}\frac{n}{r}\Lambda f_{\alpha\beta\gamma}, \tag{106}$$

where Λ is a real number characteristic of the representation and r is the dimensionality of the Lie algebra [10].

Equations (102) to (106) imply

$$[\gamma^{\mu},\, G_{\alpha}] = 0, \tag{107}$$

$$\{\gamma^{\mu},\, \gamma^{\nu}\} = 2\,\eta^{\mu\nu}\,\mathbb{1}_{n\cdot 2^{N/2}}, \qquad [G_{\alpha},\, G_{\beta}] = G_{\gamma} f^{\gamma}{}_{\alpha\beta}, \tag{108}$$

$$\eta\gamma^{\mu}\eta^{-1} = -\gamma^{\mu\sim}, \qquad \eta G_{\alpha}\eta^{-1} = -G_{\alpha}{}^{\sim}, \qquad \eta^{\sim} = -\eta, \tag{109}$$

$$\left.\begin{aligned} &\gamma^{\alpha\beta} G_{\alpha} G_{\beta} = -\Lambda \mathbb{1}_{n\cdot 2^{N/2}}, \qquad \mathrm{tr}(G_{\alpha} G_{\beta}) = -2^{N/2}\, \tfrac{n}{r} \Lambda \gamma_{\alpha\beta}, \\ &\mathrm{tr}(G_{\alpha} G_{\beta} G_{\gamma}) = -\tfrac{1}{2}\, 2^{N/2}\, \tfrac{n}{r} \Lambda f_{\alpha\beta\gamma}, \end{aligned}\right\} \tag{110}$$

and from these equations it is easy to verify that S_{ψ} is invariant under the gauge transformation (60), (61) provided ψ suffers simultaneously the transformation

$$\delta\psi^{i} = \int Q^{i}{}_{\alpha'}\, \delta\xi^{\alpha'}\, d^{N}x', \tag{111}$$

where

$$Q^{i}{}_{\alpha'} = g_0\, G_{\alpha}{}^{i}{}_{j}\, \psi^{j}\, \delta(x,x'). \tag{112}$$

With the spinor field present, the space of histories Φ becomes a supermanifold, and when Φ is endowed with a metric field the metric components must be antisymmetric in the fermionic sector. There is again a unique ultralocal gauge invariant metric field, which is obtained by adjoining to the components (63) also the following:

$$\gamma_{\alpha}{}^{\mu}{}_{j'} = 0, \qquad \gamma_{i\beta'}{}^{\nu'} = 0, \qquad \gamma_{ij'} = \mu^{-1}\, \eta_{ij}\, \delta(x,x'). \tag{113}$$

Although it will never again make an appearance, the factor μ^{-1} is inserted into the last of these expressions in order to make the mass dimension of the associated arc length uniformly equal to -1. Without loss of generality μ^{-1} may, if one likes, be assumed equal to the inverse of the auxiliary mass used in dimensional regularization.

The extended metric (63) + (113), like the metric (63), is constant. Hence the Riemannian curvature vanishes, as do the Riemannian connection components, in the coordinate system $\{A^{\alpha}{}_{\mu}, \psi^{i}\}$, and once again the dot in Eq. (55) may be replaced by a comma. The operator $\mathfrak{F}_{\alpha\beta'}$ is obtained by adding to Eq. (H.6) the expression

$$\begin{aligned} &-\int d^{N}x'' \int d^{N}x'''\, \gamma_{i''j'''}\, Q^{i''}{}_{\alpha}\, Q^{j'''}{}_{\beta'} \\ &= -\mu^{-1}\, g_0{}^{2} \psi^{\sim} G_{\alpha}{}^{\sim}\, \eta\, G_{\beta}\, \psi\, \delta(x,x') \\ &= \mu^{-1}\, g_0{}^{2} \psi^{\sim} \eta G_{\alpha}\, G_{\beta}\, \psi\, \delta(x,x'), \end{aligned} \tag{114}$$

and its Fourier transform reduces to $-\gamma_{\alpha\beta}\, p^{2}$, as before, when *both* $A^{\alpha}{}_{\mu}$ and ψ are set equal to be zero. In this limit the Fourier transform of the connection 1-form on Φ becomes

$$\omega^{\alpha}{}_{\beta'}{}^{\nu'} \longrightarrow -i\delta^{\alpha}{}_{\beta}\, p^{\mu}/p^{2}, \qquad \omega^{\alpha}{}_{i'} \longrightarrow 0 \tag{115}$$

(see Eq. (H.8)), and the horizontal projection operator reduces to

$$\begin{aligned} \Pi^{\alpha}{}_{\mu\beta'}{}^{\nu'} &\longrightarrow \delta^{\alpha}{}_{\beta}(\delta_{\mu}{}^{\nu} - p_{\mu}p^{\nu}/p^{2}), \qquad \Pi^{\alpha}{}_{\mu j'} \longrightarrow 0, \\ \Pi_{i\beta'}{}^{\nu'} &\longrightarrow 0, \qquad \Pi^{i}{}_{j'} \longrightarrow \delta^{i}{}_{j}. \end{aligned} \tag{116}$$

Evidently no Π-factors need be applied to spinor prongs.

15. RENORMALIZATION OF GRAPHS WITH SPINOR LINES

Two new renormalization constants are now needed, a constant $\tilde{Z}$ that renormalizes the spinor field,

$$\psi = \tilde{Z}^{1/2}\, \psi_R, \tag{117}$$

and a constant δm that, together with $\tilde{Z}^{-1}$, renormalizes the bare mass m_0:

$$m_0 = m + \tilde{Z}^{-1}\, \delta m. \tag{118}$$

The spinor action (101) takes the form

$$\begin{aligned} S_\psi = {} & \frac{1}{2}\, i\tilde{Z} \int \psi_R{}^{\sim} \eta (\gamma^\mu \psi_{R,\mu} + m\, \psi_R)\, d^N x \\ & + \frac{1}{2}\, i\, \delta m \int \psi_R{}^{\sim}\, \eta\, \psi_R\, d^N x \\ & + \frac{1}{2}\, i\mu^{2-N/2} g \tilde{Z} Z^{-1} Y \int \psi_R{}^{\sim} \eta \gamma^\mu G_\alpha A_R{}^\alpha{}_\mu \psi_R\, d^N x, \end{aligned} \tag{119}$$

which yields the following basic structures needed in perturbation theory:

$$\begin{aligned} & \left(\frac{\overrightarrow{\delta}}{\delta \psi_R{}^{\sim}}\, S_\psi\, \frac{\overleftarrow{\delta}}{\delta \psi'_R} \right)_{\substack{\tilde{Z}=1, \quad Z=1 \\ Y=1, \quad \delta m = 0}} \\ & = i\eta \left[\gamma^\mu \left(\frac{\partial}{\partial x^\mu} + \mu^{2-N/2} g G_\alpha A_R{}^\alpha{}_\mu \right) + m \right] \delta(x, x') \\ & \longrightarrow \eta(\gamma^\mu p_\mu + im), \end{aligned} \tag{120}$$

$$\begin{aligned} & \left(\frac{\overrightarrow{\delta}}{\delta \psi_R{}^{\sim}}\, \frac{\delta S_\psi}{\delta A_R{}^{\alpha''}{}_{\mu''}}\, \frac{\overleftarrow{\delta}}{\delta \psi'_R} \right)_{\substack{\tilde{Z}=1, \quad Z=1 \\ Y=1, \quad \delta m = 0}} \\ & \longrightarrow i\, \mu^{2-N/2} g\, \eta\, \gamma^\mu\, G_\alpha. \end{aligned} \tag{121}$$

The spinor propagator is the negative inverse of expression (120), namely

$$\frac{-\gamma^\mu p_\mu + im}{p^2 + m^2}\, \eta^{-1}. \tag{122}$$

Expression (121) is the traditional spinor vertex function. In the ghost-free formalism it must be multiplied by $\Pi_\mu{}^\nu(p'')$.

Using these structures as well as standard trace and contraction formulas for the Dirac matrices [9], one easily calculates the value of the one-loop spinor "self-energy" graph:

$$\begin{aligned} & -\frac{i}{(2\pi)^N} \left(i\mu^{2-N/2} g \right)^2 \eta\, \gamma^\mu\, G_\alpha \\ & \times \int \gamma^{\alpha\beta} \left(\eta_{\mu\nu} - \frac{k_\mu k_\nu}{k^2} \right) \frac{1}{k^2}\, \frac{-\gamma^\sigma (k_\sigma + p_\sigma) + im}{(k+p)^2 + m^2}\, \gamma^\nu\, G_\beta\, d^N k \\ & = -6 \frac{g^2 \Lambda}{(4\pi)^2} \frac{1}{N-4}\, i\, \eta\, m + \text{finite part}. \end{aligned} \tag{123}$$

To one-loop order the counter-term contribution to $\tilde{Z}$ is seen to vanish, and we have

$$\tilde{Z} = 1 + O(g^4), \tag{124}$$

$$\delta m = \left[6 \frac{\Lambda}{(4\pi)^2} \frac{1}{N-4} g^2 + O(g^4) \right] m. \tag{125}$$

Graphs containing closed spinor loops induce corrections to Eqs. (83), (87) and (95) for the renormalization constants Z, Y and X. The value of the graph in which two ϕ-prongs are inserted into a single fermion loop is given by

$$\begin{aligned} &\frac{i}{2} \frac{1}{(2\pi)^N} \left(i\mu^{2-N/2} g\right)^2 \Pi_\sigma{}^\mu(p)\, \Pi_\tau{}^\nu(p) \\ &\times \int \mathrm{tr} \left(G_\alpha \gamma^\sigma \frac{-\gamma^\rho (k_\rho + p_\rho) + im}{(k+p)^2 + m^2} G_\beta \gamma^\tau \frac{-\gamma^\lambda k_\lambda + im}{k^2 + m^2} \right) d^N k \\ &= \frac{4}{3} \frac{n}{r} \frac{\Lambda}{(4\pi)^2} \frac{1}{N-4} g^2 \gamma_{\alpha\beta} \left(\eta^{\mu\nu} p^2 - p^\mu p^\nu\right) + \text{finite part}, \end{aligned} \tag{126}$$

which yields the following modification of Eq. (83):

$$Z = 1 - \left(\frac{25}{6} \lambda - \frac{4}{3} \frac{n}{r} \Lambda \right) \frac{1}{(4\pi)^2} \frac{1}{N-4} g^2 + O(g^4). \tag{127}$$

The value of the graph in which three ϕ-prongs are inserted into a fermion loop is found to be

$$-\frac{4}{3} \frac{n}{r} \frac{g^2 \Lambda}{(4\pi)^2} \frac{1}{N-4} S_3 + \text{finite part}, \tag{128}$$

which implies that Eq. (87) gets changed to

$$Y = 1 - \left(\frac{11}{4} \lambda - \frac{4}{3} \frac{n}{r} \Lambda \right) \frac{1}{(4\pi)^2} \frac{1}{N-4} g^2 + O(g^4). \tag{129}$$

Finally, the value of the graph in which four ϕ-prongs are inserted into a fermion loop is found to be

$$-\frac{4}{3} \frac{n}{r} \frac{g^2 \Lambda}{(4\pi)^2} \frac{1}{N-4} S_4 + \text{finite part}, \tag{130}$$

which implies, first, that fermions make no contribution to the nonlocal counter term (80) in one loop order and, second, that Eq. (95) gets changed to

$$X = 1 - \frac{4}{3} \left(\lambda - \frac{n}{r} \Lambda \right) \frac{1}{(4\pi)^2} \frac{1}{N-4} g^2 + O(g^4). \tag{131}$$

Comparison of Eqs. (127), (129) and (131) shows that Eq. (70) continues to be satisfied, verifying once again the gauge invariance of the theory. One also easily checks that the β-function now becomes [11]

$$\beta = -\left(\frac{7}{2} \lambda - \frac{2}{3} \frac{n}{r} \Lambda \right) \frac{1}{(4\pi)^2} g^3 + O(g^5). \tag{132}$$

Although fermions make no one-loop contribution to the nonlocal counterterm (80) they do require the introduction of another nonlocal counter term, of the form

$$\Delta S_\psi = \frac{1}{2}\, i\, \mu^{2-N/2} g \tilde{\Xi} \int \psi_R{}^{\sim} \eta\, \gamma^\mu\, G_\alpha\, \psi_R\, \bar{A}_{R\,\mu}{}^\alpha\, d^N x, \tag{133}$$

where $\bar{A}_{R\,\mu}{}^\alpha$ is given by Eq. (81). To see this one must compute the values of graphs having two external fermion prongs and one external ϕ-prong, the divergent parts of which should in principle be cancelled by counter terms coming from the factor $\tilde{Z}Z^{-1}Y$ in the last term of expression (119). In one-loop order there are two such graphs, one in which the ϕ-prong is attached to a fermion line and one in which it is attached to a ϕ-line. The value of the former turns out to be finite, while that of the latter is found to be

$$-i\frac{3}{2}\frac{g^2\lambda}{(4\pi)^2}\frac{1}{N-4}\, g\, \eta\, \gamma^\nu\, G_\alpha\, \Pi_\nu{}^\mu(p'') + \text{finite part.} \tag{134}$$

However, the factor $\tilde{Z}Z^{-1}Y$ provides only

$$i\frac{17}{12}\frac{g^2\lambda}{(4\pi)^2}\frac{1}{N-4} g\, \eta\, \gamma^\nu\, G_\alpha\, \Pi_\nu{}^\mu(p'') \tag{135}$$

as a counter term. The remainder needed to cancel the divergent part of (134) must come from the counter term (133), with $\tilde{\Xi}$ chosen to be

$$\tilde{\Xi} = \frac{1}{12}\frac{\lambda}{(4\pi)^2}\frac{1}{N-4} g^2 + O(g^4). \tag{136}$$

Note that $\tilde{\Xi}$ does not depend on the constant Λ in one-loop order. Note also that the addition of ΔS_ψ to the classical action, like the addition of ΔS, neither violates gauge invariance (since expression (133) is gauge invariant by itself), changes the particle content of the theory, nor affects the β-function.

16. DISCUSSION

The ultimate importance of the ghost-free formalism lies not in its demonstration that one can actually get along without ghosts (and the whole paraphernalia of BRST as well!), thus remaining close to the spirit of the classical theory from which one starts. What is important is the fact that one obtains a quantum effective action Γ that is independent of the choice of gauge breaking terms and of the ghost propagator (although not independent of the choice of the base point φ_*). This has implications for attempts to tackle not only problems with "in–out" boundary conditions, which constitute the usual framework for the quantum effective action, but also problems involving "in–in" boundary conditions, in which one tries to get information about expectation values. By using some variant of the well-known "two-time" or "closed-time-path" formalism, together with Vilkovisky's ideas, one should be able to construct an "in–in" effective action that is ghost and gauge-breaking independent. This would lead to the possibility of finding gauge-covariant, ghost-independent nonlocal dynamical equations that govern, at least approximately, the evolution of field expectation values.

The most interesting application of such a development would be in quantum gravity. Here, of course, there are formidable obstacles : (1) The algebraic complexity that always arises in gravitational calculations. (2) The fact that the number of basic vertices is no longer finite. (3) The fact that γ is no longer flat. (4) The fact that gravity is not perturbatively renormalizable. Modern computing machines are making it increasingly possible to confront obstacle 1. They make it also possible to begin dealing with obstacle 2 and the fact that the higher vertex functions quickly run to millions of terms. Obstacle 3 produces only a minor perturbation on obstacle 2. Obstacle 4 is the most difficult to deal with.

The best strategy here is perhaps to treat Einsteinian gravity as an effective field theory arising from some deeper theory, and to deal with the infinities generated by quantization simply by throwing them away as they arise, through dimensional regularization and minimal subtraction. Such a procedure leaves the particle content of the theory intact in every loop order and preserves unitarity. Its chief weakness is that there are no running coupling constants (i.e., no β-functions) and that each choice of auxiliary mass μ leads to a different theory. A reasonable choice for μ would seem to be simply the Planck mass, or something close to it. One could then examine both the "in–out" and "in–in" effective actions, at least in one-loop order (which corresponds to the highly useful WKB approximation of ordinary quantum mechanics). If one could do this in ordinary space and not momentum space, with a few selected background metrics, one would be poised to undertake a stably-based, ghost-independent, nontrivial attack on the black-hole back-reaction problem.

ACKNOWLEDGMENTS

The authors are grateful for student and travel support from the University of Texas through an endowed professorship established by Jane and Roland Blumberg and through the Center for Relativity.

APPENDICES

A. DERIVATION OF THE VILKOVISKY CONNECTION

Let the symbol $\doteq$ denote equality modulo linear combinations of the $Q^i{}_\alpha$ or, what is the same thing, of the $\varphi^i{}_{,\alpha}(= -Q^i{}_\alpha\, \mathfrak{G}^\alpha{}_\beta)$. Then, making use of Eqs. (28) and the well known coordinate transformation law for connections on frame bundles, and discarding terms involving the $\varphi^i{}_{,\alpha}$, obtain

$$\Gamma^i{}_{jk} \doteq \varphi^i{}_{,A}\, I^B{}_{,j}\, I^C{}_{,k}\, \Gamma^A{}_{BC} + \varphi^i{}_{,A}\, I^A{}_{,jk}. \tag{A.1}$$

Insert the explicit form (26) into this, and use Eqs. (18) and (24) to get

$$\begin{aligned}\Gamma^i{}_{jk} \doteq{}& \frac{1}{2}\varphi^i{}_{,A}\, I^B{}_{,j}\, I^C{}_{,k}\, I^A{}_{,l}\, I^D{}_{,m}\, \gamma^{lm} \\ &\times[(\bar{\gamma}_{rs}\varphi^r{}_{,D}\varphi^s{}_{,B})_{,C} + (\bar{\gamma}_{rs}\varphi^r{}_{,D}\varphi^s{}_{,C})_{,B} - (\bar{\gamma}_{rs}\varphi^r{}_{,B}\varphi^s{}_{,C})_{,D}] \\ &+\varphi^i{}_{,A} I^A{}_{,jk}.\end{aligned} \tag{A.2}$$

Noting that

$$\varphi^i{}_{,A}\, I^A{}_{,l} = \delta^i{}_l - \varphi^i{}_{,\alpha}\, K^\alpha{}_{,l} \doteq \delta^i{}_l \tag{A.3}$$

and

$$g_{AB,C}\, I^C{}_{,k} = g_{AB,k}, \tag{A.4}$$

rewrite Eq. (A.2) in the form

$$\begin{aligned}
\Gamma^i{}_{jk} &\doteq \frac{1}{2}\,\gamma^{im}\,[I^D{}_{,m}\,I^B{}_{,j}\,(\bar{\gamma}_{rs}\,\varphi^r{}_{,D}\,\varphi^s{}_{,B})_{,k} \\
&\quad + I^D{}_{,m}\,I^C{}_{,k}\,(\bar{\gamma}_{rs}\,\varphi^r{}_{,D}\,\varphi^s{}_{,C})_{,j} \\
&\quad - I^B{}_{,j}\,I^C{}_{,k}\,(\bar{\gamma}_{rs}\,\varphi^r{}_{,B}\,\varphi^s{}_{,C})_{,m}] + \varphi^i{}_{,A}\,I^A{}_{,jk} \\
&= \frac{1}{2}\,\gamma^{im}\,[I^D{}_{.m}\,I^B{}_{.j}\,(\bar{\gamma}_{rs}\,\varphi^r{}_{,D}\,\varphi^s{}_{,B})_{.k} \\
&\quad + I^D{}_{.m}\,I^C{}_{.k}\,(\bar{\gamma}_{rs}\,\varphi^r{}_{,D}\,\varphi^s{}_{,C})_{.j} \\
&\quad - I^B{}_{.j}\,I^C{}_{.k}\,(\bar{\gamma}_{rs}\,\varphi^r{}_{,B}\,\varphi^s{}_{,C})_{.m}] + \varphi^i{}_{,A}\,I^A{}_{,jk},
\end{aligned} \tag{A.5}$$

where the dots denote covariant functional differentiation based on the Riemannian connection $\Gamma_\gamma{}^i{}_{jk}$.

Although the replacement of commas by dots is a trivial step since

$$I^B, I^C, I^D, (\bar{\gamma}_{rs}\varphi^r{}_{,D}\varphi^s{}_{,B}),$$

etc. are here to be viewed as scalar fields on Φ, the next step is nontrivial: Bring the I's inside the parentheses and make use of (A.3) and the fact that

$$\bar{\gamma}_{pq}\,\varphi^q{}_{,\alpha} = 0, \tag{A.6}$$

to get

$$\begin{aligned}
\Gamma^i{}_{jk} &\doteq \frac{1}{2}\,\gamma^{im}\,(\bar{\gamma}_{mj\cdot k} + \bar{\gamma}_{mk\cdot j} - \bar{\gamma}_{jk\cdot m} \\
&\quad - \bar{\gamma}_{ms}\,\varphi^s{}_{,B}\,I^B{}_{\cdot jk} - \bar{\gamma}_{rj}\,\varphi^r{}_{,D}\,I^D{}_{\cdot mk} \\
&\quad - \bar{\gamma}_{ms}\,\varphi^s{}_{,C}\,I^C{}_{\cdot kj} - \bar{\gamma}_{rk}\,\varphi^r{}_{,D}\,I^D{}_{\cdot mj} \\
&\quad + \bar{\gamma}_{rk}\,\varphi^r{}_{,B}\,I^B{}_{\cdot jm} + \bar{\gamma}_{js}\,\varphi^s{}_{,C}\,I^C{}_{\cdot km}) + \varphi^i{}_{,A}\,I^A{}_{,jk}.
\end{aligned} \tag{A.7}$$

The fifth and ninth terms inside the parentheses cancel, as do the seventh and eighth. Combining the fourth and sixth terms, noting that $\gamma^{im}\,\bar{\gamma}_{ms} \doteq \delta^i{}_s$, and using the explicit form (19) in the first three terms, get

$$\begin{aligned}
\Gamma^i_{jk} &\doteq -\frac{1}{2}(Q^i{}_\alpha\,\omega^\alpha{}_j)_{\cdot k} - \frac{1}{2}(Q^i{}_\alpha\,\omega^\alpha{}_k)_{\cdot j} + \frac{1}{2}(Q_{j\alpha}\,\omega^\alpha{}_k)^{\cdot i} \\
&\quad - \varphi^i{}_{,A}(I^A{}_{\cdot jk} - I^A{}_{,jk}) \\
&\doteq -\frac{1}{2}Q^i{}_{\alpha\cdot k}\,\omega^\alpha{}_j - \frac{1}{2}Q^i{}_{\alpha\cdot j}\,\omega^\alpha{}_k + \frac{1}{2}(Q_{j\alpha}\mathfrak{G}^{\alpha\beta}Q_{k\beta})^{\cdot i} \\
&\quad + \varphi^i{}_{,A}\,I^A{}_{,l}\,\Gamma_\gamma{}^l{}_{jk} \\
&\doteq \Gamma_\gamma{}^i{}_{jk} - Q^i{}_{\alpha\cdot j}\,\omega^\alpha{}_k - Q^i{}_{\alpha\cdot k}\,\omega^\alpha{}_j + \frac{1}{2}Q_{j\alpha}\mathfrak{G}^{\alpha\beta}{}_{\cdot}{}^{i}Q_{k\beta},
\end{aligned} \tag{A.8}$$

in which Eqs. (6), (8) and (A.3) are used to arrive at the last line. Finally, make use of

$$\mathfrak{G}^{\alpha\beta}{}_{\cdot}{}^{i} = \mathfrak{G}^{\alpha\gamma}\,\mathfrak{F}_{\gamma\delta\cdot}{}^{i}\,\mathfrak{G}^{\delta\beta} \tag{A.9}$$

together with Eqs. (6), (8) and (9) to arrive at Eq. (29) of the text.

B. HORIZONTAL GEODESICS

Let $\lambda : \mathbb{R} \longrightarrow \Phi$ be an affinely parametrized geodesic in Φ based on the Riemannian connection ${\Gamma_\gamma}^i{}_{jk}$. Let $\varphi^i(s)$ be the coordinates of $\lambda(s)$ in a chart $\{\varphi^i\}$ in Φ.

Theorem. $Q_{i\alpha}\,[\varphi(s)]d\varphi^i(s)/ds$ is constant along λ.

Proof: Denoting the covariant derivative along λ by D/ds and suppressing arguments, write

$$0 = \frac{D}{ds}\frac{d\varphi^i}{ds} = \frac{d^2\varphi^i}{ds^2} + {\Gamma_\gamma}^i{}_{jk}\frac{d\varphi^j}{ds}\frac{d\varphi^k}{ds}, \tag{B.1}$$

$$\begin{aligned}\frac{d}{ds}\left(Q_{i\alpha}\frac{d\varphi^i}{ds}\right) &= \frac{D}{ds}\left(Q_{i\alpha}\frac{d\varphi^i}{ds}\right)\\ &= Q_{i\alpha\cdot j}\frac{d\varphi^i}{ds}\frac{d\varphi^j}{ds} + Q^i_\alpha\frac{D}{ds}\frac{d\varphi^i}{ds}\\ &= \frac{1}{2}\left(Q_{i\alpha\cdot j} + Q_{j\alpha\cdot i}\right)\frac{d\varphi^i}{ds}\frac{d\varphi^j}{ds},\end{aligned} \tag{B.2}$$

which vanishes by Eq. (6).

If the tangent to λ is horizontal at one point then $Q_{i\alpha}\,d\varphi^i/ds = 0$ at that point. Since $Q_{i\alpha}\,d\varphi^i/ds$ is constant along λ it follows that λ is horizontal everywhere. If λ is horizontal then also $\omega^\alpha{}_i\,d\varphi^i/ds = 0$, and one sees from the form of the Vilkovisky connection (31) that

$$\frac{d^2\varphi^i}{ds^2} + \Gamma^i{}_{jk}\frac{d\varphi^j}{ds}\frac{d\varphi^k}{ds} = \frac{d^2\varphi^i}{ds^2} + {\Gamma_\gamma}^i{}_{jk}\frac{d\varphi^j}{ds}\frac{d\varphi^k}{ds} = 0, \tag{B.3}$$

so λ is also a geodesic based on $\Gamma^i{}_{jk}$. Since a geodesic through a given point is completely determined as soon as the tangent at that point is given it follows that if the tangent vector to a geodesic based on $\Gamma^i{}_{jk}$ is horizontal at one point then it is horizontal everywhere.

C. DERIVATION OF EQ. (30)

Write

$$\Gamma^i{}_{jk} = {\Gamma_\gamma}^i{}_{jk} + \Delta\Gamma^i{}_{jk} \tag{C.1}$$

where (see Eq. (29))

$$\begin{aligned}\Delta\Gamma^i{}_{jk} = &-Q^i{}_{\alpha\cdot j}\,\omega^\alpha{}_k - Q^i{}_{\alpha\cdot k}\,\omega^\alpha{}_j\\ &+\frac{1}{2}\,\omega^\alpha{}_j\,Q^i{}_{\alpha\cdot l}\,Q^l{}_\beta\,\omega^\beta{}_k + \frac{1}{2}\,\omega^\alpha{}_k\,Q^i{}_{\alpha\cdot l}\,Q^l{}_\beta\,\omega^\beta{}_j.\end{aligned} \tag{C.2}$$

Then

$$\begin{aligned}Q^i{}_{\alpha;j} &= Q^i{}_{\alpha,j} + \Gamma^i{}_{jk}\,Q^k{}_\alpha\\ &= Q^i{}_{\alpha,j} + {\Gamma_\gamma}^i{}_{jk}\,Q^k{}_\alpha + \Delta\Gamma^i{}_{jk}\,Q^k{}_\alpha\\ &= Q^i{}_{\alpha\cdot j} - Q^i{}_{\beta\cdot j}\,\omega^\beta{}_k\,Q^k{}_\alpha - Q^i{}_{\beta\cdot k}\,\omega^\beta{}_j\,Q^k{}_\alpha\\ &\quad+\frac{1}{2}\omega^\beta{}_j Q^i{}_{\beta\cdot l} Q^l{}_\gamma \omega^\gamma{}_k\,Q^k{}_\alpha + \frac{1}{2}\,\omega^\beta{}_k Q^i{}_{\beta\cdot l} Q^l{}_\gamma \omega^\gamma{}_j Q^k{}_\alpha.\end{aligned} \tag{C.3}$$

Now make use of Eq. (10) together with the component form of Eq. (2), namely

$$Q^i{}_{\alpha,j}Q^j{}_\beta - Q^i{}_{\beta,j}Q^j{}_\alpha = Q^i{}_{\alpha\cdot j}Q^j{}_\beta - Q^i{}_{\beta\cdot j}Q^j{}_\alpha = Q^i{}_\gamma c^\gamma{}_{\alpha\beta}, \tag{C.4}$$

to reexpress (C.3) in the form

$$\begin{aligned} Q^i{}_{\alpha;j} &= -Q^i{}_{\beta\cdot k}\,\omega^\beta{}_j\,Q^k{}_\alpha + \frac{1}{2}Q^i{}_{\beta\cdot k}\,\omega^\beta{}_j\,Q^k{}_\alpha + \frac{1}{2}Q^i{}_{\alpha\cdot l}\,Q^l{}_\gamma\,\omega^\gamma{}_j \\ &= \frac{1}{2}\left(Q^i{}_{\alpha\cdot k}Q^k{}_\beta - Q^i{}_{\beta\cdot k}Q^k{}_\alpha\right)\omega^\beta{}_j = \frac{1}{2}Q^i{}_\gamma\,c^\gamma{}_{\alpha\beta}\,\omega^\beta{}_j. \end{aligned} \tag{C.5}$$

D. VERTICAL GEODESICS

Let $\lambda : \mathbb{R} \longrightarrow \Phi$ be a curve with coordinates φ^i satisfying

$$\frac{d\varphi^i}{ds} = Q^i{}_\alpha[\varphi]\xi^\alpha, \tag{D.1}$$

where the ξ's are constants. Since the $\mathbf{Q}_\alpha$ are vertical vector fields λ is tangent to and, in fact, lies in a single fibre. λ is also a geodesic based on $\Gamma^i{}_{jk}$, as follows from

$$\begin{aligned} \frac{d^2\varphi^i}{ds^2} + \Gamma^i{}_{jk}\,\frac{d\varphi^j}{ds}\,\frac{d\varphi^k}{ds} &= Q^i{}_{\alpha,j}\,\frac{d\varphi^j}{ds}\,\xi^\alpha + \Gamma^i{}_{jk}\,Q^k{}_\alpha\,\frac{d\varphi^j}{ds}\,\xi^\alpha \\ &= Q^i{}_{\alpha;j}\,Q^j{}_\beta\,\xi^\alpha\,\xi^\beta = \frac{1}{2}Q^i{}_\gamma\,c^\gamma{}_{\alpha\delta}\,\omega^\delta{}_j\,Q^j{}_\beta\,\xi^\alpha\,\xi^\beta \\ &= \frac{1}{2}Q^i{}_\gamma\,c^\gamma{}_{\alpha\beta}\,\xi^\alpha\,\xi^\beta \end{aligned} \tag{D.2}$$

which vanishes by the antisymmetry of $c^\gamma{}_{\alpha\delta}$. Since a geodesic through a given point is completely determined as soon as its tangent at that point is given, and since all vertical tangents are expressible in the form (D.1) for suitable ξ^α, it follows that a geodesic that is tangent to a fibre at one point lies in that fibre.

E. DERIVATION OF EQ. (31)

First note that

$$\omega^\alpha{}_i\,\Pi^i{}_j = 0, \qquad \Pi^i{}_j\,Q^j{}_\alpha = 0, \tag{E.1}$$

whence it follows, in particular, that

$$\Delta\Gamma^i{}_{lm}\,\Pi^l{}_j\,\Pi^m{}_k = 0 \tag{E.2}$$

(see Eq. (C.2)). Next note that

$$\begin{aligned} \omega^\alpha{}_{i\cdot j} &= (\mathfrak{G}^{\alpha\beta}Q_{i\beta})_{\cdot j} = -\mathfrak{G}^{\alpha\gamma}(Q_{k\gamma}\,Q^k{}_\delta)_{\cdot j}\mathfrak{G}^{\delta\beta}Q_{i\beta} + \mathfrak{G}^{\alpha\beta}Q_{i\beta\cdot j} \\ &= -\mathfrak{G}^{\alpha\gamma}\,Q_{k\gamma\cdot j}\,Q^k{}_\delta\,\omega^\delta{}_i - \omega^\alpha{}_k\,Q^k{}_{\delta\cdot j}\,\omega^\delta{}_i + \mathfrak{G}^{\alpha\beta}Q_{i\beta\cdot j} \\ &= \mathfrak{G}^{\alpha\beta}Q_{k\beta\cdot j}\Pi^k{}_i - \omega^\alpha{}_k\,Q^k{}_{\delta\cdot j}\,\omega^\delta{}_i. \end{aligned} \tag{E.3}$$

Therefore

$$(\omega^{\alpha}{}_{k\cdot l}+\omega^{\alpha}{}_{l\cdot k})\Pi^{k}{}_{i}\,\Pi^{k}{}_{j}=\mathfrak{G}^{\alpha\beta}(Q_{k\beta\cdot l}+Q_{l\beta\cdot k})\Pi^{k}{}_{i}\,\Pi^{l}{}_{j}=0 \tag{E.4}$$

(see Eq. (6)).

If B is gauge invariant then

$$B_{,i}Q^{i}{}_{\alpha}=0, \tag{E.5}$$

which implies

$$B_{;i}=B_{,i}=B_{,j}\,\Pi^{j}{}_{i}=B_{\cdot j}\,\Pi^{j}{}_{i}. \tag{E.6}$$

Equation (31) evidently holds for $n=1$. Its validity for all n follows by induction:

$$\begin{aligned}
B_{;(i_1\ldots i_{n+1})} &= \frac{1}{(n+1)!}\sum_{\text{perm}} B_{;i_1\ldots i_{n+1}} \\
&= \frac{1}{(n+1)!}\sum_{\text{perm}} B_{;j_1\ldots j_{n+1}}\Pi^{j_1}{}_{i_1}\ldots\Pi^{j_{n+1}}{}_{i_{n+1}} \\
&= \frac{1}{(n+1)!}\sum_{\text{perm}} \left(B_{\cdot k_1\ldots k_n}\Pi^{k_1}{}_{j_1}\ldots\Pi^{k_n}{}_{j_n}\right)_{;j_{n+1}}\Pi^{j_1}{}_{i_1}\ldots\Pi^{j_{n+1}}{}_{i_{n+1}} \\
&= \frac{1}{(n+1)!}\sum_{\text{perm}} [(B_{\cdot k_1\ldots k_n}\Pi^{k_1}{}_{j_1}\ldots\Pi^{k_n}{}_{j_n})_{\cdot j_{n+1}} \\
&\quad -B_{\cdot k_1\ldots k_n}(\Pi^{k_1}{}_{l}\,\Pi^{k_2}{}_{j_2}\ldots\Pi^{k_n}{}_{j_n}\,\Delta\Gamma^{l}{}_{j_1 j_{n+1}}+\ldots \\
&\quad +\Pi^{k_1}{}_{j_1}\ldots\Pi^{k_{n-1}}{}_{j_{n-1}}\Pi^{k_n}{}_{l}\,\Delta\Gamma^{l}{}_{j_n j_{n+1}})] \\
&\quad \times\Pi^{j_1}{}_{i_1}\ldots\Pi^{j_{n+1}}{}_{i_{n+1}} \\
&= \frac{1}{(n+1)!}\sum_{\text{perm}} [B_{\cdot k_1\ldots k_n j_{n+1}}\Pi^{k_1}{}_{j_1}\ldots\Pi^{k_n}{}_{j_n} \\
&\quad +B_{\cdot k_1\ldots k_n}(\Pi^{k_1}{}_{j_1\cdot j_{n+1}}\Pi^{k_2}{}_{j_2}\ldots\Pi^{k_n}{}_{j_n}+\ldots \\
&\quad +\Pi^{k_1}{}_{j_1}\ldots\Pi^{k_{n-1}}{}_{j_{n-1}}\Pi^{k_n}{}_{j_n\cdot j_{n+1}})] \\
&\quad \times\Pi^{j_1}{}_{i_1}\ldots\Pi^{j_{n+1}}{}_{i_{n+1}} \\
&= B_{\cdot(j_1\ldots j_{n+1})}\,\Pi^{j_1}{}_{i_1}\ldots\Pi^{j_{n+1}}{}_{i_{n+1}},
\end{aligned} \tag{E.7}$$

where Eqs. (27) and (E.2) have been used to arrive at the penultimate expression and the final expression follows from that by virtue of

$$\begin{aligned}
\Pi^{i}{}_{l\cdot m}(\Pi^{l}{}_{j}\Pi^{m}{}_{k}+\Pi^{l}{}_{k}\Pi^{m}{}_{j}) &= (\Pi^{i}{}_{l\cdot m}+\Pi^{i}{}_{m\cdot l})\Pi^{l}{}_{j}\Pi^{m}{}_{k} \\
&= -Q^{i}{}_{\alpha}(\omega^{\alpha}{}_{l\cdot m}+\omega^{\alpha}{}_{m\cdot l})\Pi^{l}{}_{j}\Pi^{m}{}_{k}=0
\end{aligned} \tag{E.8}$$

(see Eqs. (E.1) and (E.4)).

F. DERIVATION OF EQ. (34)

Expression (32) defines a point in Φ lying on the geodesic through φ_* tangent to the vector $(\partial/\partial s)_{\lambda(s_*)}$. Denote the coordinates of this point in an arbitrary chart by $\varphi^i(s)$. These coordinates vary as one moves along the geodesic (by changing s) but they satisfy the geodesic equation

$$\frac{D}{ds}\frac{d\varphi^i}{ds}=0 \tag{F.1}$$

where D/ds now denotes the covariant derivative based on Vilkovisky's connection. Repeated use of (F.1) allows one to convert the standard Taylor expansion to a covariant Taylor expansion:

$$\begin{aligned} A[\varphi(s)] &= \sum_{n=0}^{\infty} \frac{(s-s_*)^n}{n!} \left\{ \left(\frac{d}{ds}\right)^n A[\varphi(s)] \right\}_{s=s_*} \\ &= \sum_{n=0}^{\infty} \frac{(s-s_*)^n}{n!} \left\{ \left(\frac{D}{ds}\right)^n A[\varphi(s)] \right\}_{s=s_*} \\ &= \sum_{n=0}^{\infty} \frac{(s-s_*)^n}{n!} \left\{ A_{;i_1 \dots i_n}[\varphi(s)] \frac{d\varphi^{i_1}(s)}{ds} \cdots \frac{d\varphi^{i_n}(s)}{ds} \right\}_{s=s_*} \\ &= \sum_{n=0}^{\infty} \frac{(s-s_*)^n}{n!} \left\{ A_{;(i_1 \dots i_n)}[\varphi(s)] \frac{d\varphi^{i_1}(s)}{ds} \cdots \frac{d\varphi^{i_n}(s)}{ds} \right\}_{s=s_*} . \end{aligned} \tag{F.2}$$

Equation (34) follows immediately upon the identification

$$\phi^a = (s-s_*) \left[\frac{d\varphi^a(s)}{ds} \right]_{s=s_*} . \tag{F.3}$$

G. INVARIANCE OF THE LOOP GRAPHS

Begin with the fact that $S[\varphi_*, \bar{\phi}]$ is gauge invariant, the gauge transformation law for the $\bar{\phi}^a$ being understood to be

$$\delta\bar{\phi}^a = Q^a{}_\alpha[\varphi_*, \bar{\phi}]\, \delta\xi^\alpha . \tag{G.1}$$

Since

$$S_{,a}[\varphi_*, \bar{\phi}] = S_{,A}[I_*, \bar{I}]\, \bar{I}^A{}_{,a} = S_{,A}[I_*, \bar{I}]\, P^A{}_a[\varphi_*], \tag{G.2}$$

it follows, from the $\bar{\phi}$-independence of $P^A{}_a$ and the obvious gauge invariance of $S_{,A}[I_*, \bar{I}]$, that $S_{,a}[\varphi_*, \bar{\phi}]$ too is gauge invariant. Indeed, $S_{,a_1 \dots a_n}[\varphi_*, \bar{\phi}]$ is gauge invariant for all n, and

$$S_{,a_1 \dots a_n}[\varphi_*, \bar{\phi}] Q^{a_n}{}_\alpha[\varphi_*, \bar{\phi}] = 0, \tag{G.3}$$

a relation that does not require either φ_* or $\bar{\phi}$ to be "on shell."

Now suppress the arguments φ_* and $\bar{\phi}$, as well as all indices, and write Eq. (54) in the form

$$F = S_2 + P^{\sim} \kappa P, \tag{G.4}$$

where "$\sim$" denotes the transpose. Since $S_2 Q = 0$ (see Eq. (G.3)) it follows that

$$F Q = P^{\sim} \kappa P Q = P^{\sim} \kappa \hat{\mathfrak{F}} . \tag{G.5}$$

Multiply this equation on the left by the Green's function G of F and on the right by the Green's function $\hat{\mathfrak{G}}$ of $\hat{\mathfrak{F}}$. Then change sign, getting

$$Q \hat{\mathfrak{G}} = G P^{\sim} \kappa . \tag{G.6}$$

The one-loop contribution to Σ is $-\frac{i}{2}\ln\det G$. Under infinitesimal changes $\delta\kappa_{\alpha\beta}$, $\delta P^{\alpha}{}_{a}$ in the κ's and P's this contribution suffers the change

$$
\begin{aligned}
-\frac{i}{2}\delta\ln\det G &= -\frac{i}{2}G^{ba}\delta F_{ab} \\
&= -iG^{ba}\,P^{\alpha}{}_{a}\,\kappa_{\alpha\beta}\,\delta P^{\beta}{}_{b} - \frac{i}{2}G^{ba}\,P^{\alpha}{}_{a}\,\delta\kappa_{\alpha\beta}\,P^{\beta}{}_{b} \\
&= -iQ^{b}{}_{\alpha}\,\hat{\mathfrak{G}}^{\alpha}{}_{\beta}\,\delta P^{\beta}{}_{b} - \frac{i}{2}Q^{b}{}_{\gamma}\,\hat{\mathfrak{G}}^{\gamma}{}_{\delta}\,\kappa^{-1\delta\alpha}\,\delta\kappa_{\alpha\beta}\,P^{\beta}{}_{b} \\
&= -i\hat{\mathfrak{G}}^{\alpha}{}_{\beta}\,\delta\hat{\mathfrak{F}}^{\beta}{}_{\alpha} - \frac{i}{2}\hat{\mathfrak{F}}^{\beta}{}_{\gamma}\,\hat{\mathfrak{G}}^{\gamma}{}_{\delta}\,\kappa^{-1\delta\alpha}\,\delta\kappa_{\alpha\beta} \\
&= i\delta(\ln J + \frac{1}{2}\ln\det\kappa)
\end{aligned}
\qquad (G.7)
$$

in which Eqs. (47) and (G.6) have been used to arrive at the final form. If $(\det\kappa)^{1/2}$ and J are restored to the integrand of expression (53) they contribute an amount $-i(\ln J + \frac{1}{2}\ln\det\kappa)$ to Σ, the change of which precisely cancels (G.7).

Under the infinitesimal gauge transformation (G.1) the one-loop contribution to Σ suffers the change

$$
-\frac{i}{2}\delta\ln\det G = -\frac{i}{2}G^{bc}F_{cb,a}Q^{a}{}_{\alpha}\delta\xi^{\alpha} = -\frac{i}{2}G^{bc}S_{,cba}Q^{a}{}_{\alpha}\delta\xi^{\alpha}, \qquad (G.8)
$$

which vanishes by virtue of Eq. (G.3).

The two-loop contribution to Σ consists of two terms,

$$
-\frac{1}{8}G^{ab}S_{,abcd}G^{cd} \qquad \text{and} \qquad -\frac{1}{12}S_{,abc}G^{ad}G^{be}G^{cf}S_{,def}
$$

respectively. The gauge invariance of these terms is trivial because of Eq. (G.3). Under a change in the P's the first suffers the change

$$
\begin{aligned}
-\frac{1}{8}\delta(G^{ab}S_{,abcd}G^{cd}) &= -\frac{1}{2}G^{ab}S_{,abcd}\,G^{ce}\,P^{\alpha}{}_{e}\,\kappa_{\alpha\beta}\,\delta P^{\beta}{}_{f}\,G^{fd} \\
&= -\frac{1}{2}G^{ab}\,S_{,abcd}\,Q^{c}{}_{\alpha}\,\hat{\mathfrak{G}}^{\alpha}{}_{\beta}\,\delta P^{\beta}{}_{f}\,G^{fd},
\end{aligned}
\qquad (G.9)
$$

which again vanishes by (G.3). The change in the second term vanishes similarly. The same kind of analysis shows, moreover, that both terms are invariant under changes in $\kappa_{\alpha\beta}$. One easily sees that these arguments extend to all loop graphs. The vertex functions are already P-, κ- and gauge invariant. The lines (i.e., propagators G^{ab}) are gauge invariant and only suffer changes under changes in the P's and κ's. But these changes are always of a form that allows Eq. (G.6) to be invoked, bringing Q's to bear on vertices so that the change in the graph vanishes by virtue of Eq. (G.3).

H. VERTEX FUNCTIONS AND FOURIER TRANSFORMS

The standard vertex functions are well known. Starting from Eq. (72) and carrying out some slightly tedious computations one finds

$$
\left(\frac{\delta^2 S}{\delta A_R{}^{\alpha}{}_{\mu}\delta A_R{}^{\beta'}{}_{\nu'}}\right)_{Z=1} \longrightarrow -\gamma_{\alpha\beta}(\eta^{\mu\nu}p^2 - p^{\mu}p^{\nu}), \qquad (H.1)
$$

$$\left(\frac{\delta^3 S}{\delta A_R{}^\alpha{}_\mu \delta A_R{}^{\beta'}{}_{\nu'} \delta A_R{}^{\gamma''}{}_{\sigma''}}\right)_{Y=1}$$
$$\longrightarrow i\mu^{2-N/2} g f_{\alpha\beta\gamma}[\eta^{\mu\nu}(p^\sigma - p'^\sigma) + \eta^{\nu\sigma}(p'^\mu - p''^\mu) + \eta^{\sigma\mu}(p''^\nu - p^\nu)], \tag{H.2}$$
$$\left(\frac{\delta^4 S}{\delta A_R{}^\alpha{}_\mu \delta A_R{}^{\beta'}{}_{\nu'} \delta A_R{}^{\gamma''}{}_{\sigma''} \delta A_R{}^{\delta'''}{}_{\tau'''}}\right)_{X=1}$$
$$\longrightarrow -\mu^{4-N} g^2 [f_{\alpha\beta\varepsilon} f^\varepsilon{}_{\gamma\delta}(\eta^{\mu\sigma}\eta^{\nu\tau} - \eta^{\mu\tau}\eta^{\nu\sigma})$$
$$+ f_{\alpha\gamma\varepsilon} f^\varepsilon{}_{\delta\beta}(\eta^{\mu\tau}\eta^{\sigma\nu} - \eta^{\mu\nu}\eta^{\sigma\tau})$$
$$+ f_{\alpha\delta\varepsilon} f^\varepsilon{}_{\beta\gamma}(\eta^{\mu\nu}\eta^{\sigma\tau} - \eta^{\mu\sigma}\eta^{\nu\tau})]. \tag{H.3}$$

Here all momenta are *incoming*; i.e., one affixes factors $e^{ip\cdot x}$, $e^{ip'\cdot x'}$, $e^{ip''\cdot x''}$, $e^{ip'''\cdot x'''}$ and integrates over x, x', etc. The momenta in Eq. (H.2) are subject to the conservation law $p + p' + p'' = 0$.

From Eq. (74) one has

$$Q^\alpha{}_{\mu\beta'} \longrightarrow i\delta^\alpha{}_\beta p_\mu, \tag{H.4}$$

and the function $Q_{i\alpha}$ of Eq. (6) takes, for Yang–Mills fields, the form

$$Q_\alpha{}^\mu{}_{\beta'} = \int \gamma_\alpha{}^\mu{}_{\gamma''}{}^{\nu''} Q^{\gamma''}{}_{\nu''\beta'} \, d^N x''$$
$$= \left(-\gamma_{\alpha\beta}\frac{\partial}{\partial x_\mu} + \mu^{2-N/2} Z^{-1} Y g f_{\alpha\beta\gamma} A_R{}^{\gamma\mu}\right)\delta(x,x') \longrightarrow i\gamma_{\alpha\beta} p^\mu. \tag{H.5}$$

The operator $\mathfrak{F}_{\alpha\beta}$ of Eq. (9) becomes

$$\mathfrak{F}_{\alpha\beta'} = -\int Q_{\gamma''}{}^{\mu''}{}_\alpha Q^{\gamma''}{}_{\mu''\beta'} \, d^N x''$$
$$= \left(\gamma_{\alpha\beta}\frac{\partial^2}{\partial x_\mu \partial x^\mu} - \mu^{2-N/2} Z^{-1} Y g f_{\alpha\beta\gamma}\left\{A_R{}^{\gamma\mu}, \frac{\partial}{\partial x^\mu}\right\}\right.$$
$$\left. + \mu^{4-N} Z^{-3} Y^2 g^2 f_{\alpha\gamma\varepsilon} f^\varepsilon{}_{\delta\beta} A_R{}^\gamma{}_\mu A_R{}^{\delta\mu}\right)\delta(x,x')$$
$$\longrightarrow -\gamma_{\alpha\beta} p^2, \tag{H.6}$$

which implies, for its Green's function,

$$\mathfrak{G}^{\alpha\beta'} \longrightarrow \gamma^{\alpha\beta}/p^2. \tag{H.7}$$

From this the connection 1-form on Φ follows:

$$\omega^\alpha{}_{\beta'}{}^{\mu'} = \int \mathfrak{G}^{\alpha\gamma''} Q_{\beta'}{}^{\mu'}{}_{\gamma''} \, d^N x'' \longrightarrow -i\delta^\alpha{}_\beta \, p^\mu / p^2, \tag{H.8}$$

and this in turn yields the horizontal projection operator

$$\Pi^\alpha{}_{\mu\beta'}{}^{\nu'} = \delta^\alpha{}_{\beta'}\delta_\mu{}^{\nu'} - \int Q^\alpha{}_{\mu\gamma''}\,\omega^{\gamma''}{}_{\beta'}{}^{\nu'} \, d^N x''$$
$$\longrightarrow \delta^\alpha{}_\beta(\delta_\mu{}^\nu - p_\mu p^\nu / p^2). \tag{H.9}$$

Finally, from Eq. (64) one has

$$P^\alpha{}_{\beta'}{}^{\mu'} \longrightarrow i\delta^\alpha{}_\beta \, p^\mu \tag{H.10}$$

which, together with Eqs. (65) and (H.1), yield for the operator (54),

$$F_{\alpha}{}^{\mu}{}_{\beta'}{}^{\nu'} = \frac{\delta^2 S}{\delta A^{\alpha}{}_{\mu}\delta A^{\beta'}{}_{\nu'}} + \int d^N x'' \int d^N x''' \, \kappa_{\gamma''\delta'''} P^{\gamma''}{}_{\alpha}{}^{\mu} P^{\delta'''}{}_{\beta'}{}^{\nu'}$$
$$\longrightarrow -\gamma_{\alpha\beta}\, \eta^{\mu\nu} p^2. \tag{H.11}$$

I. GHOST VERTEX FUNCTIONS

Since the ghost operator $\hat{\mathfrak{F}}^{\alpha}{}_{\beta}$ is equal to $P^{\alpha}{}_{a}Q^{a}{}_{\beta}$, and since the $P^{\alpha}{}_{a}$ are constants, to get the ghost vertex functions one must compute derivatives of $Q^{a}{}_{\beta}$ with respect to the ϕ's. Because the **Q**'s are vector fields one has

$$Q^{a}{}_{\alpha}[\varphi_*, \phi] = \phi^{a}{}_{,i}[\varphi_*, \varphi] Q^{i}{}_{\alpha}[\varphi]. \tag{I.1}$$

The $\phi^{a}[\varphi_*, \varphi]$ may be viewed as scalar functions on Φ. Therefore Eq. (I.1), with arguments suppressed, may be rewritten in the form

$$Q^{a}{}_{\alpha} = \phi^{a}{}_{;i}\, Q^{i}{}_{\alpha}. \tag{I.2}$$

What is desired is an expansion in powers of the ϕ^{a}:

$$Q^{a}{}_{\alpha}[\varphi_*, \phi] = [Q^{a}{}_{\alpha}] + [Q^{a}{}_{\alpha,b}]\phi^{b} + \frac{1}{2}[Q^{a}{}_{\alpha,bc}]\phi^{b}\phi^{c} + \ldots, \tag{I.3}$$

where "[ ]" on the right means "set $\phi = 0$." Equate the right hand sides of (I.2) and (I.3) and covariantly differentiate with respect to the φ's:

$$[Q^{a}{}_{\alpha}] + [Q^{a}{}_{\alpha,b}]\phi^{b} + \frac{1}{2}[Q^{a}{}_{\alpha,bc}]\phi^{b}\phi^{c} + \ldots = \phi^{a}{}_{;i}Q^{i}{}_{\alpha}, \tag{I.4}$$

$$[Q^{a}{}_{\alpha,b}]\phi^{b}{}_{;j} + [Q^{a}{}_{\alpha,bc}]\phi^{b}\phi^{c}{}_{;j} + \ldots = \phi^{a}{}_{;ij}\, Q^{i}{}_{\alpha} + \phi^{a}{}_{;i}\, Q^{i}{}_{\alpha;j}, \tag{I.5}$$

$$[Q^{a}{}_{\alpha,b}]\phi^{b}{}_{;jk} + [Q^{a}{}_{\alpha,bc}]\phi^{b}{}_{;j}\phi^{c}{}_{;k} + [Q^{a}{}_{\alpha,bc}]\phi^{b}\phi^{c}{}_{;jk} + \ldots = \phi^{a}{}_{;ijk}\, Q^{i}{}_{\alpha} + \phi^{a}{}_{;ij}\, Q^{i}{}_{\alpha;k} + \phi^{a}{}_{;ik}\, Q^{i}{}_{\alpha;j} + \phi^{a}{}_{;i}\, Q^{i}{}_{\alpha;jk}. \tag{I.6}$$

Now the ϕ's closely approximate the $\varphi - \varphi_*$'s in the vicinity of $\phi = 0$. Therefore

$$[\phi^{a}{}_{;i}] = \delta^{a}{}_{i}, \tag{I.7}$$

where "[ ]" means "take the *coincidence limit* $\varphi \to \varphi_*$ or $(\phi \to 0)$." With the aid of special techniques (which there is no space to describe here) one can also show

$$[\phi^{a}{}_{;ij}] = 0, \tag{I.8}$$

$$[\phi^{a}{}_{;ijk}] = \frac{1}{3}(R^{a}{}_{ijk} + R^{a}{}_{jik}), \tag{I.9}$$

where the $R^{a}{}_{ijk}$ are the components of the curvature tensor for Vilkovisky's connection, evaluated at $\varphi = \varphi_*$. Taking the coincidence limit of Eqs. (I.4) to (I.6) and remembering that indices on tensor components evaluated at the base point φ_* may be taken from either the first or the middle of the alphabet, one gets

$$[Q^a{}_\alpha] = Q^a{}_\alpha[\varphi_*], \tag{I.10}$$

$$[Q^a{}_{\alpha,b}] = Q^a{}_{\alpha;b}[\varphi_*] = \frac{1}{2}Q^a{}_\gamma[\varphi_*]c^\gamma{}_{\alpha\beta}\,\omega^\beta{}_b[\varphi_*], \tag{I.11}$$

$$\begin{aligned}[Q^a{}_{\alpha,bc}] &= \frac{1}{3}(R^a{}_{dbc}[\varphi_*] + R^a{}_{bdc}[\varphi_*])Q^d{}_\alpha[\varphi_*] + Q^d{}_{\alpha;bc}[\varphi_*] \\ &= \frac{1}{6}(R^a{}_{bdc}[\varphi_*] + R^a{}_{cdb}[\varphi_*])Q^d{}_\alpha[\varphi_*] \\ &\quad + \frac{1}{2}(Q^a{}_{\alpha;bc}[\varphi_*] + Q^a{}_{\alpha;cb}[\varphi_*]).\end{aligned} \tag{I.12}$$

By a lengthy computation, which makes use of every conceivable identity satisfied by the Q's, expression (I.12) can be reduced to

$$\begin{aligned}[Q^a{}_{\alpha,bc}] &= \frac{1}{12}Q^a{}_\delta\, c^\delta{}_{\gamma\varepsilon}\, c^\varepsilon{}_{\beta\alpha}\,(\omega^\beta{}_b\,\omega^\gamma{}_c + \omega^\beta{}_c\,\omega^\gamma{}_b) \\ &\quad - \frac{1}{12}Q^a{}_\beta\, c^\beta{}_{\alpha\gamma}\,(\omega^\gamma{}_{b\cdot c} + \omega^\gamma{}_{b\cdot c}) \\ &\quad - \frac{1}{6}Q^a{}_\beta\, c^\beta{}_{\alpha\gamma}\,\omega^\gamma{}_d\,(Q^d{}_{\delta\cdot b}\,\omega^\delta{}_c + Q^d{}_{\delta\cdot c}\,\omega^\delta{}_b) \\ &\quad + \frac{1}{12}Q^a{}_\beta\, c^\beta{}_{\alpha\gamma}\,\omega^\gamma{}_d\, Q^d{}_{\delta\cdot e}\, Q^e{}_\varepsilon\,(\omega^\delta{}_b\,\omega^\varepsilon{}_c + \omega^\delta{}_c\,\omega^\varepsilon{}_b)\end{aligned} \tag{I.13}$$

all quantities on the right being evaluated at φ_*. Note that Eqs. (I.11) and (I.13) confirm the general structure of the derivatives of the Q's described in the text, namely, being proportional to the Q's themselves, with coefficients proportional to the δ-function or its derivative coming from the structure constants.

The ghost vertex functions are obtained from expressions (I.11) and (I.13) through multiplication by the P's. In Appendix J use will be made only of the ghost vertex bearing a single non-ghost prong:

$$P^\alpha{}_b[\varphi_*][Q^b{}_{\beta,a}] = \frac{1}{2}\hat{\mathfrak{F}}^\alpha{}_\gamma[\varphi_*]c^\gamma{}_{\beta\delta}\,\omega^\delta{}_a[\varphi_*]. \tag{I.14}$$

For the Yang–Mills field one gets, from Eqs. (H.4) and (H.10),

$$\hat{\mathfrak{F}}^\alpha{}_{\beta'} = \int P^\alpha{}_{\gamma''}{}^{\mu''}\, Q^{\mu''}{}_{\gamma''\beta'}\, d^Nx'' \longrightarrow -\delta^\alpha{}_\beta p^2. \tag{I.15}$$

By virtue of Eqs. (62), (71) and (H.8), expression (I.14) therefore becomes

$$\begin{aligned}&\frac{1}{2}\int d^Nx''' \int d^Nx''''\, \hat{\mathfrak{F}}^\alpha{}_{\delta'''}\, c^{\delta'''}{}_{\beta'\varepsilon''''}\,\omega^{\varepsilon''''}{}_{\gamma''}{}^{\mu''} \\ &\longrightarrow \frac{1}{2}(-\delta^\alpha{}_\delta p^2)(\mu^{2-N/2}Z^{-3/2}Ygf^\delta{}_{\beta\varepsilon})(i\delta^\varepsilon{}_\gamma\, p''^\mu/p''^2) \\ &= -\frac{1}{2}i\mu^{2-N/2}Z^{-3/2}Ygf^\alpha{}_{\beta\gamma}(p^2/p''^2)\, p''^\mu.\end{aligned} \tag{I.16}$$

The basic vertex function that one would use in computing counter terms is obtained from this by setting $Y = 1$ and $Z = 1$:

$$-\frac{1}{2}i\mu^{2-N/2}gf^\alpha{}_{\beta\gamma}(p^2/p''^2)\, p''^\mu. \tag{I.17}$$

This is to be contrasted with the corresponding vertex function in the traditional formalism:

$$-\mu^{2-N/2} g f^{\alpha}{}_{\beta\gamma}\, p^{\mu} \qquad \text{(traditional)}. \tag{I.18}$$

J. GHOST GRAPHS

Consider first the ghost loop with two non-ghost prongs attached. If the prongs emanate from the same point then the value of the graph is trivially zero under dimensional regularization. If the prongs emanate from different points then its value is

$$\frac{i}{(2\pi)^N}\left(\frac{-i}{2}\right)^2 \mu^{4-N} g^2 f^{\gamma}{}_{\delta\alpha} f^{\varepsilon}{}_{\zeta\beta} \int \frac{k^2}{p^2} p^{\mu} \frac{(k+p)^2}{p^2} p^{\nu} \frac{\delta^{\zeta}{}_{\gamma}}{k^2} \frac{\delta^{\delta}{}_{\varepsilon}}{(k+p)^2} d^N k \tag{J.1}$$

Note the cancellation of the internal propagator factor $1/(k+p)^2$ brought about by the special form of the vertex function (I.17). Expression (J.1) immediately reduces to

$$\frac{-i\lambda}{4(2\pi)^N} \mu^{4-N} g^2 \gamma_{\alpha\beta} \frac{p^{\mu} p^{\nu}}{p^4} \int d^N k, \tag{J.2}$$

which vanishes under dimensional regularization, leaving not even a finite part. Thus the ghost makes no contribution to the renormalization constant Z in one-loop order. By evaluating expression (I.13) for the Yang–Mills field, one can show that ghost loops with three non-ghost prongs attached also have vanishing values. Hence the ghost makes no one-loop contributions to the renormalization constant Y. By virtue of gauge invariance one infers that it also makes no one-loop contribution to the renormalization constant X.

In the traditional formalism one needs to calculate counter terms contributing to a renormalization constant for the ghost field itself. In the present formalism the one-loop graph that would give rise to such a counter term has the value

$$\begin{aligned} -\frac{i}{(2\pi)^N} &\left(\frac{-i}{2}\right)^2 \mu^{4-N} g^2 f^{\alpha}{}_{\gamma\delta} f^{\varepsilon}{}_{\beta\zeta} \int \frac{p^2}{k^2} k^{\mu} \frac{(k+p)^2}{k^2} (-k^{\nu}) \\ &\times \frac{\delta^{\delta}{}_{\varepsilon}}{(k+p)^2} \frac{\gamma^{\gamma\zeta} \eta_{\mu\nu}}{k^2} d^N k \\ &= \frac{-i\lambda}{4(2\pi)^N} \mu^{4-N} g^2 \delta^{\alpha}{}_{\beta}\, p^2 \int \frac{d^N k}{k^4}, \end{aligned} \tag{J.3}$$

and this too vanishes under dimensional regularization. Note again the cancellation of the internal propagator factor $1/(k+p)^2$.

There are three one-loop graphs that could potentially require counter terms contributing to a renormalization of the vertex function (I.17) itself. One of these, in which the non-ghost prong is attached to a non-ghost internal line, vanishes trivially by virtue of the presence of horizontal projection operators and the fact that the momentum p''^{μ} in the vertex function (I.17) is the momentum associated with the non-ghost prong. Another, in which the non-ghost prong is attached to an internal ghost line, has the value

$$-\frac{i}{(2\pi)^N}\left(\frac{-i}{2}\right)^3 \mu^{6-3N/2} g^3 f^{\alpha}{}_{\delta\kappa} f^{\varepsilon}{}_{\zeta\gamma} f^{\theta}{}_{\beta\lambda}$$

$$\times \int \frac{p^2}{k^2} k^\nu \frac{(k+p)^2}{p''^2} p''^\mu \frac{(k-p')^2}{k^2} (-k^\sigma) \frac{\delta^\delta{}_\varepsilon}{(k+p)^2} \frac{\delta^\zeta{}_\theta}{(k-p')^2} \frac{\gamma^{\kappa\lambda}\eta_{\nu\sigma}}{k^2} d^N k$$
$$= \frac{\lambda}{16(2\pi)^N} \mu^{6-3N/2} g^3 f^\alpha{}_{\beta\gamma} \frac{p^2}{p''^2} p''^\mu \int \frac{d^N k}{k^4}, \tag{J.4}$$

which again vanishes under dimensional regularization. The third which requires use of expression (I.13), can also be shown to vanish.

REFERENCES

[1] R. P. Feynman, *Acta Physica Polonica* **24**, 697 (1963). The work presented in this paper was carried out in 1962.

[2] Throwing away the noncausal loops can be shown to be equivalent to throwing away nonvanishing contributions from arcs at infinity in the Wick rotation procedure. See B. DeWitt, *Supermanifolds (Second Edition)* (Cambridge University Press, 1992) chapter 6 (especially exercise **6.11**, pp. 393–395) where rejection of the noncausal loops is also shown to be intimately connected to the choice of measure for the functional integral that defines the loop expansion.

[3] The question of Lorentz invariance arises because Feynman was assuming spacetime to be asymptotically Minkowskian.

[4] B. DeWitt, *Dynamical Theory of Groups and Fields* (Gordon and Breach, 1965).

[5] The linear independence of the $\mathbf{Q}_\alpha$ and the invertibility and ultralocality of γ_{ij} guarantee that $\mathfrak{F}_{\alpha\beta}$ will be an invertible operator, typically differential or pseudodifferential.

[6] G. A. Vilkovisky in *Quantum Theory of Gravity*, ed. S. M. Christensen (Adam Hilger, 1984).

[7] This is not true in gravity theory, for which γ_{ij} is *not* flat.

[8] This would not be true if the invariant metric $\gamma_\alpha{}^\mu{}_{\beta'}{}^{\nu'}$ had not been chosen to be ultralocal. In fact the renormalization program almost certainly could not be carried out with any choice but (63).

[9] J. C. Collins, *Renormalization* (Cambridge University Press, 1984).

[10] In the case of the adjoint representation, generated by the f_α of Eq. (59) these equations reduce to Eqs. (58) together with $\gamma^{\alpha\beta} f_\alpha f_\beta = -\lambda \mathbb{1}_r$.

[11] If more than one Majorana field is present (a complex spinor field counts as two) each will contribute its own $-\frac{2}{3}\left(\frac{n}{r}\right)\Lambda$ term to expression (132). If too many are present the theory will cease to be asympotically free.

12

LOCALIZATION AND DIAGONALIZATION: A REVIEW OF FUNCTIONAL INTEGRAL TECHNIQUES FOR LOW-DIMENSIONAL GAUGE THEORIES AND TOPOLOGICAL FIELD THEORIES

M. Blau[1*] and G. Thompson[2†]

[1]LPTHE-ENSLAPP[‡]
ENS-Lyon, 46 Allée d'Italie
F-69364 Lyon CEDEX 07, France
[2]ICTP
P.O. Box 586
34014 Trieste, Italy

ABSTRACT

We review localization techniques for functional integrals which have recently been used to perform calculations in and gain insight into the structure of certain topological field theories and low-dimensional gauge theories. These are the functional integral counterparts of the Mathai–Quillen formalism, the Duistermaat–Heckman theorem, and the Weyl integral formula respectively. In each case, we first introduce the necessary mathematical background (Euler classes of vector bundles, equivariant cohomology, topology of Lie groups), and describe the finite dimensional integration formulae. We then discuss some applications to path integrals and give an overview of the relevant literature. The applications we deal with include supersymmetric quantum mechanics, cohomological field theories, phase space path integrals, and two-dimensional Yang–Mills theory.

(Extended version of lectures given by M.B. This article originally appeared in the *Special Issue on Functional Integration* of the *Journal of Mathematical Physics*, Jour. Math. Physics 36 No. 5 (1995) 2192-2236. We thank the American Institute of Physics (AIP) for

*e-mail: mblau@enslapp.ens-lyon.fr; supported by EC Human Capital and Mobility Grant ERB-CHB-GCT-93-0252
†e-mail: thompson@ictp.trieste.it
‡URA 14-36 du CNRS, associée à l'E.N.S. de Lyon, et à l'Université de Savoie

Functional Integration: Basics and Applications
Edited by Cécile DeWitt-Morette, Plenum Press, New York, 1997

permission to reproduce this article here. It has been slightly updated for inclusion in these Proceedings.)

CONTENTS

1. INTRODUCTION

In recent years, the functional integral has become a very popular tool in a branch of physics lying on the interface between string theory, conformal field theory and topological field theory on the one hand and topology and algebraic geometry on the other. Not only has it become popular but it has also, because of the consistent reliability of the results the functional integral can produce when handled with due care, acquired a certain degree of respectability among mathematicians.

Here, however, our focus will not be primarily on the results or predictions obtained by these methods, because an appreciation of these results would require a rather detailed understanding of the mathematics and physics involved. Rather, we want to explain some of the general features and properties the functional integrals appearing in this context have in common. Foremost among these is the fact that, due to a large number of (super-)symmetries, these functional integrals essentially represent finite-dimensional integrals.

The transition between the functional and finite dimensional integrals can then naturally be regarded as a (rather drastic) localization of the original infinite dimensional integral. The purpose of this article is to give an introduction to and an overview of some functional integral tricks and techniques which have turned out to be useful in understanding these properties and which thus provide insight into the structure of topological field theories and some of their close relatives in general.

More specifically, we focus on three techniques which are extensions to functional integrals of finite dimensional integration and localization formulae which are quite interesting in their own right, namely

1. the Mathai–Quillen formalism [1], dealing with integral representations of Euler classes of vector bundles;
2. the Duistermaat–Heckman theorem [2] on the exactness of the stationary phase approximation for certain phase space path integrals, and its generalizations [3, 4];
3. the classical Weyl integral formula, relating integrals over a compact Lie group or Lie algebra to integrals over a maximal torus or a Cartan subalgebra.

We will deal with these three techniques in sections 2–4 of this article respectively. In each case, we will first try to provide the necessary mathematical background (Euler classes of vector bundles, equivariant cohomology, topology of Lie groups). We then explain the integration formulae in their finite dimensional setting before we move on boldly to apply these techniques to some specific infinite dimensional examples like supersymmetric quantum mechanics or two-dimensional Yang–Mills theory.

Of course, the functional integrals we deal with in this article are very special, corresponding to theories with no field theoretic degrees of freedom. Moreover, our treatment of functional integrals is completely formal as regards its functional analytic aspects. So lest we lose the reader interested primarily in honest quantum mechanics or field theory functional integrals at this point, we should perhaps explain why we believe that it is, nevertheless, worthwhile to look at these examples and techniques.

First of all, precisely because the integrals we deal with are essentially finite dimensional integrals, and the path integral manipulations are frequently known to produce the correct (meaning e.g., topologically correct) results, any definition of, or approach to, the functional integral worth its salt should be able to reproduce these results and to incorporate these techniques in some way. This applies in particular to the infinite dimensional analogue of the Mathai–Quillen formalism and to the Weyl integral formula.

Secondly, the kinds of theories we deal with here allow one to study kinematical (i.e., geometrical and topological) aspects of the path integral in isolation from their dynamical aspects. In this sense, these theories are complementary to, say, simple interacting theories. While the latter are typically kinematically linear but dynamically non-linear, the former are usually dynamically linear (free field theories) but kinematically highly non-linear (and the entire non-triviality of the theories resides in this kinematic non-linearity). This is a feature shared by all the three techniques we discuss.

Thirdly, in principle the techniques we review here are also applicable to theories with field theoretic degrees of freedom, at least in the sense that they provide alternative approximation techniques to the usual perturbative expansion. Here we have in mind primarily the Weyl integral formula and the generalized WKB approximation techniques based on the Duistermaat–Heckman formula.

Finally, we want to draw attention to the fact that many topological field theories can be interpreted as 'twisted' versions of ordinary supersymmetric field theories, and that this relation has already led to a dramatic improvement of our understanding of both types of theories, see e.g., [5]–[9] for some recent developments.

Hoping to have persuaded the reader to stay with us, we will now sketch briefly the organization of this paper. As mentioned above, we will deal with the three different techniques in three separate sections, each one of them beginning with an introduction to the mathematical background. The article is written in such a way that these three sections can be read independently of each other. We have also tried to set things up in such a way that the reader primarily interested in the path integral applications should be able to skip the mathematical introduction upon first reading and to move directly to the relevant section, going back to the more formal considerations with a solid example at hand. Furthermore, ample references are given so that the interested reader should be able to track down most of the applications of the techniques that we describe.

Originally, we had intended to include a fifth section explaining the interrelations among the three seemingly rather different techniques we discuss in this article. These relations exist. A nice example to keep in mind is two-dimensional Yang–Mills theory which can be solved either via a version of the Duistermaat–Heckman formula or using the Weyl integral formula, but which is also equivalent to a certain two-dimensional cohomological field theory which provides a field theoretic realization of the Mathai–Quillen formalism. Unfortunately, we haven't been able to come up with some satisfactory general criteria for when this can occur and we have therefore just added some isolated remarks on this in appropriate places in the body of the text.

NOTE ADDED FOR THE PROCEEDINGS

This is a (significantly) extended version of lectures given by M.B. at the Cargèse Summer School. During the lectures I focussed on two-dimensional Yang–Mills theory in order to complement the more quantum mechanical aspects and applications of path integrals that prevailed in most of the other lectures.

This article originally appeared in the *Special Issue on Functional Integration* of the *Journal of Mathematical Physics*, Jour. Math. Physics 36 No. 5 (1995) 2192–2236. I am grateful to the American Institute of Physics (AIP) for permission to reproduce this article here. It has been slightly updated for inclusion in these Proceedings.

The list of references is already quite long and detailed and I have not attempted to add to it articles that have appeared in the meantime. However, I want to draw attention to one article:

Richard Szabo has written an excellent and extensive review of equivariant localization techniques, partially overlapping with but also significantly extending the material covered in section 3 of these lectures. It is available from the preprint archives as `hep-th/9608068`.

I want to thank the organizers of the School, Pierre Cartier, Cecile DeWitt-Morette, Antoine Folacci, and Chris Isham, for inviting me to lecture in Cargèse and for all their efforts that made this meeting a truly memorable one.

2. LOCALIZATION VIA THE MATHAI–QUILLEN FORMALISM

The sort of localization we are interested in in this section is familiar from classical differential geometry and topology. Namely, let us recall that classically there exist two quite different prescriptions for calculating the Euler number $\chi(X)$ (number of vertices minus number of edges plus ...) of some manifold X. The first is topological in nature and instructs one to choose a vector field V on X with isolated zeros and to count these zeros with signs (this is the Hopf theorem). The second is differential geometric and represents $\chi(X)$ as the integral over X of a density (top form) e_∇ constructed from the curvature of some connection ∇ on X (the Gauss–Bonnet theorem).

The integral over X of e_∇ can hence be localized to (in the sense of: evaluated in terms of contribution from) the zero locus of V. This localization can be made quite explicit by using a more general formula for $\chi(X)$, obtained by Mathai and Quillen [1], which interpolates between these two classical prescriptions. It relies on the construction of a form $e_{V,\nabla}(X)$ which depends on both a vector field V and a connection ∇. Its integral over X represents $\chi(X)$ for all V and ∇ and can be shown to reduce to an integral over the zero locus X_V of V in the form

$$\chi(X)=\int_X e_{V,\nabla}(X)=\int_{X_V} e_{\nabla'}(X_V)=\chi(X_V). \tag{2.1}$$

Here ∇' is the induced connection on X_V.

What makes this construction potentially interesting for functional integrals is the following observation of Atiyah and Jeffrey [10]. What they pointed out was that, although e_∇ and $\int_X e_\nabla$ do not make sense for infinite dimensional E and X, the Mathai–Quillen form $e_{V,\nabla}$ can be used to formally define *regularized Euler numbers* $\chi_V(X)$ of such bundles by

$$\chi_V(X):=\int_X e_{V,\nabla}(X) \tag{2.2}$$

for certain choices of V (e.g., such that their zero locus is finite dimensional so that the integral on the right hand side of (2.1) makes sense). Although not independent of V, these numbers $\chi_V(X)$ are naturally associated with X for natural choices of V and are therefore likely to be of topological interest. Furthermore, this construction (and its generalization to Euler classes $e(E)$ of arbitrary real vector bundles over X) provides one with an interesting class of functional integrals which are expected to localize to, and hence be equivalent to, ordinary finite dimensional integrals, precisely the phenomenon we want to study in these notes.

We should add that it is precisely such a representation of characteristic classes or numbers by functional integrals which is the characteristic property of topological field theories. This suggests, that certain topological field theories can be interpreted or obtained in this way which is indeed the case. We will come back to this in the examples at the end of this section.

A familiar theory in which this scenario is realized, and which preceded and inspired both the Mathai–Quillen formalism and topological field theory is supersymmetric quantum mechanics, an example we will discuss in some detail in section 2.2.

2.1. The Mathai–Quillen Formalism

The Euler class of a finite dimensional vector bundle. Consider a real vector bundle $\pi : E \to X$ over a manifold X. We will assume that E and X are orientable, X is compact without boundary, and that the rank (fibre dimension) of E is even and satisfies $rk(E) = 2m \leq \dim(X) = n$.

The *Euler class* of E is an integral cohomology class $e(E) \in H^{2m}(X,\mathbb{R}) \equiv H^{2m}(X)$ and there are two well-known and useful ways of thinking about $e(E)$.

The first of these is in terms of sections of E. In general, a twisted vector bundle will have no nowhere-vanishing non-singular sections and one defines the Euler class to be the homology class of the zero locus of a generic section of E. Its Poincaré dual is then a cohomology class $e(E) \in H^{2m}(X)$.

The second makes use of the Chern–Weil theory of curvatures and characteristic classes and produces an explicit representative $e_\nabla(E)$ of $e(E)$ in terms of the curvature Ω_∇ of a connection ∇ on E. Thinking of Ω_∇ as a matrix of two-forms one has

$$e_\nabla(E) = (2\pi)^{-m}\mathrm{Pf}(\Omega_\nabla) \tag{2.3}$$

where $\mathrm{Pf}(A)$ denotes the Pfaffian of the real antisymmetric matrix A. Standard arguments show that the cohomology class of e_∇ is independent of the choice of ∇. For later use we note here that Pfaffians can be written as fermionic (Berezin) integrals. More precisely, if we introduce real Grassmann odd variables χ^a, then

$$e_\nabla(E) = (2\pi)^{-m} \int d\chi\, e^{\frac{1}{2}\chi_a\Omega_\nabla^{ab}\chi_b}. \tag{2.4}$$

If the rank of E is equal to the dimension of X (e.g., if $E = TX$, the tangent bundle of X) then $H^{2m}(X) = H^n(X) = \mathbb{R}$ and nothing is lost by considering, instead of $e(E)$, its evaluation on (the fundamental class $[X]$ of) X, the *Euler number* $\chi(E) = e(E)[X]$. In terms of the two descriptions of $e(E)$ given above, this number can be obtained either as the number of zeros of a generic section s of E (which are now isolated) counted with signs,

$$\chi(E) = \sum_{x_k : s(x_k)=0} (\pm 1), \tag{2.5}$$

or as the integral

$$\chi(E) = \int_X e_\nabla(E). \tag{2.6}$$

If $2m < n$, then one cannot evaluate $e(E)$ on $[X]$ as above. One can, however, evaluate it on homology $2m$-cycles or (equivalently) take the product of $e(E)$ with elements of $H^{n-2m}(X)$ and evaluate this on X. In this way one obtains *intersection numbers* of X associated with the vector bundle E or, looked at differently, intersection numbers of the zero locus of a generic section. The Mathai–Quillen form to be introduced below also takes care of the situation when one chooses a non-generic section s. In this case, the Euler number or class of E can be expressed as the Euler number or class of a vector bundle over the zero locus of s (see [11] for an argument to that effect).

For $E = TX$, the Euler number can be defined as the alternating sum of the Betti numbers of X,

$$\chi(X) = \sum_{k=0}^{n}(-1)^k b_k(X), \qquad b_k(X) = \dim H^k(X,\mathbb{R}). \tag{2.7}$$

In this context, equations (2.5) and (2.6) are the content of the *Poincaré–Hopf theorem* and the *Gauss–Bonnet theorem* respectively. For $E = TX$ there is also an interesting generalization of (2.5) involving a possibly non-generic vector field V, i.e., with a zero locus X_V which is not necessarily zero-dimensional, namely

$$\chi(X) = \chi(X_V). \tag{2.8}$$

This reduces to (2.5) when X_V consists of isolated points and is an identity when V is the zero vector field.

The Mathai–Quillen representative of the Euler class. One of the beauties of the Mathai–Quillen formalism is that it provides a corresponding generalization of (2.6), i.e., an explicit differential form representative $e_{s,\nabla}$ of $e(E)$ depending on both a section s of E and a connection ∇ on E such that

$$e(E) = [e_{s,\nabla}(E)], \tag{2.9}$$

or (for $m = 2n$)

$$\chi(E) = \int_X e_{s,\nabla}(E) \tag{2.10}$$

for any s and ∇. The construction of $e_{s,\nabla}$ involves what is known as the *Thom class* $\Phi(E)$ of E [12] or — more precisely — the explicit differential form representative $\Phi_\nabla(E)$ of the Thom class found by Mathai and Quillen [1]. The construction of $\Phi_\nabla(E)$ is best understood within the framework of equivariant cohomology. What this means is that one does not work directly on E, but that one realizes E as a vector bundle associated to some principal G-bundle P as $E = P\times_G V$ (V the standard fibre of E) and works G-equivariantly on $P\times V$. We will not go into the details here and refer to [10, 13, 14, 15] for discussions of various aspects of that construction. Given $\Phi_\nabla(E)$, $e_{s,\nabla}$ is obtained by pulling back $\Phi_\nabla(E)$ to X via a section $s: X\to E$ of E, $e_{s,\nabla}(E) = s^*\Phi_\nabla(E)$. It is again most conveniently represented as a Grassmann integral, as in (2.4), and is explicitly given by

$$e_{s,\nabla}(E) = (2\pi)^{-m}\int d\chi\, e^{-\frac{1}{2}|s|^2 + \frac{1}{2}\chi_a\Omega_\nabla^{ab}\chi_b + i\nabla s^a\chi_a}. \tag{2.11}$$

Here the norm $|s|^2$ refers to a fixed fibre metric on E and ∇ is a compatible connection. Integrating out χ one sees that $e_{s,\nabla}(E)$ is a $2m$-form on X. That $e_{s,\nabla}(E)$ is closed is reflected in the fact that the exponent in (2.11) is invariant under the transformation

$$\delta s = \nabla s, \qquad \delta\chi = is. \tag{2.12}$$

To make contact with the notation commonly used in the physics literature it will be useful to introduce one more set of anti-commuting variables, ψ^μ, corresponding to the one-forms dx^μ

of a local coordinate basis. Given any form ω on X, we denote by $\omega(\psi)$ the object that one obtains by replacing dx^μ by ψ^μ,

$$\omega = \frac{1}{p!}\omega_{\mu_1\cdots\mu_p}dx^{\mu_1}\dots dx^{\mu_p} \to \omega(\psi) = \frac{1}{p!}\omega_{\mu_1\cdots\mu_p}\psi^{\mu_1}\dots\psi^{\mu_p}. \tag{2.13}$$

With the usual rules of Berezin integration, the integral of a top-form $\omega^{(n)}$ can then be written as

$$\int_X \omega^{(n)} = \int_X dx \int d\psi\, \omega^{(n)}(\psi). \tag{2.14}$$

As the measures dx and $d\psi$ transform inversely to each other, the right hand side is coordinate independent (and this is the physicist's way of saying that integration of differential forms has that property). Anyway, with this trick, the exponent of (2.11) and the transformations (2.12) (supplemented by $\delta x^\mu = \psi^\mu$) then take the form of a supersymmetric 'action' and its supersymmetry.

We now take a brief look at $e_{s,\nabla}(E)$ for various choices of s. The first thing to note is that for s the zero section of E (2.11) reduces to (2.4), $e_{s=0,\nabla}(E) = e_\nabla(E)$. As $\Phi_\nabla(E)$ is closed, standard arguments now imply that $e_{s,\nabla}(E)$ will be cohomologous to $e_\nabla(E)$ for any choice of section s. If $n = 2m$ and s is a generic section of E transversal to the zero section, we can calculate $\int_X e_{s,\nabla}(E)$ by replacing s by γs for $\gamma \in \mathbb{R}$ and evaluating the integral in the limit $\gamma \to \infty$. In that limit the curvature term in (2.11) will not contribute and one can use the stationary phase approximation (see section 3) to reduce the integral to a sum of contributions from the zeros of s, reproducing equation (2.5). The calculation is straightforward and entirely analogous to similar calculations in supersymmetric quantum mechanics (see e.g., [16]) and we will not repeat it here. The important thing to remember, though, is that one can use the s-(and hence γ-)independence of $[e_{s,\nabla}]$ to evaluate (2.10) in terms of local data associated with the zero locus of some conveniently chosen section s.

Finally, if $E = TX$ and V is a non-generic vector field on X with zero locus X_V, the situation is a little bit more complicated. In this case, $\int_X e_{\gamma V,\nabla}$ for $\gamma \to \infty$ can be expressed in terms of the Riemann curvature $\Omega_{\nabla'}$ of X_V with respect to the induced connection ∇'. Here $\Omega_{\nabla'}$ arises from the data Ω_∇ and V entering $e_{V,\nabla}$ via the classical *Gauss–Codazzi equations*. Then equation (2.8) is reproduced in the present setting in the form (cf. (2.2))

$$\chi(X) = \int_X e_{V,\nabla} = (2\pi)^{-\dim(X_V)/2}\int_{X_V} \mathrm{Pf}(\Omega_{\nabla'}). \tag{2.15}$$

Again the manipulations required to arrive at (2.15) are exactly as in supersymmetric quantum mechanics, the Gauss–Codazzi version of which has been introduced and discussed in detail in [17].

That the content of (2.11) is quite non-trivial even in finite dimensions where, as mentioned above, all the forms $e_{s,\nabla}(E)$ are cohomologous, can already be seen in the following simple example, variants of which we will use throughout the paper to illustrate the integration formulae (see e.g., (3.4,3.13,3.47)).

Let us take $X = S^2$, $E = TX$, and equip S^2 with the standard constant curvature metric $g = d\theta^2 + \sin^2\theta d\phi^2$ and its Levi-Civita connection ∇. To obtain a representative of the Euler class of S^2 which is different from the Gauss–Bonnet representative, we pick some vector field V on S^2. Let us e.g., choose V to be the vector field ∂_ϕ generating rotations about an axis of S^2. Then the data entering (2.11) can be readily computed. In terms of an orthonormal frame $e^a = (d\theta, \sin\theta d\phi)$ one finds

$$\begin{aligned} |V|^2 &= \sin^2\theta, & \nabla V^a &= (\sin\theta\cos\theta d\phi, -\cos\theta d\theta), \\ \Omega_\nabla^{12} &= \sin\theta d\theta d\phi, & \tfrac{1}{2}\chi_a\Omega_\nabla^{ab}\chi_b &= \chi_1\chi_2\sin\theta d\theta d\phi. \end{aligned} \tag{2.16}$$

To make life more interesting let us also replace V by γV. The χ-integral in (2.11) is easily done and one obtains

$$e_{V,\nabla}(TS^2) = (2\pi)^{-1}\mathrm{e}^{-\frac{\gamma^2}{2}\sin^2\theta}(1+\gamma^2\cos^2\theta)\sin\theta d\theta d\phi. \tag{2.17}$$

The quintessence of the above discussion, as applied to this example, is now that the integral of this form over S^2 is the Euler number of S^2 and hence in particular independent of γ. We integrate over ϕ (as nothing depends on it) and change variables from θ to $x = \sin\theta$. Then the integral becomes

$$\chi(S^2) = \int_{-1}^{1} \mathrm{e}^{\frac{\gamma^2}{2}(x^2-1)}(1+\gamma^2x^2)dx. \tag{2.18}$$

For $\gamma = 0$, the integrand is simply dx and one obtains $\chi(S^2) = 2$ which is (reassuringly) the correct result. A priori, however, it is far from obvious that this integral is really independent of γ. One way of making this manifest is to note that the entire integrand is a total derivative,

$$d(\mathrm{e}^{\frac{\gamma^2}{2}(x^2-1)}x) = \mathrm{e}^{\frac{\gamma^2}{2}(x^2-1)}(1+\gamma^2x^2)dx. \tag{2.19}$$

This is essentially the statement that (2.17) differs from the standard representative

$$(1/2\pi)\sin\theta d\theta d\phi$$

at $\gamma = 0$ by a globally defined total derivative or, equivalently, that the derivative of (2.17) with respect to γ is exact. This makes the γ-independence of (2.18) somewhat less mysterious. Nevertheless, already this simple example shows that the content of the Mathai–Quillen formula (2.11) is quite non-trivial.

The Mathai–Quillen formalism for infinite dimensional vector bundles. Let us recapitulate briefly what we have achieved so far. We have constructed a family of differential forms $e_{s,\nabla}(E)$ parametrized by a section s and a connection ∇, all representing the Euler class $e(E) \in H^{2m}(X)$. In particular, for $E = TX$, the equation $\chi(X) = \int_X e_{V,\nabla}(X)$ interpolates between the classical Poincaré–Hopf and Gauss–Bonnet theorems. It should be borne in mind, however, that all the forms $e_{s,\nabla}$ are cohomologous so that this construction, as nice as it is, is not very interesting from the cohomological point of view (although, as we have seen, it has its charm also in finite dimensions).

To be in a situation where the forms $e_{s,\nabla}$ are not necessarily all cohomologous to e_∇, and where the Mathai–Quillen formalism thus 'comes into its own' [10], we now consider infinite dimensional vector bundles where e_∇ (an 'infinite-form') is not defined at all. In that case the added flexibility in the choice of s becomes crucial and opens up the possibility of obtaining well-defined, but s-dependent, 'Euler classes' of E. To motivate the concept of a regularized Euler number of such a bundle, to be introduced below, recall equation (2.8) for the Euler number $\chi(X)$ of a manifold X. When X is finite dimensional this is an identity, while its left hand side is not defined when X is infinite dimensional. Assume, however, that we can find a vector field V on X whose zero locus is a finite dimensional submanifold of X. Then the right hand side of (2.8) *is* well defined and we can use it to tentatively define a *regularized Euler number* $\chi_V(X)$ as

$$\chi_V(X) := \chi(X_V). \tag{2.20}$$

By (2.10) and the same localization arguments as used to arrive at (2.15), we expect this number to be given by the (functional) integral

$$\chi_V(X) = \int_X e_{V,\nabla}(X). \tag{2.21}$$

This equation can (formally) be confirmed by explicit calculation. A rigorous proof can presumably be obtained in some cases by probabilistic methods as used e.g., by Bismut [18, 19] in related contexts. We will, however, content ourselves with verifying (2.21) in some examples below.

More generally, we are now led to define the regularized Euler number $\chi_s(E)$ of an infinite dimensional vector bundle E as

$$\chi_s(E) := \int_X e_{s,\nabla}(E). \tag{2.22}$$

Again, this expression turns out to make sense (for a physicist) when the zero locus of s is a finite dimensional manifold X_s, in which case $\chi_s(E)$ is the Euler number of some finite dimensional vector bundle over X_s (a quotient bundle of the restriction $E|_{X_s}$, cf. [11, 20]).

Of course, there is no reason to expect $\chi_s(E)$ to be independent of s, even if one restricts one's attention to those sections s for which the integral (2.22) exists. However, if s is a section of E naturally associated with E (we will see examples of this below), then $\chi_s(E)$ is also naturally associated with E and can be expected to carry interesting topological information. This is indeed the case.

2.2. The Regularized Euler Number of Loop Space, or: Supersymmetric Quantum Mechanics

Our first application and illustration of the Mathai–Quillen formalism for infinite dimensional vector bundles will be to the *loop space* $X = LM$ of a finite dimensional manifold M and its tangent bundle $E = T(LM)$. We will see that this is completely equivalent to supersymmetric quantum mechanics [21, 22] whose numerous attractive and interesting properties now find a neat explanation within the present framework. As the authors of [1] were certainly in part inspired by supersymmetric quantum mechanics, deriving the latter that way may appear to be somewhat circuitous. It is, nevertheless, instructive to do this because it illustrates all the essential features of the Mathai–Quillen formalism in this non-trivial, but manageable, quantum-mechanical context.

Geometry of loop space. We denote by M a smooth orientable Riemannian manifold with metric g and by $LM = \mathrm{Map}(S^1, M)$ the loop space of M. Elements of LM are denoted by $x = \{x(t)\}$ or simply $x^\mu(t)$, where $t \in [0,1]$, x^μ are (local) coordinates on M and $x^\mu(0) = x^\mu(1)$. In supersymmetric quantum mechanics it is sometimes convenient to scale t such that $t \in [0,T]$ and $x^\mu(0) = x^\mu(T)$ for some $T \in \mathbb{R}$, and to regard T as an additional parameter of the theory.

A tangent vector $V(x) \in T_x(LM)$ at a loop $x \in LM$ can be regarded as a deformation of the loop, i.e., as a section of the tangent bundle TM restricted to the loop $x(t)$ such that $V(x)(t) \in T_{x(t)}(M)$. In more fancy terms this means that

$$T_x(LM) = \Gamma(x^* TM) \tag{2.23}$$

is the infinite-dimensional space of sections of the pull-back of the tangent bundle TM to S^1 via the map $x : S^1 \to M$.

There is a canonical vector field on LM generating rigid rotations $x(t) \to x(t+\varepsilon)$ of the loop, given by $V(x)(t) = \dot{x}(t)$ (or $V = \dot{x}$ for short). A metric g on M induces a metric $\hat{g}$ on LM through

$$\hat{g}_x(V_1, V_2) = \int_0^1 dt\, g_{\mu\nu}(x(t)) V_1^\mu(x)(t) V_2^\nu(x)(t). \tag{2.24}$$

Likewise, any differential form α on M induces a differential form $\hat{\alpha}$ on LM via

$$\hat{\alpha}_x(V_1, \ldots, V_p) = \int_0^1 dt\, \alpha(x(t))(V_1(x(t)), \ldots, V_p(x(t))). \tag{2.25}$$

Because of the reparametrization invariance of these integrals, $\hat{g}$ and $\hat{\alpha}$ are invariant under the flow generated by $V = \dot{x}$, $L(V)\hat{g} = L(V)\hat{\alpha} = 0$. We will make use of this observation in section 3.2.

Supersymmetric quantum mechanics. We will now apply the formalism developed in the previous section to the data $X = LM$, $E = T(LM)$, and $V = \dot{x}$ (and later on some variant thereof). The zero locus $(LM)_V$ of V is just the space of constant loops, i.e., M itself. If we therefore define the regularized Euler number of LM via (2.20) by

$$\chi_V(LM) := \chi((LM)_V) = \chi(M), \tag{2.26}$$

we expect the functional integral $\int_{LM} e_{V,\nabla}(LM)$ to calculate the Euler number of LM. Let us see that this is indeed the case.

The anticommuting variables χ_a parametrize the fibres of $T(LM)$ and we write them as $\chi_a = e_a{}^\mu \bar{\psi}_\mu$ where $e_a{}^\mu$ is the inverse vielbein corresponding to $g_{\mu\nu}$. This change of variables produces a Jacobian $\det[e]$ we will come back to below. Remembering to substitute $dx^\mu(t)$ by $\psi^\mu(t)$, the exponent of the Mathai–Quillen form $e_{V,\nabla}(LM)$ (2.11) becomes

$$S_M = \int_0^T dt\, [-\tfrac{1}{2} g_{\mu\nu} \dot{x}^\mu \dot{x}^\nu + \tfrac{1}{4} R^{\mu\nu}{}_{\rho\sigma} \bar{\psi}_\mu \psi^\rho \bar{\psi}_\nu \psi^\sigma - i \bar{\psi}_\mu \nabla_t \psi^\mu], \tag{2.27}$$

where ∇_t is the covariant derivative along the loop $x(t)$ induced by ∇. This is precisely the standard action of de Rham (or $N = 1$) supersymmetric quantum mechanics [21, 22, 23, 24, 25] (with the conventions as in [16]), the supersymmetry given by (2.12). In order to reduce the measure to the natural form $[dx][d\psi][d\bar{\psi}]$, one can introduce a multiplier field B^μ to write the bosonic part of the action as

$$\int_0^T dt\, g_{\mu\nu}(B^\mu \dot{x}^\nu + B^\mu B^\nu) \approx \int_0^T dt\, [-\tfrac{1}{2} g_{\mu\nu} \dot{x}^\mu \dot{x}^\nu], \tag{2.28}$$

because integration over B will give rise to a determinant $\det[g]^{-1/2}$ that cancels the $\det[e]$ from above. With this understood, the integral we are interested in is the partition function of this theory,

$$\int_{LM} e_{V,\nabla}(LM) = Z(S_M) \tag{2.29}$$

Now it is well known that the latter indeed calculates $\chi(M)$ and this is our first non-trivial confirmation that (2.21) makes sense and calculates (2.20). The explicit calculation of $Z(S_M)$ is not difficult, and we will sketch it below.

It may be instructive, however, to first recall the standard *a priori* argument establishing $Z(S_M) = \chi(M)$. One starts with the definition (2.7) of $\chi(M)$. As there is a one-to-one correspondence between cohomology classes and harmonic forms on M, one can write $\chi(M)$ as a trace over the kernel Ker Δ of the Laplacian $\Delta = dd^* + d^*d$ on differential forms,

$$\chi(M) = \mathrm{tr}_{\mathrm{Ker}\,\Delta}(-1)^F \tag{2.30}$$

(here $(-1)^F$ is $+1$ (-1) on even (odd) forms). As the operator $d+d^*$ commutes with Δ and maps even to odd forms and vice-versa, there is an exact pairing between 'bosonic' and 'fermionic' eigenvectors of Δ with non-zero eigenvalue. It is thus possible to extend the trace in (2.30) to a trace over the space of all differential forms,

$$\chi(M) = \mathrm{tr}_{\Omega^*(M)}(-1)^F \mathrm{e}^{-T\Delta}. \tag{2.31}$$

As only the zero modes of Δ will contribute to the trace, it is evidently independent of the value of T. Once one has put $\chi(M)$ into this form of a statistical mechanics partition function, one can use the Feynman–Kac formula to represent it as a supersymmetric path integral [24] with the action (2.27).

This Hamiltonian way of arriving at the action of supersymmetric quantum mechanics should be contrasted with the Mathai–Quillen approach. In the former one starts with the operator whose index one wishes to calculate (e.g., $d+d^*$), constructs a corresponding Hamiltonian, and then deduces the action. On the other hand, in the latter one begins with a finite dimensional topological invariant (e.g., $\chi(M)$) and represents that directly as an infinite dimensional integral, the partition function of a supersymmetric action.

What makes such a path integral representation of $\chi(M)$ interesting is that one can now go ahead and try to somehow evaluate it directly, thus possibly obtaining alternative expressions for $\chi(M)$. Indeed, one can obtain path integral 'proofs' of the Gauss–Bonnet and Poincaré–Hopf theorems in this way. This is just the infinite dimensional analogue of the fact discussed above that different choices of s in $\int_X e_{s,\nabla}(E)$ can lead to different expressions for $\chi(E)$.

Let us, for example, replace the section $V = \dot{x}$ in (2.27) by $\gamma\dot{x}$ for some $\gamma \in \mathbb{R}$. This has the effect of multiplying the first term in (2.27) by γ^2 and the third by γ. In order to be able to take the limit $\gamma \to \infty$, which would localize the functional integral to $\dot{x} = 0$, i.e., to M, we proceed as follows. First of all, one expands all fields in Fourier modes. Then one scales all non-constant modes of $x(t)$ by γ and all the non-constant modes of $\psi(t)$ and $\bar{\psi}(t)$ by $\gamma^{1/2}$. This leaves the measure invariant. Then in the limit $\gamma \to \infty$ only the constant modes of the curvature term survive in the action, while the contributions from the non-constant modes cancel identically between the bosons and fermions because of supersymmetry. The net effect of this is that one is left with a finite dimensional integral of the form (2.4) leading to

$$Z(S_M) = (2\pi)^{-\dim(M)/2} \int_M \mathrm{Pf}(\Omega_\nabla(M)) = \chi(M). \tag{2.32}$$

This provides a path integral derivation of the Gauss–Bonnet theorem (2.6). The usual supersymmetric quantum mechanics argument to this effect [23, 25] makes use of the T-independence of the partition function to suppress the non-constant modes as $T \to 0$. But the Mathai–Quillen formalism now provides an understanding and explanation of the mechanism by which the path integral (2.29) over LM localizes to the integral (2.32) over M.

Other choices of sections are also possible, e.g., a vector field of the form $V = \dot{x}^\mu + \gamma g^{\mu\nu}\partial_\nu W$, where W is some function on M and $\gamma \in \mathbb{R}$ a parameter. This introduces a potential into the supersymmetric quantum mechanics action (2.27). It is easy to see that the zero locus of this vector field is the zero locus of the gradient vector field $\partial_\mu W$ on M whose Euler number is the same as that of M by (2.8). Again this agrees with the explicit evaluation of the path integral of the corresponding supersymmetric quantum mechanics action which is, not unexpectedly, most conveniently performed by considering the limit $\gamma \to \infty$. In that limit, because of the term $\gamma^2 g^{\mu\nu}\partial_\mu W \partial_\nu W$ in the action, the path integral localizes around the critical points of W. Let us assume that these are isolated. By supersymmetry, the fluctuations around the critical points will, up to a sign, cancel between the bosonic and fermionic contributions and one finds that the partition function is

$$\chi(M) = Z(S_M) = \sum_{x_k : dW(x_k=0} \text{sign}(\det H_{x_k}(W)), \tag{2.33}$$

where

$$H_{x_k}(W) = (\nabla_\mu \partial_\nu W)(x_k) \tag{2.34}$$

is the *Hessian* of W at x_k. This is the Poincaré–Hopf theorem (2.5) (for a gradient vector field).

If the critical points of W are not isolated then, by a combination of the above arguments, one recovers the generalization $\chi(M) = \chi(M_{W'})$ (2.8) of the Poincaré–Hopf theorem (for gradient vector fields) in the form (2.15).

Finally, by considering a section of the form $V = \dot{x} + v$, where v is an arbitrary vector field on M, one can also, to complete the circle, rederive the general finite dimensional Mathai–Quillen form $e_{v,\nabla}(M)$ (2.11) from supersymmetric quantum mechanics by proceeding exactly as in the derivation of (2.32) [26, 15].

This treatment of supersymmetric quantum mechanics has admittedly been somewhat sketchy and we should perhaps, summarizing this section, state clearly what are the important points to keep in mind as regards the Mathai–Quillen formalism and localization of functional integrals:

1. Explicit evaluation of the supersymmetric quantum mechanics path integrals obtained by formally applying the Mathai–Quillen construction to the loop space LM confirms that we can indeed represent the regularized Euler number $\chi_V(LM)$, as defined by (2.20), by the functional integral (2.21).

2. In particular, it confirms that functional integrals arising from or related to the Mathai–Quillen formalism (extended to infinite dimensional bundles) localize to finite dimensional integrals.

3. This suggests that a large class of other quantum mechanics and field theory functional integrals can be constructed which also have this property. These integrals should be easier to understand in a mathematically rigorous fashion than generic functional integrals arising in field theory.

2.3. Other Examples and Applications — an Overview

The purpose of this section is to provide an overview of other applications and appearances of the Mathai–Quillen formalism in the physics literature. None too surprisingly, all of these are related to topological field theories or simple (non-topological) perturbations thereof.

It was first shown by Atiyah and Jeffrey [10] that the rather complicated looking action of Witten's four-dimensional topological Yang–Mills theory [27] (Donaldson theory) had a neat explanation in this framework. Subsequently, this interpretation was also shown to be valid and useful in topological sigma models coupled to topological gravity [11]. In [26, 15] this strategy was turned around to construct topological field theories from scratch via the Mathai–Quillen formalism. A nice explanation of the formalism from a slightly different perspective can be found in the recent work of Vafa and Witten [9], and other recent applications include [28, 29].

As these applications thus range from topological gauge theory (intersection numbers of moduli spaces of connections) over topological sigma models (counting holomorphic curves) to topological gravity (intersection theory on some moduli space of metrics) and the string theory interpretation of 2d Yang–Mills theory, this section will not and cannot be self-contained. What we will do is to try to provide a basic understanding of why so-called cohomological topological field theories can generally be understood within and constructed from the Mathai–Quillen formalism.

The basic strategy — illustrated by Donaldson theory. On the basis of what has been done in the previous section, one possible approach would be to ask if topological field theories can in some sense be understood as supersymmetric quantum mechanics models with perhaps infinite-dimensional target spaces. For topological field theory models calculating the Euler number of some moduli space this has indeed been shown to be the case in [17]. In principle, it should also be possible to extend the supersymmetric quantum mechanics formalism of the previous section to general vector bundles E and their Mathai–Quillen forms $e_{s,\nabla}$. This generalization would then also cover other topological field theories.

This is, however, not the point of view we are going to adopt in the following. Rather, we will explain how the Mathai–Quillen formalism can be used to construct a field theory describing intersection theory on some given moduli space (of connections, metrics, maps, ...) of interest.

Before embarking on this, we should perhaps point out that, from a purely pragmatic point of view, an approach to the construction of topological field theories based on, say, BRST quantization may occasionally be more efficient and straightforward. What one is gaining by the Mathai–Quillen approach is geometrical insight.

Assume then, that the finite dimensional moduli space on which one would like to base a topological field theory is given by the following data:

1. A space Φ of fields ϕ and (locally) a defining equation or set of equations $\mathbb{F}(\phi) = 0$;
2. A group $\mathcal{G}$ of transformations acting on Φ leaving invariant the set $\mathbb{F}(\phi) = 0$, such that the moduli space is given by

$$\mathcal{M}_{\mathbb{F}} = \{\phi \in \Phi : \mathbb{F}(\phi) = 0\}/\mathcal{G} \subset \Phi/\mathcal{G} \equiv X. \tag{2.35}$$

This is not the most general set-up one could consider but will be sufficient for what follows.

The prototypical example to keep in mind is the moduli space of anti-self-dual connections (instantons), where Φ is the space $\mathcal{A}$ of connections A on a principal bundle P on a Riemannian four-manifold (M,g), $\mathbb{F}(A) = (F_A)_+$ is the self-dual part of the curvature of A, and $\mathcal{G}$ is the group of gauge transformations (vertical automorphisms of P). In other cases either $\mathbb{F}$ or $\mathcal{G}$ may be trivial — we will see examples of that below.

Now, to implement the Mathai–Quillen construction, one needs to be able to interpret the moduli space $\mathcal{M}_{\mathbb{F}}$ as the zero locus of a section $s_{\mathbb{F}}$ of a vector bundle E over X. In practice, it is often more convenient to work 'upstairs', i.e., equivariantly, and to exhibit the set $\{\mathbb{F} = 0\}$ as the zero locus of a $\mathcal{G}$-equivariant vector bundle over Φ. This may seem to be rather abstract, but in practice there are usually moderately obvious candidates for E and s. One also needs to choose some connection ∇ on E.

Let us again see how this works for anti-self-dual connections. In that case, $\mathbb{F}$ can be regarded as a map from $\mathcal{A}$ to the space $\Omega^2_+(M,\mathrm{ad}P)$, i.e., roughly speaking the space of Lie algebra valued self-dual two-forms on M. The kernel of this map is the space of anti-self-dual connections. To mod out by the gauge group we proceed as follows. For appropriate choice of $\mathcal{A}$ or $\mathcal{G}$, $\mathcal{A}$ can be regarded as the total space of a principal $\mathcal{G}$-bundle over $X = \mathcal{A}/\mathcal{G}$. As $\mathcal{G}$ acts on $\Omega^2_+(M,\mathrm{ad}P)$, one can form the associated infinite dimensional vector bundle

$$\mathcal{E}_+ = \mathcal{A}\times_{\mathcal{G}}\Omega^2_+(M,\mathrm{ad}P). \tag{2.36}$$

Clearly, the map $\mathbb{F}$ descends to give a well-defined section $s_{\mathbb{F}}$ of $\mathcal{E}_+$ whose zero-locus is precisely the finite dimensional moduli space of anti-self-dual connections and we can choose $E = \mathcal{E}_+$. This bundle also has a natural connection coming from declaring the horizontal subspaces in $\mathcal{A}\to\mathcal{A}/\mathcal{G}$ to be those orthogonal to the $\mathcal{G}$-orbits in $\mathcal{A}$ with respect to the metric on $\mathcal{A}$ induced by that on M.

Once one has assembled the data X, E and $s = s_{\mathbb{F}}$ one can — upon choice of some connection ∇ on E — plug these into the formula (2.11) for the Mathai–Quillen form. In this way one obtains a functional integral which localizes onto the moduli space $\mathcal{M}_{\mathbb{F}}$. A few remarks may help to clarify the status and role of the 'field theory action' one obtains in this way.

1. *A priori* this functional integral can be thought of as representing a regularized Euler number $\chi_s(E)$ or Euler class of the infinite dimensional vector bundle E. As mentioned in section 2.1 (and as we have seen explicitly in section 2.2), this can also be interpreted as the Euler number (class) of a vector bundle over $\mathcal{M}_{\mathbb{F}}$. If the rank of this vector bundle is strictly less than the dimension of $\mathcal{M}_{\mathbb{F}}$, than one needs to consider the pairing of this Euler class with cohomology classes on $\mathcal{M}_{\mathbb{F}}$.

   From a mathematical point of view the necessity of this is clear while in physics parlance this amounts to inserting operators (observables) into the functional integral to 'soak up the fermionic zero modes'.

2. Frequently, the connection and curvature of the bundle E are expressed in terms of Green's functions on the underlying 'space-time' manifold. As such, the data entering the putative field theory action (the exponent of (2.11)) are non-local in space-time — an undesirable feature for a fundamental action. The ham-handed way of eliminating this non-locality is to introduce auxiliary fields. However, what this really amounts to is to working 'upstairs' and constructing an equivariant Mathai–Quillen form (cf. the remarks before (2.11)). In practice, therefore, it is often much simpler to start upstairs directly and to let Gaussian integrals do the calculation of the curvature and connection terms of E entering (2.11).

Let us come back once again to the moduli space of instantons. Note that, among all the possible sections of $\mathcal{E}_+$ the one we have chosen is really the only natural one. Different sections could of course be used in the Mathai–Quillen form. But in the infinite dimensional case there is no reason to believe that the result is independent of the choice of s and for other choices of s it would most likely not be of any mathematical interest. With the choice $s_{\mathbb{F}}$, however, the resulting action is interesting and, not unexpectedly, that of Donaldson theory [30, 27]. We refer to [16, pp. 198–247] for a partial collection of things that can and should be said about this theory.

A topological gauge theory of flat connections. With the Mathai–Quillen formalism at one's disposal, it is now relatively easy to construct other examples. In the following we will sketch two of them, one for a moduli space of flat connections and the other related to holomorphic curves.

If one is, for instance, interested in the moduli space $\mathcal{M}^3$ of flat connections $F_A = 0$ on a principal G-bundle on a three-manifold M, one could go about constructing an action whose partition function localizes onto that moduli space as follows. The flatness condition expresses the vanishing of the two-form F_A or — by duality — the one-form $*F_A \in \Omega^1(M, \text{ad}P)$. Now, as the space $\mathcal{A}$ is an affine space modelled on $\Omega^1(M, \text{ad}P)$, we can think of $*F_A$ as a section of the tangent bundle of $\mathcal{A}$, i.e., as a vector field on $\mathcal{A}$. In fact, it is the gradient vector field of the *Chern–Simons functional*

$$S_{CS}(A) = \int_M \operatorname{Tr} A\,dA + \tfrac{2}{3}A^3. \tag{2.37}$$

This vector field passes down to a vector field on $\mathcal{A}/\mathcal{G}$ whose zero locus is precisely the moduli space $\mathcal{M}^3$ of flat connections and our data are therefore $X = \mathcal{A}/\mathcal{G}$, $E = TX$, and $V(A) = *F_A$ (in accordance with section 2.1 we denote sections of the tangent bundle by V). Following the steps outlined above to construct the action form the Mathai–Quillen form (2.11), one finds that it coincides with that constructed e.g., in [31, 32, 33, 26]. Again, as in supersymmetric quantum mechanics, one finds full agreement of

$$\chi_V(\mathcal{A}/\mathcal{G}) \equiv \chi((\mathcal{A}/\mathcal{G})_V) = \chi(\mathcal{M}^3) \tag{2.38}$$

with the partition function of the action which gives $\chi(\mathcal{M}^3)$ in the form (2.15), i.e., via the Gauss–Codazzi equations for the embedding $\mathcal{M}^3 \subset \mathcal{A}/\mathcal{G}$ [26]. We conclude this example with some remarks.

1. In [10] the partition function of this theory was first identified with a regularized Euler number of $\mathcal{A}/\mathcal{G}$. We can now identify it more specifically with the Euler number of $\mathcal{M}^3$.

2. In [34] it was shown that for certain three-manifolds (homology spheres) $\chi_V(\mathcal{A}^3/\mathcal{G}^3)$ is the Casson invariant. This has led us to propose the Euler number of $\mathcal{M}^3$ as a generalization of the Casson invariant for more general three-manifolds. See [26] for a preliminary investigation of this idea.

3. Note that the chosen moduli space $\mathcal{M}^3$ itself, via the defining equation $F_A = 0$, determined the bundle E to be used in the Mathai–Quillen construction. Hence it also determined the fact that we ended up with a topological field theory calculating the Euler number of $\mathcal{M}^3$ rather than allowing us to model intersection theory on $\mathcal{M}^3$. While this may appear to be a shortcoming of this procedure, there are also other reasons to believe that intersection theory on $\mathcal{M}^3$ is not a particularly natural and meaningful thing to study.

4. Nevertheless, in two dimensions one can construct both a topological gauge theory corresponding to intersection theory on the moduli space $\mathcal{M}^2$ of flat connections (this is just the 2d analogue of Donaldson theory) and a topological gauge theory calculating the Euler number of $\mathcal{M}^2$. For the former, the relevant bundle is $\mathcal{E} = \mathcal{A} \times_{\mathcal{G}} \Omega^0(M, \text{ad}P)$ (cf. (2.36)) with section $*F_A$. In the latter case, life turns out not be so simple. The base space X cannot possibly be $\mathcal{A}/\mathcal{G}$ as $*F_A$ does not define a section of its tangent bundle. Rather, the base space turns out to be $\mathcal{E}$ with $E = T\mathcal{E}$ and an appropriate section.

Because the fibre directions of $\mathcal{E}$ are topologically trivial, this indeed then calculates the Euler number of $\mathcal{M}^2$ — see the final remarks in [15].

Topological sigma models. Finally, we will consider an example of a topological field theory which is not a gauge theory but rather, in a sense, the most obvious field theoretic generalization of supersymmetric quantum mechanics, namely the topological sigma model [35, 36]. In its simplest form this is a theory of maps from a Riemann surface Σ to a Kähler manifold M localizing either onto holomorphic maps (in the so-called A-model) or onto constant maps (in the B-model). As an obvious generalization of supersymmetric quantum mechanics, the interpretation of the A-model in terms of the Mathai–Quillen formalism is completely straightforward. It is already implicit in the Langevin equation approach to the construction of the model[37] and has recently been spelled out in detail in [28, 29]. We will review this construction below. The Mathai–Quillen interpretation of the B-model is slightly more subtle and we will discuss that elsewhere.

In terms of the notation introduced at the beginning of this section (cf. (2.35)), the space of fields is the space $\Phi = \mathrm{Map}(\Sigma, M)$ of maps from a Riemann surface Σ (with complex structure j) to a Kähler manifold M (with complex structure J and hermitian metric g). We will denote by $T^{(1,0)}\Sigma$ etc. the corresponding decomposition of the complexified tangent and cotangent bundles into their holomorphic and anti-holomorphic parts.

The defining equation is the condition of holomorphicity of $\phi \in \Phi$ with respect to both j and J, $\mathbb{F}(\phi) = \bar{\partial}_J\phi = 0$. In terms of local complex coordinates on Σ and M, ϕ can be represented as a map $(\phi^k(z,\bar{z}), \phi^{\bar{k}}(z,\bar{z}))$ and the holomorphicity condition can be written as

$$\partial_{\bar{z}}\phi^k = \partial_z\phi^{\bar{k}} = 0. \tag{2.39}$$

Solutions to (2.39) are known as *holomorphic curves* in M. The moduli space of interest is just the space $\mathcal{M}_{\bar{\partial}_J}$ of holomorphic curves,

$$\mathcal{M}_{\bar{\partial}_J} = \{\phi \in \mathrm{Map}(\Sigma, M) : \bar{\partial}_J\phi = 0\}, \tag{2.40}$$

so that this is an example where the symmetry group $\mathcal{G}$ is trivial.

Let us now determine the bundle E over $X = \mathrm{Map}(\Sigma, M)$. In the case of supersymmetric quantum mechanics, the map $x(t) \to \dot{x}(t)$ could be regarded as a section of the tangent bundle (2.23) of $LM = \mathrm{Map}(S^1, M)$. Here the situation is, primarily notationally, a little bit more involved. First of all, as in supersymmetric quantum mechanics, the tangent space at a map ϕ is

$$T_\phi(\mathrm{Map}(\Sigma, M)) = \Gamma(\phi^*TM). \tag{2.41}$$

The map $\phi \to d\phi$ can hence be regarded as a section of a vector bundle $\mathcal{E}$ whose fibre at ϕ is $\Gamma(\phi^*TM \otimes T^*\Sigma)$. The map

$$s_{\mathbb{F}} : \phi \to (\partial_{\bar{z}}\phi^k, \partial_z\phi^{\bar{k}}) \tag{2.42}$$

we are interested in can then be regarded as a section of a subbundle $\mathcal{E}^{(0,1)}$ of that bundle spanned by $T^{(1,0)}M$-valued $(0,1)$-forms $\psi^k_{\bar{z}}$ and $T^{(0,1)}M$-valued $(1,0)$-forms $\psi^{\bar{k}}_z$ on Σ. Evidently, the moduli space $\mathcal{M}_{\bar{\partial}_J}$ is the zero locus of the section $s_{\mathbb{F}}(\phi) = \bar{\partial}_J\phi$ of $E = \mathcal{E}^{(0,1)}$.

Since the bundle $\mathcal{E}^{(0,1)}$ is contructed from the tangent and cotangent bundles of M and Σ, it inherits a connection from the Levi-Cività connections on M and Σ. With the convention of replacing one-forms on X by their fermionic counterparts (cf. (2.13,2.27)), the exterior covariant derivative of $s_{\mathbb{F}}$ can then be written as

$$\nabla s_{\mathbb{F}}(\psi) = (D_{\bar{z}}\psi^k, D_z\psi^{\bar{k}}). \tag{2.43}$$

One can now obtain the action S_{TSM} for the topological sigma model from the Mathai–Quillen form (2.11),

$$S_{TSM}(\Phi) = \int_\Sigma d^2z g_{k\bar{l}}(\tfrac{1}{2}\partial_{\bar{z}}\phi^k\partial_z\phi^{\bar{l}} + i\bar{\psi}_z^{\bar{l}}D_{\bar{z}}\psi^k + i\bar{\psi}_{\bar{z}}^k D_z\psi^{\bar{l}}) - R_{k\bar{k}l\bar{l}}\bar{\psi}_{\bar{z}}^k\bar{\psi}_z^{\bar{k}}\psi^l\psi^{\bar{l}}. \tag{2.44}$$

For M a Ricci-flat Kähler manifold (i.e. a Calabi–Yau manifold), this action and its B-model partner have recently been studied intensely in relation with *mirror symmetry*, see e.g., [36, 20] and the articles in [38]. Usually, one adds a topological term S_{TOP} to the action which is the pullback of the Kähler form ω of M,

$$S_{TOP} = \int_\Sigma \phi^*\omega = \int_\Sigma d^2z g_{k\bar{l}}(\partial_z\phi^k\partial_{\bar{z}}\phi^{\bar{l}} - \partial_{\bar{z}}\phi^k\partial_z\phi^{\bar{l}}). \tag{2.45}$$

This term keeps track of the instanton winding number. Its inclusion is also natural from the conformal field theory point of view as the 'topological twisting' of the underlying supersymmetric sigma model automatically gives rise to this term. We refer to [35] and [16] to background information on the topological sigma model in general and to [36, 20] for an introduction to the subject of mirror manifolds and holomorphic curves close in spirit to the point of view presented here.

The topological sigma model becomes even more interesting when it is coupled to topological gravity [39], as such a coupled model can be regarded as a topological string theory. A similar model, with both Σ and M Riemann surfaces, and with localization onto a rather complicated (Hurwitz) space of branched covers, has recently been studied as a candidate for a string theory realization of 2d Yang–Mills theory [28].

BRST fixed points and localization. This has in part been an extremely brief survey of some applications of the Mathai–Quillen formalism to topological field theory. The main purpose of this section, however, was to draw attention to the special properties the functional integrals of these theories enjoy. We have seen that it is the interpretation of the geometrical and topological features of the functional integral in terms of the Mathai–Quillen formalism which allows one to conclude that it is *a priori* designed to represent a finite dimensional integral.

In order to develop the subject further, and directly from a functional integral point of view, it may also be useful to know how physicists read off this property as a sort of fixed point theorem from the action or the 'measure' of the functional integral and its symmetries. The key point is that all these models have a Grassmann odd scalar symmetry δ of the kind familiar from BRST symmetry. This is essentially the field theory counterpart of (2.12), which we repeat here in the form

$$\delta\chi = is_{\mathbb{F}}(x), \qquad \delta s_{\mathbb{F}}(x) = \nabla s_{\mathbb{F}}(\psi). \tag{2.46}$$

The 'essentially' above refers to the fact that this may be the δ symmetry of the complete action only modulo gauge transformations (equivariance) and equations of motion.

Now, roughly one can argue as follows (see [11, 36]). If the δ action were free, then one could introduce an anti-commuting collective coordinate θ for it. The path integral would then contain an integration $\int d\theta \ldots$ But δ-invariance implies θ-independence of the action, and hence the integral would be zero by the rules of Berezin integration. This implies that the path integral only receives contributions from some arbitrarily small δ-invariant tubular neighborhood of the fixed point set of δ. While the integral over the fixed point set itself has to be calculated exactly, the integral over the 'tranverse' directions can then (for generic fixed points) be calculated in a stationary phase (or one-loop) approximation. If one wants to make something obvious look difficult, one can also apply this reasoning to the ordinary BRST symmetry of gauge fixing. In that case the above prescription says that one can set all the ghost and multiplier fields to zero provided that one takes into account the gauge fixing condition and the Faddeev–Popov determinant (arising form the quadratic ghost-fluctuation term) — a truism.

From (2.46) one can read off that, in the case at hand, this fixed point set is precisely the desired moduli space described by the zero locus of $s_{\mathbb{F}}$ and its tangents ψ satisfying the linearized equation $\nabla s_{\mathbb{F}}(\psi) = 0$. In this way the functional integral reduces to an integral of differential forms on $\mathcal{M}_{\mathbb{F}}$.

This point of view has the virtue of being based on easily verifiable properties of the BRST like symmetry. It also brings out the analogy with the functional integral generalization of the Duistermaat–Heckman theorem, the common theme being a supersymmetry modelling an underlying geometric or topological structure which is ultimately responsible for localization.

3. EQUIVARIANT LOCALIZATION AND THE STATIONARY PHASE APPROXIMATION — THE DUISTERMAAT–HECKMAN THEOREM

In this section we will discuss a localization formula which, roughly speaking, gives a criterion for the stationary phase approximation to an oscillatory integral to be exact.

To set the stage, we will first briefly recall the ordinary stationary phase approximation. Thus, let X be a smooth compact manifold of dimension $n = 2l$, f and dx a smooth function and a smooth density on X. Let $t \in \mathbb{R}$ and consider the function (integral)

$$F(t) = \int_X e^{itf} dx. \tag{3.1}$$

The stationary phase approximation expresses the fact that for large t the main contributions to the integral come from neighborhoods of the critical points of f. In particular, if the critical points of f are isolated and non-degenerate (so that the determinant of the Hessian $H_f = \nabla df$ of f is non-zero there), then for $t \to \infty$ the integral (3.1) can be approximated by

$$F(t) = \left(\frac{2\pi}{t}\right)^l \sum_{x_k : df(x_k)=0} e^{\frac{i\pi}{4}\sigma(H_f(x_k))} |\det(H_f(x_k)|^{-1/2} e^{itf(x_k)} + O(t^{-l-1}), \tag{3.2}$$

where $\sigma(H_f)$ denotes the signature of H_f (the number of positive minus the number of negative eigenvalues).

Duistermaat and Heckman [2] discovered a class of examples where this approximation gives the exact result and the error term in the above vanishes. To state these examples and the Duistermaat–Heckman formula in their simplest form, we take X to be a symplectic manifold with symplectic form ω and Liouville measure $dx(\omega) = \omega^l/l!$. We will also assume that the function f generates a circle action on X via its Hamiltonian vector field V_f defined by

$i(V_f)\omega = df$, and denote by $L_f(x_k)$ the infinitesimal action induced by V_f on the tangent space $T_{x_k}X$ (essentially, L_f can be represented by the matrix H_f). Then the Duistermaat–Heckman formula reads

$$\int_X e^{itf} dx(\omega) = (\frac{2\pi i}{t})^l \sum_{x_k : df(x_k)=0} \det(L_f(x_k))^{-1/2} e^{itf(x_k)}. \tag{3.3}$$

Being careful with signs and the definition of the square root of the determinant, it can be seen that (3.3) expresses nothing other than the fact that the stationary phase approximation to $F(t)$ is exact for all t.

As a simple example consider a two-sphere of unit radius centered at the origin of $\mathbb{R}^3$, with $f(x,y,z) = z$ (or, more generally, $z + a$ for some constant a) the generator of rotations about the z-axis and ω the volume form. This integral can of course easily be done exactly, e.g., by converting to polar coordinates, and one finds

$$\begin{aligned}\int_{S^2} dx(\omega) e^{itf} &= \int \sin\theta d\theta d\phi e^{it(\cos\theta + a)} \\ &= \frac{2\pi i}{t}\left(e^{-it(1-a)} - e^{it(1+a)}\right).\end{aligned} \tag{3.4}$$

We see that the result can be expressed as a sum of two terms, one from $\theta = \pi$ and the other form $\theta = 0$. This is just what one expects form the Duistermaat–Heckman formula, as these are just the north and south poles, i.e., the fixed points of the $U(1)$-action. The relative sign between the two contributions is due to the fact that f has a maximum at the north pole and a minimum at the south pole. This example can be considered as the classical partition function of a spin system. Thinking of the two-sphere as $SU(2)/U(1)$, it has a nice generalization to homogeneous spaces of the form G/T, where G is a compact Lie group and T a maximal torus. The corresponding quantum theories are also all given exactly by their stationary phase (WKB) approximation and evaluate to the Weyl character formula for G. For an exposition of these ideas see the beautiful paper [40] by Stone.

It is now clear that the Duistermaat–Heckman formula (3.3) (and its generalization to functions with non-isolated critical points) can be regarded as a localization formula as it reduces the integral over X to a sum (integral) over the critical point set. As the stationary phase approximation is very much like the semi-classical or WKB approximation investigated extensively in the physics (and, in particular, the path integral) literature, it is of obvious interest to inquire if or under which circumstances a functional integral analogue of the formula (3.3) can be expected to exist. In this context, (3.3) would now express something like the exactness of the one-loop approximation to the path integral.

The first encouraging evidence for the existence of such a generalization was pointed out by Atiyah and Witten [41]. They showed that a formal application of the Duistermaat–Heckman theorem to the infinite-dimensional loop space $X = LM$ of a manifold M and, more specifically, to the partition function of $N = \frac{1}{2}$ supersymmetric quantum mechanics representing the index of the Dirac operator on M, reproduced the known result correctly. This method of evaluating the quantum mechanics path integral has been analyzed by Bismut [19] within a mathematically rigorous framework. Subsequently, the work of Stone [40] brought the Duistermaat–Heckman theorem to the attention of a wider physicists audience. Then, in [42, 43], a general supersymmetric framework for investigating Duistermaat–Heckman- (or WKB-)like localization theorems for (non-supersymmetric) phase space path integrals was introduced, leading to a fair amount of activity in the field — see e.g., [44, 45, 46, 47, 48, 49, 50]. We will review some of these matters below.

While Duistermaat and Heckman originally discovered their formula within the context of symplectic geometry, it turns out to have its most natural explanation in the setting of equivariant cohomology and equivariant characteristic classes [51, 3, 4]. This point of view also suggests some generalizations of the Duistermaat–Heckman theorem, e.g., the Berline–Vergne formula [3] valid for Killing vectors on general compact Riemannian manifolds X. Strictly speaking, the example considered by Atiyah and Witten falls into this category as the free loop space of a manifold is not quite a symplectic manifold in general. We will discuss the interpretation of these formulae in terms of equivariant cohomology below. An excellent survey of these and many other interesting matters can be found in [4].

There is also a rather far-reaching generalization of the localization formula which is due to Witten [52]. It applies to non-Abelian group actions and to 'actions' (the exponent on the lhs of (3.3)) which are polynomials in the 'moment map' f rather than just linear in the moment map as in (3.3). This generalization is potentially much more interesting for field-theoretic applications, and we give a brief sketch of the set-up in section 3.3, but as a detailed discussion of it would necessarily go far beyond the elementary and introductory scope of this paper we refer to [52, 53, 54] for further information.

3.1. Equivariant Cohomology and Localization

As mentioned above, the localization theorems we are interested in, in this section, are most conveniently understood in terms of equivariant cohomology and we will first introduce the necessary concepts. Actually, all we will need is a rather watered down version of this, valid for Abelian group actions. The following discussion is intended as a compromise between keeping things elementary in this simple setting and retaining the flavor and elegance of the general theory. We will see that the localization formulae of Berline–Vergne [3] and Duistermaat–Heckman [2] are then fairly immediate consequences of the general formalism. Everything that is said in this section can be found in [4].

Equivariant cohomology. Let X be a compact smooth $(n = 2l)$-dimensional manifold, and let G be a compact group acting smoothly on X (the restriction to even dimension is by no means necessary — it will just allow us to shorten one of the arguments below). We denote by $\mathfrak{g}$ the Lie algebra of G, by $\mathbb{C}[\mathfrak{g}]$ the algebra of complex valued polynomial functions on $\mathfrak{g}$, and by $\Omega^*(X)$ the algebra of complex valued differential forms on X. For any $V \in \mathfrak{g}$ there is an associated vector field on X which we denote by V_X. We note that any metric h on X can be made G-invariant by averaging h over G. Hence we can assume without loss of generality that G acts on the Riemannian manifold (X,h) by isometries and that all the vector fields V_X are Killing vectors of h.

The G-equivariant cohomology of X, a useful generalization of the cohomology of X/G when the latter is not a smooth manifold, can be defined by the following construction (known as the Cartan model of equivariant cohomology). We consider the tensor product $\mathbb{C}[\mathfrak{g}] \otimes \Omega^*(X)$. As there is a natural (adjoint) action of G on $\mathbb{C}[\mathfrak{g}]$ and the G-action on X induces an action on $\Omega^*(X)$, one can consider the space $\Omega^*_G(X)$ of G-invariant elements of $\mathbb{C}[\mathfrak{g}] \otimes \Omega^*(X)$. Elements of $\Omega^*_G(X)$ will be called equivariant differential forms. They can be regarded as G-equivariant maps $\mu : V \to \mu(V)$ from $\mathfrak{g}$ to $\Omega(X)$. The $\mathbb{Z}$-grading of $\Omega^*_G(X)$ is defined by assigning degree two to the elements of $\mathfrak{g}$. G-equivariance implies that the operator

$$\begin{aligned} d_{\mathfrak{g}} &: \mathbb{C}[\mathfrak{g}] \otimes \Omega^*(X) \to \mathbb{C}[\mathfrak{g}] \otimes \Omega^*(X) \\ &(d_{\mathfrak{g}}\mu)(V) = d(\mu(V)) - i(V_X)(\mu(V)), \end{aligned} \tag{3.5}$$

which raises the degree by one, squares to zero on $\Omega_G^*(X)$. The G-equivariant cohomology $H_G^*(X)$ of X is then defined to be the cohomology of d_g acting on $\Omega_G^*(X)$. In analogy with the ordinary notions of cohomology an equivariant differential form μ is said to be equivariantly closed (respectively equivariantly exact) if $d_{\mathfrak{g}}\mu = 0$ ($\mu = d_{\mathfrak{g}}\lambda$ for some $\lambda \in \Omega_G^*(X)$).

It follows from these definitions that $H_G^*(X)$ coincides with the ordinary cohomology if G is the trivial group, and that the G-equivariant cohomology of a point is the algebra of G-invariant polynomials on $\mathfrak{g}$, $H_G^*(pt) = \mathbb{C}[\mathfrak{g}]^G$.

Integration of equivariant differential forms can be defined as a map from $\Omega_G^*(X)$ to $\mathbb{C}[\mathfrak{g}]^G$ by

$$(\int_X \mu)(V) = \int_X \mu(V), \tag{3.6}$$

where it is understood that the integral on the right hand side picks out the top-form component $\mu(V)^{(n)}$ of $\mu(V)$ and the integral is defined to be zero if μ contains no term of form-degree n.

A final thing worth pointing out is that one can also generalize the standard notions of characteristic classes, Chern–Weil homomorphism etc. to the equivariant case. The equivariant Euler form of a vector bundle E over X to which the action of G lifts, with ∇ a G-invariant connection on E, will make a brief appearance later. It is defined in the standard way by (2.3), with the curvature Ω_∇ replaced by the equivariant curvature

$$\Omega_\nabla^{\mathfrak{g}}(V) = \Omega_\nabla + (L^E(V) - \nabla_V) \tag{3.7}$$

of E. Here $L^E(V)$ denotes the infinitesimal G-action on E and the term in brackets is also known as the equivariant moment of the action — in analogy with the moment map of symplectic geometry. It can be checked directly that the differential form one obtains in this way is equivariantly closed and the usual arguments carry over to this case to establish that its cohomology class does not depend on the choice of G-invariant connection ∇.

For E the tangent bundle of X and ∇ the Levi-Civitá connection one finds that, as a consequence of $\nabla_V W - \nabla_W V = [V, W]$ (no torsion), the Riemannian moment map is simply $-(\nabla V)_{\mu\nu} = -(\nabla_\mu h_{\nu\lambda} V^\lambda)$. Note that this matrix is anti-symmetric because V is Killing. Hence the Riemannian equivariant curvature is simply

$$\Omega_\nabla^{so(n)}(V) = \Omega_\nabla - \nabla V. \tag{3.8}$$

The equivariant Â-genus $\hat{A}(\Omega_\nabla^{so(n)})$ of the tangent bundle appears in the localization theorem of [47] which we will discuss below.

This is as far as we will follow the general story. In our applications we will be interested in situations where we have a single Killing vector field $V_X \equiv V$ on X, perhaps corresponding to an action of $G = S^1$. This means that we will be considering equivariant differential forms $\mu(V)$ evaluated on a fixed $V \in \mathfrak{g}$. In those cases, nothing is gained by carrying around $\mathbb{C}[\mathfrak{g}]$ and we will simply be considering the operator $d(V) = d - i(V)$ on $\Omega^*(X)$. By the Cartan formula, the square of this operator is (minus) the Lie derivative $L(V)$ along V,

$$(d(V))^2 = (d - i(V))^2 = -(i(V)d + di(V)) = -L(V), \tag{3.9}$$

so that it leaves invariant the space $\Omega_V^*(X)$ of V-invariant forms on X (the kernel of $L(V)$) and squares to zero there. Thus it makes sense to consider the cohomology of $d(V)$ on $\Omega_V^*(X)$ and we will call forms on X satisfying $d(V)\alpha = 0$ $d(V)$-closed or equivariantly closed etc. The final observation we will need is that if a differential form is $d(V)$-exact, $\alpha = d(V)T$, then its top-form component is actually exact in the ordinary sense. This is obvious because the $i(V)$-part of $d(V)$ lowers the form-degree by one so there is no way that one can produce a top-form by acting with $i(V)$.

The Berline–Vergne and Duistermaat–Heckman localization formulae. We now come to the localization formulae themselves. We will see that integrals over X of $d(V)$-closed forms localize to the zero locus of the vector field V. There are two simple ways of establishing this localization and as each one has its merits we will present both of them. It requires a little bit more work to determine the precise contribution of each component of X_V to the integral, and here we will only sketch the required calculations and quote the result.

The essence of the localization theorems is the fact that equivariant cohomology is determined by the fixed point locus of the G-action. The first argument for localization essentially boils down to an explicit proof of this fact at the level of differential forms. Namely, we will show that any equivariantly closed form α, $d(V)\alpha = 0$, is equivariantly exact away from the zero locus $X_V = \{x \in X : V(x) = 0\}$ of V. To see this, we will construct explicitly a differential form β on $X \setminus X_V$ satisfying $d(V)\beta = \alpha$. As this implies that the top-form component of α is exact, it then follows from Stokes theorem that the integral $\int_X \alpha$ only receives contributions from an arbitrarily small neighborhood of X_V in X.

It is in the construction of β that the condition enters that V be a Killing vector. Using the invariant metric h we can construct the metric dual one-form $h(V)$. It satisfies $d(V)h(V) = d(h(V)) - |V|^2$ and (since V is a Killing vector) $L(V)h(V) = 0$. Away from X_V, the zero-form part of $d(V)h(V)$ is non-zero and hence $d(V)h(V)$ is invertible. Here the inverse of an inhomogeneous differential form with non-zero scalar term is defined by analogy with the formula $(1+x)^{-1} = \sum_k (-x)^k$. We now define the (inhomogeneous) differential form Θ by

$$\Theta = h(V)(d(V)h(V))^{-1}. \tag{3.10}$$

It follows immediately that

$$d(V)\Theta = 1, \qquad L(V)\Theta = 0. \tag{3.11}$$

Hence, 1 is equivariantly exact away from X_V and we can choose $\beta = \Theta\alpha$,

$$d(V)\alpha = 0 \qquad \Rightarrow \qquad \alpha = (d(V)\Theta)\alpha = d(V)(\Theta\alpha) \text{ on } X \setminus X_V, \tag{3.12}$$

so that any equivariantly closed form is equivariantly exact away from X_V. In particular, the top-form component of an equivariantly closed form is exact away from X_V.

Let us see what this amounts to to in the example (3.4) of the introduction to this section (cf. also (2.17,3.47)). Using the symplectic form $\omega = \sin\theta d\theta d\phi$, the Hamiltonian vector field corresponding to $\cos\theta$ is $V = \partial_\phi$. This vector field generates an isometry of the standard metric $d\theta^2 + \sin^2\theta d\phi^2$ on S^2, and the corresponding Θ is $\Theta = -d\phi$. This form is, as expected, ill defined at the two poles of S^2. Now the integral (3.4) can be written as

$$\int_{S^2} \omega e^{itf} = (it)^{-1} \int_{S^2} d(e^{itf} d\phi), \tag{3.13}$$

thus receiving contributions only from the critical points $\theta = 0, \pi$, the endpoints of the integration range for $\cos\theta$, in agreement with the explicit evaluation.

Alternatively, to establish localization, one can use the fact that the integral $\int_X \alpha$ of a $d(V)$-closed form only depends on its cohomology class $[\alpha]$ to evaluate the integral using a particularly suitable representative of $[\alpha]$ making the localization manifest. Again with the help of an invariant metric, such a representative can be constructed. Consider the inhomogeneous form

$$\alpha_t = \alpha e^{td(V)h(V)}, \qquad t \in \mathbb{R}. \tag{3.14}$$

Clearly, α_t is cohomologous to α for all t. Equally clearly, as $d(V)h(V) = -|V|^2 + d(h(V))$, the form α_t is increasingly sharply Gaussian peaked around $X_V \subset X$ as $t \to \infty$. Hence, evaluating the integral as

$$\int_X \alpha = \lim_{t\to\infty} \int_X \alpha_t \tag{3.15}$$

reestablishes the localization to X_V.

Perhaps it is good to point out here that there is nothing particularly unique about the choice α_t to localize the integral of α over X. Indeed, instead of $h(V)$ one could choose some other $L(V)$-invariant form β in the exponent of (3.14) and try to evaluate $\int_X$ as some limit of

$$\int_X \alpha = \int_X \alpha e^{td(V)\beta}. \tag{3.16}$$

This can potentially localize the integral to something other than X_V (we will see an example of this in the context of path integrals later on), and so lead to a seemingly rather different expression for the integral. In principle this argument for localization could also work without the assumption that V is a Killing vector, but it appears to be difficult to make any general statements in that case. However, as everything in sight is clearly $d(V)$-closed, it is possible to reduce the resulting expression further to X_V by applying the above localization arguments once more, now to the localized expression.

One *caveat* to keep in mind is that the above statements require some qualifications when the manifold X is not compact. In that case, t-independence of the rhs of (3.16) is only ensured if the asymptotic behavior of α is not changed by the replacement $\alpha \to \alpha \exp d(V)\beta$. For a version of the Duistermaat–Heckman formula for non-compact manifolds see [55]. The extension to non-compact groups is also not immediate. For example [4], consider the nowhere vanishing vector field

$$V = (1 + \tfrac{1}{2}\sin x)\partial_y \tag{3.17}$$

on the two-torus ($x, y \in \mathbb{R}/2\pi\mathbb{Z}$). Then one can easily check that

$$\alpha(V) = -\tfrac{1}{2}(7\cos x + \sin 2x) + (1 - 4\sin x)dxdy \tag{3.18}$$

is equivariantly closed. Its integral over the torus, on the other hand, is equal to $(2\pi)^2$, so that the two-form component of $\alpha(V)$ cannot possibly be exact.

Another ambiguity can arise when α is equivariantly closed with resepct to two vector fields V and W, $d(V)\alpha = d(W)\alpha = 0$. In this case one can get seemingly different expressions by localizing to either X_V or X_W or $X_V \cap X_W$. Of course, in whichever way one chooses to evaluate the integral, the results are guaranteed to agree, even if not manifestly so.

Having established that the integral will indeed localize, to obtain a localization formula one needs to determine the contributions to the integral from the connected components of X_V. This is most readily done when X_V consists of isolated points x_k, $V(x_k) = 0$. In that case the Lie derivative $L(V)$ induces invertible linear transformations $L_V(x_k)$ on the tangent spaces $T_{x_k}X$. One can introduce local coordinates in a G-invariant neighborhood of these points such that the localized integrals become essentially Gaussian. Then one finds that the contribution of x_k to the integral is just $(-2\pi)^l \det(L_V(x_k))^{-1/2}$. Restricting the equivariantly closed form α to X_V, only the scalar part $\alpha^{(0)}(x_k)$ survives and one finds the Berline–Vergne localization formula

$$d(V)\alpha = 0 \quad \Rightarrow \quad \int_X \alpha = (-2\pi)^l \sum_{x_k \in X_V} \det(L_V(x_k))^{-1/2} \alpha^{(0)}(x_k). \tag{3.19}$$

If X_V has components of non-zero dimension, things become a little bit more complicated. In that case, in the above discussion the tangent space $T_{x_k}X$ has to be replaced by the normal bundle N to X_V in X and $L_V(x_k)$ by its equivariant curvature (3.7). As $V=0$ on X_V, the scalar part of (3.7) reduces to $L^N(V)$, thought of as an endomorphism of N. The localization formula one obtains in this case is then

$$d(V)\alpha = 0 \quad \Rightarrow \quad \int_X \alpha = \int_{X_V} \det(L^N(V)+\Omega_\nabla)^{-1/2}\alpha|_{X_V} \tag{3.20}$$

(a sum over the components of X_V being understood). As pointed out before, the form appearing in the denominator of this formula can be regarded as the equivariant Euler form of the normal bundle N. We refer to [4] and [56] for details and some of the beautiful applications of this formula.

We will now consider a particular case of the Berline–Vergne formula. Namely, let (X,ω) be a symplectic manifold and assume that there is a Hamiltonian action of G on X. This means that the vector fields V_X generating the action of G on X are Hamiltonian. That is, there is a moment map $\mu : X \to \mathfrak{g}^*$ such that $\langle \mu, V\rangle \equiv \mu(V) \in C^\infty(X)$ generates the action of G on X via the Hamiltonian vector fields V_X. Since ω is closed, the defining equation

$$i(V_X)\omega = d(\mu(V)) \tag{3.21}$$

implies that $\mu+\omega$ is an equivariantly closed form,

$$d_{\mathfrak{g}}(\mu+\omega)(V) = d(\mu(V)) - i(V_X)\omega = 0. \tag{3.22}$$

In fact, finding an equivariantly closed extension of ω is equivalent to finding a moment map for the G-action [51].

If ω is integral, it can be regarded as the curvature of a line bundle E on (M,ω), the prequantum line bundle of geometric quantization[57]. In that case, the equivariant curvature form as defined in (3.7) agrees with the equivariant extension of ω given above. Indeed in geometric quantization it is well known that the 'prequantum operator' $L^E(V)$ can be realized as $L^E(V) = \nabla_V + \mu(V)$.

To produce from this an equivariantly closed form on X which we can integrate (which has a top-form component), we consider the form $\exp it(\mu(V)+\omega)$ which is annihilated by $d(V)$. Its integral is

$$\begin{aligned}\int_X e^{it(\mu(V)+\omega)} &= \int_X \frac{(it\omega)^l}{l!} e^{it\mu(V)} \\ &= (it)^l \int_X dx(\omega) e^{it\mu(V)}.\end{aligned} \tag{3.23}$$

Assuming that the zeros of V_X are isolated, we can apply the Berline–Vergne formula (3.19) to obtain

$$\int_X dx(\omega) e^{it\mu(V)} = (it)^{-l}(-2\pi)^l \sum_{x_k \in X_V} \det(L_V(x_k))^{-1/2} e^{it\mu(V)(x_k)}. \tag{3.24}$$

As the critical points of $\mu(V)$ are precisely the zeros of V_X (ω is non-degenerate) this is the Duistermaat–Heckman formula (3.3) if V is a generator of a circle action for some $U(1) \subset G$. For an example we refer back to the classical spin system discussed in the introduction to this section. Of course there is a corresponding generalization of the Duistermaat–Heckman formula for the case that the zeros of V_X are not isolated which follows from applying (3.20) to (3.23).

3.2. Localization Formulae for Phase Space Path Integrals

In quantum mechanics there are not too many path integrals that can be evaluated explicitly and exactly, while the semi-classical approximation can usually be obtained quite readily. It is therefore of obvious interest to investigate if there is some path integral analogue of the Duistermaat–Heckman and Berline–Vergne formulae.

One class of quantum mechanics models for which the Duistermaat–Heckman formula clearly seems to make sense is $N = \frac{1}{2}$ supersymmetric quantum mechanics. We will come back to this in section 3.3. What one would really like, however, is to have some version of the equivariant localization formulae available which can be applied to non-supersymmetric models and when the partition functions cannot be calculated directly (or only with difficulty) by some other means. One non-trivial field theoretic example in which (a non-Abelian version of) the Duistermaat–Heckman theorem has been applied successfully is two-dimensional Yang–Mills theory [52]. This theory has a natural symplectic interpretation because the space of gauge fields in two dimensions is symplectic and the Yang–Mills action is just the square of the moment map generating gauge transformations of the gauge fields. Hence one is in principle in the right framework to apply equivariant localization. We will sketch an Abelian version of the localization formula for the square of the moment map in section 3.3.

A large class of examples where one also has an underlying equivariant cohomology which could be responsible for localization is provided by phase space path integrals, i.e., the direct loop space analogues of the lhs of (3.3). Now phase space path integrals are of course notoriously awkward objects, and it is for this reason that the localization formulae we will obtain in this way should not be regarded as definite predictions but rather as suggestions for what kind of results to expect. Because of the lack of rigour that goes into the derivation of these localization formulae it is perhaps surprising that nevertheless some of the results that have been obtained are not only conceptually interesting but also physically reasonable.

Equivariant cohomology for phase space path integrals. In this section we choose our manifold X to be the loop space $X = LM$ (see section 2.2) of a finite-dimensional symplectic manifold (M, ω), the phase space of a classical system. LM is also a symplectic manifold in the sense that $\hat{\omega}$ (cf. (2.25)) defines a closed and non-degenerate two-form on LM. We denote by $S = S[x]$ the phase space action and by H its Hamiltonian. Thus, if we denote by θ a local symplectic potential for ω, $d\theta = \omega$, then the action is essentially of the form

$$S[x] = \int_0^T dt\, (\theta_\mu(x(t))\dot{x}^\mu(t) - H(x(t))). \tag{3.25}$$

Of course, when ω is not globally exact this has to be suitably defined by regarding the first term as a Wess–Zumino like term for ω, but this is a standard procedure which will be tacitly adopted in the following.

Consider now the phase space path integral

$$Z(T) = \int_{LM} [dx(\hat{\omega})] e^{\frac{i}{\hbar} S[x]}. \tag{3.26}$$

As in (3.23), we will lift the symplectic form into the exponent which will then take the form $S + \hat{\omega}$. Clearly, therefore, the integrand is equivariantly closed with respect to the flow generated on the loop space by the Hamiltonian vector field V_S of the action S,

$$d(V_S)(S + \hat{\omega}) = 0. \tag{3.27}$$

The zeros of this vector field are precisely the critical points of the action, i.e., the classical solutions. This is the reason why this set-up has the potential to produce WKB-like localization formulae [42, 43].

Let us make the above formulae a little bit more explicit. First of all, if we denote by H the Hamiltonian corresponding to S, then the components of V_S can be written as

$$V_S^\mu(x(t)) = \dot{x}^\mu(t) - \omega^{\mu\nu}(x(t))\partial_\nu H(x(t)), \tag{3.28}$$

so that clearly $V_S(x) = 0$ when $x(t)$ satisfies the classical equations of motion. The first term is just the canonical vector field $V = \dot{x}$ on LM, while the second is the Hamiltonian vector field V_H of H on M, regarded as a vector field on LM.

Using also the trick of section 2 (2.14) to replace the loop space one-forms $dx(t)$ by anticommuting variables $\psi(t)$, we can write the partition function (3.26) as

$$Z(T) = \int_{LM}[dx][d\psi]e^{\frac{i}{\hbar}(S+\hat{\omega}(\psi))}. \tag{3.29}$$

Then the fact (3.27) that $S+\hat{\omega}$ is equivariantly closed translates into the statement that the augmented action $S[x,\psi]$ appearing in (3.29) is invariant under the supersymmetry

$$\delta x^\mu = \psi^\mu, \qquad \delta\psi^\mu = -V_S^\mu \qquad \Rightarrow \qquad \delta(S+\hat{\omega}(\psi)) = 0. \tag{3.30}$$

We will mostly set $\hbar = 1$ in the following.

Localization formulae. The above is of course not yet sufficient to establish that (formally) the path integral localizes. As we know from the previous section, at the very least we need a metric on LM with respect to which V_S is Killing. To investigate when this is the case, we choose a metric $\hat{g}$ on LM of the ultra-local form (2.24). As any such metric is invariant under $V = \dot{x}$, $L(V)\hat{g} = 0$, the condition $L(V_S)\hat{g} = 0$ reduces to the condition that the Hamiltonian vector field V_H of H on M be a Killing vector of the metric g,

$$L(V_S)\hat{g} = 0 \quad \Leftrightarrow \quad L(V_H)g = 0. \tag{3.31}$$

This is exactly the same condition we encountered in the finite-dimensional case. For these examples, then, for which there are localization formulae for the classical partition function, can we hope to find analogous formulae for the quantum partition function as well.

We now proceed as in the finite-dimensional case and localize the integral by adding some appropriate $d(V_S)$-exact term $d(V_S)\beta$ with $L(V_S)\beta = 0$ to the action. First of all we need to show that this does not change the value of the path integral. In the finite dimensional case this (or the t-independence of integrals like (3.15,3.16)) relied, at least implicitly, on Stokes' theorem which is not directly available for integrals over LM. However, instead of that one has a kind of Stokes' theorem in the form of a Ward identity associated with the supersymmetry δ (or, in more elementary terms, a change of variables argument [43]) to reach the desired conclusion provided that the supersymmetry is non-anomalous. We will assume this and proceed with fingers crossed.

The obvious candidate for β is, as in the finite dimensional case, the metric dual one-form $\hat{g}(V_S)$ of the Hamiltonian vector field V_S itself [43] which, by our assumption $L(V_H)g = 0$, satisfies $L(V_S)\hat{g}(V_S) = 0$. This choice will formally lead to a localization onto the critical points of S and hence onto classical trajectories $x_c(t)$. If these are isolated and non-degenerate this will lead to the exact analogue of the Duistermaat–Heckman formula (3.3), namely

$$Z(T) \sim \sum_{x_c(t)} \det[L_S(x_c)]^{-1/2} e^{\frac{i}{\hbar}S[x_c]}. \tag{3.32}$$

Here the determinant is essentially the functional determinant of the Jacobi operator (the Hessian of the action). This formula can be interpreted as saying that the WKB approximation to the path integral is exact, the only difference to the usual WKB approximation being that here one is to sum over all critical points of the action and not just the local minima.

The validity of (3.32) has been well investigated in the path integral literature (although more for configuration space path integrals), and we have little to add to that discussion. We only want to point out that in those examples we know of where the semi-classical approximation is known to be correct, the assumptions that went into the derivation of (3.32) here, like the existence of an invariant phase space metric, are indeed satisfied (see e.g., [58, 59, 44] for the propagator of a particle moving on a group manifold).

Nevertheless, (3.32) should be taken with a grain of salt. First of all, as (3.32) is a sum over classical periodic trajectories with period T, these will typically either occur only at the critical points of the Hamiltonian (in which case (3.32) resembles even more closely the classical Duistermaat–Heckman formula) or on a non-zero dimensional submanifold of M (in which case one should really use the degenerate version of the Duistermaat–Heckman formula). Moreover, more care has to be exercised (already in the finite dimensional case) when one is dealing with non-compact phase spaces, which is the rule in systems of physical interest.

However, even if this formula or its obvious modifications are not correct as they stand, it would be interesting to uncover the reason for that. At the very least this could then provide one with a systematic geometric method for analyzing corrections to the WKB approximation. We are not aware of any work that has been done in this direction yet.

One of the strengths of the present setting is the ability to produce other, different, localization formulae, with perhaps different ranges of validity, by exploiting the flexibility in the choice of β. For example, one can easily convince oneself that the two terms of $\hat{g}(V_S)$ (3.28), namely $\hat{g}(V=\dot{x})$ and $\widehat{g(V_H)}$, are separately invariant under V_S so that one can choose β to be a general linear combination of them [45, 46],

$$\beta = s\hat{g}(\dot{x}) - r\widehat{g(V_H)}. \tag{3.33}$$

For $s=r$ this reduces to what we did above. But, by choosing e.g., $r=0$ and $s\to\infty$ one can localize the path integral (3.29) to an integral over the classical phase space M. Alternatively, by choosing $s=0$ and $r\to\infty$, one can obtain an expression for $Z(T)$ in terms of the critical points of the Hamiltonian. We will only consider the first of these possibilities here, as it leads to a nice expression in terms of equivariant characteristic classes. The action one obtains in this way is

$$\begin{aligned} S^s[x,\psi] &= S[x] + \hat{\omega}(\psi) + sd(V_S)\hat{g}(\dot{x}) \\ &= S[x] + \hat{\omega}(\psi) + sd(\hat{g}(\dot{x}))(\psi) - s\hat{g}(\dot{x},\dot{x}) + s\hat{g}(\dot{x},V_H) \\ &= S[x] + \hat{\omega}(\psi) + s\int_0^T dt\; \psi(t)\nabla_t\psi(t) \\ &\quad + g(\dot{x}(t),\dot{x}(t)) - g(\dot{x}(t),V_H(x(t))) \end{aligned} \tag{3.34}$$

As in (2.27), ∇_t denotes the covariant derivative along the loop $x(t)$ induced by the Riemannian connection ∇ on M. This action can be evaluated as in supersymmetric quantum mechanics. One possibility is to expand all fields in Fourier modes and to scale the non-constant modes by $s^{-1/2}$ so as to eliminate positive powers of s from the action. Then the limit $s \to \infty$ can be taken with impunity, the integration over the non-constant modes gives rise to a ratio of determinants and in the exponent one is left with the contribution from the constant modes, namely $T(\omega(\psi) - H)$. The result one obtains is

$$Z(T) = \int_M dx\, d\psi \det\left[\frac{\frac{T}{2}(\Omega_\nabla + \nabla V_H)}{\sinh[\frac{T}{2}(\Omega_\nabla + \nabla V_H)]}\right]^{1/2} e^{iT(\omega(\psi) - H)} \tag{3.35}$$

(see (3.42) and (3.43) in section 3.3 for a brief explanation of the appearance of this particular determinant in supersymmetric path integrals). This is the Niemi–Tirkkonen localization formula [47]. In this expression we recognize the $d(-V_H)$-equivariant Riemannian curvature (3.8) and its Â-genus (the occurrence of minus signs at awkward places is unavoidable in symplectic geometry...). The exponent can also be interpreted in equivariant terms. Namely, as we have seen in the discussion of the Duistermaat–Heckman theorem, as the $(-V_H)$-equivariant extension of the symplectic form ω. Roughly speaking (modulo the fact that ω need not be integral) the exponential can then be regarded as an equivariant Chern character and the result can then be written more succinctly and elegantly in the notation of (3.6) as

$$Z(T) = \left(\int_M \hat{A}(T\Omega_\nabla^{so(n)})\mathrm{Ch}(T\omega^{so(n)})\right)(-V_H). \tag{3.36}$$

This differs from the classical partition function for the dynamical system described by H by the Â-term which can be thought of as encoding the information due to the quantum fluctuations.

As the integrand is clearly $d(-V_H)$-closed, this integral can be localized further to an integral over the critical points of H provided that the zero locus of V_H is non-degenerate. The possible advantage of the present formula is that no such assumption appears to have been necessary in the derivation of (3.35) which thus potentially applies to examples where the ordinary WKB approximation breaks down (e.g., when classical paths coalesce).

This possibility has been analyzed in [48] for some simple one-dimensional quantum mechanics examples, the harmonic oscillator and the 'hydrogen atom', i.e., a particle moving in a $1/|x|$-potential. The latter example, in particular, is interesting because there the classical paths are known to coalesce so that the traditional WKB formula is not applicable. The results obtained in [48] as well as some further considerations in [49] illustrate the potential usefulness of these generalized localization formulae. However, in order to establish to what extent formulae like the above are trustworthy and can be turned into reliable calculational tools, other (higher dimensional) examples will need to be worked out.

3.3. Other Examples and Applications — an Overview

In this section we shall sketch some of the other applications of equivariant localization and, in particular, the Duistermaat–Heckman formula. Most prominent among these are applications to supersymmetric quantum mechanics and index theorems.

Supersymmetric quantum mechanics and index theorems. As mentioned in the introduction to this section, the first infinite dimensional application of the Duistermaat–Heckman theorem is due to Atiyah and Witten [41] (see also [19] and [60]) who applied it to the loop space LM of a Riemannian manifold and $N=\frac{1}{2}$ supersymmetric quantum mechanics. This is the model which represents the index of the Dirac operator[23, 24, 25] in the same way that the $N=1$ model we discussed in section 2.2 represents the index of the de Rham complex. Just like the Mathai–Quillen formalism provides the appropriate framework for understanding the topological and localization properties of $N=1$ (de Rham) supersymmetric quantum mechanics, those of $N=\frac{1}{2}$ (Dirac) supersymmetric quantum mechanics find their natural explanation within the framework of loop space equivariant localization. And, just as in the case of the Mathai–Quillen formalism, this way of looking at the $N=\frac{1}{2}$ models provides some additional insights and flexibility in the evaluation of the path integrals[46].

Roughly speaking, the action of the $N=\frac{1}{2}$ model can be obtained from that of the $N=1$ model (2.27) by setting $\psi=\bar{\psi}$. Then the four-fermi curvature term drops out by the Bianchi identity $R_{\mu[\nu\rho\sigma]}=0$ and (with a suitable rescaling of the fermi fields) one is left with

$$S_M[x,\psi]=\int_0^\beta dt\, g_{\mu\nu}\left(\dot{x}^\mu(t)\dot{x}^\nu(t)-\psi^\mu(t)\nabla_t\psi^\nu(t)\right). \tag{3.37}$$

This action has the supersymmetry

$$\delta x^\mu(t)=\psi^\mu(t),\qquad \delta\psi^\mu(t)=-\dot{x}^\mu(t)\qquad\Rightarrow\qquad \delta S_M[x,\psi]=0, \tag{3.38}$$

which we can now recognize immediately as the action of the equivariant exterior derivative $d(V=\dot{x})=d-i(\dot{x})$ on LM. Hence the action of $N=\frac{1}{2}$ supersymmetric quantum mechanics defines an equivariant differential form on LM and on the basis of the general arguments we expect its integral to localize to an integral over M, the zero locus of $\dot{x}$. This is of course well known to be the case [23, 24, 25].

One new feature of this model is that the action is actually equivariantly exact,

$$S_M=d(V)\hat{g}(V)\qquad\Leftrightarrow\qquad S_M[x,\psi]=\delta\int_0^\beta dt\, g_{\mu\nu}\dot{x}^\mu(t)\psi^\nu(t), \tag{3.39}$$

so that the partition function will not depend on the coefficient of the action (and thus manifestly localizes onto $\dot{x}=0$). As one can think of this coefficient as $\hbar$, this is another way of seeing that the semi-classical approximation to this model is exact. Nevertheless one can of course not simply set the coefficient to zero as the integral is ill-defined in this case (infinity from the x-integral times zero from the ψ-integral).

There are now at least three different ways of calculating the partition function

$$Z[M]=\int[dx][d\psi]e^{i\int_0^\beta dt\, g_{\mu\nu}\left(\dot{x}^\mu(t)\dot{x}^\nu(t)-\psi^\mu(t)\nabla_t\psi^\nu(t)\right)}. \tag{3.40}$$

The traditional method is to make use of the presumed β-independence of the theory to evaluate Z in the limit $\beta\to 0$ using a normal coordinate expansion [23, 24, 25]. One can also apply directly the Berline–Vergne formula (3.20), valid when the zero locus is not zero-dimensional. In that case one has to calculate the equivariant Euler form of the normal bundle N to M in LM (this is the bundle spanned by the non-constant modes) and evaluate the determinant appearing in (3.20) using e.g., a zeta-function prescription. This was the approach adopted in [41]. Finally, one can scale the action (3.39) by some parameter s, scale the non-constant modes of x and ψ by $s^{-1/2}$ and take the limit $s \to \infty$. The remaining integral is then Gaussian. In whichever way one proceeds one obtains the result

$$Z[M] = \int_M \hat{A}(M), \tag{3.41}$$

which is the index of the Dirac operator on M if M is a spin-manifold.

The reason for the ubiquitous appearance of the Â-character in supersymmetric quantum mechanics path integrals is, that it arises whenever one calculates the determinant of a first-order differential operator on the circle. Let $D_a = \partial_t + a$ be such an operator. We can without loss of generality assume that a is constant as the non-constant modes of a could always be removed by conjugating the entire operator by $\exp if(t)$ for some function f, an operation that does not change the determinant of D_a. Acting on periodic functions on S^1, the eigenvalues of D_a are $a + 2\pi in$ for $n \in \mathbb{Z}$ (and $n \neq 0$ if one excludes the constant mode). The determinant of D_a is now formally defined as the product of the eigenvalues. Of course, this requires some regularization and a suitable prescription is zeta-function regularization. Then the determinant $\det' D_a$ over the non-constant modes can be written as

$$\begin{aligned} \det' D_a &= \prod_{n\neq 0}(2\pi in + a) = \prod_{n>0}(a^2 + (2\pi n)^2) \\ &= (\prod_{n>0}(2\pi n)^2)\prod_{n>0}(1 + a^2/(2\pi n)^2) = \prod_{n>0}(1 + a^2/(2\pi n)^2), \end{aligned} \tag{3.42}$$

as formally the infinite prefactor is equal to one by zeta-function regularization. The function defined by the infinite product in the last line has zeros for $a \in 2\pi i\mathbb{Z}$, $a \neq 0$, and is equal to one at $a = 0$. It is nothing other than the function $(\sinh a/2)/(a/2)$, so that we can write

$$\det' D_a = \frac{\sinh a/2}{a/2} = \hat{A}(a)^{-1}, \tag{3.43}$$

where we define the function $\hat{A}(x) = (x/2)/\sinh(x/2)$.

This can of course be generalized in various ways to include the coupling of fermions to a gauge field on a vector bundle E, yielding the result

$$Z[M,E] = \int_M \mathrm{Ch}(E)\hat{A}(M). \tag{3.44}$$

The actual calculations involved in these three different approaches are practically identical and it is really only in the interpretation of what one is doing that they differ. However, there seems to be one instance where method three appears to be superior to method one [61], namely when one is dealing with odd-dimensional open-space index theorems like that of Callias and Bott. In that case it is simply not true that the partition function is independent of β (the index is obtained for $\beta \to \infty$) so that one cannot simplify matters by going to the $\beta \to 0$ limit. In [61] it was shown that the correct result can be obtained by introducing the parameter s and calculating the partition function as $s \to \infty$.

A localization formula for the square of the moment map. In [52], Witten introduced a new non-Abelian localization formula for finite dimensional integrals and applied it to path integrals. He was able in this way to deduce the intersection numbers of the moduli space of flat connections on a two-surface Σ from the solution of Yang–Mills theory on Σ.

Subsequently, certain cases of this localization formula have been derived rigorously by Jeffrey and Kirwan[54] and Wu [53].

In its simplest version, this formula applies to integrals over symplectic manifolds of the form

$$Z(\varepsilon)=(2\pi\varepsilon)^{-m/2}\int_X e^{-I/2\varepsilon+\omega}, \tag{3.45}$$

where $I=(\mu,\mu)$ is the square with resepct to some invariant scalar product on $\mathfrak{g}^*$ of the moment map $\mu: X\to\mathfrak{g}^*$ of a Hamiltonian G-action on M and $\dim G=m$. By introducing an auxiliary integral over $\mathfrak{g}$, (3.45) can be put into a form very similar to the Duistermaat–Heckman integral (3.3,3.24), namely

$$Z(\varepsilon)=(2\pi)^{-m}\int_{\mathfrak{g}} d\phi e^{-\varepsilon(\phi,\phi)/2}\int_M e^{\omega+i(\mu,\phi)}. \tag{3.46}$$

Here we recognize again the equivariant extension of the symplectic form (the factor of i is irrelevant). Hence, as the integrand is equivariantly closed, one can again attempt to localize it by adding a $d_{\mathfrak{g}}$-exact term $d_{\mathfrak{g}}\beta$ to the exponent, where $\beta\in\Omega_G^*(X)$. A reasonably canonical choice for β is $\beta=J(dI)/2$, where J is some positive G-invariant almost complex structure on X. In that case it can be shown that the integral localizes to the critical points of I. These are either critical points of μ, as in the Duistermaat–Heckman formula, or zeros of μ. The latter are a new feature of this localization theorem and are interesting for a number of reasons. For one, as the absolute minima of I they give the dominant contribution to the integral in the 'weak coupling' limit $\varepsilon\to 0$, the contributions form the other points being roughly of order $\exp(-1/\varepsilon)$. Furthermore, the contribution from $\mu=0$ is, by G-equivariance, related to the Marsden–Weinstein reduced phase space (or symplectic quotient) $X//G=\mu^{-1}(0)/G$. Hence, integrals like (3.45) can detect the cohomology of $X//G$.

This is of interest in 2d Yang–Mills theory, where $X=\mathcal{A}$ is the space of gauge potentials A, and $\mu(A)=F_A$ is the curvature of A. The Yang–Mills action is therefore just the square of the moment map and the symplectic quotient is the moduli space of flat connections.

In general, the contribution from the non-minimal critical points of I can be quite complicated, even for G Abelian. In that case the integral over M in (3.46) can also be evaluated via the Duistermaat–Heckman formula and a comparison of the expressions one obtains by following either route has been performed in [53]. To illustrate the complicated structure that arises, the example of the spin system (3.4,3.13) will suffice. In that case we have to study the integral

$$\begin{aligned}Z(\varepsilon)&=(2\pi)^{-1}\int_{-\infty}^{\infty} d\phi e^{-\varepsilon\phi^2/2}\int_X e^{\omega+i\phi(\cos\theta+a)}\\&=(2\pi/\varepsilon)^{1/2}\int_{-1}^{1} dx e^{-(x+a)^2/2\varepsilon}.\end{aligned} \tag{3.47}$$

For $|a|<1$ this can be written as

$$Z(\varepsilon)=2\pi(1-I_+-I_-), \tag{3.48}$$

where

$$I_\pm=\pm(2\pi\varepsilon)^{-1/2}\int_{\pm 1}^{\pm\infty} dx e^{-(x+a)^2/2\varepsilon}. \tag{3.49}$$

These terms correspond to the contributions from the three critical points of $I=\mu^2$, the absolute minimum at $\cos\theta=-a$ contributing the simple first term, and the other two contributions coming from the critical points of μ at $\theta = 0$ and $\theta = \pi$. The appearance of the error function in this example is in marked contrast with the elementary functions that appear as the contributions from the critical points in the Duistermaat–Heckman formula. The above integral should be compared with the (closely related) integral (2.17) which we discussed in the context of the Mathai–Quillen formalism. There is much more that should be said about these localization formulae, but for this we refer to [52].

Character Formulae and other applications. In this section we will just mention some other applications of the Duistermaat–Heckman formula and other equivariant localization theorems in the physics literature. We have already mentioned the path integral derivation of the Weyl character formula by Stone[40] and the application of the Duistermaat–Heckman theorem to the particle on a group manifold by Picken [44] (recovering old results by Schulman [58] and Dowker [59] in this way). In a similar setting, the action invariant of Weinstein [62], an invariant probing the first cohomology group of the symplectomorphism group of a symplectic manifold, has been related to what is known as Chern–Simons quantum mechanics using the Duistermaat–Heckman formula in [42]. The derivation of Stone has been generalized by Perret [63] to the Weyl–Kac character formula for Kac–Moody algebras.

Because of their classical properties, coherent states are particularly well suited for studying semi-classical properties of quantum systems, and in [50] the predictions of the Duistermaat–Heckman theorem have been verified for coherent state path integrals associated with $SU(2)$ and $SU(1,1)$. A very careful analysis of the WKB approximation for these coherent state path integrals has recently been performed in [64].

It is a longstanding conjecture that the quantum theories of classically integrable systems are given approximately by their semi-classical approximation. An application of the finite-dimensional Duistermaat–Heckman formula to certain integrable models can be found in [65] and some suggestive formulae for the phase space path integrals of integrable models have been obtained in [66].

Finally, we have recently found [67] that a localization formula of Bismut[68] for equivariant Kähler geometry has a field theoretic realization in the G/G gauged Wess–Zumino–Witten model and can be used to shed some light on and give an alternative derivation to [69] of the Verlinde formula from the G/G model.

4. GAUGE INVARIANCE AND DIAGONALIZATION — THE WEYL INTEGRAL FORMULA

In this section we discuss a technique for solving or simplifying path integrals which is quite different in spirit to those we encountered in sections 2 and 3. Both the Mathai–Quillen formalism and the Duistermaat–Heckman and Berline–Vergne localization theorems are fundamentally cohomological in nature and can be understood in terms of a supersymmetry allowing one to deform the integrand without changing the integral. The technique we will discuss here, on the other hand, requires not a supersymmetry but an ordinary non-Abelian symmetry of the integrand (as in non-Abelian gauge theories), the idea being to reduce such an integral to one with a (much more tractable) Abelian symmetry.

More precisely, the classical integration formula, which we wish to generalize to functional integrals, is what is known in group theory and harmonic analysis as the Weyl integral formula. To state this formula, we need some notation (and refer to the next section for details). We denote by G a compact Lie group and by T a maximal torus of G. As every element of

G is conjugate to some element of T (in other words, every $g \in G$ can be 'diagonalized'), a conjugation invariant function f on G, $f(g) = f(h^{-1}gh)$ for all $h \in G$, is determined by its restriction to T. In particular, therefore, the integral of f over G can be expressed as an integral over T, and the Weyl integral formula gives an expression for the integrand on T,

$$\int_G dg f(g) = \int_T dt \det \Delta_W(t) f(t). \tag{4.1}$$

Here $\Delta_W(t)$ is the Weyl determinant whose precise form we will give below. One can read this formula as expressing the fact that an integral of a function with a non-Abelian (conjugation) symmetry can be reduced to an integral of a function with an Abelian symmetry. In physics parlance one would say that one has integrated over the G/T part of the gauge volume or partially fixed the gauge using an abelianizing gauge condition.

It is this formula that we wish to generalize to functional integrals, i.e., to integrals over spaces of maps $\mathrm{Map}(M,G)$ from some manifold M into G. What is interesting about this generalization is the fact that the naive extension of (4.1) to this case is definitely wrong for basic topological reasons. This is in marked contrast with the other path integral formulae we have discussed above which appear to correctly capture the topological aspects of the situation without any modifications. The correct formula in this case (correct in the sense that the topology comes out right) turns out to include a summation over isomorphism classes of non-trivial T-bundles on M on the right hand side of (4.1), so that (very roughly) one has

$$\text{``}\int_{\mathrm{Map}(M,G)} [dg] F[g] = \sum_{T-\text{bundles}} \int_{\mathrm{Map}(M,T)} [dt] \det \Delta_W[t] F[t]\text{''} \tag{4.2}$$

(see (4.21) for a less outrageous rendition of this formula). To see how this comes about, suffice it to note here that in order to apply (4.1) to spaces of maps, one first needs to establish that maps from M to G can be diagonalized. It turns out that there are topological obstructions to doing this globally (again, in physics parlance, to choose this abelianizing gauge globally), and the summation over topological sectors reflects this fact.

The classical finite dimensional version of the Weyl integral formula (4.1) and its various ramifications have played an important role in physics in the context of matrix models for a long time[70], and in particular recently in view of the connections between matrix models and quantum gravity (see e.g., [71] for a review). Some of these integration formulae [72] can also be understood in terms of the Duistermaat–Heckman theorem applied to integrals over coadjoint orbits [40], which provides an intriguing link between these two types of localization.

The path integral version (4.2) of the Weyl integral formula was first used in [69], where we calculated the partition function and correlation functions of some low-dimensional gauge theories (Chern–Simons theory and the G/G gauged Wess–Zumino–Witten model) from the path integral. Subsequently, we also applied it to 2d Yang–Mills theory [73], rederiving the results which had been previously obtained by other methods[74, 75, 76, 77]. The method has been applied by Witten[78] to solve the Grassmannian $N = 2$ sigma model, and the topological obstructions to diagonalization have been analyzed in detail in [79].

In the following we will first recall the necessary background from the theory of Lie groups and Lie algebras and then discuss the topological aspects to the extent that the more precise version of (4.2) acquires some degree of plausibility. To keep the presentation as simple as possible, we will here only deal with simply connected groups. We then discuss 2d Yang–Mills theory as an example. For more details, the reader is referred to the review [69] and to [79].

4.1. The Weyl Integral Formula

Background from the theory of Lie groups. Let G be a compact connected and simply-connected Lie group. We denote by T a maximal torus of G, i.e., a maximal compact connected Abelian subgroup of G, and by r the rank of G, $r = \mathrm{rk}(G) = \dim T$. We also denote by G_r the set of regular elements of G, i.e., those lying in one and only one maximal torus of G, and set $T_r = G_r \cap T$. The set of non-regular elements of G is of codimension 3 and $\pi_1(G) = 0$ implies $\pi_1(G_r) = 0$.

The crucial information we need is that any element g of G lies in some maximal torus and that any two maximal tori are conjugate to each other. We will henceforth choose one maximal torus T arbitrarily and fix it. It follows that any element g of G can be conjugated into T, $h^{-1}gh \in T$ for some $h \in G$. Such an h is of course not unique. First of all, h can be multiplied on the right by any element of T, $h \to ht, t \in T$ as T is Abelian. To specify the residual ambiguity in h, we need to introduce the Weyl group W. It can be defined as the quotient $W = N(T)/T$, where $N(T) = \{g \in G : g^{-1}tg \in T\ \forall t \in T\}$ denotes the normalizer of T in G. If $h^{-1}gh = t \in T$, then $(hn)^{-1}g(hn) = n^{-1}tn \in T$, $n \in N(T)$, is one of the finite number of images $w(t)$ of t under the action of the Weyl group W. It follows that for regular elements $g \in G_r$ the complete ambiguity in h is $h \to hn$. For non-regular elements this ambiguity is larger (e.g., for g the identity element h is completely arbitrary).

A useful way of summarizing the above result for regular elements, one which will allow us to pose the question of diagonalizability of G_r-valued maps in a form amenable to topological considerations, is to say that the conjugation map

$$\begin{aligned} q : G/T \times T_r &\to G_r \\ ([h],t) &\mapsto hth^{-1} \end{aligned} \tag{4.3}$$

is a $|W|$-fold covering onto G_r (see e.g., [80, 81]). As G_r is simply-connected, this covering is trivial and we can think of $G/T \times T_r$ as the total space of a trivial W-bundle over G_r. Thus any element of G_r can be lifted to $G/T \times T_r$ (with a $|W|$-fold ambiguity). As coverings induce isomorphisms on the higher homotopy groups, it also follows that $\pi_2(G_r) = \pi_2(G/T) = \pi_1(T) = \mathbb{Z}^r$, to be contrasted with $\pi_2(G) = 0$.

All of the above results are also true for Lie algebras of compact Lie groups. In particular, if we denote by $\mathfrak{g}$ the Lie algebra of G and by $\mathfrak{t}$ a Cartan subalgebra of $\mathfrak{g}$, there is a trivial W-fibration

$$\begin{aligned} q : G/T \times \mathfrak{t}_r &\to \mathfrak{g}_r \\ ([h],\tau) &\mapsto h\tau h^{-1}. \end{aligned} \tag{4.4}$$

In order to state the Weyl integral formula, which can be thought of as a formula relating an integral over G_r to an integral over $G/T \times T_r$ via (4.3), we will need some more information. First of all, we choose an invariant metric $(.,.)$ on $\mathfrak{g}$ and will use it to identify $\mathfrak{g}$ with its dual $\mathfrak{g}^*$ whenever convenient. This metric also induces natural Haar measures dg and dt on G and T, normalized to $\int_G dg = \int_T dt = 1$.

For the purpose of integration over G we may restrict ourselves to G_r and we can thus use (4.3) to pull back the measure dg to $G/T \times T$. To calculate the Jacobian of q, we need to know the infinitesimal conjugation action of T on G/T. Corresponding to a choice of T we have an orthogonal direct sum decomposition of the Lie algebra $\mathfrak{g}$ of G, $\mathfrak{g} = \mathfrak{t} \oplus \mathfrak{k}$. G acts on $\mathfrak{g}$ via the adjoint representation Ad. This induces an action of T which acts trivially on $\mathfrak{t}$

and leaves $\mathfrak{k}$ invariant (the isotropy representation $\mathrm{Ad}_{\mathfrak{k}}$ of T on $\mathfrak{k}$, the tangent space to G/T). Therefore the Jacobian matrix is

$$\Delta_W(t) = 1 - \mathrm{Ad}_{\mathfrak{k}}(t)) \tag{4.5}$$

and one finds the Weyl integral formula

$$\int_G dg\, f(g) = \frac{1}{|W|}\int_T dt\ \det\Delta_W(t)\int_{G/T} dh\, f(hth^{-1}). \tag{4.6}$$

In particular, if f is conjugation invariant, this reduces to

$$\int_G dg\, f(g) = \frac{1}{|W|}\int_T dt\ \det\Delta_W(t) f(t), \tag{4.7}$$

which is the version of the Weyl integral formula which we will make use of later on. The infinitesimal version of (4.5) is the determinant of $\Delta_W(\tau) = \mathrm{ad}_{\mathfrak{k}}(\tau)$, where $\tau \in \mathfrak{t}$. It appears in the corresponding formulae for integration over Lie algebras which are otherwise the exact analogues of (4.6,4.7). A more explicit version of (4.7) for $G = SU(2)$ is given in (4.25).

The determinant can be calculated by using the Cartan decomposition of $\mathfrak{g}$. The complexified Lie algebra $\mathfrak{g}_{\mathbb{C}}$ splits into $\mathfrak{t}_{\mathbb{C}}$ and the one-dimensional eigenspaces $\mathfrak{g}_\alpha$ of the isotropy representation, labelled by the roots α. It follows that the Jacobian (Weyl determinant) can be written as

$$\begin{aligned} \det\Delta_W(t) &= \prod_\alpha (1 - e^\alpha(t)), \\ \det\Delta_W(\tau) &= \prod_\alpha (\alpha, \tau). \end{aligned} \tag{4.8}$$

For later use we note that $\Delta_W(t)$ vanishes precisely on the non-regular elements of T, this being the mechanism by which non-regular maps (i.e. maps taking values also in the non-regular elements of G) should be suppressed in the path integral.

Topological obstructions to diagonalization. We have seen above that the crucial ingredient in the Weyl integral formula is the fact that any element of G is conjugate to some element of T. In [79] we investigated to which extent this property continues to hold for spaces of (smooth) maps $\mathrm{Map}(M, G)$ from a manifold M to a compact Lie group G. Here we will summarize the relevant results for simply-connected G, as they will be essential for the generalization of the Weyl integral formula to integrals over $\mathrm{Map}(M, G)$ and $\mathrm{Map}(M, \mathfrak{g})$.

Thus the question we need to address is if a smooth map $g \in \mathrm{Map}(M, G)$ can be written as

$$g(x) = h(x)t(x)h(x)^{-1} \tag{4.9}$$

for $t \in \mathrm{Map}(M, T)$ and $h \in \mathrm{Map}(M, G)$. Of course there is no problem with doing this pointwise, so the question is really if it can be done consistently in such a way that the resulting maps are smooth. Intuitively it is clear that problems can potentially arise from the ambiguities in h and t. And it is indeed easy to see (by examples) that (4.9) cannot be achieved globally and smoothly on M in general. In fact, the situation turns out to be manageable only for regular maps, i.e., for maps taking values in G_r. The reason for this is that non-regular maps may not be smoothly diagonalizable in any open neighborhood of a preimage of a non-regular element even if the map takes on non-regular values only at isolated points (we will see an example of this below). Clearly then, methods of (differential) topology do not suffice to deal with the problems posed by non-regular maps. Henceforth we will deal almost exclusively with maps taking values in the dense set G_r of regular elements of G. For these regular maps, the problem can be solved completely and for simply-connected groups the results can be summarized as follows:

1. Conjugation into T can always be achieved locally, i.e., in open contractible neighborhoods of M.

2. The diagonalized map t can always be chosen to be smooth globally.

3. Non-trivial T-bundles on M are the obstructions to finding smooth functions h which accomplish (4.9) globally. In particular, when M and G are such that there are no non-trivial G bundles on M (e.g., for G simply-connected and M two-dimensional), all isomorphism classes of torus bundles appear as obstructions.

We actually want to go a little bit further than that. Namely, as we want to use (4.9) as a change of variables (gauge transformation) in the path integral, we need to inquire what the effect of the occurrence of these obstructions will be on the other fields in the theory. In particular, when gauge fields are present, a transformation $g \to h^{-1}gh$ has to be accompanied by a gauge transformation $A \to A^h = h^{-1}Ah + h^{-1}dh$. E.g., in [69] it was found that the Weyl integral formula leads to the correct results for the gauge theories investigated there only when the gauge field integral includes a sum over T-connections on all isomorphism classes of T-bundles on M (a two-manifold in those examples), even though the original bundle was trivial. One would therefore like to know if the above obstructions to diagonalization can account for this. It turns out that the results concerning the behavior of gauge fields are indeed in complete agreement with what one expects on the basis of the results of [69, 73], namely:

4. If P_T is the (non-trivial) principal T-bundle which is the obstruction to the global smoothness of h, then the t-component of A^h defines a connection on P_T.

5. Hence the Weyl integral formula, when applied to gauge theories, should contain a sum over all those isomorphism classes of T-bundles on M which arise as obstructions to diagonalization.

Rather than present proofs of all these statements (which can be found in [79]), we will illustrate the above in two different ways. On the one hand we will present a very hands-on example which allows one to see explicitly the obstruction and how it is related to connections on non-trivial T-bundles. On the other hand we will set up the general problem in such a way that the reader with a background in topology will be able to read off the answers to at least the questions (1)-(3) almost immediately.

Our example will be a particular map from S^2 to $SU(2)$. We parametrize elements of $SU(2)$ as

$$x_4\mathbf{1} + \sum_{k=1}^{3} x_k\sigma_k = \begin{pmatrix} x_4 + ix_3 & x_1 + ix_2 \\ -x_1 + ix_2 & x_4 - ix_3 \end{pmatrix}, \tag{4.10}$$

subject to $\sum_{k=1}^{4}(x_k)^2 = 1$, and consider the identity map $g(x) = \sum_{k=1}^{3} x_k\sigma_k$ from the two-sphere to the equator $x_4 = 0$ of $SU(2)$. As the only non-regular elements of $SU(2)$ are $\pm\mathbf{1}$, g is a smooth regular map. To detect a possible obstruction to diagonalizing g smoothly, we proceed as follows. To any map f from S^2 to S^2 we can assign an integer, its winding number $n(f)$. It can be realized as the integral of the pull-back of the normalized volume-form ω on the target-S^2, $n(f) = \int_{S^2} f^*\omega$. This winding number is invariant under smooth and continuous homotopies of f. Clearly for the identity map g we have $n(g)=1$. Representing (as above) f as $f=\sum_k f_k\sigma_k$ with $\sum_k(f_k)^2=1$, a more explicit realization of $n(f)$ is

$$n(f)=-\tfrac{1}{32\pi}\int_{S^2}\operatorname{tr} f[df,df]. \tag{4.11}$$

Here and in the following expressions like $[df,df]=df_k df_l[\sigma_k,\sigma_l]$ denote the wedge product of forms combined with the commutator in the Lie algebra. Now suppose that one can smoothly conjugate the map g into $U(1)$ via some h, $h^{-1}gh=t$. As the space of maps from S^2 to $SU(2)$ is connected ($\pi_2(G)=0$), g is homotopic to t and one has $n(g)=n(t)$. But, since $g^2=-\mathbf{1}$, t is a constant map (in fact, $t=\pm\sigma_3$, the two being related by a Weyl transformation) so that $n(t)=0$, a contradiction. This shows that there can be no smooth or continuous h satisfying $h^{-1}gh=t$. As both g and its diagonalization $\pm\sigma_3$ may just as well be regarded as Lie algebra valued maps, this example establishes that obstructions to diagonalization will also arise in the (seemingly topologically trivial) case of Lie algebra valued maps.

To see how connections on non-trivial bundles arise, we consider a slight generalization $n(f,A)$ of $n(f)$,

$$n(f,A)=-\tfrac{1}{32\pi}\int_{S^2}\operatorname{tr} f[df,df]-\tfrac{1}{2\pi}\int_{S^2}\operatorname{tr}[d(fA)], \tag{4.12}$$

depending on both f and an $SU(2)$-connection A. Because the second term is a total derivative, (4.12) obviously coincides with (4.11) when A is globally defined. The crucial property of $n(f,A)$ is that it is gauge invariant, i.e., invariant under simultaneous transformation of f and A,

$$n(h^{-1}fh,A^h)=n(f,A), \tag{4.13}$$

even for discontinuous h (the key point being that no integration by parts is required in establishing (4.13). The advantage of a formula with such an invariance is that it allows one to relate maps which are not necessarily homotopic. In particular, let us now choose some (possibly discontinuous) h such that it conjugates g into $U(1)$, say $g=h\sigma_3h^{-1}$ (since this can be done pointwise, some such h will exist). Using (4.13) we find

$$n(g,A)=1=-\tfrac{1}{2\pi}\int_{S^2}\operatorname{tr}\sigma_3 d(A^h). \tag{4.14}$$

In particular, if we introduce the Abelian gauge field $a=-\operatorname{tr}\sigma_3A^h$, we obtain

$$n(g,A)=1=\tfrac{1}{2\pi}\int_{S^2}da. \tag{4.15}$$

We now see the price of conjugating into the torus. The first Chern class of the $U(1)$ component of the gauge field A^h is equal to the winding number of the original map! We have picked up the sought for non-trivial torus bundles. In this case it is just the pull-back of the $U(1)$-bundle $SU(2)\to SU(2)/U(1)\sim S^2$ via g and this turns out to be more or less what happens in general.

Finally, we will show that a certain non-regular extension of this map provides us with an example of a map which cannot be smoothly diagonalized in any open neighborhood of a non-regular point. As a preparation, we make the obvious observation that if two maps from M to G_r are regularly homotopic, i.e., homotopic in $\mathrm{Map}(M,G_r)$, then if there are no obstructions to diagonalizing one of them, there are also none for the other. Incidentally, this also shows why winding numbers play a role at all in this discussion although the space of maps from S^2 to $SU(2)$ is connected. As we have seen in our discussion of G_r, $\pi_2(G_r)=\mathbb{Z}^r$, so that the space of regular maps is not connected but decomposes into disjoint sectors labelled by an r-tuple of winding numbers.

Consider now the extension $\tilde{g}$ of g to the identity map from the three-sphere to $SU(2)$, $\tilde{g}(x) = x_4\mathbf{1} + \sum_k x_k\sigma_k$. This map takes on non-regular values only at $x_4 = \pm 1$. There is clearly no smooth diagonalization of the restriction of this map to any open set containing the north-pole $\{x_4 = 1\}$. If there were, this would in particular imply the existence of a global smooth diagonalization of a map from S^2 to $SU(2)_r$ which is regularly homotopic to g, as any neighborhood of the north pole contains a surrounding two-sphere — a contradiction.

We will now return to the general situation and briefly describe how to set up the problem concerning obstructions to diagonalization in such a way that it can be solved by standard topological arguments.

Being able to (locally) conjugate smoothly into the maximal torus is the statement that one can (locally) find smooth maps h and t such that $g_U = hth^{-1}$. In other words, one is looking for a (local) lift of the map $g \in \mathrm{Map}(M, G_r)$ to a map $(h,t) \in \mathrm{Map}(M,G) \times \mathrm{Map}(M,T_r)$. It will be convenient to break this problem up into parts and to establish the (local) existence of this lift in a two-step procedure, as indicated in the diagram below.

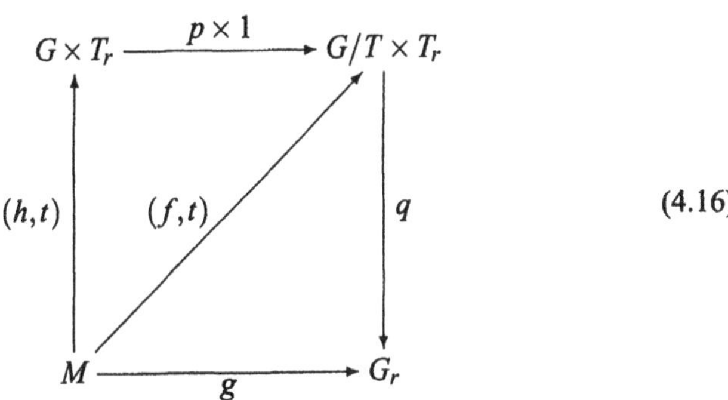

(4.16)

Here, on the right hand side of this diagram we recognize the conjugation map (4.3) and its trivial W-fibration. The top arrow, on the other hand, is essentially the non-trivial fibration $G \to G/T$. In the first step one attempts to lift g along the diagonal, i.e., to construct a pair (f,t), where $f \in \mathrm{Map}(M, G/T)$, which projects down to g via the projection q. By a standard argument the obstruction to doing this globally is the possible non-triviality of the W-bundle $g^*(G/T \times T_r)$ on M. Happily, when G is simply connected, (4.3) is trivial and this obstruction is absent. Thus we have just established the existence of a globally smooth diagonalization t of g.

In the second step, dealing with the upper triangle, one needs to lift f locally to $\mathrm{Map}(M,G)$. The obstruction to doing this globally is the possible non-triviality of the principal T-bundle $f^*(G)$ over M (as in the above example). If this bundle is trivial (e.g., when $H^2(M,\mathbb{Z}) = 0$, so that there are no non-trivial torus bundles at all on M), a smooth h accomplishing the diagonalization will exist globally. Locally, a lift h can of course always be found.

The principal T-bundles which appear as obstructions are precisely those that can be obtained from the trivial principal G-bundle P_G by restriction of the structure group from G to T (or, more informally, that sit inside P_G). Under such a restriction, the $\mathfrak{t}$-part of a connection on P_G becomes a connection on P_T, while the $\mathfrak{k}$-part becomes a one-form with values in the bundle associated to P_T via the isotropy representation of T on $\mathfrak{k}$.

Because of the similarity of (4.3) and (4.4), the situation for Lie algebra valued maps is exactly the same as above (differences only occur for non-simply connected groups).

The Weyl integral formula for path integrals. We have now collected all the results we need to state the Weyl integral formula for integrals over $\mathrm{Map}(M,G)$ which correctly takes into account the topological considerations of the previous section.

For concreteness, consider a local functional $S[g;A]$ (the 'action') of maps $g \in \mathrm{Map}(M,G)$ and gauge fields A on a trivial principal G-bundle P_G on M (a dependence on other fields could of course be included as well). Assume that $\exp iS[g,A]$ is gauge invariant,

$$\mathrm{e}^{iS[g,A]} = \mathrm{e}^{iS[h^{-1}gh,A^h]} \qquad \forall\, h \in \mathrm{Map}(M,G), \tag{4.17}$$

at least for smooth h. If e.g., a partial integration is involved in establishing the gauge invariance (as in Chern–Simons theory), this may fail for non-smooth h's and more care has to be exercised when such a gauge transformation is performed. Then the functional $F[g]$ obtained by integrating $\exp iS[g;A]$ over A,

$$F[g] := \int [dA]\,\mathrm{e}^{iS[g,A]}, \tag{4.18}$$

is conjugation invariant,

$$F[h^{-1}gh] = F[g]. \tag{4.19}$$

It is then tempting to use a formal analogue of (4.7) to reduce the remaining integral over g to an integral over maps taking values in the Abelian group T. In field theory language this amounts to using the gauge invariance (4.17) to impose the 'gauge condition' $g(x) \in T$. The first modification of (4.7) will then be the replacement of the Weyl determinant $\det \Delta_W(t)$ by a functional determinant $\det \Delta_W[t]$ of the same form which needs to be regularized appropriately (see the Appendix of [73]).

However, the crucial point is, of course, that this is not the whole story. We already know that this 'gauge condition' cannot necessarily be achieved smoothly and globally. Insisting on achieving this 'gauge' nevertheless, albeit via non-continuous field transformations, turns the $\mathfrak{t}$-component $A^{\mathfrak{t}}$ of the transformed gauge field A^h into a gauge field on a possibly non-trivial T bundle P_T (and the $\mathfrak{k}$-component turns out to transforms as a section of the associated bundle $P_T \times_T \mathfrak{k}$). Moreover, we know that all those T bundles will contribute which arise as restrictions of the (trivial) bundle P_G. Let us denote the set of isomorphism classes of these T bundles by $[P_T;P_G]$. Hence the 'correct' (meaning correct modulo the analytical difficulties inherent in making any field theory functional integral rigorous) version of the Weyl integral formula, capturing the topological aspects of the situation, is one which includes a sum over the contributions from the connections on all the isomorphism classes of bundles in $[P_T;P_G]$.

Let us denote the space of connections on P_G and on a principal T bundle P_T^l representing an element $l \in [P_T;P_G]$ by $\mathcal{A}$ and $\mathcal{A}[l]$ respectively and the space of one-forms with values in the sections of $P_T^l \times_T \mathfrak{k}$ by $\mathcal{B}[l]$. Then, with

$$Z[P_G] = \int [dA] \int [dg]\mathrm{e}^{iS[g,A]}, \tag{4.20}$$

the Weyl integral formula for functional integrals reads

$$Z[P_G] = \sum_{l\in[P_T;P_G]} \int_{\mathcal{A}[l]} [dA^{\mathfrak{t}}] \int_{\mathcal{B}[l]} [dA^{\mathfrak{k}}] \int [dt] \Delta_W[t] \mathrm{e}^{iS[t,A^{\mathfrak{t}},A^{\mathfrak{k}}]} \tag{4.21}$$

(modulo a normalization constant on the right hand side). The t-integrals carry no l-label as the spaces of sections of $\mathrm{Ad}P_T^l$ are all isomorphic to the space of maps into T. There is an exactly analogous formula generalizing the Lie algebra version of the Weyl integral formula. Namely, assuming that we have a gauge invariant action $S[\phi,A]$, where $\phi \in \mathrm{Map}(M,\mathfrak{g})$, we can use the Weyl integral formula to reduce its partition function to

$$Z[P_G] = \sum_{l\in[P_T;P_G]} \int_{\mathcal{A}[l]}[dA^{\mathfrak{t}}] \int_{\mathcal{B}[l]}[dA^{\mathfrak{k}}] \int [d\tau]\Delta_W[\tau] e^{iS[\tau,A^{\mathfrak{t}},A^{\mathfrak{k}}]} \tag{4.22}$$

In the examples considered in [69, 73], the fields $A^{\mathfrak{k}}$ entered purely quadratically in the reduced action $S[t,A^{\mathfrak{t}},A^{\mathfrak{k}}]$ and could be integrated out directly, leaving behind an effective Abelian theory depending on the fields t and $A^{\mathfrak{t}}$ (respectively τ and $A^{\mathfrak{t}}$). The information on the non-Abelian origin of this theory is contained entirely in the measure determined by $\Delta_W[t]$ and the (inverse) functional determinant coming from the $A^{\mathfrak{k}}$-integration. In all the examples considered so far, e.g., in [69, 73, 78], this procedure simplified the path integral to the extent that it could be calculated explicitly. One does of course not expect to be able to get that far in general. Nevertheless, the simplification brought about by replacing an interacting non-Abelian theory with a theory with only an Abelian gauge symmetry should allow one to obtain at least some new information from the path integral. We will see the formula (4.21) at work in the next section, where we will use it to solve 2d Yang–Mills theory.

4.2. Solving 2d Yang–Mills Theory via Abelianization

In this section we will apply the Weyl integral formula (4.21) to Yang–Mills theory on a two-dimensional closed surface Σ [73]. As mentioned before, this theory has recently been solved by a number of other techniques as well [75, 74, 76, 77], but it appears to us that none of them is as simple and elementary as the one based on Abelianization. As the partition function of Yang–Mills theory for a gauge group G can be expressed entirely in terms of the representation theory (more precisely the dimensions and the quadratic Casimirs of the representations) of G, we will first briefly recall the relevant formulae below. Then we will calculate the partition function. To keep the Lie algebra theory as simple as possible (as it is only of tangential interest here) we will assume that G is not only simply-connected but also simply-laced. For the relation and application of these results to topological Yang–Mills theory see [52, 73].

Some Lie algebra theory. It turns out that the partition function of Yang–Mills theory for a group G can be expressed in terms of the dimensions and quadratic Casimirs of the unitary irreducible representations of G. We will recall here the relevant formulae, see e.g., [80].

First of all, we choose a positive Weyl chamber C^+ and introduce a corresponding set of simple roots $\{\alpha_a, a=1,\ldots,r\}$, a dual set of fundamental weights $\{\lambda^a, a=1,\ldots,r\}$ and denote the weight lattice by $\Lambda = \mathbb{Z}[\lambda^a]$. Then the highest weights of the unitary irreducible representations can be identified with the elements of $\Lambda^+ = \Lambda\cap\bar{C}^+$, $\bar{C}^+$ denoting the closure of the Weyl chamber. For $\mu\in\Lambda^+$ we denote by $d(\mu)$ and $c(\mu)$ the dimension and the quadratic Casimir of the corresponding representation. The formulae for these are

$$d(\mu) = \prod_{\alpha>0} \frac{(\alpha,\mu+\rho)}{(\alpha,\rho)} \tag{4.23}$$

$$c_\mu = (\mu+\rho,\mu+\rho)-(\rho,\rho), \tag{4.24}$$

where $\rho = \frac{1}{2}\sum_{\alpha>0}\alpha$ is the Weyl vector.

Let us illustrate this and the formulae obtained in our discussion of the Weyl integral formula in the case $G = SU(2)$ and $T = U(1)$. We use the trace to identify the Lie algebra $\mathfrak{t}$ of T with its dual. Then we can choose the positive root α and the fundamental weight λ to be $\alpha = \mathrm{diag}(1,-1)$ and $\lambda = \alpha/2$, so that $\rho = \lambda$. We also parametrize elements of T as $t = \exp i\lambda\phi = \mathrm{diag}(\exp i\phi/2, \exp -i\phi/2)$. It follows that the Weyl determinant $\det\Delta_W(t)$ (4.5) is $\det\Delta_W(t) = 4\sin^2\phi/2$. Hence the Weyl integral formula for class functions (conjugation invariant functions) is

$$\int_G dg\, f(g) = \frac{1}{2\pi}\int_0^{4\pi} d\phi\, \sin^2(\phi/2) f(\phi). \tag{4.25}$$

For the spin j representation of $SU(2)$ (with highest weight $\mu(j) = 2j\lambda \in \Lambda^+$), one finds $d(\mu(j)) = 2j+1$ and $c(\mu(j)) = 2j(j+1)$.

2d Yang–Mills theory. The action of Yang–Mills theory on a two-dimensional surface Σ_g of genus g is

$$S[A] = \tfrac{1}{2\varepsilon}\int_{\Sigma_g} \mathrm{tr} F_A * F_A. \tag{4.26}$$

Here $F_A = dA + \frac{1}{2}[A,A]$ is the curvature of a gauge field A on a trivial principal G-bundle P_G on Σ_g, $*$ denotes the Hodge duality operator with respect to some metric on Σ_g, tr refers to the trace in the fundamental representation for $G = SU(n)$. ε represents the coupling constant of the theory. A tr will henceforth be understood in integrals of Lie algebra valued forms. It is readily seen that, since $*F_A$ is a scalar, the action does not depend on the details of the metric but only on the area $A(\Sigma_g) = \int_{\Sigma_g} *1$ of the surface. As a change in the metric can thus be compensated by a change in the coupling constant, we can without loss of generality choose a metric and a corresponding volume two-form ω with unit area. ω is a symplectic form on Σ_g representing the generator of $H^2(\Sigma_g, \mathbb{Z}) \sim \mathbb{Z}$.

The action is invariant under gauge transformations $A \to A^g$ and the path integral that we would like to compute is

$$Z_{\Sigma_g}(\varepsilon) = \int_{\mathcal{A}} [dA]\, \mathrm{e}^{\frac{1}{2\varepsilon}\int_{\Sigma_g} F_A * F_A}. \tag{4.27}$$

It will be convenient to rewrite this in first order form by introducing an additional $\mathfrak{g}$-valued scalar field $\phi \in \mathrm{Map}(\Sigma_g, \mathfrak{g})$ (cf. (3.46)), so that the partition function becomes

$$Z_{\Sigma_g}(\varepsilon) = \int_{\mathcal{A}} [dA] \int_{\mathrm{Map}(\Sigma_g,\mathfrak{g})} [d\phi] \mathrm{e}^{\int_{\Sigma_g} i\phi F_A + \frac{\varepsilon}{2}\phi * \phi}. \tag{4.28}$$

The gauge invariance of the action $S[\phi, A]$ appearing in (4.28) is

$$S[g^{-1}\phi g, A^g] = S[\phi, A], \tag{4.29}$$

and we are thus precisely in a situation where we can apply the considerations of the previous section. Before embarking on that, it will however be useful to recall some properties of the quantum field theory defined by the classical action (4.26).

Yang–Mills theory in two-dimensions is super-renormalisable and there are no ultraviolet infinities associated with diagrams involving external A or ϕ fields. Furthermore, on a compact manifold Σ_g there are no infrared divergences associated with these diagrams either. However, because of the coupling of ϕ to the metric, there are diagrams which do not involve external ϕ or

A legs but do have external background graviton legs and which require regularization. These terms arise in the determinants that are being calculated and they depend only on the area and topology of Σ_g, i.e., they have the form $\alpha_1(\varepsilon)A(\Sigma_g)+\alpha_2(\varepsilon)\chi(\Sigma_g)$, where $\chi(\Sigma_g)=2-2g$ is the Euler number of Σ_g. These may be termed area and topological standard renormalizations. We wish to ensure that the scaling invariance that allowed us to move all of the metric dependence into ε is respected. Hence only those regularization schemes which preserve this symmetry are to be considered. Then the dependence of α_1 on ε is fixed to be $\alpha_1(\varepsilon)=\varepsilon\beta$ and α_2 and β are independent of ε.

With these remarks in mind, we can now proceed by using the gauge freedom to conjugate ϕ into $\mathrm{Map}(\Sigma_g,\mathfrak{t})$ (i.e., to impose $\phi^{\mathfrak{k}}=0$). This simply amounts to plugging $S[\phi,A]$ into (4.22). Alternatively, one can of course implement this gauge condition by following the standard Faddeev–Popov procedure. The ghost determinant one obtains in this way is precisely the Weyl determinant $\det\Delta_W[\tau]$ appearing in (4.22).

We should perhaps stress once more that in choosing this gauge we are assuming that ϕ is regular. We have discussed this issue at some length in [69, 73] and have nothing to add to this here. We just want to point out that only constant modes of ϕ contribute to the path integral and that a non-regular constant ϕ would give a divergent contribution to the partition function from the gauge fields (and a zero from the Weyl determinant). Thus any way of regularizing this divergence will set the contribution from the non-regular points to zero (by the ghost contribution) and is therefore tantamount to discarding the non-regular maps which is what we will do.

Let us now determine the ingredients that go into (4.22). The Weyl determinant $\det\Delta_W[\tau]$ is the functional determinant of the adjoint action by τ on the space $\mathrm{Map}(\Sigma_g,\mathfrak{k})=\Omega^0(\Sigma_g,\mathfrak{k})$,

$$\det\Delta_W[\tau]=\det\mathrm{ad}(\tau)|_{\Omega^0(\Sigma_g,\mathfrak{k})}. \tag{4.30}$$

This determinant of course requires some regularization. We will say more about it once we have combined this with another determinant that will arise below.

We now look at the reduced action $S[\tau,A^{\mathfrak{t}},A^{\mathfrak{k}}]$. Because of the orthogonality of $\mathfrak{t}$ and $\mathfrak{k}$ with respect to the trace, the only terms that survive in the action are

$$S[\tau,A^{\mathfrak{t}},A^{\mathfrak{k}}]=\int_{\Sigma_g}\tau dA^{\mathfrak{t}}+\tfrac{1}{2}[\tau,A^{\mathfrak{k}}]A^{\mathfrak{k}}+\tfrac{\varepsilon}{2}\tau*\tau. \tag{4.31}$$

Integrating out $A^{\mathfrak{k}}$, one obtains the inverse square root of the determinant of $\mathrm{ad}(\tau)$, this time acting on the space $\Omega^1(\Sigma_g,\mathfrak{k})$ of $\mathfrak{k}$-valued one-forms,

$$\int[dA^{\mathfrak{k}}]\quad\Rightarrow\quad\det{}^{-1/2}\mathrm{ad}(\tau)|_{\Omega^1(\Sigma_g,\mathfrak{k})}. \tag{4.32}$$

It is clear that this almost cancels against the Weyl determinant as, modulo zero-modes, a one-form in 2d has as many degrees of freedom as two scalars. The zero mode surplus is one constant scalar mode minus $(2g/2)=g$ harmonic one-form modes so that the combined determinant would simply reduce to a finite dimensional determinant,

$$\frac{\det\mathrm{ad}(\tau)|_{\Omega^1(\Sigma_g,\mathfrak{k})}}{\det{}^{1/2}\mathrm{ad}(\tau)|_{\Omega^1(\Sigma_g,\mathfrak{k})}}=(\det\mathrm{ad}(\tau)|_{\mathfrak{k}})^{\chi(\Sigma_g)/2}, \tag{4.33}$$

if τ were constant. As we will see shortly that only the constant modes of τ contribute to the path integral, we will just leave it at (4.33) and refer to [73] for a more careful discussion using a heat kernel regularization of the determinants involved.

Putting all the above together we see that we are left with an Abelian functional integral over τ and $A^{\mathfrak{t}}$ and a sum over the topological sectors,

$$Z_{\Sigma_g}(\varepsilon) = \sum_{l\in[P_T;P_G]} \int_{\mathcal{A}[l]} [dA^{\mathfrak{t}}] \int [d\tau] (\det \mathrm{ad}(\tau)|_{\mathfrak{k}})^{\chi(\Sigma_g)/2} \mathrm{e}^{\int_{\Sigma_g} (i\tau dA^{\mathfrak{t}} + \frac{\varepsilon}{2}\tau * \tau)}. \tag{4.34}$$

To evaluate this, we will use one more trick, namely to change variables from $A^{\mathfrak{t}}$ to $F^{\mathfrak{t}} = dA^{\mathfrak{t}}$. To see how to do that let us first note that the theory we have arrived at has still got local Abelian gauge invariance (as well as W-invariance as a remnant of the non-Abelian gauge invariance of the original action). Let us therefore choose some gauge fixing condition $E(A^{\mathfrak{t}}) = 0$ and change variables from $A^{\mathfrak{t}}$ to $(F^{\mathfrak{t}}, E(A^{\mathfrak{t}}))$. This change of variables is fine away from the gauge equivalence classes of flat connections (i.e., the moduli space T^{2g} of flat T-connections on Σ_g). As these flat connections do not appear in the action, they just give a finite volume factor (depending on the normalization of the metric on T) which we will not keep track of. On the remaining modes of the gauge field this change of variables is then well defined. What makes it so useful is the fact, demonstrated in [82], that the corresponding Jacobian cancels precisely against the Faddeev–Popov determinant arising from the gauge fixing.

As a consequence, after this change of variables and integration over the ghosts, the multiplier field and $E(A^{\mathfrak{t}})$ all derivatives disappear from (4.38) and the remaining path integral over τ and $F^{\mathfrak{t}}$ is completely elementary. The only thing to keep track of is that we are not integrating over arbitrary two-forms $F^{\mathfrak{t}}$ but only over those which arise as the curvature two-form on some T-bundle over Σ which arises as the restriction of the trivial G-bundle P_G. As there are no non-trivial G-bundles on Σ_g, the sum over topological sectors $l \in [P_T;P_G]$ will extend over all isomorphism classes of T-bundles on Σ_g. These are classified by their first Chern class in $H^2(\Sigma_g, \mathbb{Z}^r) \sim \mathbb{Z}^r$, and hence the summation over l in (4.22) can be thought of as a summation over r-tuples of integers. In particular, if we expand $F^{\mathfrak{t}}$ in a basis of simple roots $\{\alpha_a\}$ of $\mathfrak{g}$ as $F^{\mathfrak{t}} = i\alpha_a F^a$, then in the topological sector characterized by $l = (n^1, \ldots, n^r)$ we have

$$\int_{\Sigma_g} F^a = 2\pi n^a. \tag{4.35}$$

As all topological sectors appear, one way to enforce (4.35) in the path integral is to impose a periodic delta-function constraint on the curvature $F^{\mathfrak{t}}$ of the torus gauge field, i.e. to insert

$$\delta^P(\int F^{\mathfrak{t}}) = \prod_{a=1}^{r} \sum_{m_a\in\mathbb{Z}} \mathrm{e}^{im_a \int_{\Sigma_g} F^a} \tag{4.36}$$

into the path integral. It will be convenient to write this as a sum over the weight lattice Λ dual to the lattice spanned by the simple roots (or Chern classes),

$$\delta^P(\int F^{\mathfrak{t}}) = \sum_{\lambda\in\Lambda} \mathrm{e}^{\int_{\Sigma_g} \lambda F^{\mathfrak{t}}}. \tag{4.37}$$

Then the path integral to be evaluated reads

$$Z_{\Sigma_g}(\varepsilon) = \sum_{\lambda\in\Lambda} \int [F^{\mathfrak{t}}] \int [d\tau] (\det \mathrm{ad}(\tau)|_{\mathfrak{k}})^{\chi(\Sigma_g)/2} \mathrm{e}^{\int_{\Sigma_g} (i\tau + \lambda) F^{\mathfrak{t}} + \frac{\varepsilon}{2}\tau * \tau}. \tag{4.38}$$

Let us do the $F^{\mathfrak{t}}$-integral first. This imposes the delta-function constraint $\tau = i\lambda$ on τ, so that in particular only the constant modes of τ ever contribute to the partition function. We are then left with just the sum over the (regular) elements of Λ. Eliminating the Weyl group invariance is then the same as summing only over λ's in the interior of the positive Weyl chamber C^+. Writing λ as $\lambda = \mu + \rho$ one sees that the sum is now precisely over all the highest weights $\mu \in \Lambda^+ = \Lambda \cap \bar{C}^+$ of the unitary irreducible representations of G,

$$Z_{\Sigma_g}(\varepsilon) = \sum_{\mu \in \Lambda^+} \prod_{\alpha > 0} (\alpha, \mu + \rho)^{\chi(\Sigma_g)} e^{-\frac{\varepsilon}{2}(\mu + \rho, \mu + \rho)}. \tag{4.39}$$

Using (4.23) and (4.24), we can rewrite this as

$$Z_{\Sigma_g}(\varepsilon) = e^{-\frac{\varepsilon}{2}(\rho,\rho)} \prod_{\alpha>0} (\alpha,\rho)^{\chi(\Sigma_g)} \sum_{\mu \in \Lambda^+} d(\mu)^{\chi(\Sigma_g)} e^{-\frac{\varepsilon}{2}c(\mu)}. \tag{4.40}$$

This differs from the expression obtained in e.g., [74, 76, 75, 77] only by the first two terms which are precisely of the form of a standard area and topological renormalization respectively. What is interesting, though, is that the form of the partition function obtained here (with $(\mu+\rho,\mu+\rho)$ in the exponential instead of c_μ) agrees with what one obtains from non-Abelian localization [52] and is the correct one to use for determining the intersection numbers of the moduli space of flat G-connections on Σ_g.

The uses of the Weyl integral formula are of course not restricted to Yang–Mills theory (and its non-linear cousin, the G/G-model). For instance, in principle it can also be applied to gauge theories on manifolds of the form $M \times S^1$, where the 'temporal gauge' $\partial_0 A_0 = 0$ can profitably be augmented by using time-independent gauge transformations to achieve the gauge condition $A_0^{\mathfrak{k}} = 0$ (see e.g., [69] for an application to Chern–Simons theory). In other words, the temporal gauge reduces A_0 to a map from M to $\mathfrak{g}$ to which the considerations of this section regarding diagonalizability and the ensuing form of the path integral can be applied.

In practice, however, this method has its limits, not only because in higher dimensions the issue of non-regular maps raises its ugly head but also because simplifications brought about by a suitable choice of gauge alone will not be sufficient to make the theory solvable without recourse to more traditional techniques of quantum field theory as well. On a more optimistic note one may hope that the technique developed here can shed some light on the global aspects of the Abelian projection technique introduced in [83], currently popular in lattice gauge theories.

Finally, we want to mention that arguments as in section 4.1 permit one to also deduce a path integral analogue of the Weyl integral formula (4.6) valid for functions which are not conjugation invariant. This formula is then applicable to theories with less or no gauge symmetry, and can potentially help to disentangle the degrees of freedom of the theory. This appears to be the case e.g., in the ordinary (ungauged) Wess–Zumino–Witten model.

REFERENCES

[1] V. Mathai and D. Quillen, *Superconnections, Thom classes and equivariant differential forms*, Topology 25 (1986) 85–110.

[2] J. J. Duistermaat and G. Heckman, *On the variation in the cohomology of the symplectic form of the reduced phase space*, Inv. Math. 69 (1982) 259–269, ibid. 72 (1983) 153–158.

[3] N. Berline and M. Vergne, *Classes caractéristiques équivariantes. Formule de localisation en cohomologie équivariante*, C. R. Acad. Sci. Paris 295 (1982) 539–541; and *Zéros d'un champ decteurs et classes caractéristiques équivariantes*, Duke Math. J. 50 (1983) 539–549.

[4] N. Berline, E. Getzler and M. Vergne, *Heat Kernels and Dirac Operators*, Springer Verlag (1992).

[5] M. Bershadsky, S. Cecotti, H. Ooguri and C. Vafa, *Kodaira–Spencer theory of gravity and exact results for quantum string amplitudes*, Commun. Math. Phys. 165 (1994) 311.

[6] E. Witten, *Supersymmetric Yang–Mills theory on a four-manifold*, J. Math. Phys. 35 (1994) 5101–5135.

[7] N. Seiberg and E. Witten, *Electro-magnetic duality, monopole condensation, and confinement in $N = 2$ supersymmetric Yang–Mills theory*, Nucl. Phys. B426 (1994) 19–52 (Erratum Nucl. Phys. B430 (1994) 485–486); *Monopoles, duality and chiral symmetry breaking in $N = 2$ supersymmetric QCD*, Nucl. Phys. B431 (1994) 484–550.

[8] E. Witten, *Monopoles and four-manifolds*, Math. Research Lett. 1 (1994) 769.

[9] C. Vafa and E. Witten, *A strong coupling test of S-duality*, Nucl. Phys. B431 (1994) 3–77.

[10] M. F. Atiyah and L. Jeffrey, *Topological Lagrangians and cohomology*, J. Geom. Phys. 7 (1990) 119–136.

[11] E. Witten, *The N matrix model and gauged WZW models*, Nucl. Phys. B371 (1992) 191–245.

[12] R. Bott and L. W. Tu, *Differential Forms in Algebraic Topology*, Springer, New York, 1986.

[13] J. Kalkman, *BRST model for equivariant cohomology and representatives for the equivariant Thom class*, Commun. Math. Phys. 153 (1993) 447–463.

[14] V. Mathai, *Heat kernels and Thom forms*, J. Funct. Anal. 104 (1992) 34–46.

[15] M. Blau, *The Mathai–Quillen formalism and topological field theory*, J. Geom. Phys. 11 (1993) 95–127.

[16] D. Birmingham, M. Blau, M. Rakowski and G. Thompson, *Topological Field Theory*, Physics Reports 209 Nos. 4&5 (1991) 129–340.

[17] M. Blau and G. Thompson, *Topological gauge theories from supersymmetric quantum mechanics on spaces of connections*, Int. J. Mod. Phys. A8 (1993) 573–585.

[18] J.-M. Bismut, *The Atiyah–Singer theorems. A probabilistic approach. I. The index theorem*, J. Funct. Anal. 57 (1984) 56–99; *II. The Lefschetz fixed point formulas*, ibid. 57 (1984) 329–348.

[19] J.-M. Bismut, *Index theorem and equivariant cohomology on the loop space*, Commun. Math. Phys. 98 (1985) 213–237; *Localization formulas, superconnections, and the index theorem for families*, ibid. 103 (1986) 127–166.

[20] P. S. Aspinwall and D. R. Morrison, *Topological field theory and rational curves*, Commun. Math. Phys. 150 (1993) 245–262.

[21] E. Witten, *Supersymmetry and Morse theory*, J. Diff. Geom. 17 (1982) 661–692.

[22] E. Witten, *Constraints on supersymmetry breaking*, Nucl. Phys. B202 (1982) 253–316.

[23] L. Alvarez-Gaumé, *Supersymmetry and the Atiyah–Singer index theorem*, Commun. Math. Phys. 90 (1983) 161–173.

[24] L. Alvarez-Gaumé, *Supersymmetry and index theory*, in *Supersymmetry*, NATO Adv. Study Institute, eds. K. Dietz et al. (Plenum, New York, 1985).

[25] D. Friedan and P. Windey, *Supersymmetric derivation of the Atiyah–Singer index theorem*, Nucl. Phys. B235 (1984) 395–416.

[26] M. Blau and G. Thompson, *$N = 2$ topological gauge theory, the Euler characteristic of moduli spaces, and the Casson invariant*, Commun. Math. Phys. 152 (1993) 41–71.

[27] E. Witten, *Topological quantum field theory*, Commun. Math. Phys. 117 (1988) 353–386.

[28] S. Cordes, G. Moore and S. Ramgoolam, *Large N 2D Yang–Mills theory and topological string theory*, Yale preprint YCTP-P23–93, 97 p., hep-th/9402107; Nucl. Phys. Proc. Suppl. 41 (1995) 184.

[29] S. Wu, *On the Mathai–Quillen formalism of topological sigma models*, 12 p., hep-th/9406103.

[30] S. K. Donaldson, *A polynomial invariant for smooth four-manifolds*, Topology 29 (1990) 257–315.

[31] E. Witten, *Topology changing amplitudes in $2+1$ dimensional gravity*, Nucl. Phys. B323 (1989) 113–140.

[32] D. Birmingham, M. Blau and G. Thompson, *Geometry and quantization of topological gauge theories*, Int. J. Mod. Phys. A5 (1990) 4721–4752.

[33] D. Montano and J. Sonnenschein, *The topology of moduli space and quantum field theory*, Nucl. Phys. B324 (1989) 348–370; J. Sonnenschein, *Topological quantum field theories, moduli spaces and flat gauge connections*, Phys. Rev. D42 (1990) 2080–2094.

[34] C. H. Taubes, *Casson's invariant and Riemannian geometry*, J. Diff. Geom. 31 (1990) 547–599.

[35] E. Witten, *Topological sigma models*, Commun. Math. Phys. 118 (1988) 411–449.

[36] E. Witten, *Mirror manifolds and topological field theory*, in [38, p. 120–159].

[37] D. Birmingham, M. Rakowski and G. Thompson, *Topological field theories, Nicolai maps, and BRST quantization*, Phys. Lett. B214 (1988) 381–386.

[38] S. T. Yau (editor), *Essays on Mirror Manifolds*, International Press, Hong Kong (1992).

[39] E. Witten, *Two dimensional gravity and intersection theory on moduli space*, Surveys in Diff. Geom. 1 (1991) 243–310.

[40] M. Stone, *Supersymmetry and the quantum mechanics of spin*, Nucl. Phys. B314 (1989) 557–586.

[41] M. F. Atiyah, *Circular symmetry and stationary phase approximation*, Astérisque 131 (1985) 43–59.

[42] M. Blau, *Chern–Simons quantum mechanics, supersymmetry, and symplectic invariants*, Int. J. Mod. Phys. A (1991) 365–379.

[43] M. Blau, E. Keski-Vakkuri and A. J. Niemi, *Path integrals and geometry of trajectories*, Phys. Lett. B246 (1990) 92–98.

[44] R. F. Picken, *The propagator for quantum mechanics on a group manifold from an infinite-dimensional analogue of the Duistermaat–Heckman integration formula*, J. Phys. A. (Math. Gen.) 22 (1989) 2285–2297.

[45] A. J. Niemi and P. Pasanen, *Orbit geometry, group representations and topological quantum field theories*, Phys. Lett. B253 (1991) 349–356; E. Keski-Vakkuri, A. J. Niemi, G. Semenoff and O. Tirkkonen, *Topological field theories and integrable models*, Phys. Rev. D44 (1991) 3899–3905; A. J. Niemi and O. Tirkkonen, *On exact evaluation of path integrals*, Uppsala preprint UU-ITP 3/93, 40 p., hep-th/9301059, Ann. Phys. (in press).

[46] A. Yu. Morozov, A. J. Niemi and K. Palo, *Supersymmetry and loop space geometry*, Phys. Lett. B271 (1991) 365–371; A. Hietamaki, A. Yu. Morozov, A. J. Niemi and K. Palo, *Geometry of $N = 1/2$ supersymmetry and the Atiyah–Singer index theorem*, Phys. Lett. B263 (1991) 417–424.

[47] A. J. Niemi and O. Tirkkonen, *Cohomological partition functions for a class of bosonic theories*, Phys. Lett. B293 (1992) 339–343.

[48] H. M. Dykstra, J. D. Lykken and E. J. Raiten, *Exact path integrals by equivariant localization*, Phys. Lett. B302 (1994) 223–229.

[49] R. J. Szabo and G. W. Semenoff, *Phase space isometries and equivariant localization of path integrals in two dimensions*, Nucl. Phys. B421 (1994) 391–412; and *Equivariant localization, spin systems, and topological quantum theory on Riemann surfaces*, University of British Columbia preprint UBCTP-94–005 (June 1994).

[50] S. G. Rajeev, S. K. Rama and S. Sen, *Symplectic manifolds, coherent states, and semiclassical approximation*, Jour. Math. Phys. 35 (1994) 2259–2269.

[51] M. F. Atiyah and R. Bott, *The moment map and equivariant cohomology*, Topology 23 (1984) 1–28.

[52] E. Witten, *Two dimensional gauge theories revisited*, J. Geom. Phys. 9 (1992) 303–368.

[53] S. Wu, *An integration formula for the square of the moment map of circle actions*, Lett. Math. Phys. 29 (1993) 311–328.

[54] L. C. Jeffrey and F. C. Kirwan, *Localization for non-Abelian group actions*, Oxford preprint 36 p., alg-geom/9307001.

[55] E. Prato and S. Wu, *Duistermaat–Heckman measures in a non-compact setting*, alg-geom/9307005.

[56] M. Audin, *The topology of torus actions on symplectic manifolds*, Progress in Mathematics Vol. 93, Birkhäuser, Basel (1991).

[57] N. Woodhouse, *Geometric Quantization* (2nd ed.), Clarendon Press, Oxford (1992).

[58] L. S. Schulman, *A path integral for spin*, Phys. Rev. 176 (1968) 1558–1569.

[59] J. S. Dowker, *When is the 'sum over classical paths' exact?*, J. Phys. A3 (1970) 451–461.

[60] J. D. S. Jones and S. B. Petrack, *The fixed point theorem in equivariant cohomology*, Trans. Amer. Math. Soc. 322 (1990) 35–49.

[61] A. Hietamäki and A. J. Niemi, *Index theorems and loop space geometry*, Phys. Lett. B288 (1992) 331–341.

[62] A. Weinstein, *Cohomology of symplectomorphism groups and critical values of Hamiltonians*, Math. Z. 201 (1989) 75–82.

[63] R. E. Perret, *Path integral derivation of characters for Kac–Moody group*, Nucl. Phys. B356 (1991) 229–244.

[64] K. Funahashi, T. Kashiwa, S. Sakoda and K. Fujii,, *Coherent states, path integral, and semiclassical approximation*, J. Math. Phys. 36 (1995) 3232–3253.

[65] J. P. Francoise, *Canonical partition functions of Hamiltonian systems and the stationary phase formula*, Commun. Math. Phys. 117 (1988) 34–47.

[66] A. J. Niemi and K. Palo, *On quantum integrability and the Lefschetz number*, Mod. Phys. Lett. A8 (1993) 2311–2322; and *Equivariant Morse theory and quantum integrability*, Uppsala preprint UU-ITP 10/94, hep-th/9406068; T. Karki and A. J. Niemi, *On the Duistermaat–Heckman formula and integrable models*, UU-ITP 02/94, 18 p., hep-th/9402041.

[67] M. Blau and G. Thompson, *Equivariant Kähler geometry and localization in the G/G model*, Nucl. Phys. B439 (1995) 367–394.

[68] J.-M. Bismut, *Equivariant Bott–Chern currents and the Ray–Singer analytic torsion*, Math. Ann. 287 (1990) 495–507.

[69] M. Blau and G. Thompson, *Derivation of the Verlinde Formula from Chern–Simons Theory and the G/G model*, Nucl. Phys. B408 (1993) 345–390.

[70] M. L. Mehta, *Random Matrices and the Statistical Theory of Energy Levels*, Academic press, New York (1967).

[71] F. David, *Matrix models and 2-d gravity*, in *String Theory and Quantum Gravity* (eds. M. Green et al.), World Scientific, Singapore (1991) 55–90.

[72] D. Altschuler and C. Itzykson, *Remarks on Integration over Lie algebras*, Ann. Inst. Henri Poincaré 54 (1991) 1–8.

[73] M. Blau and G. Thompson, *Lectures on 2d Gauge Theories: Topological Aspects and Path Integral Techniques*, in *Proceedings of the 1993 Trieste Summer School on High Energy Physics and Cosmology* (eds. E. Gava et al.), World Scientific, Singapore (1994) 175–244, hep-th/9310144.

[74] B. Rusakov, *Loop averages and partition functions in $U(N)$ gauge theory on two-dimensional manifolds*, Mod. Phys. Lett. A5 (1990) 693–703.

[75] E. Witten, *On quantum gauge theories in two dimensions*, Commun. Math. Phys. 141 (1991) 153–209.

[76] D. S. Fine, *Quantum Yang–Mills on the two-sphere*, Commun. Math. Phys. 134 (1990) 273–292; and *Quantum Yang–Mills theory on a Riemann surface*, Commun. Math. Phys. 140 (1991) 321–338.
[77] M. Blau and G. Thompson, *Quantum Yang–Mills theory on arbitrary surfaces*, Int. J. Mod. Phys. A7 (1992) 3781–3806.
[78] E. Witten, *The Verlinde Algebra and the Cohomology of the Grassmannian*, IAS preprint IASSNS-HEP-93/41, 78 p., hep-th/9312104.
[79] M. Blau and G. Thompson, *On diagonalization in Map*(M, G), Commun. Math. Phys. 171 (1995) 639–660.
[80] T. Bröcker and T. tom Dieck, *Representations of Compact Lie Groups*, Springer, New York (1985).
[81] S. Helgason, *Differential Geometry, Lie Groups, and Symmetric Spaces*, Academic Press, Orlando (Fl.) (1978).
[82] M. Blau and G. Thompson, *A new class of topological field theories and the Ray–Singer torsion*, Phys. Lett. B228 (1989) 64–68; and *Topological gauge theories of antisymmetric tensor fields*, Ann. Phys. (NY) 205 (1991) 130–172.
[83] G. 't Hooft, *The Topology of the Gauge Condition and New Confinement Phases in Non-Abelian Gauge Theories*, Nucl. Phys. B190 (1981) 455–478.

PARTICIPANTS CONTRIBUTIONS BY TITLE AND ABSTRACT

A CONSTRUCTIVE ANALOG OF ZIMMERMANN'S FORMULA

Abdelmalek Abdesselam

Centre de Physique Théorique
Ecole Polytechnique
91128 Palaiseau Cédex, France
e-mail: abdessel@orphee.polytechnique.fr

Although some asymptotically free just renormalizable field theory models were rigorously constructed, like infrared ϕ^4 in four dimensions or the Gross-Neveu model in two dimensions, the expansions for the Schwinger functions are heavily inductive. At the perturbative level, an explicit solution of the renormalization recursion is given by Zimmermann's forest formula. However there is not yet a model independent constructive formalism which provides an analog of Zimmermann's formula for the renormalized Schwinger functions. A step in that direction would involve a thorough understanding of the combinatorics of phase cell expansions.

HEAT KERNEL AS THE BASIS FOR THE FUNCTIONAL INTEGRATION IN QUANTUM FIELD THEORY

Ivan G. Avramidi

Department of Mathematics and Computer Sciences
Ernst-Moritz-Arndt University of Greifswald
Fr.-L.-Jahnstr. 15a, D-17489 Greifswald, Germany
e-mail: avramidi@rz.uni-greifswald.de

The only way to get numbers from the functional integral in QFT is to *define* it as a series of Gaussian integrals, which are well defined in terms of the functional determinants and the Green functions, and finally, by stansard procedure, in terms of the heat kernel. Thus the heat kernel is the basis for the calculation of the functional integral in QFT. In our recent papers (see *I. G. Avramidi, J. Math. Phys.* **37** *(1996) 374;* **36** *(1995) 5055*) we studied the heat kernel in quantum gravity and gauge theories for strong slowly varying background fields and proposed a new covariant purely algebraic approach based on taking into account only a finite number of low-order covariant derivatives of the background fields. The explicit manifestly covariant closed formulas for the heat kernel expressed purely in terms of curvature invariants are obtained.

WIENER INTEGRATION FOR QUANTUM SYSTEMS: A UNIFIED APPROACH TO THE FEYNMAN-KAC FORMULA

Bernhard Bodmann and Simone Warzel

Institut für Theoretische Physik
Universität Erlangen-Nürnberg
Staudtstraße 7, D-91058 Erlangen, Germany

Gaussian linearization on the basis of Wiener integration over operator-valued functionals provides a unifying approach to the probabilistic representation of certain operator semigroups [B. Bodmann, H. Leschke, S. Warzel, in: Path integrals: Dubna '96, eds. V. S. Yarunin, M. A. Smondyrev, Dubna 1996, pp. 95-106]. Within the setting of a quantum particle in an electromagnetic field it naturally yields a basis independent version of the standard Feynman-Kac(-Ito) formula for the corresponding Schrödinger semigroup. In this framework even semigroups generated by non-standard Hamiltonians such as for a quantum particle with a spatially dependent mass can be represented by conventional Wiener integrals – in contrast to [B. Gaveau, L. S. Schulman: J. Math. Phys. **30**, (1989) 2019]. The approach offers a convenient starting point for estimations and calculations.

ANOMALIES IN BRST anti BRST APPROACH AS A FUNCTIONAL INTEGRATION PROBLEM

Violeta Calian

Department of Physics
University of Craiova
13 A. I. Cuza, Craiova 1100, Romania

The most general method for quantizing gauge theories in a manifestly covariant manner proved to be the antifields-antibrackets formalism. The resulting Ward-Takahashi identities allow an elegant proof of renormalizability and the nilpotent nature of the BRST transformations seems to be the key for the unitarity of the S-matrix. However, even if the BRST anti BRST technique and the anomalies study is well documented, the following problems need to be covered: a) the definition of a functional integral operator in the BRST anti BRST formalism as the extension of the operator obtained in standard BRST; b) identifying the differential space on which this operator is acting; c) deriving the corresponding Ward identities and the Wess-Zumino consistency conditions.

SEMICLASSICAL APPROXIMATION USING WIENER MEASURE

Fabienne Castell

CMI, Université Aix-Marseille 1
39, rue Joliot Curie
13453 Marseille Cedex 13, FRANCE
e-mail: castell@gyptis.univ-mrs.fr

To represent the wave function for the Schrödinger equation by a path integral raises the question of knowing if the path integral considered is well-defined. That is, given any potential V, is the associated oscillatory function path-integrable? This question is proved to be irrelevant regarding the semiclassical approximation problem. Indeed a probabilistic ansatz based on Wiener measure is exhibited, which under mild conditions on the potential, leads to the semiclassical approximation before the caustics. Such an anstaz is available after the caustics, but the link between this ansatz and the wave function is an open problem.

OPEN PROBLEM: THE MEASURE FOR THE FUNCTIONAL INTEGRAL

Bryce DeWitt

Department of Physics
University of Texas
Austin, TX 78703 USA

The $(\det G^{adv})^{-1/2}$ measure in the path integral for quantum mechanics is precisely what is needed to get the Van Vleck-Morette determinant in the WKB approximation (1-loop). In field theory it cancels contributions from arcs at infinity in the Wick rotation procedure for 1-loop graphs, thus justifying the Euclidean evaluation of Feynman graphs. Question: Is there a choice of measure that will cancel contributions from arcs at infinity to *all* loop orders?

EXACTNESS OF THE SEMI-CLASSICAL APPROXIMATION FOR DIFFUSION ON NON-COMPACT COSET SPACES ?

D. Endesfelder

Theoretical Physics
Oxford University
1 Keble Road
Oxford OX 1 3 NP U.K.
e-mail: d.endesfelder1@physics.oxford.ac.uk

A disordered conductor in which the mean free path for inelastic electron scattering exceeds its size is called mesoscopic. Using the Landauer-Büttiker scattering approach one can express transport quantities in terms of a transfer matrix T which belongs to a non-compact group ($Sp(2N,\mathbf{R}), SU(N,N)$, and $SO^*(4N)$ for the orthogonal, unitary, and the symplectic transfer matrix ensembles, respectively)[1]. The evolution of the probability distribution of T with the length of a quasi-one-dimensional wire can be described as diffusion on the coset spaces of these groups. It is known that the semi-classical approximation for the path integral which describes diffusion on compact group manifolds is exact. Is this also true for these non-compact coset spaces ?

(1) M. Caselle, Nucl. Phys. B S45A, 120 (1996)

THE PHASE SPACE GENERALIZATION OF THE VAN VLECK–PAULI–MORETTE APPROACH TO FEYNMAN PATH INTEGRAL

P. P. Fiziev

Department of Theoretical Physics, University of Sofia,
5 James Boucher Blvd., Sofia 1164, Bulgaria.
e-mail: fiziev@phys.uni-sofia.bg

Recently I proposed a solution of the problem of quantization of a classical system, based on a justification of the class of paths, really taken into account in a path integral on phase space (P. Fiziev, *The Class of Paths in the Feynman Path Integral on the phase space*, in Lectures on Path Integration: Trieste 1991, eds. H.Serdeira et al., World Scientific, Singapore, 1993), generalizing (C. Morette, Phys.Rev.,(1951), **81**, 848).

In the framework of the discretization procedure the general form of the different quantum Hamiltonians corresponding to a single classical one is derived using a summation on different classes of paths in the phase space. A number of simple examples with both Gaussian, and non-Gaussian path integrals are included.

INFLUENCE OF THE PATH INTEGRAL MEASURE IN QUANTUM GRAVITY

Christian Holm

Max Planck Institut für Polymerforschung
Ackermannweg 10
D-55128 Mainz, Germany
e-mail: holm@mpip-mainz.mpg.de

In Ref. 1 we applied path integral quantization to a classical 2D gravitational Lagrangian with R^2 term. This model is numerically evaluated by Monte Carlo techniques, employing Regge's simplicial calculus to reduce the infinite dimensional degrees of freedom to a finite number. In general, for quantum gravity no unique functional integration measure exists. In 2D, however, its influence can be studied by comparing the numerical results to analytic predictions coming from conformal field theory and matrix models. Special interest is given to scale invariant measures and the DeWitt measure.

1. C. Holm, W. Janke, Nucl. Phys. **B477**, 465 (1996), and references herein.

A NEW METHOD OF COMPUTATION OF FUNCTIONAL INTEGRALS

Yuri Yu. Lobanov

Lab. of Computing Techniques and Automation
Joint Institute for Nuclear Research,
141980 Dubna, Moscow Region, Russia
e-mail: lobanov@main1.jinr.dubna.su

In the framework of the rigorous definition of an integral in the abstract complete separable metric space the new method for numerical evaluation of functional integrals is elaborated. The method does not require preliminary discretization and allows to use the deterministic algorithms in computations. The convergence of approximations to an exact value of integral is proved, the estimate of the remainder is obtained. In the particular case of conditional Wiener integrals the approximation formulas with the weight functional are derived. The method is generalized to the case of computation of multiple functional integrals. Some applications of the method to the problems of quantum mechanics and quantum field theory are considered.

VARIATIONAL METHODS, PATH INTEGRALS AND QUANTUM GRAVITY

Remo Garattini

Facoltà di Ingegneria
Università degli Studi di Bergamo
Viale Marconi, 5 24044 Dalmine (Bergamo), Italy
e-mail: Garattini@mi.infn.it

We use variational methods to calculate quantum corrections in Quantum Gravity. A comparison with the effective potential calculated at quadratic order is made by means of gaussian wave functionals [R. Garattini, *Vacuum energy in ultralocal metrics for TT tensors with Gaussian wave functionals*, in General Relativity, Proceedings of the Forty Sixth Scottish Universities: Aberdeen 1995, eds. G.S.Hall and J.R. Pulham, Institute of Physics Publishing, Bristol, p. 409.]. The method is a particular case of the effective action for composite operators used in quantum field theory [J.M. Cornwall, R. Jackiw and E. Tomboulis, Phys.Rev.D, **8**, (1974), p. 2428.]. In pure gravity the method is applied for the first time. A new effective action could be extracted from the path integral and it could serve as a model for Quantum Gravity with and without matter fields.

PATH INTEGRALS AND PSEUDOCLASSICAL DESCRIPTION FOR SPINNING PARTICLES

Dmitri M. Gitman

Instituto de Física
Universidade de São Paulo
C.P. 66318 CEP 05315-970 São Paulo,SP, Brazil
e-mail: gitman@fma.if.usp.br

The propagator of a spinning particle is expressed by a path integral. The problem has different solutions in even and odd dimensions. In even dimensions the representation is a generalization from one to four dimensions (E.Fradkin, D. Gitman, Phys. Rev. **D44** (1991) 3230) and the effective action in the path integral is Berezin-Marinov pseudoclassical action. In odd dimensions the solution is presented for the first time. A new effective action can be extracted from the path integral and it can serve as a pseudoclassical model for spinning particles in odd dimensions.

FUNCTIONAL INTEGRATION AND DISORDERED SYSTEMS

Gilles Poirot

Centre de Physique Théorique
Ecole Polytechnique
91128 Palaiseau cedex, FRANCE
email: poirot@cpth.polytechnique.fr

Very often disordered systems are *self-averaging* in the thermodynamic limit. This means that any physical quantity is almost surely equal to its expectation value. But the "disorder" can be seen as a random potential, *i.e.* a random scalar field, so that computing mean values is a field theory problem. Thus, functional integration techniques can be used in addition to probabilistic ones which usually fail when the disorder is small. Question: Can rigorous functional integration allow to get results in weakly disordered systems? For instance, is it possible to prove that electrons in a metal with few impurities have extended states, completing the proof of a metal-insulator transition in semi-conductors?

PATH-INTEGRAL ANALYSIS OF WAVE PROPAGATION IN MULTIPLE-SCATTERING RANDOM MEDIA

Gregory Samelsohn

Department of Electrical and Computer Engineering,
Ben-Gurion University of the Negev,
P.O.Box 653, Beer-Sheva 84105, Israel
e-mail: gregory@newton.bgu.ac.il

In this work we implement the path integral approach to the general problem of wave propagation in random media. For this purpose we apply the method originally proposed by Fock for the integration of quantum mechanical equations. The principal idea of the method is based on the introduction of an additional pseudotime variable and the transfer to a higher-dimensional space, in which the propagation process is described by a generalized parabolic equation similar to the nonstationary Schrodinger equation in quantum mechanics [G. Samelsohn and R. Mazar, Phys. Rev. E **54**, (1996)]. We present its solution in a form of the Feynman path integral, the asymptotic evaluation of which allows us to estimate the so-called wave correction terms. These corrections are related to coherent backscattering, i.e. to the phenomenon which can not be described in the framework of conventional theories of radiative transfer or small-angle scattering.

RELATIONS BETWEEN MARKOV PROCESSES VIA LOCAL TIME AND COORDINATE TRANSFORMATIONS

Axel Pelster, Hagen Kleinert

Institut für Theoretische Physik
Freie Universität Berlin
Arnimallee 14
D-14195 Berlin, Germany

The Duru-Kleinert method of solving unknown path integrals of quantum mechanical systems by relating them to known ones does not apply to Markov processes since a Fokker-Planck equation is in general not transformed to a Fokker-Planck equation. In a contribution to Phys. Rev. Lett. (in press, cond-mat/9608120) we present a significant modification of the method, based again on a combination of path-dependent time and coordinate transformations, to obtain such relations after all. As an application we express unknown Green functions for a one-parameter family of Markov processes in terms of the known one for the Schenzle-Brand process.

ON THE MECHANISM UNDERLYING THE DIVERGENCE OF PERTURBATION THEORY

Sergio A. Pernice

Departmant of Physics and Astronomy
University of Rochester
Rochester, New York 14627 USA

It is well known that perturbation theory in quantum mechanics and quantum field theory generates a divergent series. It has been shown recently (hep-th/9609139) that such divergence can be understood in the context of Lebesgue's dominated convergence theorem. That theorem governs the validity of the formal manipulation done to generate the perturbative series in the functional integral formalism. It was shown that one of its hypothesis is violated in theories like the $\lambda\phi^4$ theory in 1, 2 or 3 space-time dimensions. Question: given that in 4 dimensions a new source of divergence appears (renormalons), can they also be understood in the context of Lebesgue's theorem?

FIVE-LOOP RENORMALIZATION GROUP FUNCTIONS OF ϕ^4-THEORY WITH $O(N)$-SYMMETRIC AND CUBIC INTERACTIONS

Verena Schulte-Frohlinde

Institut für Theoretische Physik
Freie Universität Berlin
Arnimallee 14
D-14195 Berlin, Germany

The renormalization group functions are calculated in $D = 4 - \varepsilon$ dimensions for a ϕ^4-theory with two coupling constants: $g_1 \left(\sum_\alpha^N \phi_\alpha^2\right)^2 + g_2 \sum_\alpha^N \phi_\alpha^4$. We find strong indication for the stability of the cubic fixed point for the number of components $N \geq 3$, implying that the magnetic transition of three-dimensional cubic crystals are described by critical exponents of the cubic universality class. Resummation procedures for theories with two coupling constants are still under investigation.
H. Kleinert, V. Schulte-Frohlinde, Phys. Lett. B 342 (1995) 284.
H. Kleinert, V. Schulte-Frohlinde, *Critical Behavior of ϕ^4-Theories*, to be published.

FILTERED PROPAGATOR FUNCTIONAL OF QUANTUM DISSIPATIVE PATH INTEGRATION

Eunji Sim

School of Chemical Sciences
University of Illinois
Urbana, IL 61801 USA
e-mail: e-sim1@uiuc.edu

We developed a new scheme, based on Feynman's path integral formulation of time-dependent quantum mechanics, which calculates the reduced density matrix of a one-dimensional quantum mechanical system coupled to a harmonic dissipative environment. Tracing out the bath coordinates introduces memory effects to the dynamics of the system which have finite span if the environment has a broad spectrum. By introducing a multi-time augmented reduced density vector or functional of statistically significant path segments the path integration can be reduced to matrix-vector multiplication. Monte Carlo techniques are employed to identify path segments of appreciable weight. The scheme allows iterative evaluation of the path integration with long memory kernels over long time periods. High efficiency is achieved with the aid of sorting and filtering criteria. The FORTRAN program[1] is available for either serial or shared-memory parallel execution. [1] E. Sim and N. Makri, *Comput. Phys. Commun.* (in press.)

APPLICATION OF THE FUNCTIONAL METHOD TO THE STUDY OF THE STOCHASTIC PROCESSES IN PLASMAS

Florin Spineanu and Madalina Vlad

Institute of Atomic Physics
IFTAR-L.22 P.O.Box MG-7
Magurele Bucharest
Romania
E-mail : spineanu@roifa.ifa.ro

Problems related to the transport processes in plasmas are often expressed as stochastic equations. We develop a formalism where the averages over statistical ensembles are obtained as functional integrations with appropriate measures. For a broad class of noise functions the calculation can be carried out analytically. The method has been successfully applied in the study of the particle motion in stochastic magnetic field and of the magnetic field line diffusion. We have obtained explicit results for the diffusion of particles in the presence of random trapping and for continuum time random walks with Levy distribution.

FUNCTIONAL FORMALISM FOR THE INFRARED PROBLEM

B. M. Pimentel and J. L. Tomazelli

Instituto de Física Teórica, UNESP
Rua Pamplona, 145, 01405-900 São Paulo, SP, Brazil

In the asymptotic limit for scalar QED the boson current operator can be approximated by a current operator whose eigenvalue corresponds to the classical current of a point-charge moving with constant velocity. Using this approximation in the Gell-Mann-Low formula we obtain a path-ordered phase factor which multiplies the free boson propagator (Prog. Theor. Phys. **93**, 1105). This coincides with the Bloch-Nordsieck propagator in the WKB approximation. Instead, we can substitute the classical current in the complete generating functional and first integrate over the gauge fields. This illustrates the usefulness of functional techniques for tackling the IR problem and we propose a similar construct to QCD and quantum gravity.

PROBABILITY DISTRIBUTIONS WITH INFINITE MOMENTS

Jorge F. Willemsen

Rosenstiel School of Marine and Atmospheric Science
University of Miami
Miami, FL 33149 USA
jorge@maya.rsmas.miami.edu

The functional integral representation of traditional Euclidean field theory is based on the existence of Gaussian functional integrals. Perturbation theory then consists of calculating the moments of what may be thought of as a Gaussian probability distribution functional. At the Cargese Summer School, S.K. Foong discussed functional integrals based upon the Poisson distribution and how these may be applied to problems in wave propagation. This opens the door to thinking about the possible significance of other distributions to certain physical processes. Question: Can we learn something about notoriously non-renormalizable field theories such as gravity by studying distributions such as the Levy distribution, which has infinite moments? That is, can we rephrase the questions away from what we expect from the traditional moments of a Gaussian?

NEW UNIVERSALITY CLASSES IN 1D $O(N)$-INVARIANT SPIN-MODELS WITH AN n-PARAMETRIC ACTION

Karim Yildirim

Max-Planck-Institut für Physik
Föhringer Ring 6, D-80805 Munich, Germany
e-mail : yildirim@iws186.mppmu.mpg.de

An action with n parameters (couplings), which generalizes the mixed $O(N)$-RP^{N-1}-model, is considered in one dimension for general N. The non-negativity of the parameters ensures that the transfer matrix is a positive operator due to reflection positivity and therefore allows a quantum-mechanical interpretation of the spectrum. Using asymptotic expansion techniques to determine where the model becomes critical and show that for the actions considered there exists a family of hypersurfaces whose asymptotic behaviour determines a one-parameter family of new universality classes. Reflecting the true spectral properties of the transfer matrix, they interpolate between the classes of the $O(N)$-vector-model and the RP^{N-1}-model. Furthermore continuum limits are discussed, including the exceptional case $N = 2$.

DOUBLE DELTA EXPANSIONS – AN OPEN PROBLEM

Miloslav Znojil

Ústav jaderné fyziky AV ČR
250 68 Řež, Czech Republic
e-mail: znojil@ujf.cas.cz

According to Bender et al (Phys. Rev. Lett. 58, 2615 (1987)), it is possible to treat coupling constants non-perturbatively, using an *ad hoc* new parameter δ. Inspiration: The formalism uses parametric differentiation of functional integrals. Question: Could the formalism of functional integrals also provide a similar recipe which would use a *doublet* of the expansion parameters? Hint: My preliminary, second-order study with δ replaced by (ρ, η) (Phys. Lett. A 164, 145 (1992)).

LIST OF PARTICIPANTS

Abdelmalek ABDESSELAM
Centre de Physique Théorique
Ecole Polytechnique
91128 Palaiseau Cedex, France
email: abdessel@orphee.polytechnique.fr

Stéphane ANCEY
CMCS, Université de Corse
Faculté des Sciences, B.P. 52
20250 Corte, France
email: ancey@univ-corse.fr

Ivan AVRAMIDI
Department of Mathematics
University of Greifswald
Jahnstr. 15a
D-17489 Greifswald, Germany
email: avramidi@rz.uni-greifswald.de

Gergely BANA
Eötvös University
Institute for Theoretical Physics
Puskin u. 5-7
Budapest, H-1088, Hungary
email: gbana@hal9000.elte.hu

Matthias BLAU
Laboratoire de Physique Théorique
Ecole Normale Supérieure de Lyon
46, allée d'Italie
69364 Lyon Cedex 07, France
email: mblau@enslapp.ens-lyon.fr

Bernhard BODMANN
Institut für Theoretische Physik I
Universität Erlangen-Nürnberg
Staudtstrasse 7
D-91058 Erlangen, Germany
email: bernhard@theorie1.physik.uni-erlangen.de

Federico BONETTO
Dipartimento di Matematica
Università "La Sapienza"
00185 Roma, Italia
email: bonetto@ipparco.roma1.infn.it

Maria Elisa BORELLI
Instituto de Fisica
Universidade de São Paulo
R. do Mataão, travessa R 187
Caixa Postal 66.318 05389-970
São Paulo SP, Brazil
email: mborelli@usp.br

Mark BYRD
Department of Physics, RLM 5.208
University of Texas, Austin
Austin, TX 78712-1081, USA
email: mbyrd@physics.utexas.edu

Pierre CARTIER
ENS, Département Math.-Info.
45 rue d'Ulm
75230 Paris Cedex 05, France
email: cartier@ihes.fr

Violeta CALIAN
University of Craiova
Department of Physics, 13 A.I. Cuza
Craiova - 1100, Romania

Fabienne CASTELL
LATP, Université d'Aix Marseille 1
39 rue Joliot Curie
13453 Marseille Cedex13, France
email: castell@gyptis.univ-mrs.fr

David COLLINS
Department of Physics, Center of Relativity
University of Texas, Austin
Austin, TX 78712-1081, USA
email: david@einstein.ph.utexas.edu

Samuel DE TORO ARIAS
Laboratoire de Physique de la Matière Condensée
Université de Nice-Sophia Antipolis, Parc Valrose
06108 Nice Cedex 2, France
email: sdetoro@calypso.unice.fr

Bryce DEWITT
Physics Department
University of Texas
Austin, TX 78712, USA
email: dewitt@physics.utexas.edu

Cécile DEWITT-MORETTE
Physics Department, Center for Relativity
University of Texas
Austin, TX 78712, USA
email: cdewitt@physics.utexas.edu

Bertrand DUPLANTIER
Service de Physique Théorique
Centre d'Etudes Nucléaires de Saclay
91191 Gif-sur-Yvette, France
email: bertrand@spht.saclay.cea.fr

Dirk ENDESFELDER
Oxford University
Theoretical Physics
1 Keble Road
Oxford OX1 3NP, United Kingdom
email: endes@thphys.ox.ac.uk

Sergei FEDOTOV
Department of Mathematical Physics
Ural State University, 51 Lenina St.
Jekaterinburg, 620083, Russia
email: sergei.fedotov@usu.ru

Plamen FIZIEV
Department of Theoretical Physics
Faculty of Physics
University of Sofia
Boulevard 5 James Boucher
Sofia 1164, Bulgaria
email: fiziev@phys.uni-sofia.bg

Antoine FOLACCI
CMCS, Université de Corse
Faculté des Sciences, B.P. 52
20250 Corte, France
email: folacci@univ-corse.fr

See Kit FOONG
College of Education
University of the Ryukyus
Senbaru 1, Nishihara-cho
Okinawa, Japan 903 - 01
email: foong@edu.u-ryukyu.ac.jp

Remo GARATTINI
Università degli Studi di Bergamo
Facoltà di Ingegneria
Viale Marconi, 5
24044 Dalmine (BG), Italia
email: garattini@mi.infn.it

Dmitri GITMAN
Instituto de Fisica
Universidade de São Paulo
Caixa Postal 66318
05389-970 São Paulo SP, Brazil
email: gitman@fma.if.usp.br

Yves GRANDATI
LPLI - Institut de Physique
1 bvd D.F. Arago
57070 Metz, France

Christian HOLM
Institut für Theoretische Physik
Freie Universität Berlin
Arnimallee 14
14195 Berlin, Germany
email: holm@physik.fu-berlin.de

Bruce JENSEN
CMCS, Université de Corse
Faculté des Sciences, B.P. 52
20250 Corte, France
email: jensen@univ-corse.fr

Louis H. KAUFFMAN
Dept of Mathematics
University of Illinois
851 South Morgan St.
Chicago, IL 60607-7045, USA
email: kauffman@uic.edu

Takashi KIMURA
Dep. of Mathematics, 111 Cummington Street
Boston University
Boston, MA 02275, USA
email: kimura@math.bu.edu

Hagen KLEINERT
Institut für Theoretische Physik
Freie Universität Berlin
Arnimallee 14
1000 Berlin 33, Germany
email: kleinert@physik.fu-berlin.de

Christiane KOCH
Institüt für Physik
Lehrstuhl für Statistische Physik
Invalidenstr. 110
Humboldt-Universität zu Berlin
10099 Berlin, Germany
email: koch@physik.hu-berlin.de

Flora KOUKIOU
Département de Physique
Université de Cergy-Pontoise
F-95302 Cergy-Pontoise Cedex, France
email: koukiou@u-cergy.fr

Rémi LEANDRE
Departement de Mathématiques
Université de Nancy I
54000 Vandoeuvre-les-Nancy, France
email: leandre@iecn.u-nancy.fr

Yuri LOBANOV
Joint Institute for Nuclear Research
Lab. of Computing Techniques and Automation
141980 Dubna, Moscow Region, Russia
email: lobanov@main1.jinr.dubna.su

Jacques MAGNEN
Ecole Polytechnique
Centre de Physique Théorique
91128 Palaiseau Cedex, France
email: magnen@orphee.polytechnique.fr

Nancy MAKRI
School of Chemical Sciences
University of Illinois, 505 S Mathews Ave
Urbana, IL 61801, USA
email: nancy@makri.scs.uiuc.edu

Vieri MASTROPIETRO
Università di Tor Vergata Roma
Viale della Ricerca Scientifica
Roma, Italia
email: vieri@ipparco.roma1.infn.it

Mark MIMS
Department of Physics
University of Texas, Austin
Austin, TX 78712-1081, USA
email: mmm@physics.utexas.edu

Xiaorong MORROW
Department of Physics
University of Texas, Austin
Austin, TX 78712-1081, USA
email: xiaorong@happy.cc.utexas.edu

Jan Martin PAWLOWSKI
Theoretisch-Physikalisches Institut
Universität Jena
Fröbelstieg 1
D-07743 Jena, Germany
email: pawlowski@hpfs1.physik.uni-jena.de

Axel PELSTER
Institut für Theoretische Physik
Freie Universität Berlin
Arnimallee 14
14195 Berlin, Germany
email: pelster@physik.fu-berlin.de

Sergio PERNICE
Physics and Astronomy Department
University of Rochester
Rochester, NY 14627, USA
email: sergio@charm.pas.rochester.edu

Dimitri PETRITIS
Institut de Recherche Mathématique
Université de Rennes 1, Campus de Baulieu
35042 Rennes, France
email: petritis@levy.univ-rennes1.fr

Gilles POIROT
Centre de Physique Théorique
Ecole Polytechnique
91128 Palaiseau Cedex, France
email: poirot@orphee.polytechnique.fr

Michael PRAEHOFER
LS Wagner, Theoretische Physik
Ludwig-Maximilians-Universitaet Muenchen
Theresienstr. 37
D-80333 Muenchen, Germany
email: praehofer@stat.physik.uni-muenchen.de

Johan RADE
Department of Mathematics
Lund University, Box 118
22100 Lund, Sweden
email: rade@maths.lth.se

Gregory SAMELSOHN
Ben-Gurion University of the Negev
Department of Electrical and Computer Engineering
P.O. Box 653, Beer-Sheva 84105, Israel
email: gregory@newton.bgu.ac.il

Verena SCHULTE-FROHLINDE
Institut für Theoretische Physik
Freie Universität Berlin
Arnimallee 14
14195 Berlin, Germany
email: frohlind@physik.fu-berlin.de

Alberto SCOTTI
Latrobe 122
Department of Mechanical Engineering
The Johns Hopkins University
3400 N Charles Street
Baltimore, MD 21218-2686, USA
email: scotti@jhu.edu

Eunji SIM
School of Chemical Sciences
University of Illinois
505 S. Mathews Ave.
Urbana, IL 61801, USA
email: eunji@mars.scs.uiuc.edu

Alan SOKAL
New York University
Dept. of Physics
4 Washington Place
New York, NY 10003, USA
email: sokal@nyu.edu

Florin SPINEANU
Institute of Atomic Physics
IFTAR - L.22, P.O. Box MG-7
Magurele, Bucharest, Romania
email: spineanu@roifa.ifa.ro

Guy STANDEN
University of Newcastle Upon Tyne
Department of Physics
Newcastle Upon Tyne, NEI 7RU, United Kingdom
— email: g.b.standen@ncl.ac.uk

Silke THOMS
Institut für Theoretische Physik
Freie Universität Berlin
Arnimallee 14
14195 Berlin, Germany
email: thoms@physik.fu-berlin.de

Jeferson TOMAZELLI
Instituto de Fisica Teorica, UNESP
Rua Pamplona, 145
01405-900 São Paulo -SP, Brazil
email: tomazell@axp.ift.unesp.br

Nuri UNAL
Akdeniz University
Fen-Edebiyat Fakültesi
Fizik Bölümü, P.K. 510
07200 Antalya, Turkey
email: unal@ufuk.lab.akdeniz.edu.tr

Jon URRESTILLA
Fisika Teorikoa Saila
Euskal Herriko Unibertsitatea
644 Posta Kutxa
48080 Bilbo, Espagna

Madalina Olimpia VLAD
Institute of Atomic Physics
IFTAR - L.22, P.O. Box MG-7
Magurele, Bucharest, Romania
email: madi@roifa.ifa.ro

Simone WARZEL
Institut für Theoretische Physik I
Universität Erlangen-Nürnberg
Staudtstrasse 7
D-91058 Erlangen, Germany
email: simone@theorie1.physik.uni-erlangen.de

Jorge WILLEMSEN
University of Miami
RSMAS-AMP
4600 Rickenbacker Causeway
Miami, FL 33134, USA
email: jorge@maya.rsmas.miami.edu

Alexander WURM
Department of Physics
University of Texas, Austin
Austin, TX 78712-1081, USA
email: don@utpapa.ph.utexas.edu

Karim YILDIRIM
Max-Planck-Institut für Physik
Föhringer Ring 6, Office-Room 180 A
D-80805 München, Germany
email: yildirim@mppmu.mpg.de

Miloslav ZNOJIL
Theory Group
Inst. Nuclear Physics
Czech Academy of Sciences
250 68 Rez, Czech Republic
email: znojil@ujf.cas.cz

Andreas ZOUPAS
Theoretical Physics Group
The Blackett Laboratory
Imperial College
London SW7 2BZ, United Kingdom
email: a.zoupas@ic.ac.uk

INDEX

www.ingramcontent.com/pod-product-compliance
Ingram Content Group UK Ltd.
Pitfield, Milton Keynes, MK11 3LW, UK
UKHW051131260726
13967UKWH00010B/2979
* 9 7 8 1 4 8 9 9 0 3 2 0 4 *